METHODS IN CELL BIOLOGY

VOLUME 25

The Cytoskeleton

Part B. Biological Systems and in Vitro *Models*

Advisory Board

METHODS IN CELL BIOLOGY

Prepared under the Auspices of the American Society for Cell Biology

VOLUME 25

The Cytoskeleton

Part B. Biological Systems and in Vitro *Models*

Edited by

LESLIE WILSON

DEPARTMENT OF BIOLOGICAL SCIENCES
UNIVERSITY OF CALIFORNIA, SANTA BARBARA
SANTA BARBARA, CALIFORNIA

1982

ACADEMIC PRESS

A Subsidiary of Harcourt Brace Jovanovich, Publishers

New York London
Paris San Diego San Francisco São Paulo Sydney Tokyo Toronto

ACADEMIC PRESS, INC.
111 Fifth Avenue, New York, New York 10003

United Kingdom Edition published by
ACADEMIC PRESS, INC. (LONDON) LTD.
24/28 Oval Road, London NW1 7DX

LIBRARY OF CONGRESS CATALOG CARD NUMBER: 64-14220

ISBN 0-12-564125-7

PRINTED IN THE UNITED STATES OF AMERICA

82 83 84 85 9 8 7 6 5 4 3 2 1

CONTENTS

4. *Permeabilized Cell Models for Studying Chromosome Movement in Dividing* PtK_1 *Cells*

W. Zacheus Cande

5. *Mitotic Spindles Isolated from Sea Urchin Eggs with EGTA Lysis Buffers*

E. D. Salmon

6. *Molecular and Genetic Methods for Studying Mitosis and Spindle Proteins in* Aspergillus nidulans

N. Ronald Morris, Donald R. Kirsch, and Berl R. Oakley

7. *Actin Localization within Cells by Electron Microscopy*

Arthur Forer

CONTRIBUTORS

Numbers in parentheses indicate the pages on which the authors' contributions begin.

SCOTT T. BRADY, Department of Anatomy, Case Western Reserve University School of Medicine, Cleveland, Ohio 44106 (365)

JOSEPH BRYAN, Department of Cell Biology, Baylor College of Medicine, Houston, Texas 77030 (175)

W. ZACHEUS CANDE, Department of Botany, University of California, Berkeley, California 94720 (57)

ARTHUR FORER, Biology Department, York University, Downsview, Ontario M3J 1P3, Canada (131, 227)

BARBARA H. GIBBONS, Pacific Biomedical Research Center, University of Hawaii, Honolulu, Hawaii 96822 (253)

JOHN H. HARTWIG, Hematology-Oncology Unit, Massachusetts General Hospital, and Department of Medicine, Harvard Medical School, Boston, Massachusetts 02114 (201)

JEANNE M. HEIPLE,[1] Cell and Developmental Biology, The Biological Laboratories, Harvard University, Cambridge, Massachusetts 02138 (1)

CHRISTINE L. HOWE, Department of Biology, Yale University, New Haven, Connecticut 06511 (143)

ROBERT E. KANE, Pacific Biomedical Research Center, University of Hawaii, Honolulu, Hawaii 96813 (175)

DANIEL P. KIEHART, Department of Cell Biology and Anatomy, Johns Hopkins University School of Medicine, Baltimore, Maryland 21205 (13)

DONALD R. KIRSCH, Department of Pharmacology, College of Medicine and Dentistry of New Jersey, Rutgers Medical School, Piscataway, New Jersey 08854 (107)

EDWARD D. KORN, Laboratory of Cell Biology, National Heart, Lung, and Blood Institute, National Institutes of Health, Bethesda, Maryland 20205 (313)

RAYMOND J. LASEK, Department of Anatomy, Case Western Reserve University School of Medicine, Cleveland, Ohio 44106 (365)

ELIAS LAZARIDES, Division of Biology, California Institute of Technology, Pasadena, California 91125 (333)

J. RICHARD MCINTOSH, Department of Molecular, Cellular, and Developmental Biology, University of Colorado, Boulder, Colorado 80309 (33)

MARK S. MOOSEKER, Department of Biology, Yale University, New Haven, Connecticut 06511 (143)

N. RONALD MORRIS, Department of Pharmacology, College of Medicine and Dentistry of New Jersey, Rutgers Medical School, Piscataway, New Jersey 08854 (107)

BERL R. OAKLEY, Department of Pharmacology, College of Medicine and Dentistry of New Jersey, Rutgers Medical School, Piscataway, New Jersey 08854 (107)

[1]*Present address:* Department of Biochemistry, Boston University Medical School, Boston, Massachusetts 02118.

EVE REAVEN, Department of Medicine, Stanford University School of Medicine, Stanford, California 94305, and Geriatric Research, Education, and Clinical Center, Veterans Administration Medical Center, Palo Alto, California 94304 (273)

E. D. SALMON, Department of Zoology, University of North Carolina, Chapel Hill, North Carolina 27514 (69)

MANFRED SCHLIWA,[2] Department of Molecular, Cellular, and Developmental Biology, University of Colorado, Boulder, Colorado 80309 (285)

JAMES A. SPUDICH, Department of Structural Biology, Stanford University School of Medicine, Stanford, California 94305 (359)

THOMAS P. STOSSEL, Hematology-Oncology Unit, Massachusetts General Hospital, and Department of Medicine, Harvard Medical School, Boston, Massachusetts 02114 (201)

D. LANSING TAYLOR, Cell and Developmental Biology, The Biological Laboratories, Harvard University, Cambridge, Massachusetts 02138 (1)

YU-LI WANG, Cell and Developmental Biology, The Biological Laboratories, Harvard University, Cambridge, Massachusetts 02138 (1)

[2]*Present address:* Department of Zoology, University of California, Berkeley, California 94720.

PREFACE

Knowledge of the organization of the cytoplasm of eukaryotic cells has expanded greatly during the past decade. The early view of the cytoplasm as a structureless "soup" has given way to our present understanding of it as being exquisitely ordered. This order is conferred by a three-dimensional network of fibrous structures, which include microtubules, microfilaments, and intermediate filaments, and the fiber-associated molecules that mediate the interactions and functions of the fibrous elements with each other and with cytoplasmic organelles. The three-dimensional network of fibrous structures has come to be known as the *cytoskeleton,* though the dynamic nature of the network is not captured by the term.

Early research on the cytoskeleton focused on identification of the major constituents of the cytoskeleton and their characterization, in terms of both their organization in cells and tissues and their biochemical properties. More recently, our attention has been turning strongly toward understanding the functions of cytoskeletal elements in living cells, and has focused on investigation of "less visible" cytoskeletal components (which are considered to interact functionally with the surfaces of the major types of filaments) and on the development of methods and model systems for investigating the functional interactions of cytoskeletal components with one another and with other cell components.

This volume consists of 18 chapters that describe cell systems and *in vitro* model systems presently or potentially valuable for the elucidation of the functions of the cytoskeleton or its components in living cells. The methods involved have been described in considerable detail and are often accompanied by a section of overview perspectives that should aid investigators new to this research area. The previous volume of this publication (Volume 24, The Cytoskeleton, Part A) consists of 24 chapters concerned with the isolation and characterization of cytoskeletal components, and with the development of research tools to enable the study of cytoskeletal components in living cells and *in vitro* models.

I wish to thank all of the authors who have so generously contributed chapters to this volume, and I apologize to readers who are searching for a technique that has not been included. I also wish to thank Susan Overton for her substantial help in preparing the Index.

LESLIE WILSON

Chapter 1

Fluorescent Analog Cytochemistry of Contractile Proteins

YU-LI WANG, J. M. HEIPLE, AND D. LANSING TAYLOR

Cell and Developmental Biology
Harvard University
Cambridge, Massachusetts

I. Introduction

A new approach to investigate the distribution and interaction of cytoskeletal and contractile proteins was introduced in 1978 (Taylor and Wang, 1978) by combining the techniques of microinjection, fluorescence spectroscopy, and image intensification. The general approach, which we now call fluorescent analog cytochemistry (FAC; previously referred to as molecular cytochemistry), has been defined as the incorporation of functional fluorescent analogs of cellular components into or onto living cells.

Fluorescent analog cytochemistry is specifically designed to follow the distribution and interaction of molecular components in single living cells. However, as with any technique, errors in experimental design, execution, or interpretation can effectively limit this potentially powerful approach. We have discussed some of the advantages and practical considerations of fluorescent analog cytochemis-

ISBN 0-12-564125-7

try in our early papers. This chapter is a set of guidelines and procedures for the optimal utilization of the technique. We will emphasize the most critical factors and point out common pitfalls.

II. Procedures

The technique involves four major steps: (1) purification and fluorescent labeling of cellular components; (2) biochemical and spectroscopic characterization *in vitro;* (3) incorporation into or onto living cells; and (4) image recording and interpretation. The following discussion will be limited to proteinaceous, cytoplasmic components using our results with 5-iodoacetamidofluorescein-labeled actin (5-AF-actin) as illustrations (Taylor and Wang, 1978; Wang and Taylor, 1979, 1980; Taylor *et al.*, 1980). A variety of labeled proteins have been incorporated into various cell types since we first introduced this technique (Feramisco, 1979; Feramisco and Blose, 1980; Kreis *et al.*, 1979; Wehland and Weber, 1980).

A. Purification and Fluorescent Labeling of Cellular Components

Fluorescent labeling of proteins in general has been discussed in several previous reviews (Dandliker and Portmann, 1971; Fairclough and Cantor, 1978; Stryer, 1978). The fluorescent conjugate (hereafter referred to as the "fluorescent analog" of a cellular component) can be prepared either by (a) labeling the purified protein subunits or by (b) labeling a supramolecular complex that contains the target protein, followed by fractionation of the labeled conjugates. This latter approach has the advantage that active sites for structure formation are more likely to be protected from modification.

The fluorescent-labeling reagents should be chosen based on several criteria. The fluorophore should absorb in the visible range, optimally 450–650 nm in wavelength, so that radiation damage to living cells and interference from autofluorescence can be minimized. Furthermore, the fluorophore should be stable under recording conditions. Optimally, fluorophores with high quantum yields and high extinction coefficients should be chosen, in order to maximize the signal from single cells. The reactive group should create a covalent bond that is stable inside the cell, without affecting the biochemical functions of the protein. The optimal reactive group will be determined to a large extent by the properties of the specific protein. For some well-characterized proteins such as actin, it is possible to choose site-specific reagents directed toward an apparently nonessential site (e.g., sulfhydryl reagents react predominantly with cys-373 in F-actin; Elzinga and Collins, 1975; Wang and Taylor, 1980). If nonspecific reagents are used, it is important either to isolate a functional, labeled fraction after reaction or to protect the active sites furing the reaction.

To date, several fluorescent reagents have been used for fluorescent analog cytochemistry. 5-Iodoacetamidofluorescein, rhodamine isothiocyanate (RITC), and dichlorotriazinyl aminofluorescein (DTAF) (Wang and Taylor, 1980; Feramisco, 1979; Keith *et al.*, 1980). Many other fluorescent reagents such as 7-chloro-4-nitrobenzo-2-oxa diazole (NBD-Cl), eosin isothiocyanate, and cyanine dyes fit into the above criteria and remain to be explored (Ghosh and Whitehouse, 1968; Cherry *et al.*, 1976; Waggoner, 1979). Of particular interest is the "piggyback" labeling techniques using $Rhod_7$-α-lactalbumin introduced by Shechter *et al.* (1978). This technique has great potential for preparing fluorescent analogs of minor cellular components and yielding high fluorescence intensity from a very small number of labeled molecules.

B. Biochemical and Spectroscopic Characterization *in Vitro*

It is very important to characterize various properties of the fluorescent analog *in vitro* before microinjection into living cells. These assays not only minimize the possibility of artifacts after microinjection, but also provide necessary information for interpreting results from injected cells.

The solution of the fluorescent analog should be free of unbound or noncovalently associated fluorophores; otherwise, a confusing background fluorescence will be observed in the injected cell, and the results will be impossible to interpret. The test can be performed by using either a desalting column preequilibrated with SDS or SDS-gel electrophoresis. A single fluorescent band comigrating with the protein should be detected.

The average number of fluorophores per protein molecule (F/P ratio) must be determined in order to allow proper evaluation of biochemical assays. The concentration of fluorophore can be determined most conveniently by light adsorption (Dandliker and Portmann, 1971). However, it is necessary first to determine the extinction coefficient of the bound fluorophore. A simple method to obtain approximate values has been described by Hartig *et al.* (1977). The method assumes that the extinction coefficient of the fluorophore bound to unfolded protein (in 7.5 *M* urea) is equal to that of the original fluorescent reagent in the same solvent. For those fluorescent reagents that undergo significant changes in electron distribution on reaction, model compounds prepared by reaction with small molecules (i.e., amino acids) should be used.

If a low F/P ratio is obtained, it is necessary to isolate the labeled fraction to obtain a F/P $\geqslant 0.5$, in order to allow unambiguous evaluation of biochemical assays. For reactions that alter the net charge of proteins, the fractionation can be carried out by ion-exchange chromatography (Wang and Taylor, 1980).

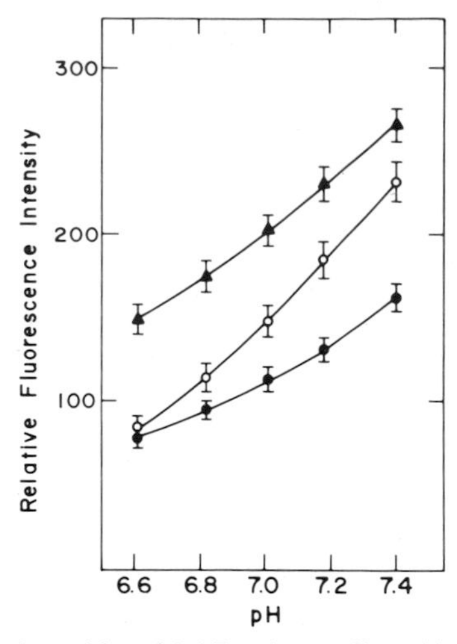

FIG. 1. Peak fluorescence intensities of 5-AF-actin are affected by both pH and the state of actin assembly. The graph indicates that the detection of G-actin (open circles) in the cell will be slightly favored over that of F-actin (solid circles). The binding of HMM (triangles) has a significant effect under saturation conditions, but given the low myosin/actin ratio in nonmuscle cells, the net effect inside cells should be limited. (From Wang and Taylor, 1980.)

Detailed biochemical assays should be carried out to assess the functional activities of the analog. For 5-AF-actin, the assays include viscosity measurement, activation of heavy meromyosin Mg^{2+} ATPase, and formation of Mg^{2+} paracrystals. The results are compared to unlabeled controls and are evaluated in relation to the percentage of protein molecules that are labeled.

Along with assays with purified proteins, additional assays can be performed using crude cell-free extracts. For example, motile extracts of *Dictyostelium discoideum* have been used to test the incorporation of 5-AF-actin into contracting fibrils and pellets (Taylor and Wang, 1978; Wang and Taylor, 1980). The extract system allows assays to be performed at a level between living cells and purified proteins and could cover functions not well characterized with purified proteins.

Spectroscopic characterization is very important for the interpretation of fluorescence signals from injected cells. It is necessary to determine whether the fluorescence intensity of spectra are affected by solvent parameters, such as pH and free Ca^{2+} ion concentration, which could vary in different regions of the cell or during different physiological states of the cell. In addition, the effect of conformational change or structural transformation on fluorescence properties should be investigated, since a particular conformation or structure could be favored at certain sites of the cell and affect the pattern of fluorescence distribution (Fig. 1).

C. Incorporation into Living Cells

The optimal target cell is determined to a large extent by the questions under investigation. For example, cultured mammalian cells are particularly amenable to whole-cell image analysis using tv-image intensifiers and computers, because of the short pathlength and distinct cytoskeletal structures. On the other hand, large cells, such as sea urchin eggs, giant amoebas, and large cultured cells (e.g., myotubes and newt lung fibroblasts), can be measured with large measuring spot sizes and yield strong fluorescence signals, and thus are more readily applicable to microspectrofluorimetric measurements.

A number of microinjection techniques, including direct-pressure microinjection, red cell ghost fusion, liposome fusion, and cell permeation, have recently been introduced or refined to deliver macromolecules into living cells (for a review, see Taylor and Wang, 1980). Two important precautions should be followed when applying these techniques.

First, the solvent condition should be compatible with both the target cell and the fluorescent analog. The cytoplasm should be kept from exposure to harmful conditions such as high pH (>7.2) or high free Ca^{2+} ion concentration ($>10^{-5}M$). In addition, appropriate solvent conditions should be chosen to stabilize the analog. For example, G-actin should be kept in the presence of nucleotides and divalent cations. Conditions that favor actin polymerization, such as high ionic strength (>10 m*M*) and high Mg^{2+} ion concentration (>0.2 m*M*), should be avoided.

Another consideration is the amount of the fluorescent analog to microinject. The final concentration of the analog in the cell should be sufficiently high to yield detectable signals, but low enough to minimize disturbance of the cell. For proteins with enzymatic activities, this precaution is especially important. We have set an upper limit at ~10% of the concentration of the corresponding endogenous component. Unfortunately, for most microinjection techniques (except direct-pressure microinjection; see Kiehart, Chapter 2, this volume), reliable ways to control the volume of delivery have yet to be developed.

An important question is whether the analog can actually be utilized by the cell after microinjection into the cytoplasm. The incorporation of analogs into cellular structures could be affected by such factors as the rate of turnover, the accessibility of incorporation sites, and the mechanism of structure formation. Using mass-incorporation methods such as ghost fusion, it should be possible to apply biochemical techniques to study the turnover of the injected analog (Rechsteiner, 1979). The question of incorporation can be studied by using model systems in which the behavior of the component has been characterized in detail. For example, the single-cell model of *Chaos carolinensis* has been used to test the incorporation of 5-AF-actin (Taylor and Wang, 1978; Taylor *et al.*, 1980). A significant fraction of 5-AF-actin is found to remain associated with the ghost

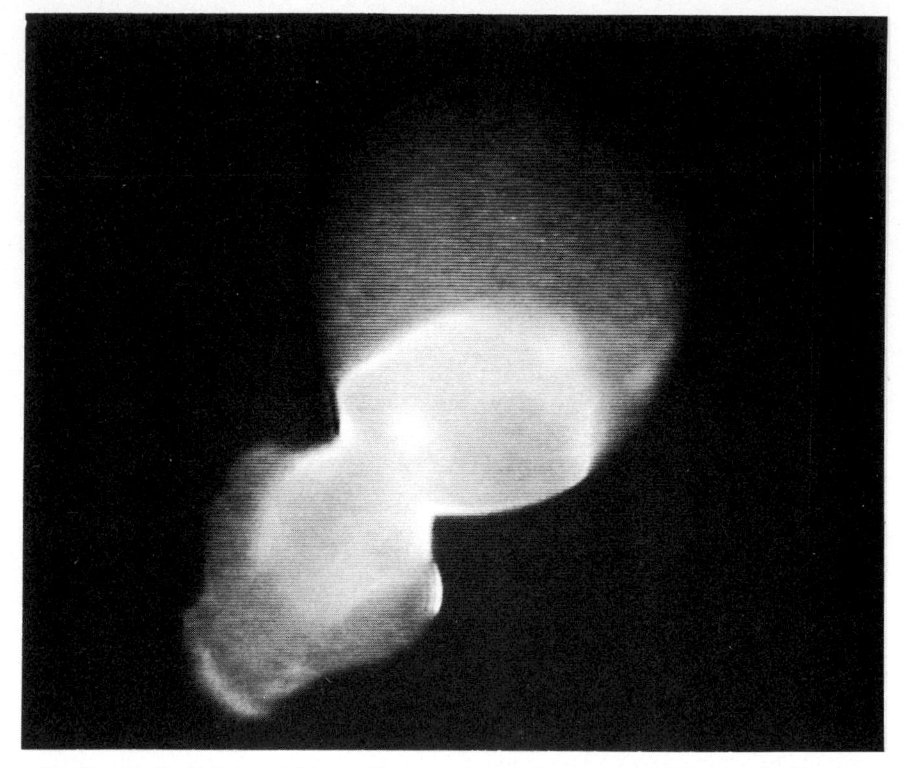

FIG. 2. Phalloidin is injected into a *Chaos carolinensis* that contains 5-AF-actin. The fluorescent fibrils contract irreversibly to one end of the cell and recruit most of the 5-AF-actin. Image recorded using a RCA SIT coupled to a NEC videotape recorder.

and membrane-free cytoplasm on rupturing a preinjected cell in a stabilizing solution. Furthermore, 5-AF-actin becomes incorporated into cytoplasmic fibrils when a cell with 5-AF-actin is postinjected with phalloidin (Fig. 2). These observations are consistent with the known characteristics of endogenous actin in amoebas (Taylor *et al.*, 1976; Stockem *et al.*, 1978). The use of model systems indicates whether the analog can be utilized by the cell under specified conditions, but does not prove that the analog can be incorporated into all possible cell structures or under all conditions. When interpretating results from injected cells, the possibility of ''false negatives'' should always be considered.

D. Image Recording and Interpretation

The fluorescence image of injected cells can be recorded using photography, and high-quality records can be obtained from nonmotile processes (under conditions similar to those used for immunofluorescence). However, the relatively long exposure required (>1–2 minutes) is not desirable for motile processes or for probes sensitive to photobleaching.

The technique of image intensification provides a powerful alternative for image recording (Reynolds, 1972; Reynolds and Taylor, 1980; Willingham and Pastan, 1978; Sedlacek *et al.*, 1976). The sensitivity is at least 3–4 orders of magnitude higher than that of photography. As a result, strong excitation light can be avoided and photobleaching can be minimized. The time resolution is 1/30 to 1/60 of a second, so most motile processes can be recorded in real time. In addition, the image can be readily digitized and processed by computers. Although the image quality becomes a limiting factor at very low light intensity, photography under those conditions is very often impossible to carry out.

The detection limit of fluorescence from single cells depends on the specific recording conditions and on the fluorophore employed. It has been demonstrated that small bundles of 5-AF-actin containing no more than ~4000 fluorophore molecules per micron (~10 filaments) can be readily detected using a 60-watt quartz-halogen lamp and a silicon-intensified target (SIT) tv camera (Wang and Taylor, 1980, Fig. 3). The actual detection limit of discrete structures in the cell will probably be more limited, since the analog will be diluted by the endogenous component, and fluorescent structures will probably be surrounded by a certain level of unorganized fluorescence. We estimate that actin bundles containing ~50 filaments should be detectable under the above conditions. However, using more sensitive image intensifiers in conjunction with computer manipulation of video images, the sensitivity can be significantly higher.

Great caution must be exercised when interpreting fluorescence images of injected cells. The two-dimensional distribution of fluorescence intensity is affected by several factors, including the plane of focus, the accessible volume of the analog, local solvent conditions, local fluorophore concentration, and the conformation or structure of the analog. Some of these factors will affect not only the image of the fluorescent analog being studied but also the image of essentially any fluorescent macromolecule (e.g., fluorescently labeled ovalbumin, fluorescent antibodies).

Accessible volume is determined by the thickness of the cell and by the distribution of membrane-bound organelles (Fig. 4). In those regions of the cell including stress fibers, mitotic spindles and hyaline cytoplasm, which are relatively free of membrane-bound organelles, high apparent fluorescence intensity should be interpreted with caution (Fig. 5). Controls for accessible volume can be carried out by incorporating a fluorescently labeled inert macromolecule with similar solubility properties. For example, labeled ovalbumin has been used as a control for 5-AF-actin. Control experiments can be performed in a separate cell (Wang and Taylor, 1979), or, using fluorophores of different excitation and emission wavelengths, performed in the cell that contains the experimental

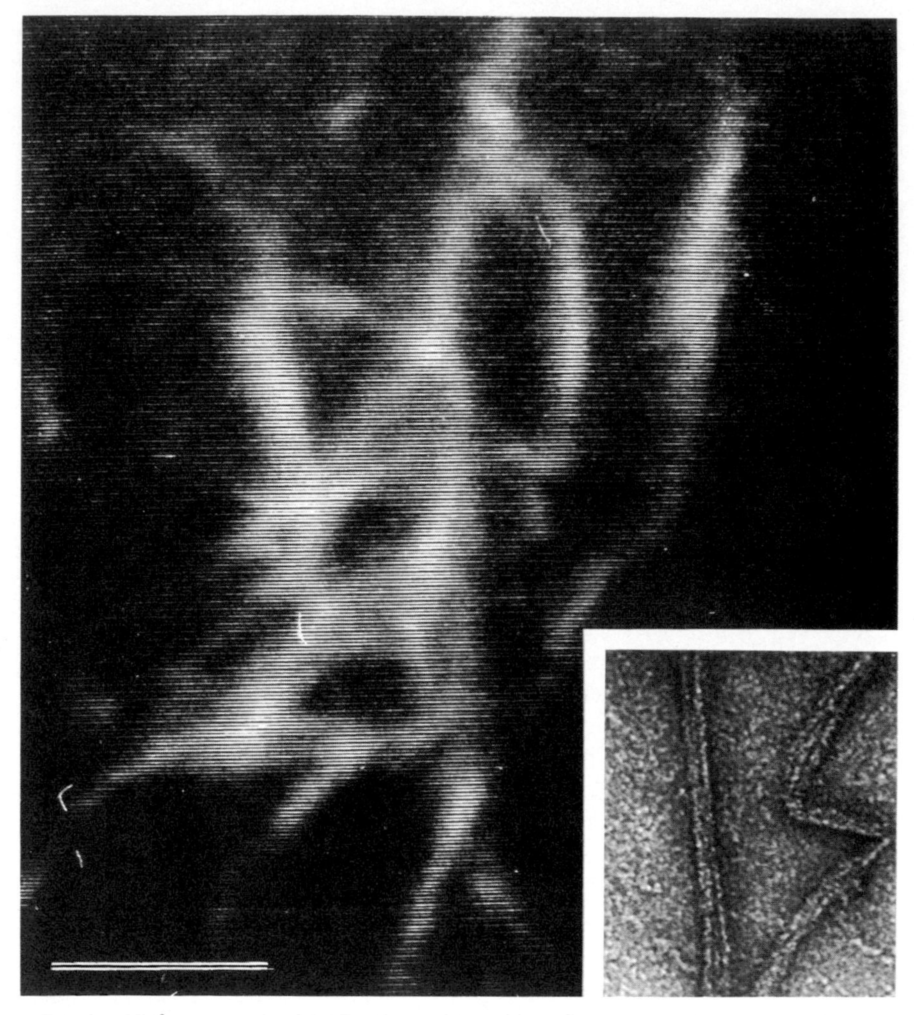

FIG. 3. Ni^{+2} paracrystals of 5-AF-actin, as detected by a fluorescence microscope coupled to a tv-image intensifier. Bar = 5 μm. Inset: electron microscopy of negatively stained Ni^{+2} paracrystals.

analog (Taylor *et al.*, 1980). This latter method is preferable for cells of irregular shape, since images of the analog and the control can be compared directly by shifting the excitation and emission filters. Such comparison can even be performed electronically by computer-aided image analysis.

As discussed previously, the sensitivity of the fluorophore to solvent conditions should be studied *in vitro* before incorporation into living cells. For those fluorophores that are sensitive to solvent conditions, experiments should be

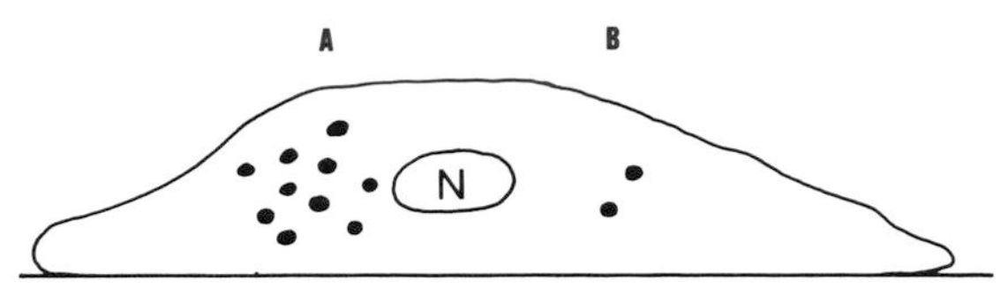

FIG. 4. Simplified diagram of a cell viewed from the side. Injected soluble macromolecules are excluded by the membrane-bound organelles. As a result, region B shows higher fluorescence intensity than region A, even if the injected fluorescent molecules are uniformly distributed.

performed to identify possible solvent effects inside the cell. The easiest approach is to inject a control molecule labeled with the same fluorophore into a separate cell, and to compare the fluorescence distribution with that of the analog. Alternatively, some solvent factors such as pH and free Ca^{2+} ion concentration can be measured by independent methods (Blinks *et al.*, 1979; Caswell, 1979; Heiple and Taylor, 1980).

The fluorescence distribution could also be affected by the conformation or the supramolecular structure of the analog. This effect will be specific for the analog and cannot be identified using a different control protein. For example, actin labeled with NBD-Cl shows dramatic increases in fluorescence intensity during polymerization (Detmers *et al.*, in 1981). Therefore, images of injected cells would be biased toward regions enriched in F-actin. This kind of analog with high microenvironmental sensitivity will be useful for studying intracellular pro-

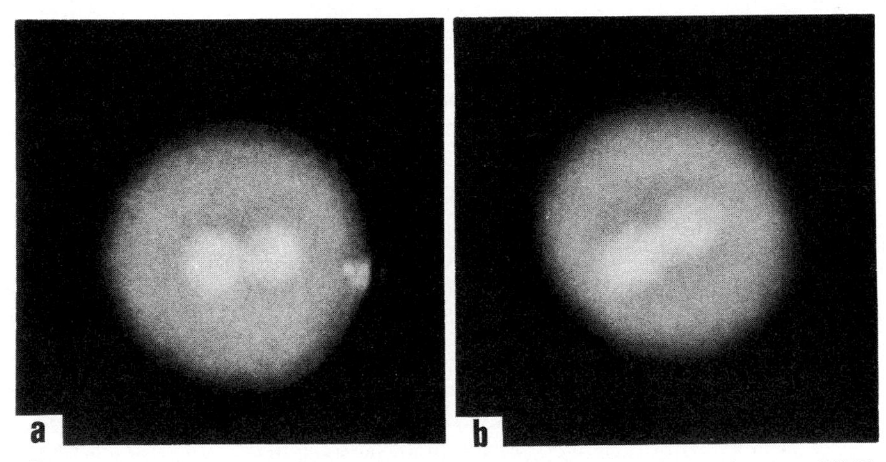

FIG. 5. Two separate sea urchin eggs during mitosis, injected with 5-AF-actin (a), and FTC-ovalbumin (b), respectively. The mitotic spindles in both cells are distinctly fluorescent, most likely because of their large accessible volume. It is impossible to tell, based on qualitative observations, whether 5-AF-actin is more concentrated in the spindle than elsewhere. (From Wang and Taylor, 1979.)

tein conformation or structure, but if it is used for distribution studies, the images should be interpreted cautiously. On the other hand, analogs relatively insensitive to microenvironment are more suitable for distribution studies. One possible way to identify specific microenvironmental effects is to compare images of different analogs labeled with fluorophores of different microenvironmental sensitivities.

III. Conclusion and Prospectus

Three levels of analysis with fluorescent analog cytochemistry are possible: (1) qualitative studies on the distribution of cellular organelles or molecules; (2) quantitative studies on the distribution of cellular organelles or molecules; and (3) microspectrofluorometric analysis of the interaction of organelles or molecules molecules.

In any given experiment, both the nature of the biological question asked and the cell type employed will dictate the level of resolution and quantitation at which this technique can be applied. To date, fluorescent analog cytochemistry has been limited to qualitative studies on the distribution of specific molecules. This approach should continue to be valuable in answering a wide variety of cellular questions in many different cell types.

Quantitative analyses of the distribution of specific proteins will be particularly important in cells with irregular shape and during dynamic cellular processes such as amoeboid movement, karyo- and cytokinesis. Computer manipulation of digitized images will enable a variety of quantitative studies to be performed, including normalization of pathlength and accessible volumes, enhancement of images, and real-time mapping of the distribution of specific molecules.

One of the most important applications of fluorescent analog cytochemistry will be the measurement of spectroscopic properties of the labeled molecules *in vivo*. Fluorescence photobleaching, polarization, and lifetime measurements, as well as resonance energy transfer, can be quantitated in living cells (see Taylor and Wang, 1980, for a review). Such data may yield information on diffusion coefficients, specific binding activities, and supramolecular structures of the labeled protein or proteins.

We have emphasized the use of fluorescent analog cytochemistry in the study of cell motility. However, these techniques can be applied to a wide variety of organelles and cellular constituents (Taylor and Wang, 1980). Careful planning and execution of the experiments and controls should make this a powerful tool in many different fields.

ACKNOWLEDGMENTS

The authors would like to thank J. Lupo, M. Rockwell, L. Meszolay, and L. Houck for excellent technical assistance. The research described in this report was supported by grants from the NIH (AM 18111) and NSF (PCM-7822499).

REFERENCES

Blinks, J. R., Mattingly, P. H., Jewell, B. R., Leeuwen, M., Harrer, G. C., and Allen, D. G. (1979). *In* "Methods in Enzymology" (W. B. Jakoby and I. H. Pastan, eds.), Vol. 58, pp. 292–328. Academic Press, New York.

Caswell, A. H. (1979). *Int. Rev. Cytol.* **56,** 145–181.

Cherry, R. J., Cogoli, A., Oppliger, M., Schneider, G., and Semenza, G. (1976). *Biochemistry* **15,** 3653–3656.

Dandliker, W. B., and Portmann, A. J. (1971). *In* "Excited States of Proteins and Nucleic Acids" (R. F. Steiner and I. Weinryb, eds.), pp. 199–275. Plenum, New York.

Detmers, P., Weber, A., Elzinga, M., and Stephens, R. (1981). In press.

Elzinga, M., and Collins, J. H. (1975). *J. Biol. Chem.* **250,** 5897–5905.

Fairclough, R. H., and Cantor, C. R. (1978). *In* "Methods in Enzymology" (S. N. Timasheff and C. H. W. Hirs, eds.). Vol. 48, pp. 347–379. Academic Press, New York.

Feramisco, J. R. (1979). *Proc. Natl. Acad. Sci. U.S.A.* **76,** 3967–3971.

Feramisco, J. R., and Blose, S. H. (1980). *J. Cell Biol.* **86,** 608–615.

Ghosh, P. B., and Whitehouse, M. W. (1968). *Biochem. J.* **108,** 155–156.

Hartig, P. R., Bertrand, N. J., and Sauer, K. (1977). *Biochemistry* **16,** 4275–4282.

Heiple, J. M., and Taylor, D. L. (1980). *J. Cell Biol.* **86,** 885–890.

Keith, C. H., Feramisco, J. R., and Shelanski, M. L. (1980). *J. Cell Biol.* **88,** 234–240.

Kreis, T. E., Winterhalter, K. H., and Birchmeier, W. (1979). *Proc. Natl. Acad. Sci. U.S.A.* **76,** 3814–3818.

Rechsteiner, M. (1979). *In* "Transfer of Cell Constituents into Eukaryotic Cells" (J. E. Celis, A. Graessmann, and A. Loyter, eds.), pp. 113–141. Plenum, New York.

Reynolds, G. T. (1972). *Q. Rev. Biophys.* **5,** 295–347.

Reynolds, G. T., and Taylor, D. L. (1980). *BioScience* **30,** 586–592.

Sedlacek, H., Gundlach, H., and Ax, W. (1976). *Behring Inst. Mitt.* **59,** 64–70.

Shechter, Y., Schlessinger, J., Jacobs, S. Chang, K.-J., and Cuatrescasas, P. (1978). *Proc. Natl. Acad. Sci. U.S.A.* **75,** 2135–2139.

Stockem, W., Weber, K., and Wehland, J. (1978). *Cytobiologic* **18,** 114–131.

Stryer, L. (1978). *Annu. Rev. Biochem.* **47,** 819–846.

Taylor, D. L., and Wang, Y.-L. (1978). *Proc. Natl. Acad. Sci. U.S.A.* **75,** 857–861.

Taylor, D. L., and Wang, Y.-L. (1980). *Nature (London)* **284,** 405–410.

Taylor, D. L., Rhodes, J. A., and Hammond, S. A. (1976). *J. Cell Biol.* **70,** 123–143.

Waggoner, A. S. (1979). *Annu. Rev. Biophys. Bioeng.* **8,** 47–68.

Wang, Y.-L., and Taylor, D. L. (1979). *J. Cell Biol.* **81,** 672–679.

Wang, Y.-L., and Taylor, D. L. (1980). *J. Histochem. Cytochem.* **28,** 1198–1206.

Wehland, J., and Weber, K. (1980). *Exp. Cell Res.* **127,** 397–408.

Willingham, M., and Pastan, I. (1978). *Cell* **13,** 501–507.

Chapter 2

Microinjection of Echinoderm Eggs: Apparatus and Procedures

DANIEL P. KIEHART

Department of Cell Biology and Anatomy
Johns Hopkins School of Medicine
Baltimore, Maryland

I. Introduction

Although biochemists have been surprisingly successful at isolating functional macromolecules and organelles from living cells, the complexity of certain processes has precluded the purification and subsequent reactivation of their isolated components. Microinjection has proved to be an invaluable technique that can be used to effect local changes in the chemical and physical milieu of these complex cellular processes *in vivo* (for recent reviews, see Baserga *et al.*, 1980; Celis *et al.*, 1980). Microinjection studies provide information on intracellular homeostasis of various ions (e.g., Ca^{2+}, H^{+}) and serve to elucidate the role and mechanism of ionic control in cell function. Further, they provide insight for subsequent *in vitro* studies on purified systems and their components.

ISBN 0-12-564125-7

The role of specific macromolecules in cell processes can also be evaluated by microinjection. The distribution of purified, fluorescently tagged proteins, microinjected into cells, can be traced through various cell functions (Taylor and Wang, 1980). Alternatively, naturally occurring or specially designed (e.g., specific antibody) inhibitors of enzymatic or other molecular functions (Mabuchi and Okuno, 1977; Kiehart *et al.*, 1977a; Furasawa *et al.*, 1976) can be microinjected to evaluate the functional relevance of particular proteins in any number of cell processes.

Although various substances can be introduced into the interior of cells by microelectrophoresis (Curtis, 1964), vesicle fusion (Poste *et al.*, 1976); Weinstein *et al.*, 1977), or red cell ghost fusion (Furasawa *et al.*, 1976), no method is as versatile as direct-pressure microinjection for rapidly introducing nonpermeable agents into different compartments of individual cells. As a consequence, a variety of pressure microinjection techniques have been devised (Barber, 1911; Chambers and Kopac, 1950; Chambers and Chambers, 1961; Diacumakos, 1973; Ellis, 1961; deFonbrune, 1949; Hiramoto, 1956, 1962, 1974; Kohen *et al.*, 1975; Kopac, 1955, 1964; Needham and Needham, 1925, 1926a,b). Precise volumes of charged or uncharged, aqueous or nonaqueous solutions or suspensions can be rapidly (in < 0.25 second) pressure-injected into local regions of the cytoplasm without harm to the target cell. Pressure microinjection is used to inject inert oils, salt and protein solutions, or even whole, living sperm into all different cell types from echinoderm gametes to cultured mammalian cells. Only with pressure microinjection can the consequences of high, experimentally induced concentration gradients of biologically active substances in local regions of the cytoplasm be studied. The dose of such pressure-injected solutions (dose $=$ volume $\times$ concentration) can be precisely determined, whereas microelectrophoresis or fusion methods require various assumptions to produce, at best, estimates of injected dose (Curtis, 1964; Poste *et al.*, 1976; Weinstein *et al.*, 1977).

This chapter provides a description of a new chamber in which marine eggs are immobilized for observation and microinjection, a new method for determining the volume of injected solutions, and a detailed description of a simple, easy-to-construct microinjection system that was devised but never fully described by Hiramoto (1962). An abbreviated version of this contribution appears elsewhere (Kiehart, 1981).

II. Injection Chamber for Marine Eggs

Sea urchin, sand dollar, starfish, and marine annelid oocytes and eggs are normally free-floating spherical cells. For microinjection, viable cells are gently flattened and immobilized between two parallel sheets of coverslip that are closely opposed in the form of a rectangular chamber. Cells in the chamber remain healthy and rapidly divide in synchrony with cells cultured under optimal conditions. The chamber allows easy access to these rapidly dividing eggs by micropipet or other microtools (Figs. 1–4) during observation with high-resolution and high-extinction polarized light, bright-field, phase, fluorescence, or differential-interference microscopy. Further, individual cells can be easily removed from the chamber for subsequent processing (preparation for electron microscopy, for example) without disturbing the other cells in the chamber.

Preparation of the injection chamber requires stocks of coverslips, coverslip fragments, and spacers. Coverslips are cleaned, dried, and stored in a dust-free cabinet as described elsewhere (Fuseler, 1975). A clean, strain-free 22 × 22 mm, #1½ coverslip is scored with a carbide or diamond glass scribe and carefully broken to make 14 rectangular coverslip fragments, each approximately 11 × 3 mm. Rectangular spacers, approximately 11 × 2 mm, are prepared from sheets or rolls of nontoxic, easy-to-cut lamena of desired thickness (see below). A clean piece of black Lucite (approximately 8 × 4 × ¼ in.) provides an excellent background and surface on which chambers are assembled. A 22 × 22 mm clean coverslip is placed on the Lucite. The long edge of the 11 × 2 mm spacer is positioned on the coverslip about 3–5 mm from, and parallel to, one edge of the coverslip (Fig. 1a). An 11 × 3 mm coverslip fragment is next centered on top of

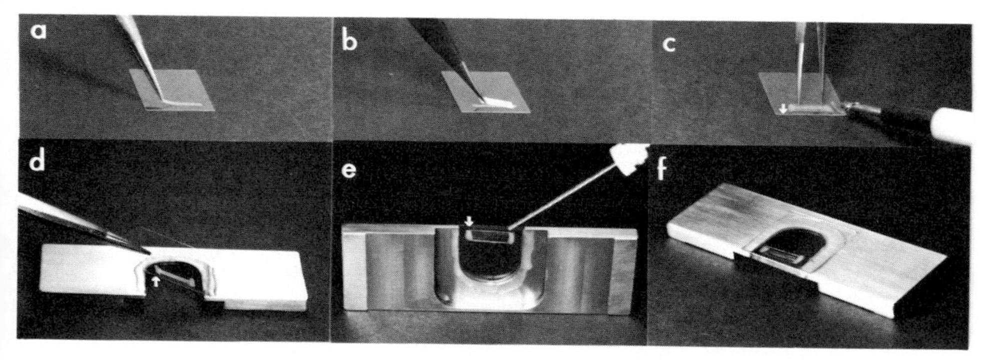

FIG. 1. Preparation and loading of an injection chamber for immobilizing and viewing cells. (a) A spacer is placed parallel to the edge of a coverslip. (b) A coverslip fragment is centered on the spacer, and (c) cemented into place. (d) Next, the chamber is sealed by the surface tension of silicone grease to a stainless-steel support slide. (e) Cells are drawn into the injection chamber by capillary action as they are pipetted into the front side of the chamber. (f) The finished chamber: A bottom coverslip is sealed into place, the preparation is filled with seawater, and the open end is capped with mineral oil. The arrows in c, d, and e indicate the edge on which a wax pencil line was drawn to keep the mineral oil from creeping back, away from the open end of the preparation. Details are given in the text.

the spacer. The short edges of the fragment are fastened into place with molten Tackiwax (Central Scientific Co., Chicago, Illinois) that is allowed to harden (Fig. 1b and c). Dental wax and valap (Fuseler, 1975) were tried but failed to hold the fragment in place reliably during use. To ensure uniform thickness of the chamber, I sandwich the spacer tightly between the coverslip and the coverslip fragment by gently pressing on the fragment with watchmakers' forceps (Fig. 1c) while the molten Tackiwax is being applied. The space underneath the overhanging fragment forms the rectangular injection chamber, a space approximately 10 mm long, 0.3–0.8 mm deep, and as high as the thickness of the spacer (Figs. 2 and 3). The long front side of the chamber remains open.

The thickness of the spacer is chosen so that when eggs are later introduced into the finished chamber they remain viable but are flattened to about 50–70% of their original diameter. For example, for *Asterias forbesi* eggs, I use Scotch Brand Magic Tape (~70 μm thick, 3M Company, St. Paul, Minnesota); for *Lytechinus pictus* and *L. variegatus* eggs, I use 100 s Mylar (~50 μm, E. I. Dupont, Wilmington, Delaware); and for *Echinarachnius parma,* I use 200 s Mylar (~100 μm) as spacer material. Flattening holds the cells immobile and improves the optical character of cytoplasmic organelles such as the mitotic spindle by displacing light-scattering granules from above and below the organelle. The design of the chamber is such that even if cells are not firmly held by flattening alone, they can be pushed back until they are immobilized against the rear wall of the chamber (i.e., the spacer).

Construction of the injection chamber is completed by drawing a line with a wax pencil along the top edge of the coverslip in front of the chamber. This line

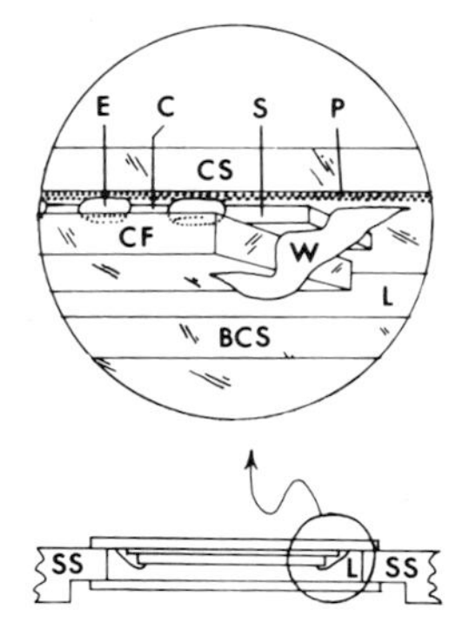

FIG. 2. Perspective schematic of the injection chamber. Lower: Front view of injection chamber mounted on the stainless-steel support slide is shown. SS, Stainless-steel support slide; L, lumen of the injection slide. Upper: Perspective view of the injection chamber, looking up and to the left. Vertical scale has been compressed to include the bottom coverslip. BCS, Bottom coverslip; C, injection chamber; CF, coverslip fragment; CS, top coverslip; E, egg; L, lumen of the injection slide; P, wax pencil line; S, spacer; W, Tackiwax. Drawing by Eleanor Fedde.

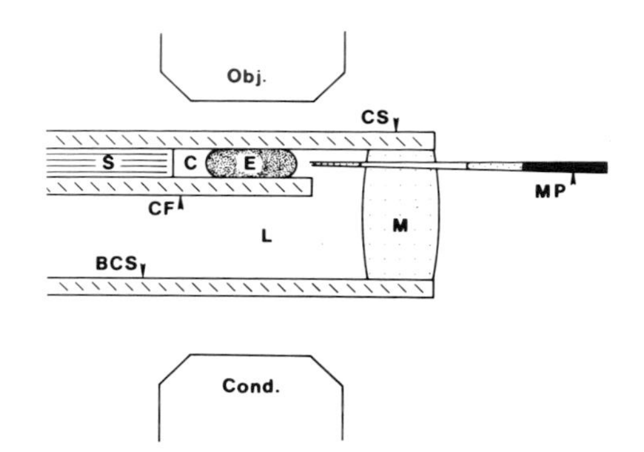

FIG. 3. Cross-sectional schematic of the injection slide. A cutaway view of the injection slide schematically depicts the relationship between egg, microinjection chamber, loaded micropipet, and microscope, just prior to an injection. Cross section is in the plane defined by the microscope's optical axis and the long axis of the micropipet. The vertical scale is somewhat compressed. Obj., Objective; Cond., condenser; BCS, bottom coverslip; CF, coverslip fragment; C, injection chamber; E, egg; S, spacer; CS, top coverslip; L, lumen of the injection slide; M, mineral oil; MP, loaded micropipet (see Fig. 9 and text).

provides a hydrophobic barrier whose purpose will later become apparent. Finished chambers are stored in a dust-free cabinet until use. For microinjection the chamber, fragment side down, is mounted with silicone grease (Dow Corning Corp., Midland, Michigan), across the U-shaped opening in a stainless-steel or

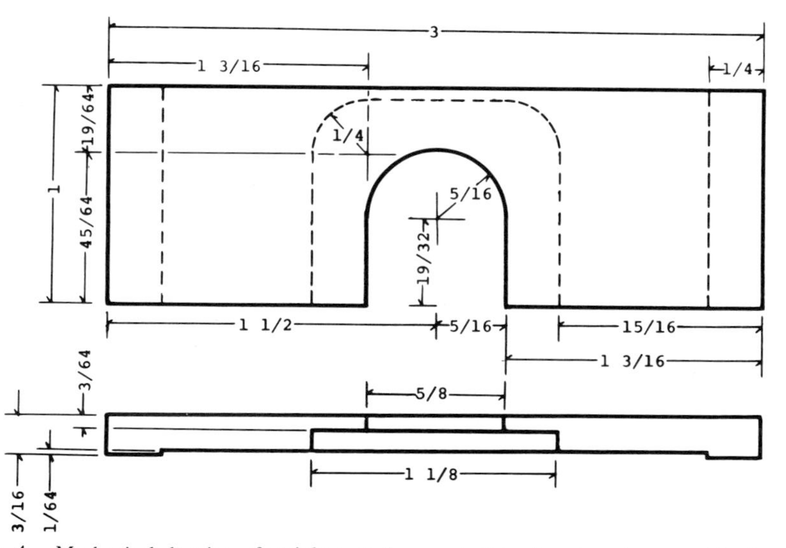

FIG. 4. Mechanical drawing of stainless-steel support slide. Drawing gives specifications for machining the metal (stainless-steel or aluminum) slide on which glass injection chambers were mounted. After Begg and Ellis, 1979.

aluminum support slide designed by Gordon Ellis (Begg and Ellis, 1979) having approximately the same dimensions as a glass microscope slide (Figs. 1, 4).

III. Loading Eggs into the Injection Chamber

The distance between the two glass surfaces of the injection chamber is less than the diameter of an echinoderm egg, so capillary action is required to draw the eggs into the chamber. Unfortunately, the eggs have a tendency to be forcibly drawn into the exceedingly narrow spaces between spacer and glass, at which time they lyse. By placing a small drop of seawater at the rear edge of the coverslip fragment just before the eggs are introduced, these narrow spaces can be filled while the chamber itself remains empty. The chamber will remain empty provided the spacer is longer than the coverslip fragment (Figs. 1b, 2). To load the chamber, an egg suspension is drawn into a 20-μl Microcap (Drummond Scientific Co., Broomall, Pennsylvania), then expelled adjacent to the lumen of the chamber (Fig. 1e). Capillary action draws the eggs and the seawater into the entire length of the injection chamber. The number of eggs in the chamber is controlled by adjusting the density of the egg suspension in the Microcap. A loose pellet of eggs formed by gentle hand centrifugation in a 1-ml conical tube yields the egg

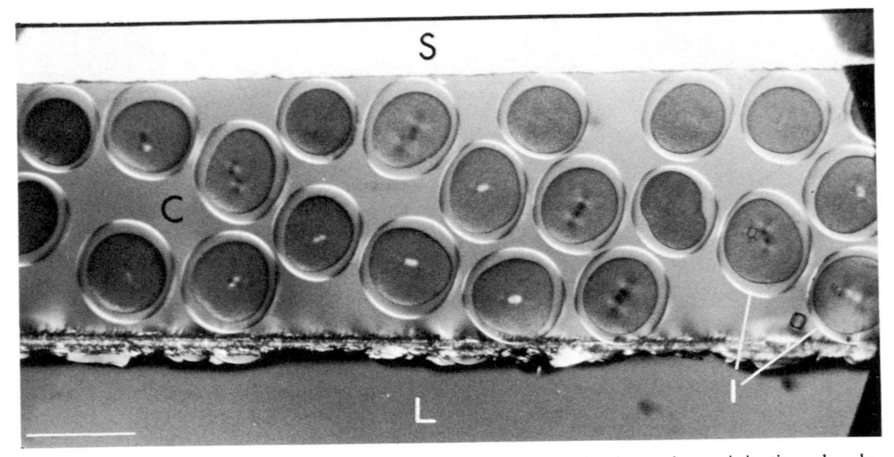

FIG. 5. Cells in an injection chamber. *Asterias forbesi* cells shown in an injection chamber viewed with a 4 × objective. Fertilized oocytes were loaded into the chamber during prometaphase of first meiosis. Polar body formation and syngamy occurred normally and synchronously throughout the chamber. Here eggs are in first mitosis approximately 2 hours and 10 minutes after they had been loaded into the chamber. C, Injection chamber; I, two injected cells; L, lumen of the injection slide; S, spacer (here a highly birefringent strip of Scotch Brand Magic Tape). Bar = 200 μm.

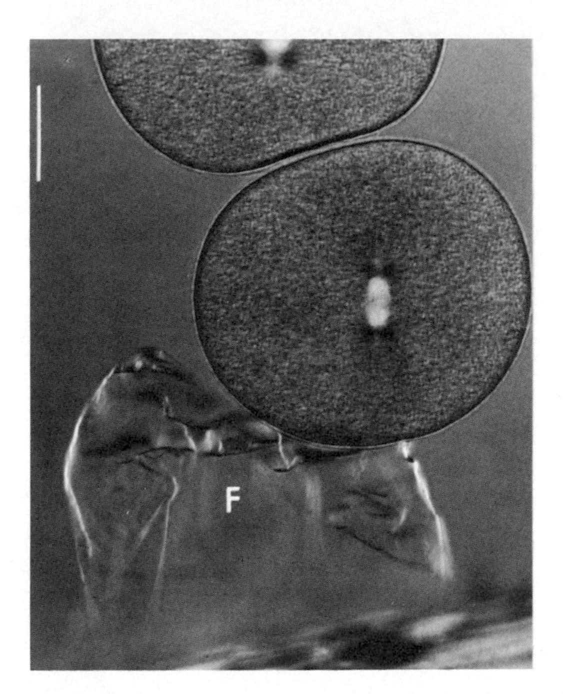

FIG. 6. Removal of fertilization envelope. Approximately 5 minutes after fertilization, *Lytechinus variegatus* eggs were loaded into an injection chamber. Capillary action drew the eggs into the narrow space between the parallel sheets of coverglass that form the top and bottom of the chamber. This process mechanically sheared the fertilization envelope off the eggs (see text). Birefringent fertilization envelope (F), is adjacent to one of the eggs. Bar = 40 μm.

density required to provide approximately 90–150 well-spaced eggs per chamber (Fig. 5). A coverslip is mounted immediately with a wax pencil line drawn across one edge on the underside of the metal support slide so that the wax pencil line faces the upper (injection chamber) coverslip (Figs. 1f, 2, and 3). Seawater is pipetted rapidly into the space between the two coverslips, up to the inside edge of the wax pencil lines. A layer of mineral oil (Squibb, drugstore grade) is added to seal the wax pencil line-thick opening between the two coverslips and bring the fluid level up to the edges of the coverslips. The two wax pencil lines provide a hydrophobic surface on the glass that prevents seawater from displacing the mineral oil from the front of the now-complete injection slide. The mineral oil acts as a cap that prevents evaporation of the seawater but allows microinstruments to pass through to the cells.

Certain eggs must be relieved of their fertilization membranes in order to allow micropipet penetration (for example, *Lytechinus variegatus,* but not *Asterias forbesi* eggs). Since the thickness of the injection chamber is considerably thinner than the diameter of the unflattened eggs, the shear between coverslip and eggs introduced when capillary action draws the eggs into the chamber can be used to remove the fertilization membrane, provided the fertilization envelope has risen fully, but has not completely hardened. Almost all *L. variegatus* eggs loaded into the injection chamber between 4 and 8 minutes after they are fertilized are demembranated. The fertilization envelope ghost can often be viewed adjacent to the demembranated egg (Fig. 6). Thus no separate demembranation technique is required.

Unfertilized eggs, flattened and immobilized in the chamber, can be fertilized

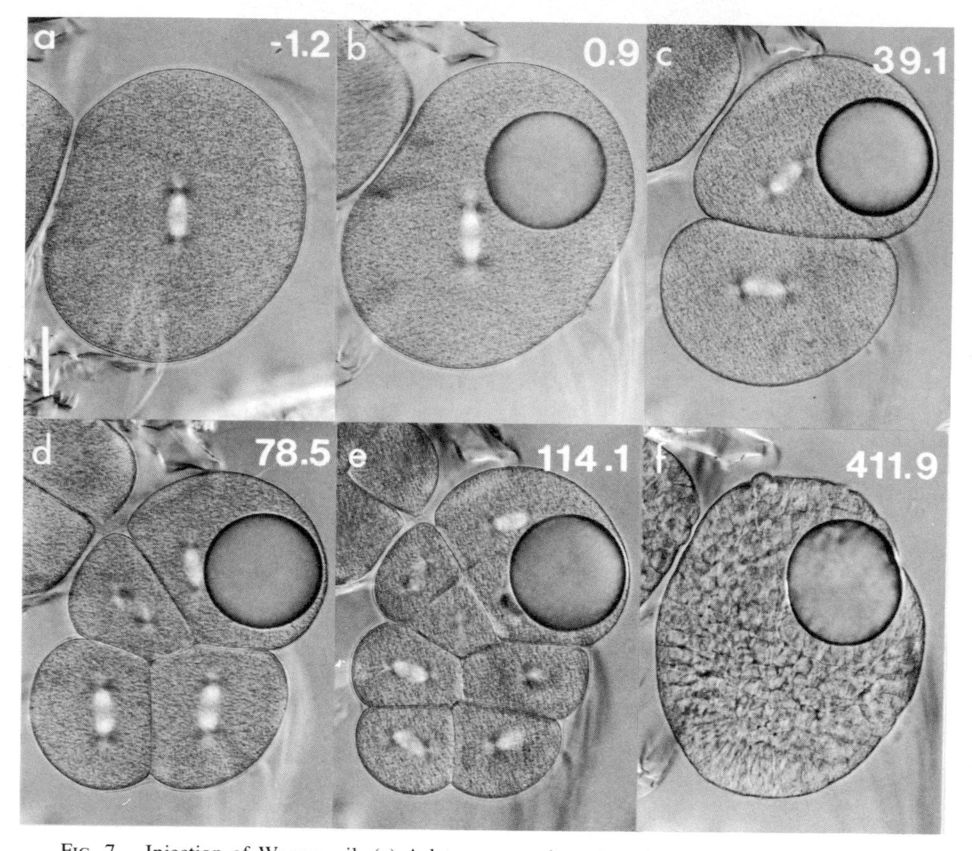

FIG. 7. Injection of Wesson oil. (a) A late-prometaphase *Lytechinus variegatus* egg, prior to injection. (b) Spindle birefringence and size are unaffected by the injection of 154 pl of Wesson oil. Cell has progressed to full metaphase. (c) The same embryo following cleavage, now in late prometaphase of second mitosis. (d and e) Later still, spindles are seen in four- and eight-cell embryos, respectively. The oil drop has no effect on spindle birefringence and size. Asymmetric cleavages are the result of the skewed position of the spindles in the oil-containing blastomeres. (f) After 7 hours, the Wesson oil-injected embryo is still dividing in perfect synchrony with uninjected controls (see embryo in upper left). Cell-formed cilia then swam out of the injection chamber into the lumen of the injection slide (not shown, see text). Time is given in minutes from time of injection. Bar = 40 μm.

by adding sperm to the seawater in the injection slide. Whether fertilized before or after being loaded into the chamber, eggs divide normally and synchronously (Fig. 5). After hatching, embryos swim out of the injection chamber and into the large lumen of the injection slide, where they mature normally.

IV. Capillary Reservoir for Injection Solutions

A 20-μl Microcap (Drummond Scientific Co., Broomall, Pennsylvania) serves as a reservoir for storing the solution to be injected. Approximately 4 μl (a column of fluid about 1 cm long) of solution is sucked by capillary action into the end of the Microcap. This solution is "capped," again by capillary action, with approximately 0.5 μl (2–3 mm) of vegetable oil. Wesson oil is used because it proves nontoxic to living cells (Fig. 7) and does not denature isolated proteins, such as aequorin (the oil is preequilibrated with 10 m*M* EDTA; Kiehart *et al.*, Eisen, 1977b) or γ-globulin (Kiehart *et al.*, 1977a). The distal end of the capillary reservoir is sealed with petroleum jelly or silicone grease. For loading pipets the reservoir is placed flat on the injection slide, perpendicular to the slide's long axis.

V. Microinjection Apparatus

The principle of the microinjection system described here was developed by Hiramoto (1962). The system employs a simple, yet elegant, pressure-volume transducer to effect the small volume changes required for gentle yet rapid pressure microinjection. The transducer consists of a screw-driven fluid-filled syringe that increases, maintains, or decreases the pressure on a small column of mercury placed in the neck of a pulled-glass micropipet (Fig. 8). Because of its high surface tension, the mercury normally assumes the form of a sphere and backs out of the necked region of the pipet. Applied pressure forces the mercury back into the neck. Changes in pressure effected by manipulating the screw-driven syringe alter the equilibrium position of the mercury, causing it to advance toward or retreat from the tip of the micropipet. Thus relatively large changes in volume in the screw-driven syringe are indirectly but precisely coupled (through the mercury pressure-volume transducer) to the minute volume changes at the pipet tip.

The rate at which pressure is applied to the mercury can be altered by varying the total volume of gas included in the microinjection system. If a large volume of air is included, advance of the piston in the screw-driven syringe develops

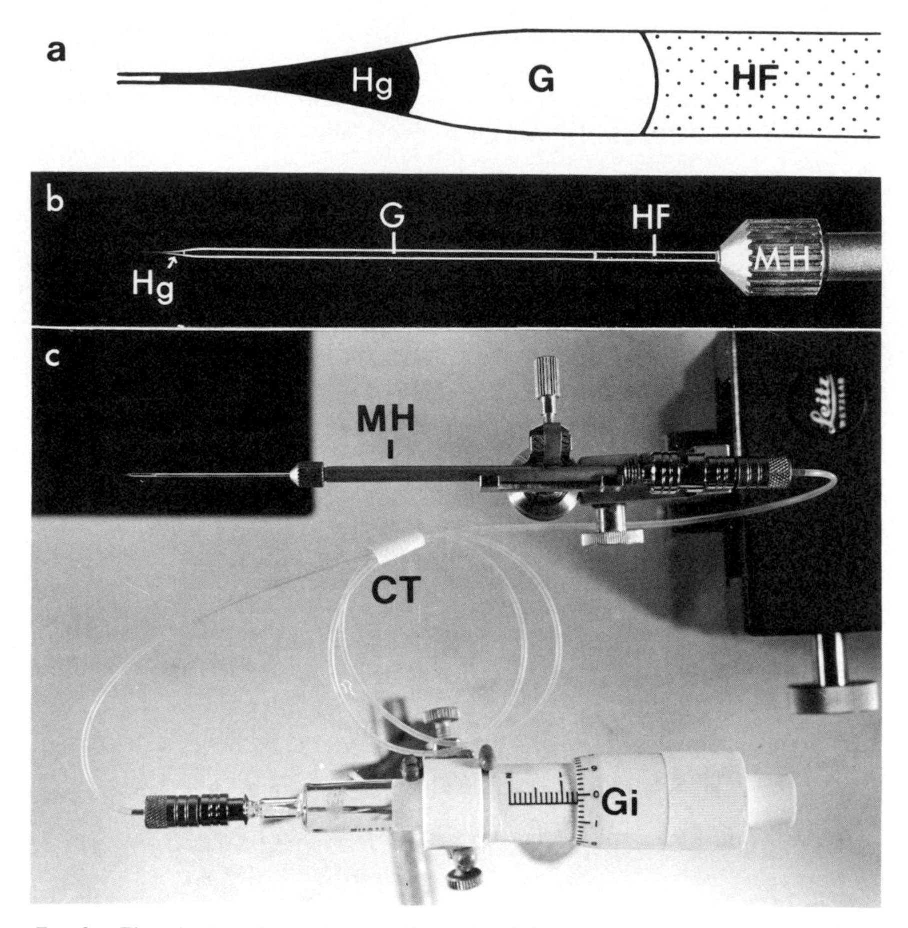

FIG. 8. The microinjection system. (a) Schematic of the mercury pressure-volume transducer. Hg, Mercury; G, gas phase; HF, hydraulic fluid. (b) Photograph of mercury pressure-volume transducer. MH, micropipet holder; HF, G, and Hg as in (a). (c) Photograph of the complete microinjection system showing a Gilmont micrometer syringe (Gi), connecting tubes (CT) fitted with Luer-lok adapters, the micropipet holder (MH) mounted in place on a Leitz manipulator, and a micropipet with mercury in place, ready for loading aqueous-injection solution.

pressure only slowly, and the volume change at the pipet tip is consequently small. If the volume of air in the syringe is decreased, pressure is rapidly developed and volume changes in the pipet tip are great. Precise control of volume changes ranging from < 0.01 pl to > 1 nl can be accommodated by choosing an appropriate volume of included air. Thus direct-pressure microinjection into a wide variety of cell types is feasible with this method.

VI. Assembly of the Microinjection Apparatus

A sufficiently pressure-tight injection system can be constructed from easily modified, commercially available components: a 1-cc glass Luer-lok tip tuberculin syringe (#2027, Becton-Dickinson Co., Rutherford, New Jersey) is mounted in a Leitz screw-advance syringe holder. The steep pitch of the syringe holder's three-start screw thread allows pressure to be applied or released very rapidly with only a small turn of the screw. The piston of the syringe is heavily greased with high-vacuum silicone grease (Dow Corning Corp., Midland, Michigan) to prevent pressure leaks. Alternatively, a 2-ml micrometer syringe (Gilmont Instruments, Great Neck, New York) fitted with a Luer tip can be used. Changes in pressure are delivered to the micropipet through a pressure-tight system of polyethylene tubing, Luer-lok fittings, and a modified Leitz microinstrument holder (Fig. 8). It is important to stress that Luer-lok fittings are required to maintain the integrity of the system against the high pressures developed during microinjection. Friction-fit connections invariably fail, often launching various parts of the apparatus across the room. For this reason, the rear end of the Leitz microinstrument holder is threaded and fit with a tapped, male Luer-lok fitting [modified from Clay-Adams No. A-1027, size A (7550)]. Similarly, the polyethylene connecting tubing (PE20, intramedic polyethylene tubing, Clay-Adams, Inc., Parsippany, New Jersey) is fit on both ends with female Luer-lok fittings [Clay-Adams No. 1-1025, size A (7540)]. To further ensure against leaks, all fittings are greased with high-vacuum silicone grease before they are joined.

Before fitting the micropipet, the system is filled with an appropriate hydraulic fluid. Cavitation under the influence of "negative" pressure is minimized by choosing a fluid with both low vapor pressure and low surface tension. Such fluids most efficiently wet internal surfaces and minimize nucleation points at which the cavitation phase change most readily occurs. For biological applications it is most convenient if the hydraulic fluid is also nontoxic. Boiled deionized water—or better, degassed fluorocarbon oil (FC47, Gibco)—prove effective. Operationally, sufficient control for injection of volumes ranging from 0.1 to 100 pl into marine gametes is afforded by completely filling the system with FC47 oil and allowing the only air in the system to be that introduced in the shaft of the micropipet.

Glass micropipets are drawn from thin-walled Pyrex glass capillary stock (0.8 mm OD, 0.6 mm ID, 100 mm long; Drummond Scientific Co., Broomall, Pennsylvania) on a DuBois-type horizontal micropipet puller (DuBois, 1931) manufactured by Leitz. Nitric acid-cleaned metallic mercury (for cleaning procedure, see Hodgmann, 1958, p. 3285) is drawn into a 10-μl Hamilton syringe, and a small cylinder of mercury about 1 mm long is deposited about 1 cm from the rear end of the micropipet.

The mercury-loaded glass micropipet is next friction-fitted into the Leitz microinstrument holder (originally designed by R. Frank; see Chambers and Kopac, 1950). This is the *only* friction-fit connection in the whole system. Inside the neck of the microinstrument holder, a silicone rubber bushing is pressed tightly around the shaft of the micropipet by a brass collar as the knurled retaining collar is tighened. The silicone rubber bushing must fit snugly around the shaft of the micropipet even before tightening for the seal to be complete and pressure tight. With the pipet in place, a continuous pressure-tight system extends from the screw-driven injection syringe, through the tubing and the instrument holder, to the tip of the micropipet (Fig. 8).

To determine whether or not the injection system is pressure tight, increase the pressure with the injection syringe so that the mercury advances toward the tip of the micropipet. If the mercury remains stationary when pressure is no longer increased, the system is pressure tight. If the mercury slowly retreats, there is a leak that has to be corrected before any microinjections are attempted.

VII. Loading the Micropipet

A Leitz mechanical micromanipulator (described in Kopac, 1964) can be used to position the micropipets for loading and microinjection. Any micromanipulator with similar features can be used. The Leitz micromanipulator provides coarse movement of the micropipet in three dimensions by screw-driven adjustments, and fine movement via a ''joystick'' in the two horizontal dimensions and via a fine-focusing knob in the vertical dimension. The microinstrument holder is clamped into a carrier mounted on a ball joint and attached directly to the micromanipulator.

A clean micropipet, loaded with mercury and mounted in the microinstrument holder, is positioned with its tip near the microscope field and angled slightly higher than horizontal. The coarse-adjustment knobs of the micromanipulator are used to bring the micropipet tip into the microscope field, and into focus. The micropipet, ready for loading, is backed off, out of the field of view.

A microscope with a movable stage and body tube for focusing is convenient, but not absolutely necessary. If the stage moves for focusing, the micropipet can be independently brought into focus, then removed on one horizontal axis while the injection slide is positioned or otherwise manipulated. When needed, the micropipet is easily brought into view by movement along this single axis, already in focus and ready for use. The advantage of a microscope with a movable body tube is that the cell and the micropipet are not moved with respect to one another during focusing.

Loading the micropipet proceeds by bringing the end of the capillary reservoir, mounted on the injection slide, into focus in the center of the microscope field. Next, the coarse adjustment on the manipulator brings the micropipet back into the field of view. Once there, coarse adjustments are rarely used and the joystick on the manipulator translates the pipet in the two dimensions parallel to the microscope stage.

Because the lumen of the pipet as pulled is often too fine for liquids, or even air, to flow through, the tip has to be broken to make its diameter approximately 1 μm. The fragile tip of the micropipet is gently touched to the glass surface at the end of the capillary reservoir. In this manner, the very tip is removed and an approximately 1 μm (ID) pipet is obtained. Tips of 1 μm allow free flow of solutions and do not damage the cells during injection. Pipets with tips up to 5–10 μm can also be used to inject suspensions of large particles.

The micropipet is backed away from the end of the reservoir. As pressure is applied with the screw-driven syringe, the advance of the mercury toward the neck of the micropipet is monitored by eye. All subsequent operations are viewed through the 20× or 40× objectives of the microscope. Pressure is applied to force the mercury to the very tip of the micropipet. The micropipet is centered in the microscope field, then advanced toward the open end of the reservoir so as to move its tip approximately halfway between the center and edge of the field. Next, the capillary reservoir is advanced so that the pipet penetrates the column of oil that caps the reservoir (Fig. 9). The pressure on the mercury is reduced by slowly unscrewing the injection syringe, causing oil from the reservoir to flow into the micropipet. The oil prevents direct contact between the aqueous solution and the mercury and is also used to determine the volume of aqueous solution injected (see below).

For an injection in which a precisely calibrated volume is desired, the distance between the pipet tip and the mercury/oil interface is measured and recorded (Fig. 9A). When aqueous solution is later loaded into the pipet, this distance is reproduced exactly so that the volume of aqueous solution injected is precisely equal to the volume of mercury-capping oil. When the position of the mercury/oil interface in the micropipet has stabilized (this requires a few seconds) at the desired distance from the pipet tip, the capillary reservoir is advanced farther so that the tip of the pipet passes through the Wesson oil cap and penetrates the aqueous solution in the reservoir (Fig. 9B). Again when pressure on the mercury is reduced within the syringe, solution is drawn into the micropipet. The distance between the tip and the oil/aqueous solution interface is made to match the recorded distance between the tip and the mercury/oil interface. When this second interface has stabilized, the capillary reservoir is partially withdrawn to leave

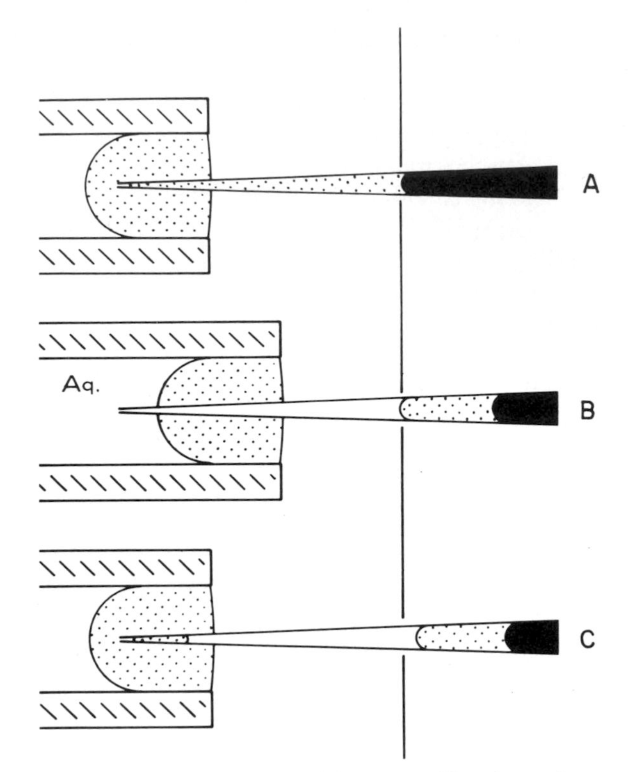

FIG. 9. Schematic for loading a micropipet with a measurable volume of aqueous solution. The stippled regions are oil filled, the black regions are mercury filled. Aq., aqueous solution in the capillary reservoir.

(A) Oil has been drawn into the pipet tip a given distance. (B) The capillary reservoir is advanced so that the tip of the micropipet is in the aqueous solution. Aqueous solution has been drawn the same distance into the tip. (C) Oil is next drawn into the tip to cap the aqueous solution. Details are given in text.

the micropipet in the oil again. A second volume of oil (exact amount unimportant) is loaded into the pipet (Fig. 9C). If one dose of aqueous solution is required, this oil acts as a cap to prevent evaporation from the tip or contamination of the contents of the micropipet by the culture medium (seawater) bathing the cells in the injection chamber. If more than one dose of aqueous solution is to be loaded, this oil acts as a spacer. If different volumes of aqueous solution are to be loaded into a single pipet, the volume of oil and aqueous solution in each pair of oil spacer (or cap) and aqueous dose are made equivalent. As described below, each spacer can be used to calibrate the volume of aqueous-injection solution with which it is paired.

The loading process can be repeated until as many as 6–10 doses are loaded into a single micropipet. The final oil drop acts as a cap. When the last interface finally stabilizes, first the reservoir, then the micropipet, is completely withdrawn from the microscope field. The reservoir, if it contained perishables, can be returned to storage.

VIII. Calibration of Volumes Injected

To determine the volume of solution injected into a cell, it is only necessary to expel into the seawater the mercury-capping oil or oil spacer (equal in volume to the injection solution), measure the diameter of the droplet, and calculate its volume. Micropipets as pulled vary in bore and taper. If it is desirable to load predetermined volumes into the micropipets, each pipet can be precalibrated with oil that is subsequently ejected into distilled water. The precalibrated pipet can then be used to inject the desired volume.

That the method for determining injection volumes is reliable is demonstrated as follows. On two separate occasions, a pipet was loaded with four "equal" doses of aqueous solution. The oil caps and spacers were also made the same size, meaning that the linear distance between the micropipet tip and interface was made equal for each oil and aqueous dose loaded. The oil drops were, in turn, expelled into seawater. The diameter of each drop was measured with a Zeiss filar micrometer and the volumes calculated. The oil drops from a given pipet were essentially equal in size: Measured diameters gave calculated volumes of the oil drops of 16.4 ± 0.8 pl for one pipet and 48.6 ± 1.5 pl for the other.

IX. Microinjection of Echinoderm Eggs

With the pipet out of the way and fully loaded, the cells in the injection chamber are positioned and focused at the very edge of the microscope field. The tip of the micropipet is brought into the center of the field. In order to bring the pipet into the injection chamber without damaging the tip, it is moved downward, out of focus, approximately 100 μm. As the chamber is advanced over the tip, the micropipet is rapidly brought back into focus. Thus the target cell and the loaded micropipet are in focus in the same microscope field (Fig. 10).

A target cell is centered in the 20× field, then its periphery is brought into focus with the 40× objective. With the vertical drive of the micromanipulator, the micropipet is also brought into focus. The shallow depth of field of the 40×

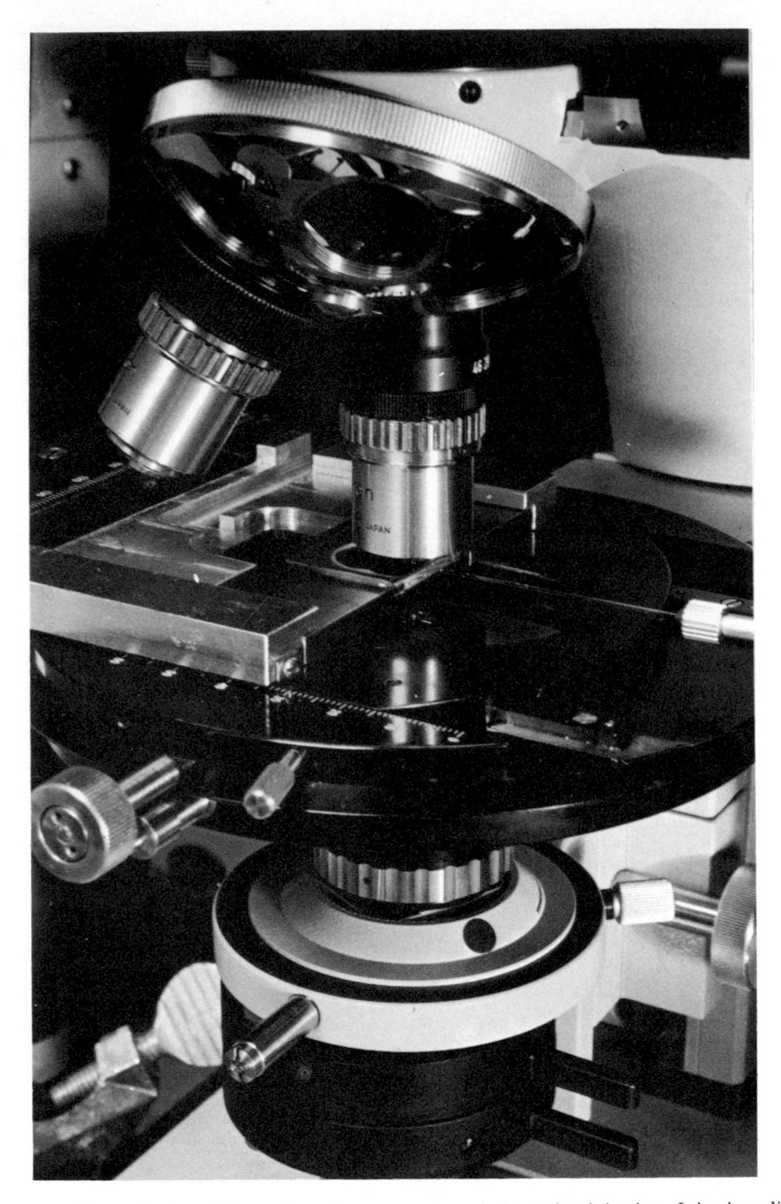

FIG. 10. Microinjection slide on the microscope stage during microinjection. Injection slide is held firmly in place with a specially machined mount.

objective ensures vertical alignment of the center of the cell and the micropipet. When large volumes of aqueous solution are to be injected (greater than 1% of the volume of the egg), it is best not to inject an equal volume of Wesson oil. Although the oil is nontoxic, even when greater than 20% of the egg volume is injected with Wesson oil (Fig. 7), such large volumes of oil may obscure structures and processes of interest. Consequently most of the oil cap is expelled into the seawater surrounding the cells.

Surface tension between the oil and seawater makes it difficult to initiate expulsion of the oil from the micropipet. Once this barrier is overcome, the oil and the aqueous solution behind it have a tendency to discharge rapidly. For this reason, it is advisable to use relatively large volumes of oil as caps or spacers, even on small-volume injections, so that once the oil begins to flow, there is enough time for the operator to react and prevent total expulsion of the oil and aqueous solution into the seawater. Once all but a tiny drop of oil (optimally $<<$ 1 pl) has been expelled, the micropipet is ready for microinjection.

Under the 40× objective, the remaining oil of the cap is advanced to the very tip of the micropipet so that no seawater remains in the tip. Driven by the manipulator's joystick, the pipet is positioned normal to the edge of the cell, then advanced further to cause a depression in the cell surface; the pipet is still not inside the cell. After the pipet has been pushed approximately one fourth to one third of the way across the cell, the deformed cell periphery snaps out around the pipet to regain its circular outline. At this time, the pipet is usually, but not always, inside the cell.[1] To ensure penetration, pressure is released with the syringe until cytoplasm flows into the pipet. After this, the pipet is assuredly in the cell.

Positive pressure is then applied to expel the small oil cap and the aqueous solution. Since the oil droplet that forms in the cell occasionally sticks to the end of the pipet, the site of the injection is not always marked by the position of the oil droplet. When the pipet is withdrawn from the cell, viscous drag usually dislodges the oil droplet to a different location. Although it may be desirable to photograph cells as soon as possible after injection, movement of cytoplasm caused by withdrawal of the pipet at times prevents a clear image for as long as 1 minute after injection.

If a pipet has been loaded with several doses of solution, most of the spacer oil is expelled into the seawater, and the microinjection process is repeated on another cell. Coordinates of the mechanical stage are recorded and, along with the injected oil droplets, serve to locate each injected cell.

[1]This can be illustrated by slowly expelling aqueous solution from the tip of the pipet. If it is inside the cell, a clear region of apparently lowered refractive index appears to displace granules in a relatively spherical region around the tip of the pipet. Otherwise, the aqueous solution passes out along the shaft of the pipet toward the periphery of the cell.

X. Summary

This chapter describes in detail the equipment and procedures I use to microinject various solutions into echinoderm eggs. The system was designed but never fully described by Hiramoto (1962). My modifications include an easily constructed rectangular chamber to immobilize echinoderm eggs for high-resolution microscopy and a method for accurately determining the volume of injected solutions. In addition to injecting salt and protein solutions into echinoderm eggs (Kiehart, 1981; Kiehart *et al.*, 1977a), I have used this system to microinject aqueous solutions into teleost eggs (Kiehart *et al.*, 1977b), annelid eggs (Inoué *et al.*, 1974), and *Acanthamoeba*. With minor modifications to the injection chamber, this simple and sensitive technique permits various substances to be microinjected into tissue culture cells as well.

Acknowledgments

I take pleasure in acknowledging the continual support of Drs. Shinya Inoué and Tom Pollard. I appreciate greatly stimulating discussions with each of them and with Dr. Gordon Ellis and Mr. Ed Horn.

Supported by NIH 5T01 HD00030 predoctoral traineeship to D.P.K. through Howard Holtzer, NIH Grant PHS 9 R01 GM23475-1 and NSF BMS 75-00473 to Shinya Inoué, Muscular Dystrophy Association Post-Doctoral Fellowship to D.P.K. and NIH GM26132-03 to Tom Pollard.

References

Barber, M. A. (1911). *J. Infect. Dis.* **8,** 348-360.

Baserga, R., Croce, C., and Rovera, G., eds. (1980). "Introduction of Macromolecules into Viable Mammalian Cells." Alan R. Liss, New York.

Begg, D. A., and Ellis, G. W. (1979). *J. Cell Biol.* **82,** 528-541.

Celis, J. E., Graessman, A., and Loyter, A., eds. (1980). "Transfer of Cell Constituents into Eukaryotic Cells." Plenum, New York.

Chambers, R. W., and Chambers, E. L. (1961). "Explorations into the Nature of the Living Cell." Harvard Univ. Press, Cambridge, Massachusetts.

Chambers, R. W., and Kopac, M. J. (1950). *In* "Handbook of Microscopical Technique" (J. McClung, ed.), 3rd ed., pp. 492-543. Harper, New York.

Curtis, D. R. (1964). *In* "Physical Techniques in Biological Research" (W. L. Nastuk, ed.), pp. 144-190. Academic Press, New York.

deFonbrune, P. (1949). "Technique de Micromanipulation." Masson, Paris.

Diacumakos, E. G. (1973). *Methods Cell Biol.* **7,** 287-311.

DuBois, D. (1931). *Science* **73,** 344-345.

Ellis, G. W. (1961). Ph.D. Thesis, University of California, Berkeley.

Furasawa, M., Yamaizumi, M., Nishimura, T., Uchida, T., and Okada, Y. (1976). *Methods Cell Biol.* **14,** 73-80.

Fuseler, J. (1975). *J. Cell Biol.* **64,** 159–171.
Hiramoto, Y. (1956). *Exp. Cell Res.* **11,** 630–636.
Hiramoto, Y. (1962). *Exp. Cell Res.* **27,** 416–426.
Hiramoto, Y. (1974). *Exp. Cell Res.* **87,** 403–406.
Hodgmann, C. D., ed. (1958). "Handbook of Chemistry and Physics," 40th ed. Chem. Rubber Publ. Co., Cleveland, Ohio.
Inoué, S., Borisy, G. G., and Kiehart, D. P. (1974). *J. Cell Biol.* **62,** 175–184.
Kiehart, D. P. (1981). *J. Cell Biol.* **88,** 604–617.
Kiehart, D. P., Inoué, S., and Mabuchi, I. (1977a). *J. Cell Biol.* **75,** 258a (abstr.).
Kiehart, D. P., Reynolds, G. T., and Eisen, A. (1977b). *Biol. Bull. Woods Hole, Mass.* **153,** 432 (abstr.).
Kohen, E., Patterson, D., Kohen, C., Bengtsson, G., and Salmon, J. M. (1975). *IEEE Trans. Biomed. Eng.* **BME-22,** 424–426.
Kopac, M. J. (1955). *Int. Rev. Cytol.* **4,** 1–29.
Kopac, M. J. (1964). *In* "Physical Techniques in Biologic Research" vol. 5., (W. L. Nastuk, ed.), Vol. 5, pp. 191–233. Academic Press, New York.
Mabuchi, I., and Okuno, M. (1977). *J. Cell Biol.* **74,** 251–263.
Needham, J., and Needham, D. M. (1925). *Proc. R. Soc. London Ser. B* **98,** 259–286.
Needham, J., and Needham, D. M. (1926a). *Proc. R. Soc. London, Ser. B* **99,** 173–199.
Needham, J., and Needham, D. M. (1926b). *Proc. R. Soc. London, Ser. B* **99,** 383–397.
Poste, G., Papahadjopoulos, D., and Vail, W. J. (1976). *Methods Cell Biol.* **14,** 33–71.
Taylor, D. L., and Wang, Y. L. (1980). *Nature (London)* **284,** 405–410.
Weinstein, J. N., Yoshikami, S., Henkart, P., Blumenthal, R., and Hagins, W. A. (1977). *Science* **195,** 489–492.

Chapter 3

Microscopic Methods for Analysis of Mitotic Spindle Structure

J. RICHARD McINTOSH

Department of Molecular, Cellular, and Developmental Biology
University of Colorado
Boulder, Colorado

I. Introduction

Much of the work directed toward understanding the mitotic spindle has been structural, because the spindle is a kind of machine, and numerous structural questions are obvious as one tries to understand a device that converts chemical energy into mechanical work. A review of the structural methods that have been successfully applied to mitosis must therefore cover a good deal of ground. This brief treatment is necessarily sketchy, but I will try to present sufficient detail on each important method so that interested readers can identify the techniques relevant to their own questions and pursue them further in the references cited. In

METHODS IN CELL BIOLOGY, VOLUME 25

ISBN 0-12-564125-7

the final section, I present my own perspective concerning promising avenues for further work.

II. Light Microscopy

A. Bright-Field Optics

The mitotic spindle in the living cell is essentially invisible in bright-field optics. There is, however, a voluminous literature on the structure of dividing cells as seen in the light microscope after various fixation and staining protocols. These techniques are not in heavy current use because of the ability of modern optical methods to reveal structural details in living cells and because of the power of the electron microscope for studying fixed material. Nonetheless, there is rich literature on the morphological diversity of mitosis in the work done with the older methods. Some of the more spectacular work of the nineteenth century, for example, Lauterborn's (1896) studies on diatoms, has recently been confirmed with the electron microscope (Tippit and Pickett-Heaps, 1977), suggesting that confidence in these descriptions is not misplaced. Useful summaries of much of the older work are available in Wilson (1925) and Schrader (1953).

B. Phase Optics

Phase optical systems generate visible contrast from objects that do not absorb light by employing a geometry that effects interference between a reference light beam and light that has passed through the specimen and suffered a retardation of phase. [Lucid treatments of phase optics are available from the manufacturers of most commercial microscopes. A more rigorous discussion is found in many optics texts, for example, Hecht and Zajec (1974). A useful combination of theory and practice is published in books on microscopic methods, such as James (1976).] Objects possessing a higher refractive index than their surroundings appear dark on a light-gray background, although "negative phase-contrast" systems are also available. Several commercial systems have been designed to obtain a direct and quantitative measure of the optical path (geometrical path $\times$ refractive index) through specific parts of the specimen. Because the refractive index is almost directly proportional to the concentration of macromolecular solute in a given region, these optical systems have been used to measure the amount of dry mass in a given region of a fully hydrated mitotic spindle (Forer and Goldman, 1972).

The phase-contrast image of the living spindle is perhaps the most familiar, in part because phase optics is so widely available and in part because the image is useful for many purposes. The chromosomes are readily visible, but the spindle is seen only indirectly because the phase-dark mitochondria and vesicles are excluded from that region of the cell (Fig. 1a). In some animal cells, the spindle poles are also visible [for crane fly spermatocytes, see Dietz (1959); for PtK_2 cells, see Rattner and Berns (1976) or Peterson and Berns (1978)], and in rare cases such as the hypermastigote flagellate *Barbulanympha,* the spindle fibers themselves may be seen in living cells (Cleveland, 1953).

These optics have proved to be the tool of choice for some important experiments on spindles. Nicklas has used phase optics for all his micromanipulation studies on chromosome behavior, in part because the chromosomes are so clearly seen, and in part because they are readily used on an inverted microscope with a long-working-distance condenser, leaving ample room for manipulations (for example, Nicklas and Koch, 1972). The same condenser will permit the use of Rose chambers (Rose *et al.*, 1958), which are convenient for long-term sterile cell culture. This route has been followed by Berns (1974) in many of his laser microsurgery experiments on chromosomes and spindles. Water-immersion phase objectives are also available, permitting experimental manipulation of cells by such techniques as microinjection while the cells are being watched in a conventional microscope with the objective above the specimen.

The strengths of phase optics derive from their flexibility of use—as detailed above—and their modest cost. The disadvantages of the method for spindle study are its general inability to display spindle fibers in living cells and the limitation on image quality imposed by objects not quite in the plane of focus. Phase optics yields diffraction images of phase-dense objects above and below the plane of good imaging, and these diffraction patterns often obscure image detail (Allen *et al.*, 1969b). An ideal specimen for phase optics is flat enough to constitute a single optical section. Many spindles are found in rounded cells that do not approximate this ideal; the resulting spindle images are often confused. For these reasons, other optical systems are sometimes preferable for the study of spindles in living cells.

C. Polarization Optics

In polarization optics, two orthogonal polarizing filters are set in the light path, both oriented normal to the optic axis, one situated above and one below the specimen. If the specimen is optically isotropic, essentially no light will pass the polarizing filters: Their extinction coefficient (incident intensity/transmitted intensity) can be as high as 10^7 with no lenses in place and 10^3–10^5 with lenses. Contrast is generated by an optically anisotropic specimen because light polarized in different orientations will travel at different velocities through the specimen as a result of the anisotropy, often called birefringence (BR). The resulting changes in phase convert the plane-polarized light generated by the first filter into elliptically polarized light, some of which will pass the second filter. Birefringent areas of the specimen then appear light on a dark background. The use of a "compensator" allows a small, uniform BR of the opposite sign to be added to the whole image, thus causing the birefringent specimen to appear dark on a light background (Fig. 1b). [For a treatment of the theory and practice of polarization optics, see Oster and Pollister (1955), Slayter (1970), or Hartshorne and Stuart (1960).]

Most fibrous materials, including spindle fibers, display BR and hence are visible in living cells with a polarizing microscope. The spindle is usually weakly birefringent, so a rather sensitive instrument is required. A standard research microscope equipped with strain-free lenses is often just adequate, but for serious work a more specialized optical system is usually required. The problem is that conventional, high-numerical aperture lenses of short focal length cause a small rotation in the plane of polarization of the light passing through them (Inoué, 1952a). The rotation varies with position over the lens surface, so no one position

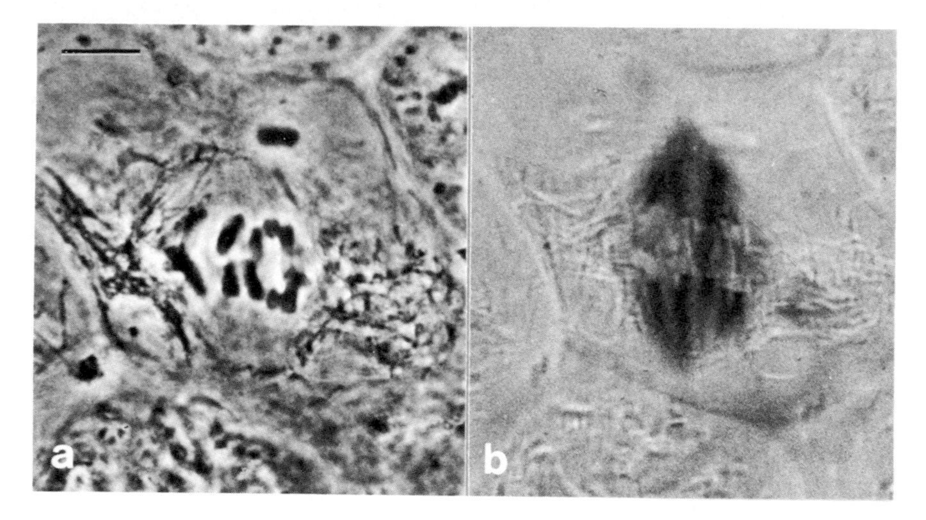

FIG. 1. Light micrographs of a grasshopper spermatocyte in anaphase of meiosis I. (a) A phase micrograph showing clearly the phase-dark chromosomes and cytoplasmic particles, but revealing the spindle only as a region lacking such material. The bright halo around the chromosomes is an example of the diffraction effects that limit image quality. (b) A polarization optical micrograph of the same live cell. A small amount of BR with sign opposite to that of the spindle has been introduced with the compensator, so that the spindle appears dark on a light background. With these optics, the spindle is obvious, but the chromosomes and many of the previously visible cytoplasmic particles are missing. Bar = 10 μm. (From Salmon and Begg, 1980. Reproduced with permission from Dr. Salmon and the *Journal of Cell Biology*.)

of the upper polarizing filter can block out all the incident light. Lenses that rectify this rotation are available (Inoué and Hyde, 1957), and there are ways to minimize the effect with conventional lenses that are beyond the scope of this chapter. The subject merits thorough study before any investments are made, especially because modern methods for contrast enhancement may make money spent on television and image-processing hardware a better investment than money spent on lenses [Inoué (1981) and Allen *et al.* (1981)]. (These gentlemen sometimes offer courses on research use of the light microscope, and an interested reader could profit greatly from their offerings.)

The power of polarizing optics lies in its capacity to reveal spindle fiber BR in living cells (Fig. 1b) and in the adaptability of the method to a quantitative approach. The amount of BR in a living cell may be measured as the retardation of the plane of polarization oriented parallel to a fiber axis relative to the light polarized in the orthogonal plane. This retardation, usually expressed in nanometers, may then be used to measure the result of an experimental perturbation of the spindle fiber BR, such as a treatment with colchicine (Inoué, 1952b), high hydrostatic pressure (Salmon, 1975), low temperature (Fuseler, 1975), or subunit dilution following lysis in microtubule polymerization buffers (Inoué *et al.*, 1974; Snyder and McIntosh, 1975). Spindle BR has also been used as an index of fiber preservation during fixation for view in fluorescence or electron microscopes (Sato *et al.*, 1975; Cande *et al.*, 1977). Polarization optics is unique in allowing this quantitation on essentially any living spindle; its importance for this purpose cannot be overstated.

A limitation of polarization optics is found, however, in the difficulty of giving an unambiguous molecular interpretation of the retardation that is measured. Inoué and his collaborators have done a thorough job of looking for fine structural correlates of BR in spindles isolated from marine eggs (Sato *et al.*, 1975). They find that microtubules can account for essentially all the BR measured in spindles isolated under conditions of constant BR. Other workers disagree with their conclusions (Forer *et al.*, 1976), and indeed, several observations on spindle BR and microtubule density as seen in the electron microscope are difficult to bring into harmony (e.g., LaFountain, 1976; Forer and Brinkley, 1977). The discrepancy is thought by some to be due to labile microfilaments in the spindle (Forer, 1978). An alternative is provided by the recent descriptions by electron microscopy of membranous vesicles among the spindle microtubules (Harris, 1975; Moll and Paweletz, 1980; Hepler, 1980). This membranous material might exist in the living cell in such a geometry that it contributes to the BR of the spindle fibers. In summary, polarization optics is a tool of importance and power for the study of spindle structure, but the difficulty of interpreting the images obtained from living cells makes the technique complementary to, not superior or inferior to other structural approaches.

D. Differential-Interference Contrast (DIC) Optics

DIC optics (often called Nomarski after one of its inventors) generates contrast by causing interference between light beams that have passed through immediately adjacent regions of the specimen. This is achieved by splitting a single beam of plane-polarized light with a modified Wolliston prism into two coherent, overlapping beams. The direction in which the beam is split, called the direction of shear, is along one chosen direction in the plane of the specimen. Small phase differences in the two beams are induced by refractive index variation in the specimen. If these local differences are superimposed on a small overall phase shift or "bias," then the method will generate bright and dark-appearing edges on opposite sides of phase-dense objects in the sample. The sign and magnitude of the contrast depends on the sign and magnitude of the phase shifts caused by the specimen. These shifts will in turn depend on the derivative with respect to position along the direction of shear of the function describing optical path through the specimen (Allen *et al.*, 1969b). The image gives the impression of a surface shown in relief, but this is an optical illusion. In reality one is seeing a thin lamina from within the specimen. Indeed, one of the advantages of the method is that objects go out of focus so rapidly that essentially no contrast is generated by material above or below a lamina about as thick as the resolving power of the microscope in the plane of the specimen. Stated another way, the resolution of this optical system is about the same in all three dimensions of space. It is therefore possible to cut "optical sections" through the spindle by making small changes in the plane of focus (Fig. 2).

The value of DIC optics to the study of spindle structure derives from the high contrast and dynamic range it offers for the study of living cells. With a good-quality DIC system, both spindle fibers (low-contrast objects) and chromosomes (high-contrast objects) can be seen in the same micrograph (Figs. 2 and 3). Although these micrographs display particularly favorable biological material and are taken with a particularly good optical setup, they show in a dramatic way how much detail can be seen in living spindles with these optics, permitting one to follow time-dependent events directly (Fig. 3). The contrast is so high that one can see small, round objects ("particles or states") along the chromosome-to-pole spindle fibers that are essentially invisible with all other optics. These structures migrate in the living metaphase spindle from wherever they are first seen toward the spindle poles, displaying a novel dynamic property of living spindle fibers. The particles move relative to the chromosomes during metaphase, but are at rest relative to the chromosomes during anaphase (Allen *et al.*, 1969a). Such enigmatic behavior stands as a challenge to theorists and as a fine example of the importance of detailed study of living cells. Although the resolving power of optical instruments is set by the physics of light, novel modes of

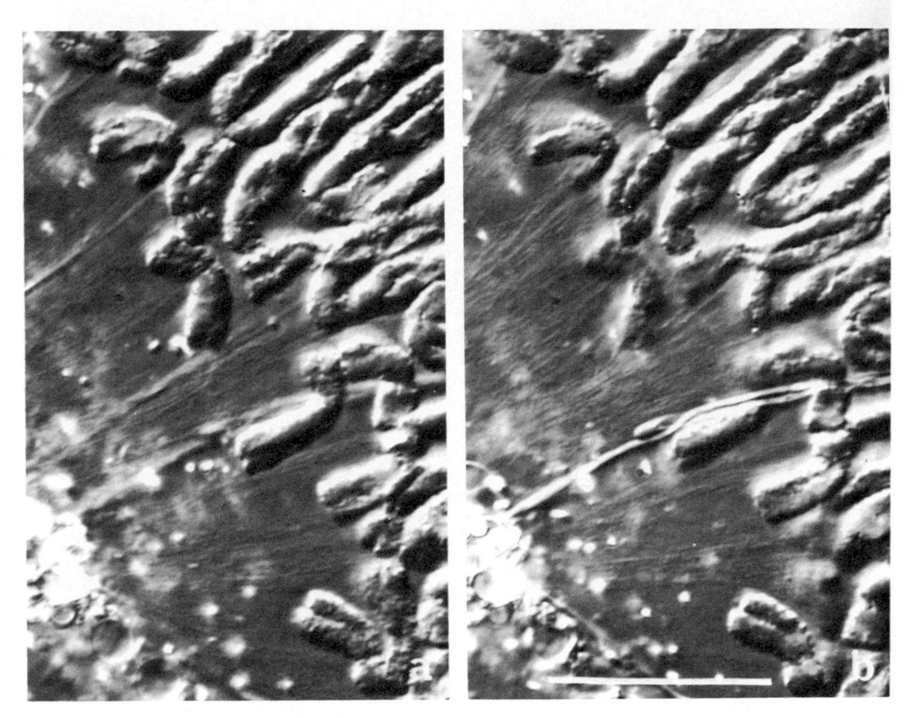

FIG. 2. DIC optical micrographs of endosperm from the plant *Haemanthus katherinae*. The two pictures differ in the plane of focus by 1 μm, showing how sharply these optics can discriminate position along the optical axis. The pictures were taken on a particularly good DIC system in the laboratory of R. D. Allen at Dartmouth College; the sharpness and clarity of the spindle fibers is unusual. (These unpublished micrographs are included through the generosity of Mr. Joel Stafstrom, University of Colorado.) Bar = 20 μm.

contrast generation, such as DIC, will sometimes show a new face on an old friend.

E. Cinematography and Television

The modes of recording the events of mitosis seen in living cells warrant mention here. Fixed specimens are so common that it is easy to overlook the value of observations made on live cells. It is often, however, difficult to perceive and appreciate the events of mitosis in a live cell because they occur slowly enough to tax our perceptions over time. Time-lapse recording that permits a speeded-up display of the events of spindle formation and function is a useful and easy way to overcome this problem. With time-lapse movies, students of mitosis

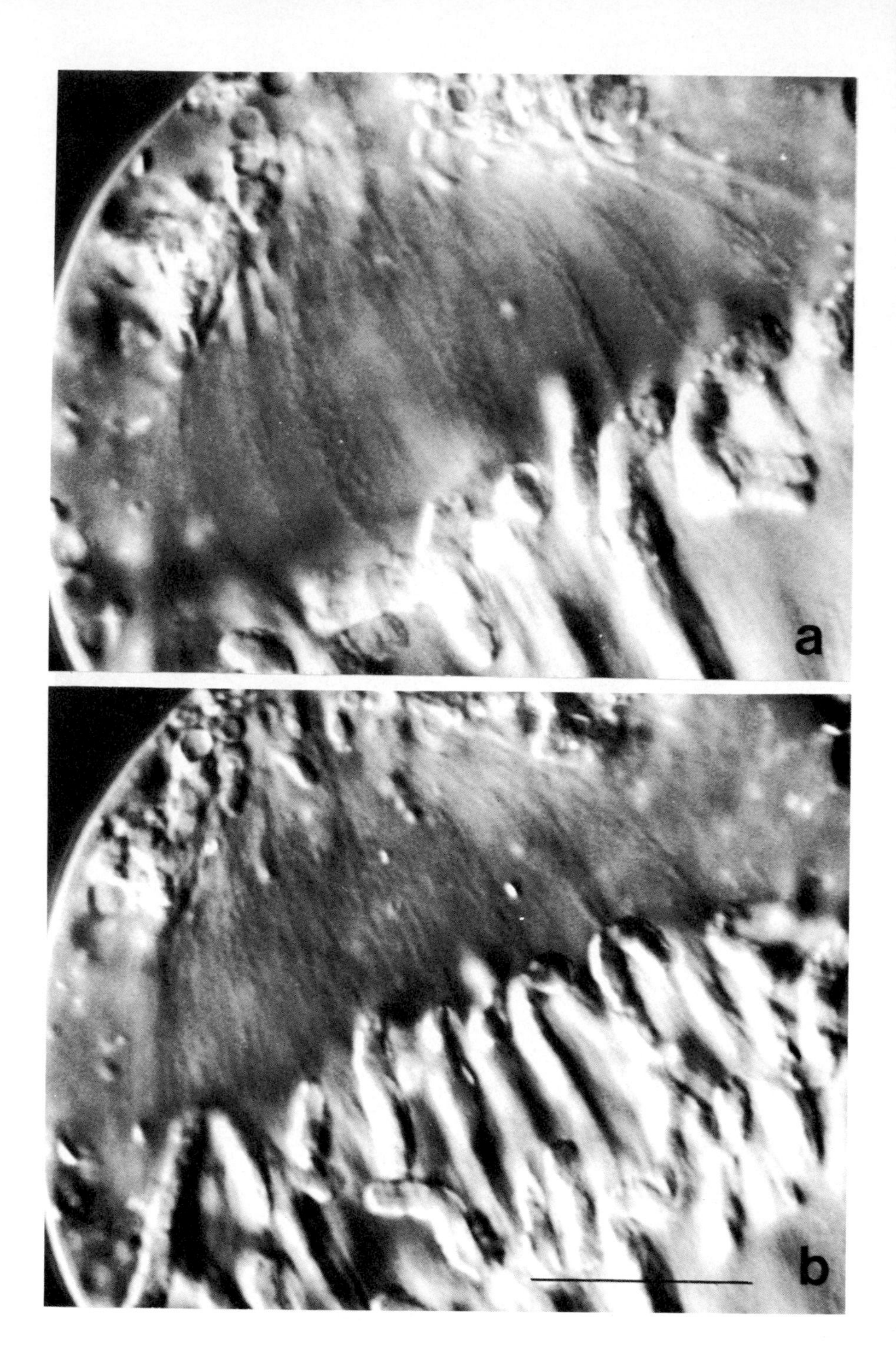
a
b

such as Andrew Bajer have recorded many examples of both normal and abnormal chromosome movement. Their films are among the most important records of mitosis that exist.

More recently, time-lapse television has become popular. Although this technology is expensive to buy, it is cheap and easy to use. It is especially valuable for experimental work, allowing one to examine at comprehensible speeds the result of an experiment just completed.

F. Fluorescence Microscopy and Immunocytochemistry

Fluorescence optics is, in theory and in practice, a technique of great sensitivity, contrast, and dynamic range. Its sensitivity is a result of displaying fluorescent objects against a black background where they are most readily visualized. It has been clearly demonstrated that a single microtubule may be seen and photographed by indirect immunofluorescence (Osborn *et al.,* 1978). The rapid bleaching during observation of some fluorochromes, such as fluorescein, and the difficulty of quantitating fluorescence detract somewhat from the value of the method, but it is nonetheless so powerful, easy, and esthetic that it has gained wide popularity. The adjoining methodology that has made fluorescence optics particularly valuable for spindle structure studies is immunochemistry. Specific cellular components can be used to elicit antibodies, which can then be tagged with a fluorescent dye and used to stain these components in fixed cells (Coons and Kaplan, 1950; Lazarides and Weber, 1974). Although there is a formidable array of immunochemical controls that should be run to minimize the likelihood of spurious immunological reactions (see, for example, Fujiwara and Pollard, 1976), and although there are numerous possibilities of fixation artifact in the methods commonly used to prepare cells for antibody staining (Cande *et al.*, 1977), the method has tremendous utility.

Studies of spindle structure with antibodies directed against tubulin have largely confirmed the sense of spindle design that grew from previous light and electron-microscopic investigations (Brinkley *et al.*, 1975). They have helped to establish the existence of microtubules in the interzone throughout metaphase and anaphase and have shown the extent of aster development in some somatic animal cells where it had not been appreciated (Figs. 4a and b). The more significant contribution of the method, however, has been in the use of immunofluorescence to study the distribution of proteins of uncertain significance for the spindle. For example, antibodies to calmodulin have revealed a surprising distri-

---

FIG. 3. The same optics and material as Fig. 2. Focus is held constant, but different times in mitosis are shown. The high contrast of DIC optics reveals tiny particles along the spindle fibers, which are seen to move poleward in the living cell. Bar = 20 μm. (From Hard and Allen, 1977. Reproduced with the permission of Dr. Hard and the *Journal of Cell Science*.)

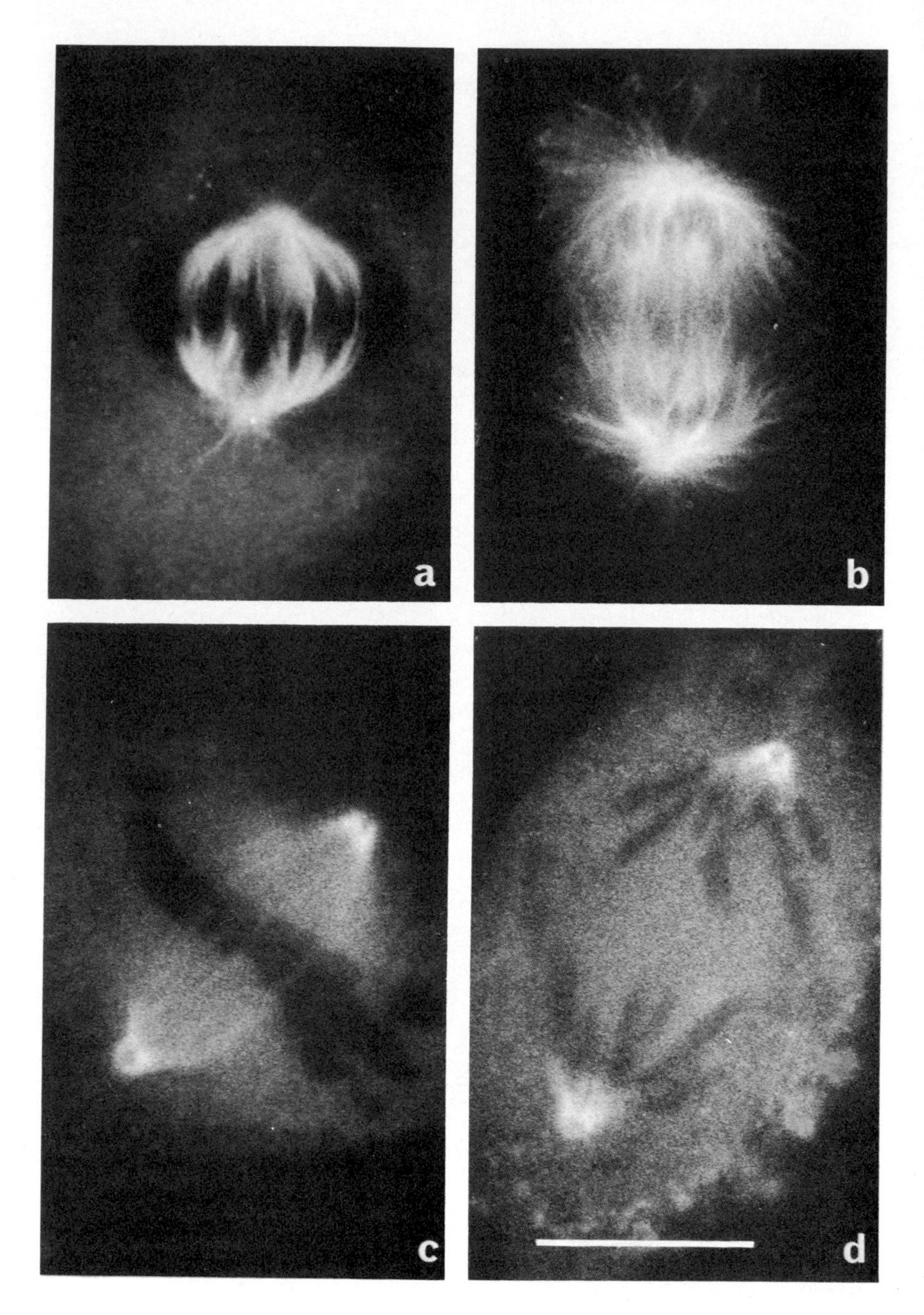
a
b
c
d

bution of this protein during mitosis. It becomes concentrated in the spindle during prometaphase, aligning with the chromosome fibers during metaphase (Welsh *et al.*, 1978) (Fig. 4c). At metaphase there is a gradient in its distribution along these fibers, a higher concentration lying near the poles (De Mey, 1980). During anaphase it is confined to the shortening chromosome fibers (Fig. 4d), but in teleophase it is found at the pole-facing ends of the midbody. Although such behavior is, at present, not interpretable in an unambiguous way, its very improbability makes it seem of considerable significance, and one is inclined to treat these observations with respect and interest.

The immunochemical observations of actin in the spindle serve to illustrate another aspect of the method. The existence of actin as a spindle component had been suggested on the basis of several light and electron-microscopic studies (Forer and Behnke, 1972; Gawadi, 1974; Sanger, 1975); immunofluorescence confirmed and extended these studies (Cande *et al.*, 1977; Herman and Pollard, 1979). More recently another look at the spindle with antiactin has thrown the matter into doubt by showing that the images with actin concentrated in the spindle may depend on exactly how the cell is fixed (Aubin *et al.*, 1979). I would emphasize that although the body of evidence suggests to me that actin is present in the living spindle, morphological evidence tells us nothing about the functional significance of actin in the spindle. One needs some sort of experimental evidence to determine whether the actin located in the spindle is used by the cell for chromosome movement.

A recent innovation may make fluorescence microscopy important for obtaining just this sort of evidence. Taylor has been developing methods for tagging cytoskeletal proteins with fluorescent dyes, injecting them into living cells, and following their behavior in the cell with fluorescence optics (Taylor and Wang, 1980). The key steps in this procedure are labeling the proteins in a nondestructive way, injecting them without important or irreversible damage to the recipient cell, and monitoring the location of the fluorescence at light levels low enough to leave the cell unperturbed. For the latter purpose, high-sensitivity television cameras are proving to be very useful (Willingham and Pastan, 1978; Fermisco, 1979). One can readily imagine that following fluorescent-tagged spindle pro-

---

FIG. 4. Fluorescence optical micrographs of mammalian cells stained by indirect immunofluorescence. (a and b) Metaphase and anaphase mouse 3T3 cells stained with antitubulin. Since the two times in division are represented by different cells, one cannot be certain that the changes seen are time dependent, but study of multiple micrographs suggests that these images are representative of the two stages. (c and d) Metaphase and anaphase PtK cells stained with antibodies to calmodulin. The change in staining reflects a surprising movement of the calmodulin during mitosis. The dissimilarity of the tubulin and calmodulin staining patterns displays the specificity of the method. Bar = 10 μm. [These unpublished micrographs are included through the generosity of Dr. B. R. Brinkley (antitubulin), Baylor College of Medicine, and Dr. M. Welsh (anticalmodulin) of the University of Michigan.]

teins injected into living cells may be an informative approach to identifying their special locations during mitosis. It is also possible that the labeling of two proteins with different fluorochromes may allow one to study their interactions *in vivo* by resonance energy transfer (Stryer, 1978). Such methods promise to make the fluorescence microscope a powerful tool for both experimental and morphological work in the coming years.

III. Electron Microscopy

A. Specimen Preparation for the Electron Microscope

The major contributions of the electron microscope (EM) to our understanding of spindle structure have resulted from the characterization of the morphological details of the mitotic apparatus. The quality of the images has depended, as one might expect, heavily on the efficacy of the fixation and staining methods employed. Spindle microtubules were discovered in OsO_4-fixed cells using the stabilizing effects of Ca^{2+} ions or low pH to support these labile filaments during osmium fixation (Roth and Daniels, 1962; Harris, 1962). The discovery of glutaraldehyde as a fixative, however, was the real turning point in studies of spindle fine structure (Sabatini *et al.*, 1963). Since the widespread use of this compound for fixation, microtubules (MTs) have been found associated with chromosome movement in every eukaryotic cell in which they have been properly sought. The operational question one asks is, therefore, what is the best way to convey the glutaraldehyde to the cells. For microorganisms, much good work has been done with glutaraldehyde added directly to the culture medium. For mammalian cells in culture, however, a buffer that supports the polymerization of microtubule protein is preferable to the culture medium as a solvent for the fixative. In surf clam eggs, good results have been achieved with mixtures of glutaraldehyde in standard fixation buffers (such as phosphate) and six-carbon glycols that stabilize spindles in lysed cells (Inoué and Sato, 1967). It seems from some observations that a cell has an intrinsic capacity to disassemble the spindle, and the MTs must be protected against this system until they are properly fixed (McIntosh *et al.*, 1975).

Some cells with particularly impenetrable surfaces require special procedures to introduce the aldehyde. Some protozoa and algae have responded well to the "semisimultaneous" application of glutaraldehyde and OsO_4 (Roth *et al.*, 1970; Tippit and Pickett-Heaps, 1977). The OsO_4 seems to help break down the permeability barriers and facilitate the entry of the glutaraldehyde. A phase-partition method has been used to improve the entry of glutaraldehyde into *Drosophila* eggs (Zalokar and Erk, 1977). Nematode eggs fix well with glutaraldehyde after their shells have been weakened by treatment with hypochlorite and digestion with chitinase (P. Bazzicalupo, personal communication, following Fairbairn and Passey, 1955).

Even in material treated with the best methods available, there is presently controversy over the existence of a fibrous spindle component other than MTs. The hypothetical fibrils are assumed to be poorly preserved with current methods of fixation (Forer, 1978). Independent evidence suggests that glutaraldehyde does not necessarily protect filaments of pure actin against the rigors of subsequent OsO_4 fixation (Maupin-Szamier and Pollard, 1978). These experiments have nutured the conjecture that actin filaments might be present in the living spindle. Certainly actin-like microfilaments are rarely seen in fixed spindles with the EM. It is too early to tell what is right on this important issue, but one must note that complete faith in the current fine structural image of spindles is unwarranted. The revitalization of permanganate fixation (Moll and Paweletz, 1980) and some improved methods for staining particular membranes (Hepler, 1980) have recently altered our understanding of the distribution of vesicles in the spindle. One can expect that further improvements through fast freezing or novel chemical treatments may serve the same role for other spindle components as well.

The scarcity of specific mitotic stages in a growing population of cells, combined with the rarity of particular mitotic structures in random sections of spindles, has led to the common use of some specific embedding techniques for microtomy. Most thorough studies of mitosis have used flat embedded cells, permitting the selection and orientation by light microscopy of a particular cell for thin sectioning. Of the several useful methods available, the one we have found most practical (Euteneuer and McIntosh, 1980) is a hybrid of those worked out by Coss and Pickett-Heaps (1974), McIntosh and Landis (1971), and Brinkley and Cartwright (1971). The fixed and embedded cell of choice can be found and marked on the stage of the light microscope, then excised from a thin epon wafer with a scalpel, glued to a dummy block, trimmed, and serial-sectioned. Alternatively, if large numbers of dividing cells can be collected and pelleted, the probability of finding in a random section a well-oriented spindle at the right stage of mitosis is acceptable. For most purposes, however, flat embedding seems to be worth the trouble.

Recently some specific staining methods for electron microscopy have been used for spindle studies with notable results. Immunocytochemistry for electron microscopy with the peroxidase–antiperoxidase method of Sternberger (1974) has been perhaps the most useful to date (De Brabander *et al.*, 1977), although the recent combination of this method with the use of colloidal gold as an EM tracer for a second antibody promises even better things to come. Rieder (1979) has used the Bernhard (1969) method for preferential staining of ribonucleoprotein (RNP) to examine the poles and kinetochores of the mammalian spindle. This method combined with RNase digestion appears to give dependable localization of RNP at the fine-structural level. The results suggest that RNP is located at both the kinetochores and the centrioles, a stimulating result given the evidence from independent sources that RNA may be functionally significant in the mitotic behavior of centrioles (Heidemann *et al.*, 1977; Peterson and Berns, 1978).

Certain aspects of spindle fine structure are not visible by simple fixation and electron microscopy. Microtubule polarity is an example. Microtubules are known to be polar fibers, because they are formed by the head-to-tail association of asymmetric dimers. Their polarity is, however, invisible in standard electron micrographs. A method for revealing MT polarity analogous to the use of heavy meromyosin for displaying actin polarity has been wanted for serveral years. Recently, such methods have been found (Haimo *et al.*, 1979; Heidemann and McIntosh, 1980), and are discussed in other chapters of this series.

B. Scanning Electron Microscope (SEM)

The SEM has not been used extensively for studies of spindle structure because it usually sees surfaces. Recently, however, this microscope has been employed to good advantage in the study of isolated spindles (Silver *et al.*, 1980). It provides a holistic image of intermediate resolution that can be useful for an overview of the isolate. It is probably most useful as a companion method to others, since details of interpretation of the images are not yet well worked out.

C. Holistic, High-Resolution Studies

The smallness of the fraction of a spindle contained within one thin section necessitates some additional action to obtain a view of the spindle that is simultaneously high in resolution yet holistic in perspective. Two methods are available: high-voltage electron microscopy and serial sectioning.

1. The High-Voltage Electron Microscope (HVEM)

The penetrating power of electrons is a monotonic, increasing function of their voltage for a given electron optical resolution, so the constraints on section thickness imposed by the nature of the electron are to some extent removed by employing a HVEM. Whole, isolated spindles have been examined after fixation and dehydration by the critical-point method, but these results are not very encouraging. Mammalian spindles contain so many microtubules that their totality is a confusing morass (McIntosh *et al.*, 1979a). Some spindle fragments are small enough to yield interpretable images, but by and large, the method has not yet proved very useful. Thick sections of fixed, embedded material have been more informative. The design of two large plant spindles has been well revealed with this tool (Coss and Pickett-Heaps, 1974; Bajer and Mole-Bajer, 1975). Comparable images of mammalian spindles lysed so as to extract a majority of the background substance, yet preserve the spindle fibers, have also been published (McIntosh *et al.*, 1975) (Fig. 5). Such images are graphic, but have not yet proved to be uniquely informative.

Peterson and Ris (1976) applied the HVEM and some serial thick sectioning to the spindle of a yeast. Here the combination of the diminutive size of the spindles and the penetrating power of the beam has led to a few views in which essentially every kinetochore MT can be seen and the central spindle is a clear MT bundle. These results show a utility of the HVEM, but also make it clear that when

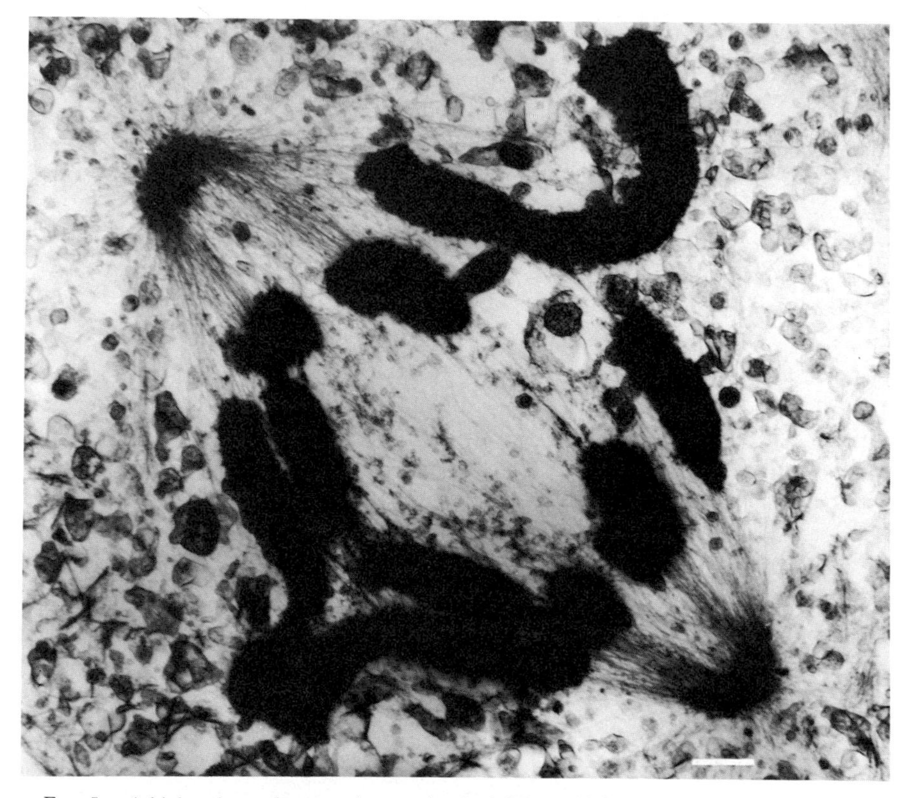

FIG. 5. A high-voltage electron micrograph of a 0.25-μm-thick section of a lysed and extracted PtK cell in anaphase. Chromosomes and spindle microtubles are evident, and the design of the spindle is made clear. The method does not, however, allow one to follow individual microtubules from one end to the other. Bar = 1 μm. (From McIntosh *et al.*, 1975. With permission.)

spindle MTs are bunched together, one cannot sort them out from one thick section. In spite of this difficulty, the HVEM has recently been used effectively to study the attaching of spindle MTs to the chromosomes after experimental manipulation. On reversal of a colcemid block, the spindle re-forms *in vivo;* the HVEM is a convenient way to identify where within the cell the MTs first reappear (Witt *et al.*, 1981). Similar studies after reversal of cold treatment have been equally informative (Rieder and Borisy, 1980). Unfortunately, the implications of the results from these two investigations are in disagreement, but the studies show clearly the power of the method for rapid and effective processing of experimental material at high space resolution.

2. Serial Thin Sectioning

Serial sectioning has become a common technique in electron microscopy. Modern microtomes, diamond knives, and embedding resins have made the cutting of multiple sections straightforward. Picking up such sections without losing any is not easy, but has been achieved by several investigators. The use of tacking wax on the upper and lower edge of the block to facilitate adherence of successive sections in a ribbon is sometimes useful. In spindle studies, one usually needs to see every section, so some completely open specimen support, such as the LKB slot grid, is necessary. In such a case, one achieves improved film stability by casting two thin plastic films rather than one thick one.

One of the first applications of serial-section technology to mitotic spindle structure was to develop a quantitative picture of the distribution of spindle MTs. Serial transverse sections were cut, and the number of MTs on every tenth to twentieth section was counted. The number of MTs plotted versus the section number constitutes a microtubule-distribution profile, which is an informative index of spindle structure (McIntosh and Landis, 1971; Brinkley and Cartwright, 1971; Fuge, 1971, 1973). The method has provided some useful insights into the changes in spindle structure with time in mitosis. The major limitation with the method has been that numerous different spindle structures could all yield the same distribution profiles, so the data are not really sufficiently detailed to define the structural constraints on spindle mechanisms.

Spindles have been successfully viewed in both serial longitudinal and serial cross sections. The longitudinal view has several advantages. Fewer sections are required to include the whole volume of the spindle, and more of each MT is

---

Fig. 6. A tracing of the positions of the microtubules from 33 adjacent longitudinal sections of a *Haemanthus* endosperm cell in anaphase superimposed upon a low-density electron micrograph of the same cell to provide a frame of reference. The capacity of the method to show the distribution of the spindle microtubules is well displayed. Bar = 10 μm. (From Jensen and Bajer, 1973. Reproduced with permission of Dr. Bajer and *Chromosoma*.)

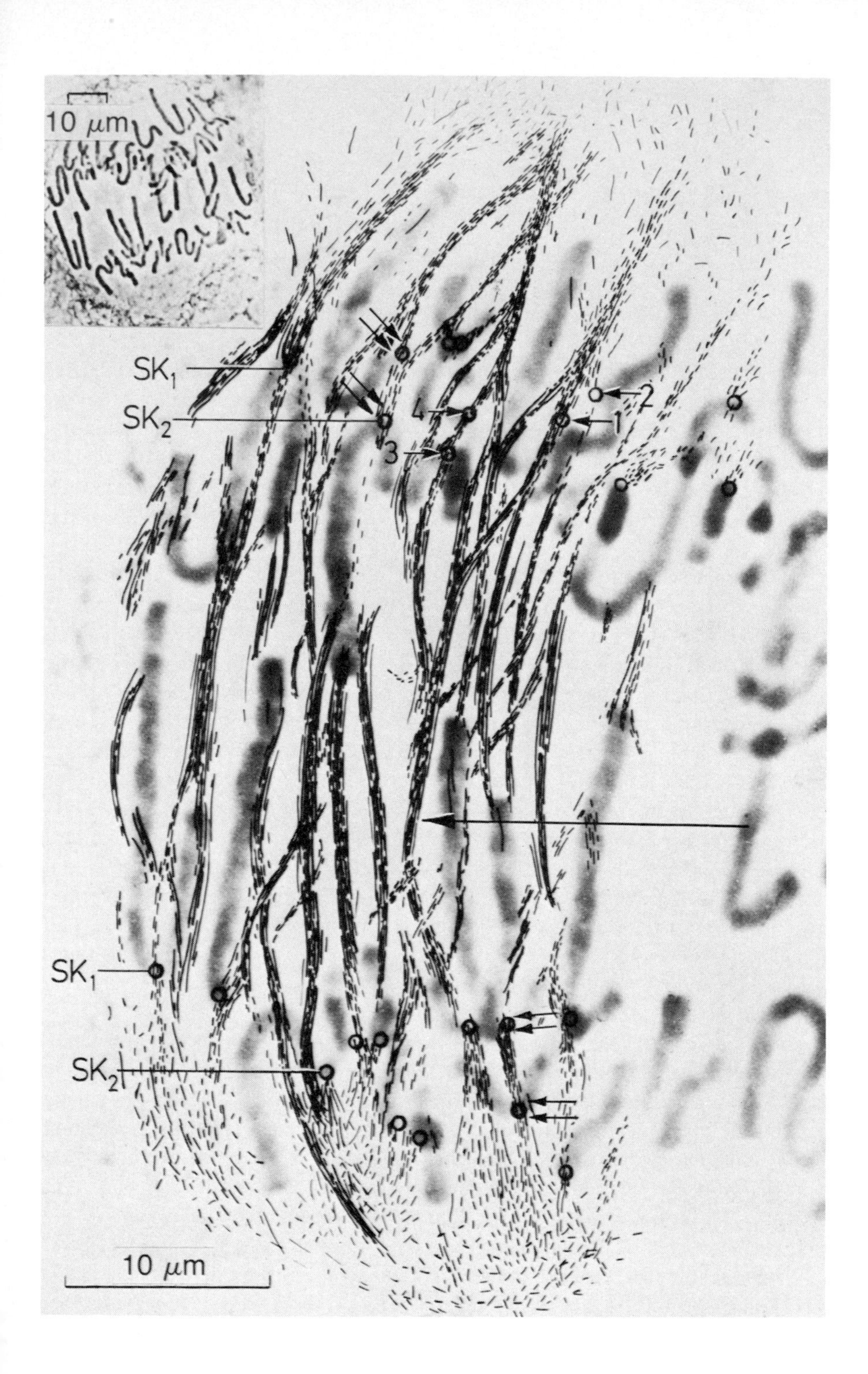
10 μm
SK1
SK2
4
3
2
1
SK1
SK2
10 μm

included in each slice. For display the MTs of several lontigudinal sections may be traced onto a single transparency to build up quite an accurate view of a portion of the spindle (Fig. 6) (Jensen and Bajer, 1973; Fuge, 1974; Rattner and Berns, 1976; Lambert and Bajer, 1977; Dietrich, 1979). The limitations of longitudinal sectioning derive from the difficulty of linking the position at which a given MT ends on one micrograph as it passes out of that section, with the place it begins on the next micrograph where it enters the subsequent section. Most longitudinal-section studies, therefore, provide representations of spindle architecture, but not true three-dimensional reconstructions. A few small spindles, however, have been studied in detail by longitudinal sections. Because of favorable spacing of the MTs in these organisms, it has been possible to follow the MTs unambiguously from one section to the next and determine their trajectories (Heath, 1974, 1978; Oakley and Heath, 1978). These studies have allowed the development of a complete reconstruction of the MTs from some or all of the spindle.

An alternative method for getting a complete spindle reconstruction is to serially cross-section the mitotic cell. Now every section contains only a short segment of each MT, but the position of the MT as it enters and leaves the section is well defined. It is therefore possible to track the MTs from one section to the next with good confidence. Typically one marks the center of each MT on a given micrograph, transfers these positions onto a transparent overlay, and uses the overlay to identify the corresponding image of each MT on the next section (Fig. 7). Such tracking studies have been accomplished for several small spindles and spindle fragments (McIntosh *et al.,* 1975; Heath and Heath, 1976; McDonald *et al.,* 1977; Tippit *et al.,* 1978). In some studies, computer graphics have been used to help with the bookkeeping and to aid in the display of the final structure (Fig. 8) (McIntosh *et al.,* 1979b). In my opinion, the cross-section tracking is easier to do with confidence than longitudinal tracking, but in certain spindles this may not be the case. Certainly the cross-section tracking is time-consuming, and the number of opportunities for human error is so large that great care must be exercised to check and recheck the work.

There are several results from these three-dimensional spindle reconstruction studies that have been informative about spindle mechanism. For example, when the positions of all the MT ends are known, the distribution of MT lengths can be determined at various times in mitosis. In *Diatoma,* this has led to the realization that at metaphase there is a bimodal distribution of MT lengths and divergent behavior of the two modes: The short MTs get shorter while the long MTs elongate (McIntosh *et al.,* 1979b). The reconstructions have also allowed the study of near-neighbor relationships between different classes of spindle fibers. In *Cryptomonas,* the spacing between kinetochore and nonkinetochore MTs is too large to permit a plausible interaction between these fibers, suggesting that

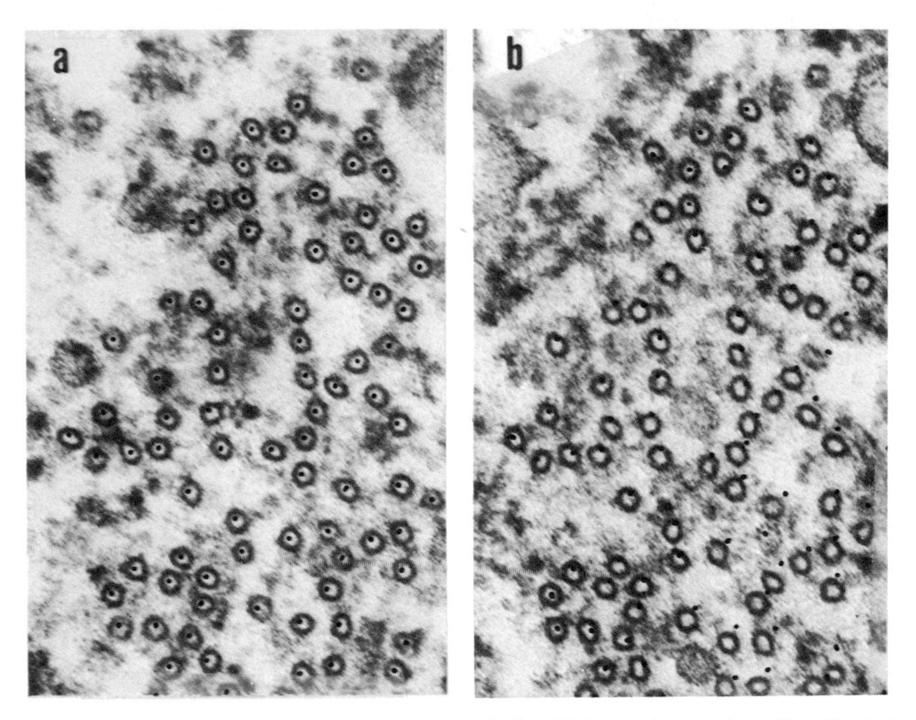

FIG. 7. Adjacent cross sections of the central spindle of *Diatoma* in metaphase. Two identical tracings of microtubule position were made from the micrograph on the left. One of these overlays fits with precision the positions of the microtubules shown in (a). The other overlay has been superimposed on (b), where it was placed so as to fit as well as possible the positions of the microtubules in the lower left portion of the micrograph. Elsewhere the dots do not necessarily lie above the centers of the tubules, but there is no ambiguity as to which dot belongs with which tubule. Further, one can imagine that a small shift in the position of the overlay would realign it with the tubules on the right side of the micrograph, losing the registration on the left. The close correspondence between the dots traced from (a) and the positions of the tubules in (b) is the reason that I am confident of three-dimensional reconstructions made from serial transverse section. (This micrograph was taken by Dr. Kent McDonald.)

the forces for chromosome movement are not a result of MT–MT interaction (Oakley and Heath, 1978). In *Diatoma* the radial density of MTs around each other has been used to study the preference of MTs for parallel or antiparallel near neighbors. The results show a marked preference for antiparallel association (McDonald *et al.*, 1979). In sum, the three-dimensional reconstructions at EM resolutions allow substantial detail in structural analysis of the spindle, both in qualitative and quantitative ways. These details can in turn clarify the structure changes that occur in the spindle as it works. One hopes that such details will ultimately contribute to our understanding of spindle mechanism.

FIG. 8. Computer drawings of central spindles of *Diatoma* in metaphase and late anaphase-telophase. Each curved line represents a microtubule. The drawing is a projection of all the spindle tubules onto a plane containing the spindle axis. The projection is shown here with a doubling in the distance between tubules to improve the visual comprehensibility of the structure. (From McIntosh *et al.*, 1979b. With permission of the *Journal of Cell Biology*.)

IV. Perspectives on the Study of Mitotic Spindle Structure

The extended period of time over which mitosis has been observed and studied has yielded a large body of descriptive material about the process. The power and limitations of most of the descriptive methods employed are detailed above, and here we use a broader brush. The simple fact is that in spite of all the work, we do not really have any clear idea about how mitosis works. It is evident that descriptive work is not enough. The question is what combination of descriptive and experimental studies is going to be most fruitful in the future.

The most hopeful avenue for descriptive work would seem to be in the characterization of phenotype of mitotic mutants. Little has yet been done to harness the power of genetic dissection for progress in mitosis research. The work from Östergren and Östergren (1966), Hartwell (1978), Byers and Goetsch (1975), and Morris (1976, 1980) has established an important beginning in this field, but much work lies ahead in which careful structural studies characterizing the behavior of particular mutants should be helpful.

Spindle biochemistry is still something of a mystery, but it is likely that immunocytochemistry will be a powerful tool in locating individual spindle components and charting their movements during mitosis. Further, the use of microscopes to observe the results of experiments with lysed cells and with exogenous material microinjected into live cells will be important for assessing the importance of particular proteins, antibodies, or other factors for chromosome movement.

In a more purely structural vein, the three-dimensional reconstruction of at least parts of spindles holds considerable promise for the immediate future. Existing morphometric methods allow an analysis of particular MT–MT interactions. If one has relatively complete structural data, it is possible to know which MTs in a bundle are associated with each pole and which are attached to kinetochores. Thorough structural analysis of mammalian chromosomal bundles should therefore help us to learn about the MT interactions that exist and how they may contribute to chromosome movement.

Finally, a word of caution about our present view of the spindle is in order. Current fixation procedures are not sacrosanct, and it is possible that a new method, such as fast freezing or an improved chemical fixation, will give us a new view of spindle-fiber composition. With the mechanism of mitosis as mysterious as it still is, the field is wide open to a major revolution of this kind.

References

Allen, R. D., Allen, N. S., and Travis, J. L. (1981). *Cell Motil.* **1,** 291–302.

Allen, R. D., Bajer, A., and LaFountain, J. (1969a). *J. Cell Biol.* **43,** 4a (abstr.).

Allen, R. D., David, G. B., and Nomarski, G. (1969b). *Z. Wiss. Mikrosk. Mikrosk. Tech.* **69,** 193–221.

Aubin, J. E., Weber, K., and Osborn, M. (1979). *Exp. Cell Res.* **124,** 79–92.

Bajer, A., and Mole-Bajer, J. (1975). *In* "Molecules and Cell Movement" (S. Inoué and R. E. Stephens, eds.), pp. 77–96. Raven, New York.

Bernard, W. (1969). *J. Ultrastruct. Res.* **27,** 250–265.

Berns, M. W. (1974). *Science* **186,** 700–705.

Brinkley, B. R., and Cartright, J. (1971). *J. Cell Biol.* **50,** 416–431.

Brinkley, B. R., Fuller, G. M., and Highfield, D. P. (1975). *Proc. Natl. Acad. Sci. U.S.A.* **72,** 4281-4285.

Byers, B., and Goetsch, L. (1975). *J. Bacteriol.* **124,** 511-523.

Cande, W. Z., Lazarides, E., and McIntosh, J. R. (1977). *J. Cell Biol.* **72,** 552-567.

Cleveland, L. R. (1953). *Trans. Am. Philos. Soc.* **43,** 848-869.

Coons, A. H., and Kaplan, M. H. (1950). *J. Exp. Med.* **91,** 1.

Coss, R. A., and Pickett-Heaps, J. D. (1964). *J. Cell Biol.* **63,** 84-98.

De Brabander, M., De Mey, J., Jonian, M., and Genens, S. (1977). *J. Cell Sci.* **28,** 283-301.

De Mey, J., Moeremans, M., Geuens, G., Nuydens, R., Van Belle, H., and De Brabander, M. (1980). *In* "Microtubules and Microtubule Inhibitors" (M. De Brabander and J. De Mey, eds.), pp. 227-242. North-Holland Publ., Amsterdam.

Dietrich, J. (1979). *Biol. Cell.* **34,** 77-82.

Dietz, R. (1959). *Z. Naturforsch.* **14,** 749-753.

Euteneuer, U., and McIntosh, J. R. (1980). *J. Cell Biol.* **87,** 509-515.

Fairbairn, D., and Passey, B. I. (1955). *Can. J. Biochem. Physiol.* **33,** 130-134.

Feramisco, J. R. (1979). *Proc. Natl. Acad. Sci. U.S.A.* **76,** 3967-3971.

Forer, A. (1978). *In* "Nuclear Division in the Fungi" (I. B. Heath, ed.), pp. 21-88. Academic Press, New York.

Forer, A., and Behke, O. (1972). *Chromosoma* **39,** 145-173.

Forer, A., and Brinkley, B. R. (1977). *Can. J. Genet. Cytol.* **19,** 503-519.

Forer, A., and Goldman, R. D. (1972). *J. Cell Sci.* **10,** 387.

Forer, A., Kalnins, V. I., and Zimmerman, A. M. (1976). *J. Cell Sci.* **22,** 115-132.

Fuge, H. (1971). *Z. Zellforsch. Mikrosk. Anat.* **120,** 579-599.

Fuge, H. (1973). *Chromosoma* **43,** 109-143.

Fuge, H. (1974). *Chromosoma* **45,** 249-260.

Fujiwara, K., and Pollard, J. (1976). *J. Cell Biol.* **71,** 848-875.

Fuseler, J. W. (1975). *J. Cell Biol.* **64,** 159-171.

Gawadi, N. (1974). *Cytobios* **10,** 17-35.

Haimo, L. T., Telzer, B. R., and Rosenbaum, J. L. (1979). *Proc. Natl. Acad. Sci. U.S.A.* **76,** 5759-5763.

Hard, R., and Allen, R. D. (1977). *J. Cell Sci.* **27,** 47-56.

Harris, P. (1962). *J. Cell Biol.* **14,** 475-487.

Harris, P. (1975). *Exp. Cell Res.* **94,** 409-425.

Hartshorne, N. H., and Stuart, A. (1960). "Crystals and the Polarizing Microscope—A Handbook for Chemists and Others," 3rd ed. Arnold, London.

Hartwell, L. H. (1978). *J. Cell Biol.* **77,** 627-637.

Heath, I. B. (1974). *J. Cell Biol.* **60,** 204-220.

Heath, I. B. (1978). *In* "Nuclear Division in the Fungi" (I. B. Heath, ed.), pp. 89-176. Academic Press, New York.

Heath, I. B., and Heath, M. C. (1976). *J. Cell Biol.* **70,** 592-607.

Hecht, E., and Zajec, A. (1974). "Optics." Addison-Wesley, Menlo Park, California.

Heidemann, S. R., and McIntosh, J. R. (1980). *Nature (London)* **286,** 517-519.

Heidemann, S. R., Sander, G., and Kirschner, M. W. (1977). *Cell* **10,** 337-350.

Hepler, P. K. (1980). *J. Cell Biol.* **86,** 490-499.

Herman, I. M., and Pollard, T. D. (1979). *J. Cell Biol.* **80,** 509-520.

Inoué, S. (1952a). *Exp. Cell Res.* **3,** 199-208.

Inoué, S. (1952b). *Exp. Cell Res., Suppl.* **2,** 305-318.

Inoué, S. (1981). *J. Cell Biol.* **89,** 346-357.

Inoué, S., and Hyde, W. L. (1957). *J. Biophys. Biochem. Cytol.* **3,** 831-838.

Inoué, S., and Sato, H. (1967). *J. Gen. Physiol., Suppl.* **50,** 259-292.

Inoué, S., Borisy, G. G., and Kiehart, D. P. (1974). *J. Cell Biol.* **62,** 175-184.
James, J. (1976). "Light Microscopic Techniques in Biology and Medicine." Nijhoff, The Hague.
Jensen, C., and Bajer, A. (1973). *Chromosoma* **44,** 73-89.
LaFountain, J. R. (1976). *J. Ultrastruct. Res.* **54,** 333-346.
Lambert, A.-M., and Bajer, A. (1977). *Cytobiologie* **15,** 1-23.
Lauterborn, R. (1896). "Untersuchungen uber Bau, Kernteilung, und Bewegung der Diatomeen." Engelmann, Leipzig.
Lazarides, E., and Weber, K. (1974). *Proc. Natl. Acad. Sci. U.S.A.* **71,** 2268-2272.
McDonald, K. L., Pickett-Heaps, J. D., McIntosh, J. R., and Tippit, D. H. (1977). *J. Cell Biol.* **74,** 377-388.
McDonald, K. L., Edwards, M. K., and McIntosh, J. R. (1979). *J. Cell Biol.* **83,** 443-461.
McIntosh, J. R., and Landis, S. C. (1971). *J. Cell Biol.* **49,** 468-497.
McIntosh, J. R., Cande, W. Z., and Snyder, J. A. (1975). *In* "Molecules and Cell Movement" (S. Inoué and R. E. Stephens, eds.), pp. 31-76. Raven, New York.
McIntosh, J. R., Sisken, J. E., and Chu, L. K. (1979a). *J. Ultrastruct. Res.* **66,** 40-52.
McIntosh, J. R., McDonald, K. L., Edwards, M. K., and Ross, B. M. (1979b). *J. Cell Biol.* **83,** 428-442.
Maupin,Szamier, P., and Pollard, T. D. (1978). *J. Cell Biol.* **77,** 837-852.
Moll, E., and Paweletz, N. (1980). *Eur. J. Cell Biol.* **21,** 280-287.
Morris, N. R. (1976). *Genet. Res.* **26,** 237-254.
Morris, N. R. (1980). *Symp. Soc. Gen. Microbiol.* **30,** 41-76.
Nicklas, R. B., and Koch, C. A. (1972). *Chromosoma* **39,** 1-26.
Oakley, B. R., and Heath, I. B. (1978). *J. Cell Sci.* **31,** 53-70.
Osborn, M., Webster, R. E., and Weber, K. (1978). *J. Cell Biol.* **77,** R27-34.
Oster, G., and Pollister, A. W. (1955). "Physical Techniques in Biological Research," Vol. 1. Academic Press, New York.
Östergren, G., and Östergren, K. (1966). *Chromosomes Today* **1,** 128-130.
Peterson, J. B., and Ris, H. (1976). *J. Cell Sci.* **22,** 219-242.
Peterson, S. P., and Berns, M. W. (1978). *J. Cell Sci.* **34,** 289-301.
Rattner, J. B., and Berns, M. W. (1976). *Cytobios* **15,** 37-43.
Rieder, C. L. (1979). *J. Cell Biol.* **80,** 1-9.
Rieder, C. L., and Borisy, G. G. (1980). *Eur. J. Cell Biol.* **22,** 312 (abstr.)
Rose, G. G., Pomerat, C. M., Shindler, T. O., and Trunnel, J. B. (1958). *J. Biophys. Biochem. Cytol.* **4,** 761-764.
Roth, L. E., and Daniels, E. W. (1962). *J. Cell Biol.* **12,** 57-78.
Roth, L. E., Pihlaja, D. J., and Shigenaka, Y. (1970). *J. Ulrastruct. Res.* **31,** 356-374.
Sabatini, D. D., Bensch, K., and Barrnett, R. J. (1963). *J. Cell Biol.* **17,** 19-58.
Salmon, E. D. (1975). *J. Cell Biol.* **65,** 603-614.
Salmon, E. D., and Begg, D. A. (1980). *J. Cell Biol.* **85,** 853-865.
Sanger, J. W. (1975). *Proc. Natl. Acad. Sci. U.S.A.* **72,** 2451-2455.
Sato, H., Ellis, G. W., and Inoué, S. (1975). *J. Cell Biol.* **67,** 501-517.
Schrader, F. (1953). "Mitosis." Columbia Univ. Press, New York.
Silver, R. B., Cole, R. D., and Cande, W. Z. (1980). *Cell* **19,** 505-516.
Slayter, E. M. (1970). "Optical Methods in Biology." Wiley (Interscience), New York.
Snyder, J. A., and McIntosh, J. R. (1975). *J. Cell Biol.* **67,** 744-760.
Sternberger, L. A. (1974). "Immunocytochemistry." Prentice-Hall, Englewood Cliffs, New Jersey.
Stryer, L. A. (1978). *Annu. Rev. Biochem.* **47,** 819-846.
Taylor, D. L., and Wang, Y. (1980). *Nature (London)* **284,** 405-410.
Tippit, D. H., and Pickett-Heaps, J. D. (1977). *J. Cell Biol.* **73,** 705-727.
Tippit, D. H., Schultz, D., and Pickett-Heaps, J. D. (1978). *J. Cell Biol.* **79,** 737-763.

Welsh, M. J., Dedman, J. R., Brinkley, B. R., and Means, A. R. (1978). *Proc. Natl. Acad. Sci. U.S.A.* **75,** 1867–1871.
Willingham, M. C., and Pastan, I. (1978). *Cell* **13,** 501–507.
Wilson, E. B. (1925). "The Cell in Development and Heredity." Macmillan, New York.
Witt, P., Ris, H., and Borisy, G. G. (1981). *Chromosoma* (in press).
Zalokar, M., and Erk, I. (1977). *Stain Technol.* **52,** 89–95.

Chapter 4

Permeabilized Cell Models for Studying Chromosome Movement in Dividing PtK_1 Cells

W. ZACHEUS CANDE

Department of Botany
University of California
Berkeley, California

I. Introduction

The breakthroughs in our understanding of how different cell-motility systems work at a molecular level have often followed a similar pattern. In the most well-understood motility systems (muscle contraction and ciliary beat), the development of an *in vitro* or demembranated model system was an initial first step. Controlled physiological experiments were performed on these model systems and the results monitored by precise quantitative electron microscopy and other techniques. Thus changes in molecular architecture were correlated with

ISBN 0-12-564125-7

changes in higher levels of organization. This sequence of experiments has led to our current understanding of the molecular mechanisms of muscle contraction and ciliary beat. It is reasonable to expect that a similar experimental strategy should lead to a better understanding of other motile processes such as anaphase chromosome movement and cleavage.

An *in vitro* system that will successfully move chromosomes would be extremely valuable (perhaps necessary) for understanding the mechanisms by which the spindle moves chromosomes. Many attempts have been made to isolate functional mitotic spindles *in vitro;* however, most have been unsuccessful (reviewed in McIntosh, 1977; Sakai, 1979). However, Sakai and colleagues (1976), using spindles isolated in bulk from sea urchin eggs, have demonstrated limited spindle elongation in the presence of ATP and exogenous tubulin, and Salmon and Segal (1979, and this volume) have shown that chromosomes move poleward when isolated spindles fall apart in the presence of micromolar free Ca^{2+}.

An alternative approach has been to use lysed cell models to study mitosis (Hoffman-Berling, 1954a,b; Cande *et al.*, 1974, 1981; McIntosh *et al.*, 1975; Cande, 1978, 1979, 1980; Cande and Wolniak, 1978). Permeabilized cell models preserve spindle function better than any spindle isolate described to date. However, the lysed cells approach unlysed cells in cytoplasmic complexity (Cande *et al.*, 1981). Two methods have been used to prepare permeabilized mitotic cells that are functional. In his pioneering studies on motility in nonmuscle cells, Hoffman-Berling (1954a,b) used glycerol to permeabilize cells and stabilize the contractile machinery. However, anaphase chromosome movement occurred only in relatively nonphysiological conditions. More recent studies have utilized low concentrations of nonionic detergents to permeabilize cells, and polymerizable tubulin or polyethylene glycol to stabilize spindle structure and function (Cande *et al.*, 1974, 1981; Cande, 1978, 1979, 1980).

In this chapter, I describe the methods used in my laboratory for preparing permeabilized cell models. After lysis, anaphase cell models maintain chromosome-to-pole movements and spindle elongation for 10 minutes at 50–75% of *in vivo* rates (Fig. 1). Movement is highly reproducible, and rates and duration of movement can be manipulated by altering the composition of the lysis medium (Cande, 1978). Recently, I have used the same lysis protocol on cleaving cells, and furrowing activity continues for 5–10 minutes after lysis at higher than 50% of *in vivo* rates (Cande, 1980). Unlike other permeabilization protocols involving the use of toluene, dextran sulfate, lysolecithin, or high detergent concentrations (reviewed in Heppel and Makin, 1977; Miller *et al.*, 1979), the conditions described here are compatible with the maintenance of labile contractile machinery such as the mitotic spindle and may be generally useful in studying the physiological requirements of many nonmuscle motile systems.

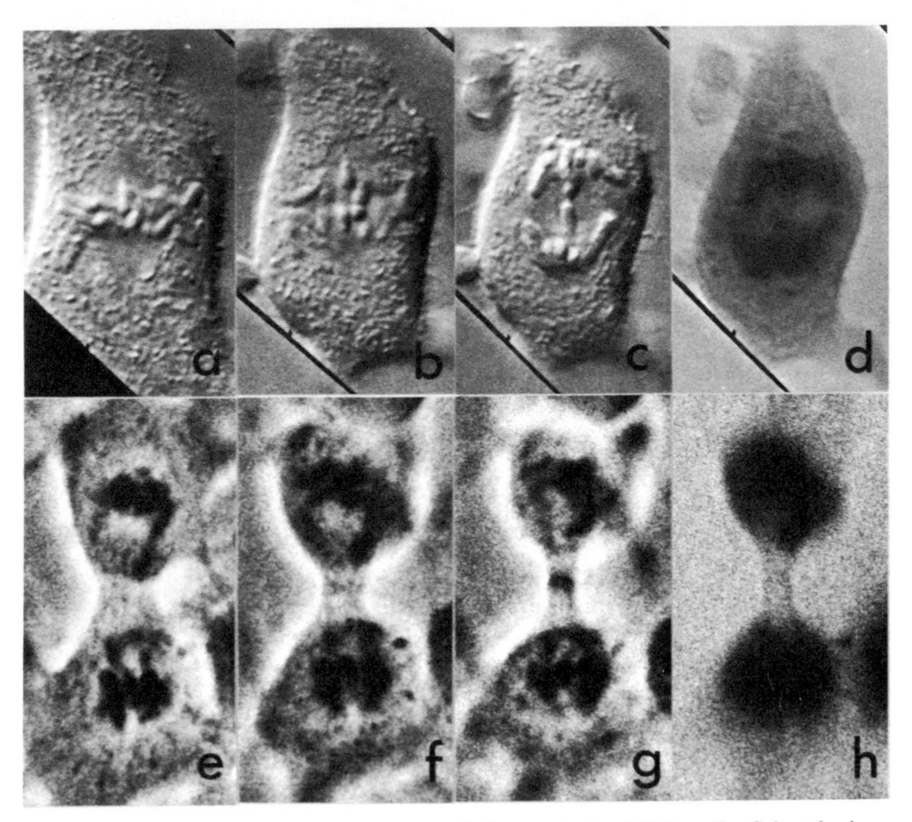

FIG. 1. Anaphase chromosome movement and cleavage in lysed PtK_1 cells. Selected micrographs taken from cinematographic record of cell undergoing anaphase (a) before lysis, (b) after 2.5 minutes in the lysis medium, (c) after 9.5 minutes, and (d) after addition of erythrosin B at 9.5 minutes. Selected micrographs of cell undergoing cleavage before lysis (e), after 2 minutes in the lysis medium (f), after 6 minutes (g), and after addition of dye at 6 minutes (h). ×2000.

II. Method of Lysis

A. Strategy of Experiments

The success of this approach is dependent on being able to monitor the activities of a single cell or a small number of cells, and know its history before and after lysis. At the appropriate stage of cell division (e.g., the beginning of anaphase), lysis medium is introduced, and the behavior of the cell after lysis is recorded by time-lapse cinematography or videotape and analyzed later. It is also possible to fix and prepare the same cell for electron microscopy at the end of the experiment (Cande *et al.*, 1981).

After lysis the interior of the permeabilized cell is accessible to physiologically important molecules such as calcium, and to exogenously supplied proteins like tubulin. Alteration of physiologically relevant components should affect the rate and extent of chromosome movement after lysis. Inhibitors directed against spindle components should interfere with spindle function. We are now in the process of using permeabilized cells to describe the physiology of chromosome movement, and preliminary accounts of these experiments have been published elsewhere (Cande, 1978, 1979; Cande and Wolniak, 1978; Cande *et al.*, 1981).

B. PtK_1 Cells

We have utilized PtK_1 cells for making permeabilized cell models because their size and morphology is well suited for studying mitosis. These cells are large for tissue culture cells, and have metaphase spindles that are over 10 μm long. Although the cells change shape and thicken as they enter mitosis, dividing cells do not round up to the same extent as most other cells. Thus they are optically ideal for monitoring chromosome movement. In addition these cells have a small number of chromosomes ($2n = 12$) that are very large. The cells cling tenaciously to the coverslip; after lysis they do not detach or move around. In contrast, other tissue culture cells (e.g., HeLa and CHO cells) may detach from the coverslip during the lysis procedure and float away.

PtK_1 cells (derived from the kidney epithelium of the rat kangaroo) can be obtained from the American Tissue Culture Collection (ATCC CCL 35). We grow these cells in plastic flasks (Corning) and petri dishes in a 4% CO_2 atmosphere at 37°C using Ham's F12 medium supplemented with 0.2 m*M* L-glutamine, 0.02% sodium ampicillin, 15 m*M* $NaHCO_3$, 10 m*M* HEPES, and 10% fetal calf serum. The cells readily form monolayers on Corning #1 glass coverslips that have been sterilized by immersion in 95% ethanol and subsequently flamed. The coverslips are normally used when the cells have reached 80% confluency and are about 2 days old.

Although the PtK_1 cell line is considered to be stable, a significant proportion (1–5%) of these cells appear to be polyploid or have abnormal mitotic spindles. The easiest way to recognize abnormal spindle morphology is to examine dividing cells with polarization optics. Abnormal cells can also be recognized with phase and Nomarski optics because they may have too many chromosomes or the chromosomes do not all line up on one metaphase plate. A dividing cell is normally oblate with a long process at either end of the cell, and the spindle is aligned parallel to the long axis of the cell. Abnormal cells are often rectangular or triangular with several processes and the spindle may be oriented perpendicular to the long axis. These spindles are often tripolar and are best avoided.

In general the easiest way to find a suitable cell is to survey the coverslip at low power with phase optics and then to record the experiment using higher magnification objectives. If the goal is to study anaphase, the cells that are ready to enter anaphase have all their chromosomes aligned tightly on the metaphase plate and the kinetochores stretched out. Such cells will usually enter anaphase within 10 minutes. Normally metaphase is only 30 minutes in duration and anaphase is completed within 15 minutes after sister chromatid separation.

C. Mounting Cells and Exchanging the Medium

For experimentation the coverslips are mounted cell side down on glass spacers to form a chamber that is open at each end (Fig. 2). The glass spacers, made from coverslips, are approximately 5 mm wide and are glued to the slide with 10% elvanol (polyvinyl alcohol, Baker). The coverslip is held in place with valap (1:1:1 mixture of Vaseline, lanolin, paraffin), but the ends are left unsealed. The total volume of this chamber is ~40 λ. Provided fresh culture medium is added periodically and the temperature remains stable, cells will continue to divide for several hours. We maintain the temperature of the stage at 35 ± 2°C, either with a Sage Air-curtain Incubator or with commercially available hairdryers (Schick Tote 'n Dry, etc.). The temperature of the stage can be monitored with a Tele-thermometer (Yellow Springs Instrument Co.) or with temperature-sensitive liquid crystals that are cut in the shape of slides and placed on the microscope stage.

Medium can be added to the preparation by placing a large drop (~200 λ) on

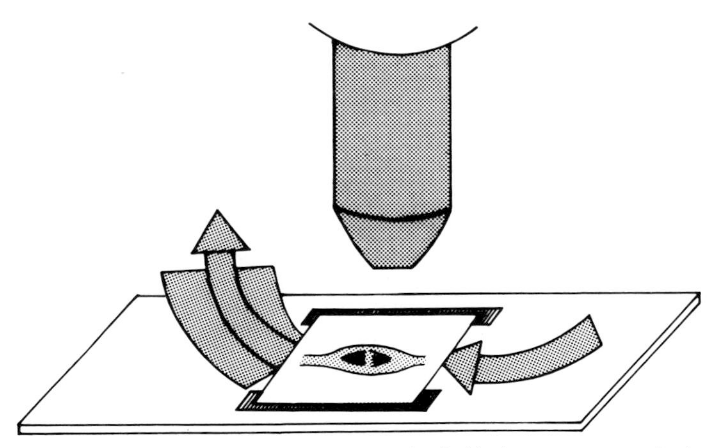

FIG. 2. Cell lysis protocol. Coverslips are mounted cell side down on spacers. Lysis medium is added to one side of the chamber and the solution is pulled past the cells using bibulous paper as a wick. The progress of the experiment is monitored by phase, polarization, or Nomarski optics and recorded by time-lapse cinematography or videotape.

one side of the chamber and simultaneously pulling the solution past the cell by using bibulous paper as a wick at the other side of the chamber (Fig. 2). Medium can be exchanged as rapidly as 20 seconds with this technique. In order to minimize problems from introduction of air bubbles and unequal mixing and exchange of medium, cells to be filmed are always selected from the center of the coverslip rather than the periphery. Since shear forces caused by flow of the medium through the chamber may contribute to the rate and extent of cell lysis, care is taken to introduce the medium to the chamber the same way in each experiment.

D. Lysis Protocol

Lysis is accomplished in two steps (Table I). First, a solution containing the nonionic detergent Brij 58 is pulled through the chamber (solution A). After the cells have been exposed for 60–70 seconds to this medium, it is replaced by a second medium (solution B). which contains in addition 2.5% polyethylene glycol (20,000 MW).

The extent of lysis and extraction of cellular components is controlled by the detergent concentration and the ratio of detergent to polyethylene glycol in the lysis medium (Table II; Cande, 1978). As the detergent level is increased with respect to the concentration of polyethylene glycol present, the duration of movement after lysis decreases. For example, in 0.5% Brij 58 and 2.5% polyethylene glycol, chromosome movement continues for only 2–3 minutes after lysis (Cande, 1978). Cells exposed to solution A or to a similar solution using Triton X-100 as the detergent move chromosomes for a brief period before the spindle falls apart, and relatively little protein remains in the cells after lysis. In contrast, after lysis in solution B, chromosome movement continues for a

TABLE I

COMPOSITION OF THE LYSIS MEDIUM

Solution A	
0.08–0.1% Brij 58	
85 m*M* PIPES, pH 6.94 (K salt)	
1.25 m*M* MgATP	
0.1 μM free Ca^{2+} 1.0 m*M* free Mg^{2+}	10 m*M* EGTA + 1.51 m*M* $CaCl_2$ + 1.28 m*M* $MgSO_4$
Solution B	
Solution A *plus*	
2.5% polyethylene glycol (20,000 MW)	

TABLE II

EFFECT OF CHANGES IN LYSIS MEDIUM ON PROPERTIES OF THE PERMEABILIZED CELL

	Percent of protein[a] on coverslip after 8 minutes in lysis medium	Duration[b] of chromosome movement after lysis
Culture medium	100	—
Solution A	40	4 minutes
Solution B	95	7 minutes
Solution A, B (normal procedure)	90	10 minutes
Solution A with 0.1% Triton X-100	35	3 minutes

[a]Average of three experiments. At the end of the lysis period, coverslips were rinsed in 0.15 *M* NaCl, then placed in a dish with 1 ml of solution containing 0.15 *M* NaCl, 1 m*M* $MgSO_4$, 5% glycerol, and 0.1% SDS. The cells were scraped off the coverslip with a rubber policeman, and the protein content of the solution was determined by the method of Lowry *et al.* (1951).

[b]Average of three separate experiments.

longer period and most of the soluble protein is retained by the cell. However, these conditions are not optimal for movement, and in addition the retained cytoplasm may act as a barrier for uptake and diffusion of molecules into the cell. We obtained optimal movement after lysis with a procedure that combines both of these lysis conditions (solution A plus B). That is, we expose the cells briefly to detergent, then in the second step add polyethylene glycol to retard extraction of cellular components.

Although we have screened a large number of detergents, Brij 58 gave the most reproducible results for making permeabilized models. Other detergents with similar properties such as Nonidet P 40 and Triton X-100, when used at a lower concentration (0.04–0.05%), yield functional permeabilized cell models, but the rates of chromosome movement after lysis were slower. One note of caution in using Brij 58 is in order. We have found that the potency of the detergent increases with the age of the aqueous stock solution. In general, we found that the most reproducible results are obtained if the aqueous stock is made up fresh daily.

We have not systematically examined the effects of manipulating the concentration of polyethylene glycol or the type of carbowax used in the second step. In general, as long as the polyethylene glycol concentration is above 2%, spindle structure and function is stabilized after lysis. Polyethylene glycol solutions above 5% are too viscous to use easily. We have tried, without success, to use BSA, glycerol, sucrose, and mannitol to stabilize spindle function. Although

25% glycerol or 0.3 *M* sucrose when added to the lysis medium will maintain a birefringent spindle as viewed with polarization optics, this stabilized spindle will not move chromosomes.

The other components of the lysis medium (Table I) are defined by the physiological requirements for maintenance of chromosome movement. In general we find that chromosome movement works best at a neutral pH, when the ionic strength of the medium is less than 350 m*M* and Mg^{2+} ions are in excess of 1 m*M* (Cande *et al.*, 1981; Cande, 1979). Due to the complexity of the lysed cell, some chromosome movement will continue in the absence of exogenous nucleotide, but chromosome movement is optimal in the presence of ATP (Cande and Wolniak, 1978; Cande, 1979). We are in the process of describing the effects of different free-calcium levels on chromosome movement (manuscript in preparation). Maximal rates of chromosome-to-pole movement are observed in 0.5 μM free calcium, but maximum spindle elongation is observed in 0.1 μM or lower free-calcium levels. Usually we lyse cells in a 10 m*M* EGTA buffer system designed to keep the cells at a calculated 0.1 μM free-Ca^{2+} and 1 m*M* free-Mg^{2+} concentration (R. Steinhardt, personal communication).

We have found that the requirements for cleavage after lysis are similar to that for chromosome movement, except optimal cleavage is seen in the presence of 5 m*M* ATP (Cande, 1980), and cleavage is not inhibited by the addition of EDTA to the lysis medium (unpublished data). As with chromosome movement, the maximum extent of cleavage is observed at a neutral pH in 0.1 μM free Ca^{2+}, with MgATP present.

III. Characteristics of the Permeabilized Cell

A. Permeability Properties

To describe the permeability properties of PtK_1 cells after lysis, cells were incubated with dyes and fluorescently labeled proteins (Cande, 1978, 1979; Cande *et al.*, 1981). After addition of solution A, cells no longer exclude dyes such as erythrosin B (MW 800) and ions like calcium (Fig. 1). Protein uptake is extensive but less rapid after exposure to lysis medium. After 1 minute, only a small number of dividing cells are stained by rhodamine-labeled immunoglobulin fragment FAB, (50,000 MW), but after 4 minutes in the lysis medium 85% of the dividing cells are stained by this fluorescently labeled protein (Cande *et al.*, 1981). By functional criteria the cortex of cleaving cells is accessible to $NEM\text{-}S_1$ within 1.5 minutes after lysis, since cleavage is inhibited by this protein after a brief incubation (Cande, 1980). Chromosome movement in anaphase cells is inhibited by high concentrations of polymerizable tubulin after addition of the lysis medium, and exogenous tubulin is incorporated into the spindle aster of the permeabilized cell (Cande *et al.*, 1981).

B. Morphology of Permeabilized Cells

The phase and Nomarski appearance of the cells after lysis are not greatly altered. Immediately after lysis, the rod-shaped mitochondria round up and appear to fragment (Fig. 1). Considerable blebbing of the cytoplasm may occur; however, the most extensive blebbing is observed in cells lysed in more than micromolar free Ca^{2+}. After addition of lysis medium, the condensed chromosomes present in mitotic cells may become less refractile and phase dense, but this change in appearance is usually brief and the chromatin reverts to its original phase-dense appearance after 1–2 minutes. After 4–5 minutes the spindles in lysed mitotic cells take on a more definite outline, and at the spindle poles the centriole complex may become visible as small dots. The nuclear envelope in

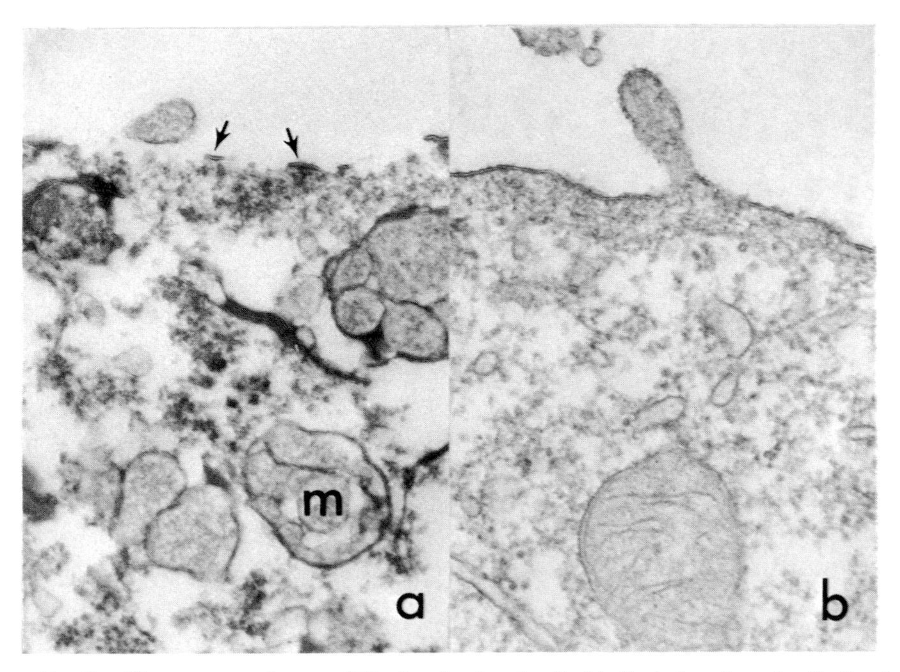

FIG. 3. Ultrastructure of permeabilized and unlysed cell. (a) Cortical region of cell after 8 minutes in lysis medium. Film records of this cell reveal that anaphase chromosome separation was proceeding at $\sim$1.0 μm/minute at the time of fixation. Except for some small patches (arrows), the plasmalemma is mostly absent from this region of the cell. Mitochondria (m) appear swollen. $\times$54,000. (b) Cortical region of a normal (unlysed) cell for comparison with (a). The cells in Fig. 3a and b were treated identically during preparation for electron microscope observation. $\times$54,000.

interphase cells characteristically stands out in relief in Nomarski optics and the chromatin becomes more stringy and condensed.

The ultrastructure of cells after lysis is similar but not identical to that of unlysed cells (Fig. 3). From our point of view, the most striking difference is that portions of the plasma membrane of the lysed cell are removed; only fragments of membrane are left attached to the cortex of the cell. However the cytoplasm underneath the disrupted membrane does not look noticeably different. The density of staining of the ground substance is similar in lysed and unlysed cells, and microfilaments, intermediate filaments, and microtubules are present in both classes of cells. The other membranous organelles in the permeabilized cell do not appear to be noticeably disrupted, although the mitochondria are swollen and the cristae distorted. The endoplasmic reticulum, noticeable especially around the spindle poles in dividing cells, appears as a reticulate network and is not broken up into small vesicles or otherwise disrupted. The chromatin in lysed cells may appear more condensed and densely stained than in unlysed cells.

C. Other Properties

Permeabilized cells retain most of their protein for at least 12 minutes after lysis (Fig. 4). After 8 minutes, over 90% of the protein remains in the cells; after 12 minutes 75% of the protein is still present in the cells. The loss of protein from cells during the lysis period is paralleled by changes in cell appearance (as described above) and cell function. After 12–15 minutes, spindle birefringence of permeabilized cells is diminished by about 50% and anaphase cells no longer move chromosomes. The cytoplasm of interphase cells looks washed out, although stress fibers remain and the swollen mitochondria are no longer visible in Nomarski optics.

All cells, including interphase cells, undergo some shape change after lysis. This is demonstrated by the occurrence of large spaces between cells and the presence of retraction fibers extending out from the periphery of the lysed cell to the old cell margins. Some rounding-up and retraction fibers are seen in all cases after lysis, even with cells lysed in cytochalasin B or phalloidin. Pretreating cells with cytochalasin B before lysis does not block retraction fiber formation. Although maximum cell shape change after lysis is observed under conditions that also favor cleavage and anaphase chromosome movement, the ubiquitous presence of retraction fibers suggests that some of this overall shape change may be due to an alteration in attachment of cells to the substrate rather than to an active contraction of the cell cortex.

By a variety of criteria, cells are not viable even after only 2 minutes of exposure to the lysis medium. Mitotic cells do not proceed through prophase; that is, nuclear envelope breakdown, chromosome condensation, and spindle formation are blocked. Saltatory movement of mitochondria does not occur after lysis.

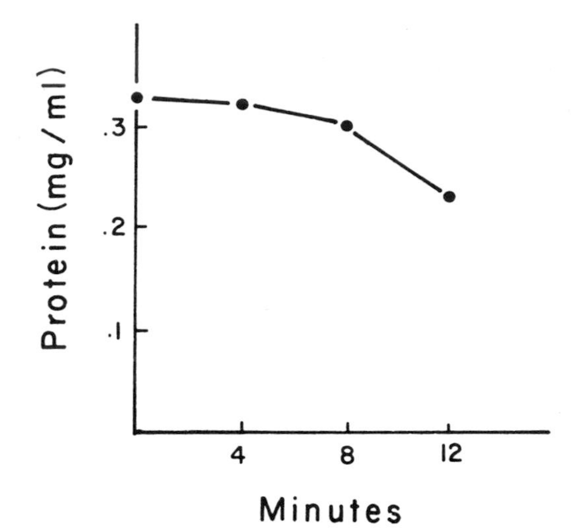

FIG. 4. Protein content of cells remaining on coverslips after lysis. Cells were lysed as described in the text and the protein content determined as described in Table II. Each point is the average of three separate experiments.

Other steps in cell division are also blocked. The transition from metaphase to anaphase, the initiation of the cleavage furrow, and chromosome decondensation also do not occur after lysis.

IV. Concluding Remarks

The mitotic spindle is an ephemeral structure that forms at one stage of the cell cycle and falls apart during the process of chromosome movement. A lysis medium designed to stabilize spindle structure perfectly would freeze the chromosomes in place and prevent further anaphase chromosome movement. To study anaphase *in vitro* it is necessary not only to preserve the motile machinery but also those parts of the cell that are involved in regulating spindle disassembly. We have solved these problems by using a gentle method of cell permeabilization that does not disrupt cellular architecture, and we have developed a lysis medium that retards but does not block spindle dissolution.

We have described in detail this strategy for preparing permeabilized cells because we feel it can also be used to study other complex motile phenomena that are not easily dissected by more conventional biochemical methods. Recently Beth Burnside (personal communication) has used a modification of this lysis protocol to reactivate teleost retinal cone contraction. We have used this procedure to describe the physiological requirements for cleavage in PtK_1 cells (Cande, 1980). Studies are in progress in this and other laboratories to prepare permeabilized models for studying other motile processes such as axoplasmic flow in neurons, cytoplasmic streaming in plant cells, and cell division in diatoms.

Acknowledgments

The advice and help of Kent McDonald and Susan Stallman are gratefully acknowledged. This research was supported by NIH grant GM 23238 and American Cancer Society Grant CD-79.

References

Cande, W. Z. (1978). *In* "Cell Reproduction" (E. R. Dirksen, D. M. Prescott, and C. F. Fox, eds.), ICN-UCLA Symp. Mol. Cell. Biol., pp. 457–464. Academic Press, New York.

Cande, W. Z. (1979). *In* "Cell Motility: Molecules and Organization" (S. Hatano, H. Ishikawa, and H. Sato, eds.), pp. 593–608. Univ. of Tokyo Press, Tokyo.

Cande, W. Z. (1980). *J. Cell Biol.* **87,** 326–335.

Cande, W. Z., and Wolniak, S. M. (1978). *J. Cell Biol.* **79,** 573–580.

Cande, W. Z., Snyder, J., Smith, D., Summers, K., and McIntosh, J. R. (1974). *Proc. Natl. Acad. Sci. U.S.A.* **71,** 1559–1563.

Cande, W. Z., Meeusen, R. L., and McDonald, K. L. (1981). *J. Cell Biol.* **88,** 618–629.

Heppel, L. A., and Makin, N. (1977). *J. Supramol. Struct.* **6,** 399–409.

Hoffman-Berling, H. (1954a). *Biochim. Biophys. Acta* **14,** 182–195.

Hoffman-Berling, H. (1954b). *Biochim. Biophys. Acta* **15,** 332–339.

Lowry, O. H., Rosenbrough, N. J., Farr, A. L., and Randall, R. J. (1951). *J. Biol. Chem.* **193,** 265–275.

McIntosh, J. R. (1977). *In* "Mitosis: Facts and Questions" (M. Little, N. Paweletz, G. Petzelt, H. Ponstingl, D. Schroeter, and H.-P. Zimmerman, eds.), pp. 167–184. Springer-Verlag, Berlin and New York.

McIntosh, J. R., Cande, W. Z., and Snyder, J. A. (1975). *In* "Molecules and Cell Movement" (S. Inoué and R. Stephens, eds.), pp. 31–76. Raven, New York.

Miller, M. R., Castellot, J. J., Jr., and Pardee, A. B. (1979). *Exp. Cell Res.* **120,** 421–425.

Sakai, H. (1979). *Int. Rev. Cytol.* **55,** 23–48.

Sakai, H., Mabuchi, I., Shimoda, S., Kuriyama, R., Ogawa, K., and Mohri, H. (1976). *Dev., Growth Differ.* **18,** 211–219.

Salmon, E. D., and Segall, R. R. (1979). *J. Cell Biol.* **83,** (pt 2), 337a.

Chapter 5

Mitotic Spindles Isolated from Sea Urchin Eggs with EGTA Lysis Buffers

E. D. SALMON

Department of Zoology
University of North Carolina
Chapel Hill, North Carolina

ISBN 0-12-564125-7

I. Introduction

This chapter describes methods used in our laboratory to isolate mitotic spindles from dividing sea urchin eggs. An obvious goal is to obtain a preparation that has a well-defined composition, that can be stored, and that can then be reactivated under controlled conditions to produce lifelike assembly characteristics and chromosome transport (Inoué, 1964; Inoué and Sato, 1967; Inoué *et al.*, 1975; Nicklas, 1971, 1975, 1977; Salmon, 1975, 1976).

The mitotic apparatus is a complex organelle, and an ideal isolated model has not yet been achieved. The spindle-fiber microtubules are labile and will generally depolymerize within several minutes if cells are lysed into normal culture media. Several spindle-isolation techniques have been developed that use glycols or purified brain microtubule protein (MTP) to stabilize the spindle microtubules on cell lysis. However, these procedures significantly alter the native assembly characteristics of the spindle microtubules, hence no activated chromosome movements have been obtained (reviewed by McIntosh, 1977; Salmon *et al.*, 1980). Significant progress toward a functional isolated model is being made, in part by recognizing that spindle-fiber microtubules are sensitive to depolymerization by micromolar concentrations of calcium ions and that calcium ions are normally sequestered below 10^{-7} *M* within the spindle microenvironment (Inoué and Kiehart, 1978; Kiehart, 1979; Kiehart and Inoué, 1976; Salmon and Jenkins, 1977; Salmon and Segall, 1980; Weisenberg, 1978).

We have found that spindles with relatively stable microtubules can be isolated from dividing sea urchin eggs using simple EGTA lysis buffers. The morphology and birefringent retardation (BR) of these isolated spindles are similar to spindles in living cells (Figs. 1–3). The isolated spindles are highly extracted, free of membranes, and made up primarily of microtubules (Figs. 4–6). These isolated spindles can be stored and, for at least one species, reactivated to display microtubule assembly-disassembly characteristics similar to living spindles and to move chromosomes in a limited manner (Salmon and Segall, 1979, 1980).

FIG. 1. An early-anaphase spindle isolated from first-division eggs of the sea urchin *Strongylocentrotus droebachiensis* using an EGTA lysis buffer containing 1% Triton X-100 to solubilize the membrane components of the mitotic apparatus. The differential-interference contrast photomicrograph shows the various spindle parts and the separated chromosomes. The central spindle includes the regions between the chromosomes and nearest pole, called the half-spindle, and the region between the separating chromosomes, termed the interzone. The spindle poles abut the centrosome (CS)-centrosphere (CSPH) complex at the center of the asters. Astral fibers extend radially outward from the periphery of the centrosphere, which grows in diameter from metaphase through teleophase and appears largely devoid of internal structure in electron micrographs, except for the centrosome complex surrounding the pair of centrioles (see Fig. 6). Bar = 10 μm.

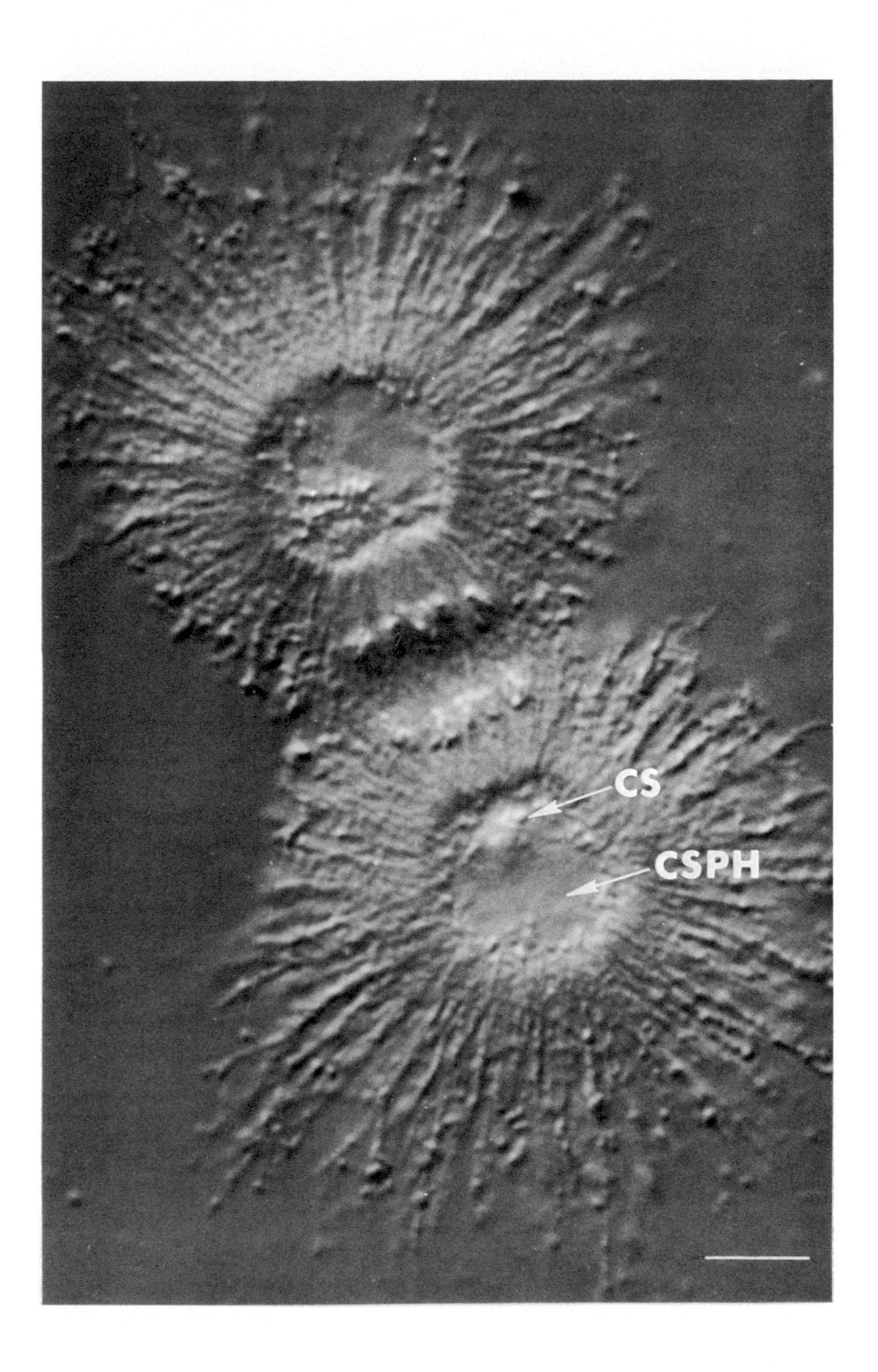
CS
CSPH

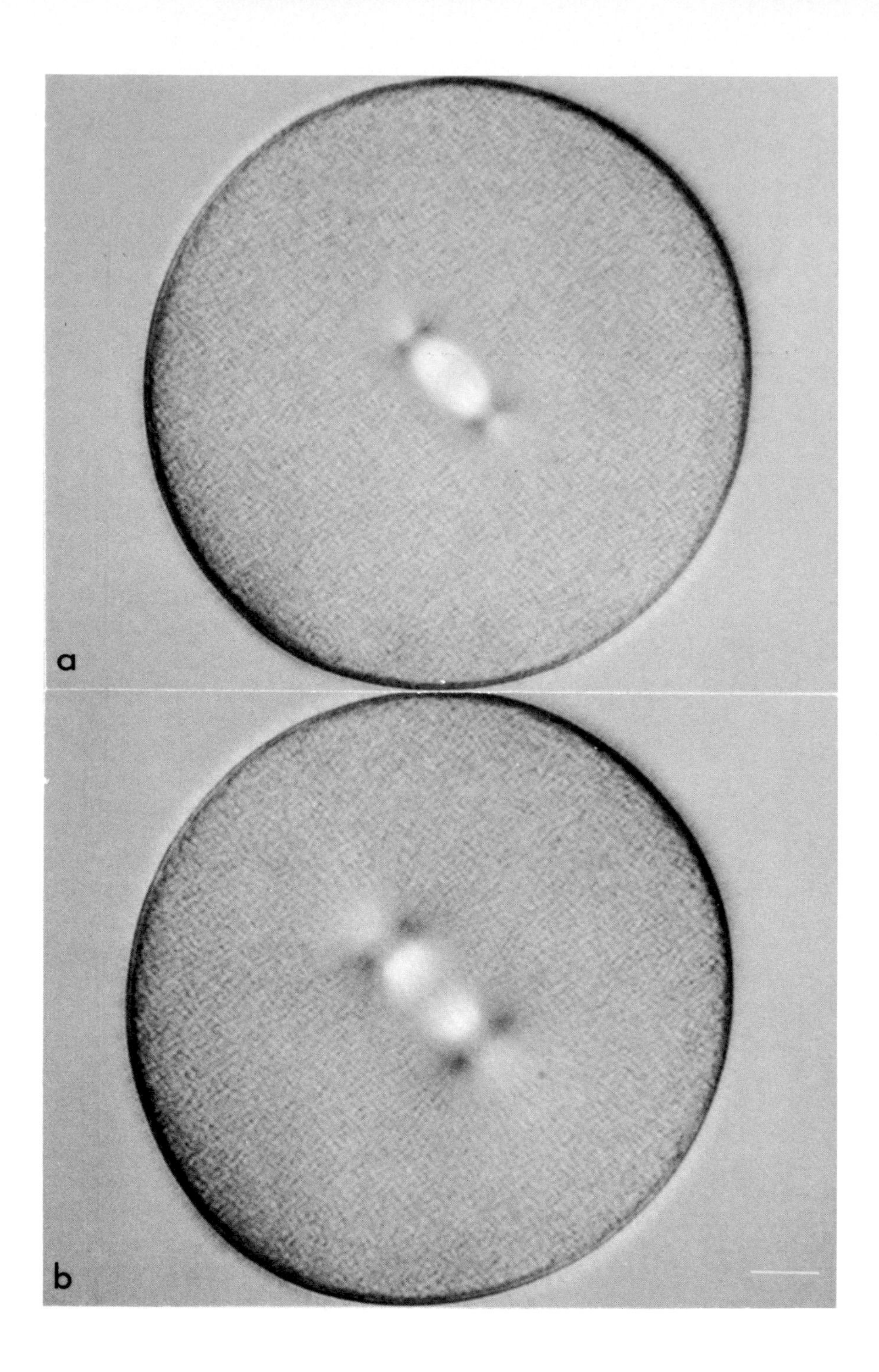
a
b

Since Mazia and Dan first isolated the mitotic apparatus from dividing sea urchin eggs in 1952, several techniques have been developed for isolating stable spindles or mitotic apparatuses. A number of excellent detailed reviews of this work exist. Although one may not wish to employ the same isolation media used by these previous investigators, their reports are worth consulting because they contain a wealth of information on optimal procedures for culturing and preparing particular species of sea urchin eggs for spindle isolation.

This chapter will first briefly describe the major *in vivo* characteristics of the mitotic spindle. A brief historical review of isolation techniques will be followed by a description of the evolution of the EGTA lysis method that I use, emphasizing the influence on microtubule stability of buffer composition and intracellular physiological conditions. First-division sea urchin eggs are an excellent source for isolated spindles because large numbers of them can be cultured to develop synchronously. However, several important procedures must be mastered in order to remove the tough extracellular layers and obtain synchronous fertilization and development, and to determine the proper time for cell lysis, which I will consider next. We have found that culture procedures, cell lysis, solubility of the egg cortex, and spindle-microtubule stability in our EGTA lysis buffers are highly species specific; this report therefore will detail the procedures and results for the two sea urchin species with which I have had the most in-depth experience, *Strongylocentrotus droebachiensis* and *Lytechinus variegatus*.

II. *In Vivo* Spindle Characteristics

A. Definition of Spindle Parts

Some definitions and considerations of the *in vivo* mitotic spindle are in order. Detailed definitions can be found in McIntosh (1977). *Mitotic apparatus* refers to the chromosomes, the central spindle and astral fibers, the membrane components, and whatever other components are indigenous to the organelle *in vivo*. The term *mitotic spindle* refers specifically to the fibrous structures of the mitotic apparatus plus attached chromosomes and the mitotic centers. Parts of the mitotic

---

FIG. 2. Polarization micrographs of a living egg of *Lytechinus variegatus* at late metaphase (a) and mid-anaphase (b). The fertilization and hyaline layers have been removed, which reveals the radially positive birefringent retardation (BR) of the microvilli of the egg cortex. Polarization micrographs of spindles are taken with the interpolar axis of the spindle oriented at 45 degrees to the analyzer-polarizer directions and the compensator set for about 3.5 nm BR from the background light extinction. Astral fibers perpendicular to the spindle's interpolar axis appear dark in contrast because of the compensator. Note that the birefringent spindle sits within a clear zone from which the yolk particles have been excluded. The egg has been slightly flattened between the slide and coverslip. Bar = 20 μm.

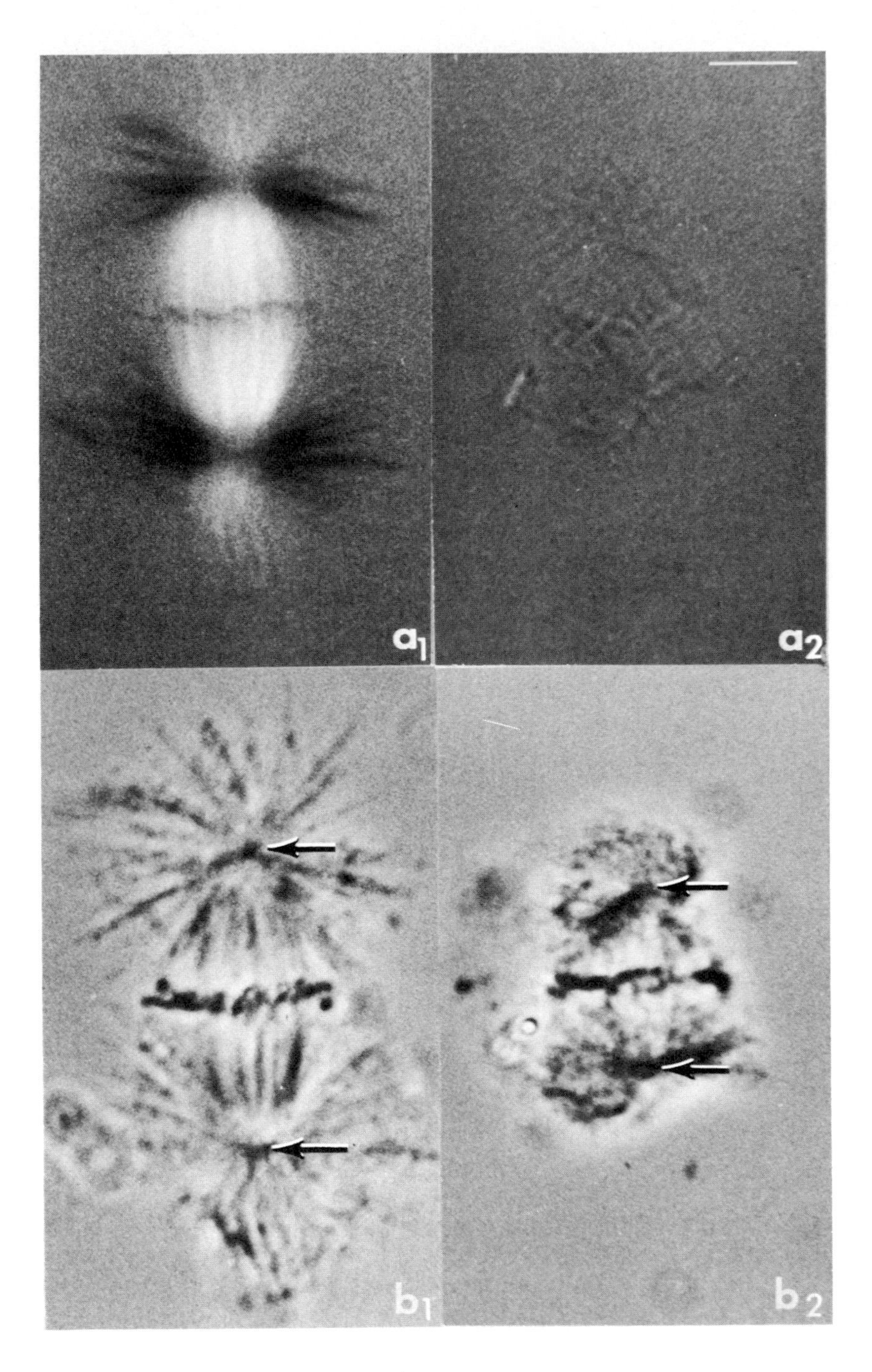
a1
a2
b1
b2

spindle include the central spindle fibers, the kinetochores, the polar and interpolar fibers, the astral fibers, and the centrosome-centrosphere complex (see Fig. 1).

B. Spindle-Fiber Microtubules

In the living cell, forces for chromosome transport are generated along the spindle fibers, and most current theories of mitosis are based on proposed properties of the microtubules and associated complexes (Bajer, 1973; Inoué, 1976; Inoué and Sato, 1967; Inoué and Ritter, 1975; Margolis, 1978; Margolis *et al.*, 1978; McIntosh, 1977; McIntosh *et al.*, 1969; Nicklas, 1971, 1975, 1977).

Microtubules are the major structural elements of the spindle fibers (Fuge, 1978; McIntosh, 1977). Spindle-fiber microtubules tend to align in parallel arrays producing weak birefringent retardation (BR), about 1 nm. Dynamic changes in the number and distribution of spindle-fiber microtubules can be monitored using polarization microscopy methods (Inoué, 1964; Sato *et al.*, 1975; Salmon and Ellis, 1976) in living cells and in isolated spindles (Figs. 2–4) (Hiramoto *et al.*, 1981; Salmon and Segall, 1980).

The first-division spindles in sea urchin eggs are large. For example, the interpolar length of a metaphase spindle from *L. variegatus* is about 22 μm (Fig. 3); in *S. droebachiensis* eggs, it is about 25 μm depending on the temperature. Sea urchin chromosomes are small, and the kinetochore regions are not distinctly differentiated as in mammalian chromosomes. Although no detailed analyses have been done of microtubule distribution throughout mitosis in sea urchin embryos, some information is available. In most species of sea urchin there are about 38–44 chromosomes (Harvey, 1956; Hiramoto *et al.*, 1981). Electron micrographs of isolated spindles have shown that the metaphase half-spindle contains 2500–3500 microtubules, depending on the conditions of fixation and the species of sea urchin (Cohen and Rebhun, 1970; Hiramoto *et al.*, 1981; E. D. Salmon, unpublished observations). The number of kinetochore microtubules per chromosome is not well established, but about 15 are reported for *Arbacia* (Cohen and Rebhun, 1970), about 30 are estimated for the heart urchin *Clypeaster japonicus* (Hiramoto *et al.*, 1981), and about 15 for *L. variegatus* E. D. Salmon, unpublished observation). Microtubules attached to the kinetochores

FIG. 3. Changes in spindle-fiber BR and morphology induced by 2 μM Ca^{2+} in isolated metaphase spindles of *L. variegatus*. Isolated spindles in EGTA buffer (no glycerol) before treatment with 2 μM Ca^{2+} (a_1 and b_1) and 5 minutes after treatment (a_2 and b_2), viewed with polarization (a_1 and a_2) and phase-contrast microscopy (b_1 and b_2). The centrosomes (arrows) can be seen as phase-dense regions abutting the spindle poles. Note the loss of spindle-fiber BR (a_2) and the shortening of the chromosomal half-spindle fibers and astral fibers (b_2) that occurred after addition of 2 μM Ca^{2+}. The width of the metaphase plate remains unchanged. Polarization micrographs were taken with the compensator set for 3.5 nm positive BR with respect to the spindle interpolar axis. (Modified from Salmon and Segall, 1980, with permission.) Bar = 10 μm; magnification, ×1200.

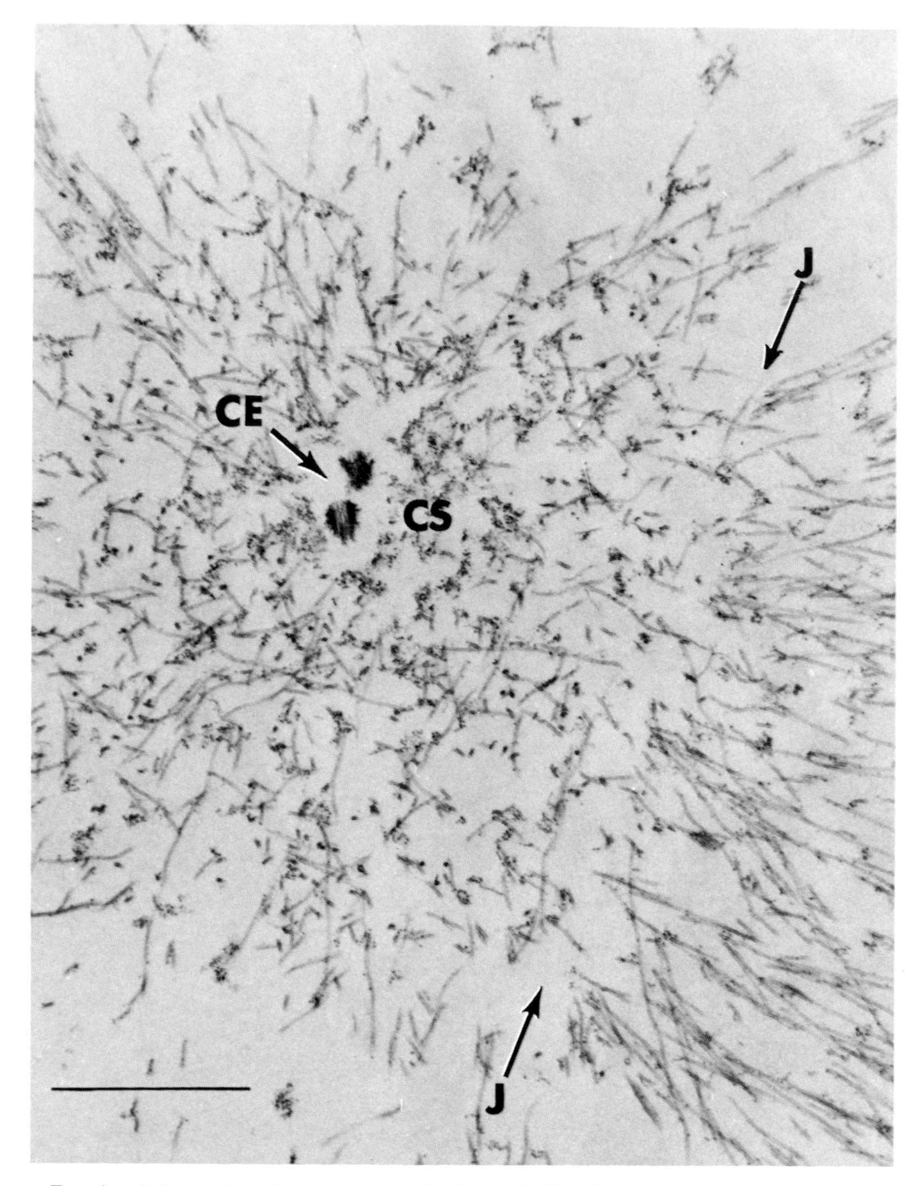

FIG. 4. A thin-section electron micrograph of the spindle pole–centrosome–aster complex of an early-metaphase spindle isolated from *L. variegatus*. Centrioles (CE) are contained within the centrosome complex (CS), which is a region where electron-dense material and ribosomelike particles are concentrated. At this stage a centrosphere structure is not apparent. Astral microtubules appear to end in the centrosome complex and extend radially away from it. The central-spindle microtubules do not appear to end in the centrosome complex, but terminate at a peripheral junction (J). (Modified from Salmon and Segall, 1980, with permission.) Bar = 1 μm.

appear to make up about 25–50% of the number of half-spindle microtubules. Most microtubules in a half-spindle are polar microtubules, which originate near the spindle pole and extend toward the chromosomes. Few extend from pole to pole; most end before reaching the chromosomes. Only a small percentage overlap the polar microtubules from the opposite pole (Hiramoto *et al.*, 1981; E. D. Salmon, unpublished observation). In first-division eggs, the aster complexes grow during anaphase to occupy a greater volume than the central spindle (Fig. 1), but by the 32-cell stage, the asters are difficult to detect (Harris, 1962, 1975). The centrosome-centrosphere complex has not been well characterized.

C. Other Components

In the living cell the spindle region appears as a clear or achromatic zone (see Fig. 2) devoid of large cellular inclusions such as mitochondria or yolk particles. Microtubules contribute only 4% or less of the spindle volume. Exactly what components besides microtubules form a functional spindle fiber has not been determined (Forer, 1969, 1974; Forer and Goldman, 1972). Clearly, the mitotic

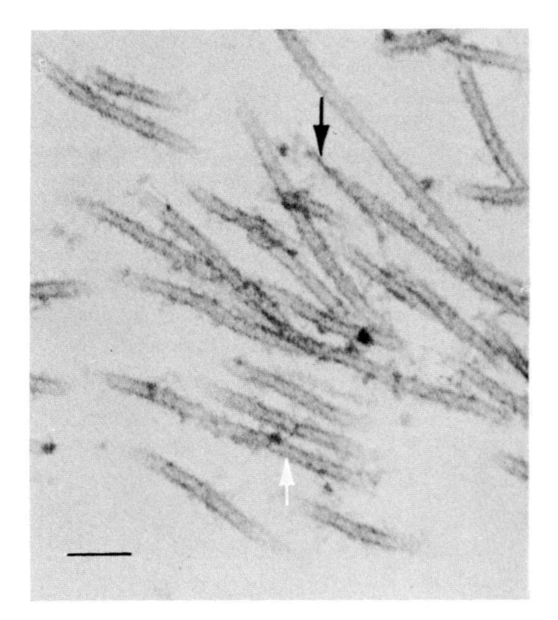

FIG. 5. Microtubules in an isolated spindle from *L. variegatus* seen in a thin-section electron nicrograph at high magnification. Spindles were fixed by standard procedures with glutaraldehyde, postfixed with osmium tetroxide, and embedded in Epon-Araldite. The microtubule walls appear to be coated with a fine, fuzzy, filamentous network (filaments about 3–4 mn diameter) that can sometimes be seen to interconnect microtubules (white arrow). Thicker filaments (about 6 nm diameter) are also occasionally seen (black arrow). (From Salmon and Segall, 1980, with permission.) Bar = 0.1 μm; magnification, ×81,700.

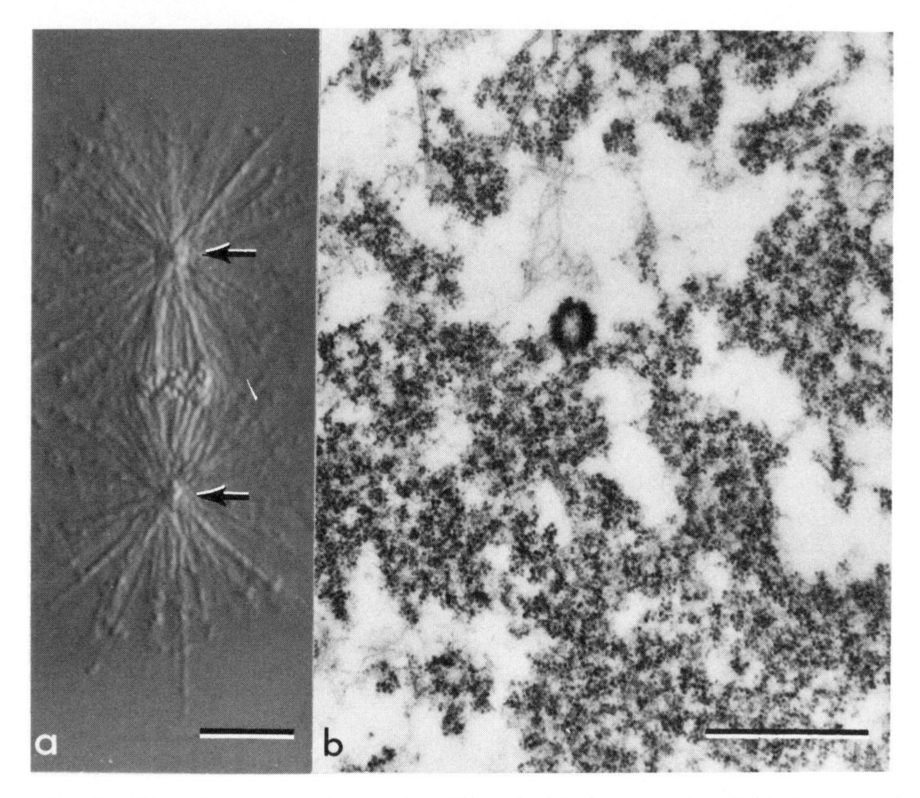

FIG. 6. The centrosome complex seen in a differential-interference contrast light micrograph of an isolated prometaphase spindle from *S. droebachiensis* (a), and seen in a corresponding electron micrograph that includes a tangential section through one of the centrioles. The phase-dense centrosome complex [arrows in (a)] is composed of a high density of ribosome-like particles clustered within an ill-defined, filamentous complex surrounding the centroles (b). The density of ribosome-like particles is much higher than seen in *L. variegatus* spindles (see Fig. 4). Clusters of filaments similar to actin are also occasionally seen within the centrosome complex. Most polar half-spindle microtubules and the astral microtubules end in the periphery of the centrosome complex, not visible in (b). (a) Bar = 10 μm; (b) bar = 1 μm.

centers in the centrosomes or centrospheres are important, but their composition and structure are still poorly defined (see Figs. 4, 6). Actin and myosin filaments have been proposed as key functional elements of spindle fibers (Forer, 1974, 1976; Fujiwara and Pollard, 1978; Herman and Pollard, 1978, 1979; Sanger, 1977; Sanger and Sanger, 1976), as has a dynein-like intermicrotubule cross-bridging protein complex (Margolis, 1978; McIntosh, 1977; McIntosh *et al.*, 1969; Pratt *et al.*, 1980). In our isolated spindles there is a matrix of 3- to 4-nm, fuzzy, ill-defined filaments enmeshed with the spindle microtubules (Fig. 5).

This filamentous network may also be an important cytoskeletal component of the spindle fibers, as may be the actin filaments (Figs. 6, 7).

Spindle microtubules are calcium labile (Fig. 3), and attention has been given recently to the endogenous network of membrane vesicles, tubules, and cisternae that has been observed in electron micrographs of fixed whole cells. The membranes appear most noticeable near the spindle poles, filling the centrosphere in first-division sea urchin spindles (Harris, 1975). Membrane tubules appear to extend upward along the kinetochore fibers toward the chromosomes, and radially outward along the astral-fiber microtubules (Harris, 1975). This network of smooth endoplasmic reticulum (SER) may function to sequester and release calcium ions as does the sarcoplasmic reticulum of muscle (Harris, 1975, 1978a; Hepler, 1977, 1979, 1980; Salmon and Segall, 1980). The calcium-binding protein, calmodulin, has also been localized by immunofluorescence techniques in the mitotic spindles of mammalian cultured cells with a distribution similar to the SER (Andersen *et al.*, 1978; Robbins and Jentzsch, 1969; Welsh *et al.*, 1979). It is a major protein of the sea urchin egg (Head *et al.*, 1979). I decided to eliminate the membrane components by using 1% Nonidet P-40 or 1% Triton X-100 in the EGTA lysis buffers.

D. Microtubule Lability

Besides deciding which components not to preserve during spindle isolation, the primary consideration of any isolation technique is to *reversibly* stabilize the spindle microtubules. As *in vivo* spindle BR studies have established, spindle microtubules are labile structures, particularly the non-kinetochore fiber microtubules. They apparently exist in a constant state of flux with a cellular pool of tubulin subunits (Cande *et al.*, 1974; Inoué, 1964; Inoué and Sato, 1967; Inoué *et al.*, 1975; Rebhun *et al.*, 1974a, 1975; Salmon, 1975; Salmon and Begg, 1980; Salmon and Segall, 1980). Spindle microtubules are reversibly depolymerized within several minutes by micromolar calcium, 4°C cooling, 400 atm pressure, or by 100 μM colcemid, colchicine, or other tubulin-binding drugs, or metabolic inhibitors. In addition, the transport of chromosomes is intimately coupled to the shortening and elongation of the spindle-fiber microtubules (Inoué and Ritter, 1975; Nicklas, 1975, 1977; Salmon, 1975, 1976).

E. Egg Cortex

By first mitosis, the sea urchin egg cortex becomes highly differentiated. Metaphase sea urchin eggs, stripped of the extracellular layers, look like fuzzy tennis balls in scanning electron micrographs (Schroeder, 1978, 1979). Densely packed microvilli cover the egg surface (see Fig. 2). The microvilli are structured

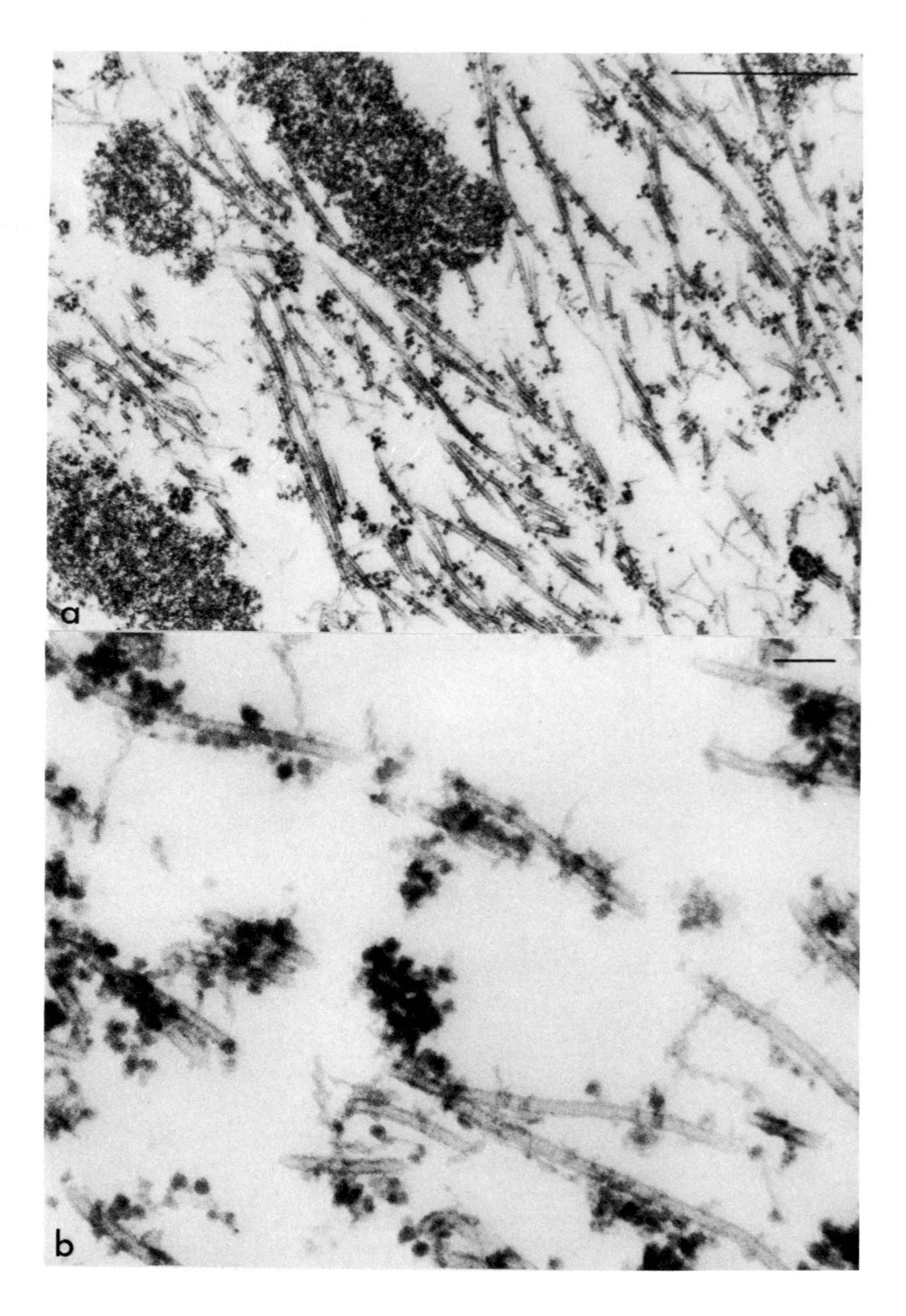
a
b

by a core of actin microfilaments that are anchored into the terminal web complex of the cortex, which is about 2 μm thick (Begg *et al.*, 1978; see also Bryan and Kane, 1982). Cytoplasmic cleavage is produced by an actin-myosin-mediated cortical contractile process (Schroeder, 1975; Inoué and Kiehart, 1978). The mitotic spindle determines the plane of cleavage, but does not produce cleavage (Rapaport, 1975). Preservation of the actin filament structure of the egg cortex during lysis with EGTA lysis buffers is species specific, as will be discussed.

III. Approaches to *in Vitro* Spindle Models

There are three general approaches for providing direct access to the spindle machinery: (1) permeabilized models of mitotic cells; (2) cells lysed into tubulin-reassembly buffers; and (3) total isolation of the mitotic spindle into nontubulin buffers. A description of the permeabilized cell models is given by Cande (1982) in this volume. Since 1972, when Weisenberg showed how to reassemble microtubules *in vitro*, several methods have been developed that use exogenous brain MTP in microtubule-reassembly buffers to stabilize spindle microtubules after cell lysis. This work, as well as the earlier spindle-isolation techniques developed by Kane using hexylene glycol stabilization medium, has been clearly described and evaluated by McIntosh (1977). For growth of exogenous MTP on mitotic centers in lysed cell preparations, see Borisy and Gould (1977), Borisy (1982), and Kirschner (1979).

The properties of mitotic spindles isolated by the early methods developed by Mazia and co-workers and by Kane are reviewed in detail by Forer (1969). See also reviews by Turner and McIntosh (1977) and Zimmerman and others (1977). Zimmerman and co-workers (1977) also describe their isolation method, which uses DMSO-glycerol medium to stabilize microtubules principally in spindles from eggs of *S. purpuratus*. Sakai (1978a) has reviewed the biochemical properties of isolated spindles. He gives a clear appraisal of the microtubule composition and assembly characteristics and spindle ATPases, as well as a description of his method of isolating spindles in glycerol-isolation buffers. See also Pratt *et al.*

FIG. 7. Identification of actin filaments in isolated spindles of *L. variegatus* (a) and *S. droebachiensis* (b). Actin filaments were labeled in pellets of spindles using myosin-S_1 and the tannic acid fixation procedures of Begg *et al.* (1978). Actin filaments are seen in much higher concentration in the *L. variegatus* spindles. No bundles of actin filaments are seen in either case. Polarity of the actin filaments with respect to the chromosome-to-pole axis appears to be random. Many actin filaments are seen to extend along and to interconnect microtubules. (a) Bar = 1 μm; (b) bar = 0.1 μm. [These unpublished electron micrographs were provided by David Weiss (a) and Paul Browne (b), University of North Carolina, Chapel Hill, North Carolina.]

(1980) and reviews by Petzelt (1979), Raff (1979), Rebhun (1976, 1977), and Stephens and Edds (1976).

To obtain isolated mitotic apparatuses that include the endogenous, calcium-sequestering membrane components, see the method recently described by Silver *et al.* (1980).

IV. Evolution of ETGA Lysis Buffer

A. Historical Considerations

In 1972 Weisenberg reported that MTP purified from mammalian brain tissue could be reversibly assembled *in vitro* by using calcium-chelating reassembly buffers. Several laboratories then demonstrated that the microtubule structures of the mitotic spindle could be stabilized by appropriate concentrations of purified brain MTP (6S tubulin plus MAPs or HMW proteins) in the *in vitro* reassembly cell lysis buffers, or by the addition of high concentrations of glycerol, which had been shown to stabilize microtubule assembly *in vitro* (reviewed by McIntosh, 1977). Spindles stabilized in tubulin buffers or with glycerol maintained sensitivity to cooling for a short period of time following cell lysis, but in several cases the spindles were not rapidly depolymerized by dilution of the tubulin concentration. Sakai and co-workers were able to maintain a limited amount of chromosome movement following lysis of anaphase eggs into a complex microtubule-reassembly buffer that contained MES, EGTA, glycerol, Mg^{2+}, GTP, ATP, tubulin, ascorbic acid, glutathione, cAMP, tubulin, and potassium fluoride (reviewed by Sakai, 1978a; see also Sakai, 1978b; Sakai *et al.*, 1979).

Rebhun *et al.* (1974b) developed a procedure in which no exogenous tubulin, glycerol, or glycols were used to isolate sea urchin and surf clam mitotic apparatuses with relatively stable microtubules. Their lysis buffer contained, at pH 6.6–6.8: 0.1 *M* MES or PIPES, 1 m*M* EGTA, 0.1 *M* TAME (a proteolysis inhibitor), plus 0.2 *M* DTT and 0.1–1.0% Triton X-100 to lyse the cells. Cande *et al.* (1974) used similar procedures to obtain partially lysed tissue culture cells using 0.1–0.2% Triton X-100 in a reassembly buffer of 0.1 *M* PIPES, 1 m*M* EGTA, 1 m*M* GTP. Cande *et al.* (1974) noted that as the detergent concentration was increased, the rate of spindle dissolution decreased in tubulin-free reassembly buffer. One might expect high concentrations of detergent, which rapidly lyse the plasma and intracellular membranes, to promote rapid spindle dissolution as a result of the dilution of the cell's tubulin concentration; however, the exact opposite occurred. Since the detergents Triton X-100 and Nonidet P-40 at 1% concentration do not alter the thermal lability of purified brain microtubules *in vitro* (Salmon and Segall, 1980), the reason for the stability of the spindle microtubules in Rebhun's and Cande's lysis buffers was not clear.

B. EGTA Lysis Buffer

In view of the wide variety of techniques published, I decided to look for *minimal* buffer conditions that would permit isolation of mitotic spindles with relatively stable microtubules. In 1977 Robert Jenkins and I found that mitotic spindles from first-mitotic sea urchin eggs of *S. droebachiensis* or *L. variegatus* could be isolated with a relatively stable normal distribution of spindle BR and organization of microtubules by rapidly lysing the eggs (plus fertilization membranes) in a simple calcium-chelating (EGTA), low-ionic-strength, Triton X-100 detergent buffer. Critical requirements for microtubule preservation during cell lysis were strong calcium chelation, low ionic strength, rapid membrane solubilization, and pH below 7.0. Adding micromolar concentrations of calcium ions or raising the KCl concentration above 0.3 *M* or the pH above 9.0 rapidly dissolved the spindle BR. Table I summarizes the stability of spindle BR as a function of various lysis buffer conditions we tried. Our standard EGTA lysis buffer now contains 5 m*M* EGTA, 0.5 m*M* $MgCl_2$, 10–50 m*M* PIPES (pH 6.8–7.0), plus 0.5–1.0% Triton X-100 or Nonidet P-40. As described below, glycerol can be added to the EGTA lysis buffer to stabilize the spindle microtubules reversibly during cell lysis and spindle storage (Sakai and Kuriyama, 1974).

C. Physiological Considerations

In the living cell, steady-state free-calcium-ion concentration is probably below $10^{-7}M$. Elevation of the free-calcium-ion concentration above 1 μM rapidly depolymerizes sea urchin spindle microtubules (Inoué and Kiehart, 1978; Kiehart, 1979, 1982; Salmon and Segall, 1980). The pH of the sea urchin may be between 7.0 and 7.6—an exact value has not yet been established (Johnson *et al.*, 1976; Shen and Steinhart, 1978; Steinhart *et al.*, 1978). Very little spindle BR was preserved with our EGTA lysis buffer at pH 7.4. Isolation at pH 6.6 preserved BR well, but with considerable cytoplasmic contamination and overstabilization of the spindle microtubules. We compromised on pH 6.8–7.0. Extraordinarily clean isolated spindles could be obtained in magnesium-free isolation buffers (10–30 m*M* EDTA added to the EGTA lysis buffer), but the microtubules were extremely labile and the spindles difficult to purify. We use 0.5 m*M* $MgCl_2$ in our isolation buffer to promote microtubule stability. Magnesium ions, however, also promote nonspecific cytoplasmic contamination and stability of the egg cortex during cell lysis, so that higher concentrations of Mg^{2+} are not useful. The osmolarity and ionic strength of the sea urchin cytoplasm is high. The ionic composition of the egg, although not well defined, appears to include more than 0.3 *M* glycine, about 0.25 *M* K^+, about 30–50 m*M* Cl^-, and other anions, perhaps 0.2 *M* glutamate (Baker and Whitaker, 1979; Kavanau, 1953;

TABLE I

Solution Parameters Affecting Preservation and Stability of Birefringent Retardation (BR) in Isolated *S. droebachiensis* Spindles after Cell Lysis[a,b]

Parameter	Buffer composition[c]	Isolation temp. (°C)	Initial BR (nm)	Half-life BR (hr)
pH	1 m*M* EGTA, 1 m*M* $MgCl_2$, 100 m*M* MES (pH 6.6)	15	5.6	2
	1 m*M* EGTA, 100 m*M* PIPES (pH 6.8)	15	5.0	1
	1 m*M* EGTA, 100 m*M* sodium cacodylate (pH 7.1)	15	3.5	2
	1 m*M* EGTA, 1 m*M* $MgCl_2$, 10 m*M* Tris (pH 8.0)	15	0	—
	1 m*M* EGTA, 10 m*M* $MgCl_2$, 0.1 *M* KCl, 10 m*M* NaH_2PO_4 (pH 7.2)	15	2.8	2
% Detergent	1 m*M* EGTA, 1 m*M* $MgCl_2$, 100 m*M* PIPES (pH 6.8), 0.025% Triton X-100	15	0	—
	As above with 0.075% Triton X-100	15	3.5	0.1
	As above with 0.25% Triton X-100	15	5.6	2
Ca^{2+}	1 m*M* $MgCl_2$, 50 m*M* PIPES (pH 6.8)	15	0	—
	5 m*M* EGTA, 10 m*M* PIPES (pH 6.8)	15	5.6	1
	10 m*M* EGTA (pH 6.8)	15	5.0	1
Mg^{2+}	5 m*M* EGTA, 10 m*M* EDTA, 10 m*M* PIPES (pH 6.8)	15	5.0	0.5
	5 m*M* EGTA, 0.5 m*M* $MgCl_2$, 10 m*M* PIPES (pH 6.8)[d]	15	5.6	2
Temperature	1 m*M* EGTA, 1 m*M* $MgCl_2$, 100 m*M* MES (pH 6.6) at	22	6.3	6
	As above at	15	5.6	2
	As above at	8	3.5	0.5
	As above at	0	0.7	0.1

[a]Spindles are from eggs grown at 8°C.

[b]*Note:* Eggs were cleaned of their jelly coat, fertilized, grown to first-division metaphase, then rapidly diluted 100-fold into lysis buffer. Preserved spindles and lysed cytoplasm were retained within the fertilization membrane, which was easily held between slide and coverslip for observation in the polarization microscope. BR measurements are averages of 10 or more spindles (S.D. = ±15%).

[c]All buffers contained 0.25% Triton X-100 unless otherwise stated.

[d]EGTA buffer referred to in text.

Rothchild and Barnes, 1953; Schatten and Mazia, 1976). The osmolarity of seawater is about 1.1, depending on local conditions. As seen in Table I, spindles were poorly preserved when lysed in media with high ionic strength. Therefore, before cell lysis, we wash cells free of growth medium with a nonionic iso-osmotic medium (we use 1 *M* glycerol, 5 m*M* Tris, pH 8.0). If this is not satisfactory, try 1 *M* dextrose (Sakai *et al.*, 1977), or an isotonic medium of 19 parts NaCl to 1 part KCl to remove divalent cations (Table II). During lysis, eggs should be rapidly diluted 100-fold (v/v) into lysis buffer to make certain that the components of the egg cytoplasm do not raise the calcium concentration or ionic strength, or lower the pH.

D. Species Specificity

I have found that the properties of the EGTA-isolated spindle preparations vary with different species. For instance, highly purified spindle preparations suitable for biochemical analysis can be obtained from *S. droebachiensis*, but the spindle microtubules from this urchin are unusually stable following lysis and washing. In contrast, spindles from *L. variegatus* are highly labile to calcium ions (Fig. 3) and can be treated to display lifelike thermal lability and a limited shortening of spindle fibers and chromosome movement. *L. variegatus* spindles are suitable for physiological or antibody-labeling studies of single spindles, but the spindle preparation also contains well-preserved egg cortices, containing actin filaments in high concentration. Thus for *L. variegatus* the EGTA lysis buffer yields a relatively stable whole-cell "cytoskeleton." In a biochemical analysis of the spindle proteins in a preparation of *L. variegatus* spindles, the spindle proteins are only a minor component. Why the spindle microtubules and egg cortices of different species have such striking differences in stability is not known, but the phenomenon is a major consideration in the choice of a biological source for isolated spindles. One can expect many other features of the isolation procedure and resulting spindles to be peculiar to individual species.

E. Other Cell Types

Over the past several years I have briefly examined several other cell types for the ability of EGTA lysis buffers to produce highly extracted spindles with relatively stable spindle-fiber BR. Successful results were obtained for the marine eggs of *Echinarachnius parma* (sand dollar), *Asterias forbesi* (starfish), *Arbacia punctulata* (sea urchin), and *Spisula* (surf clam) (see also Murphy, 1980; Rebhun *et al.*, 1974b; Sakai *et al.*, 1977). Nonmarine sources of isolated spindles have included PtK_1 tissue culture cells, *Haemanthus* endosperm cells, and the ciliate *Spirostomum* (intranuclear spindles of micronuclei) (Salmon and Jenkins, 1977).

V. Procedures for Isolating Spindles

A. General Considerations

To isolate first-mitotic spindles from developing sea urchin eggs the following steps must be accomplished (see Fig. 8).

1. Obtain ripe sea urchins.
2. Shed gametes.
3. Remove the egg jelly coat.
4. Fertilize all the eggs synchronously without inducing polyspermy.
5. Remove the tough extracellular coats—the fertilization membrane and hyaline layer.
6. Culture developing eggs to first division synchronously (generally in calcium-free seawater at constant, species-specific temperature).
7. Identify correctly the mitotic stage where isolation procedures are to be executed.

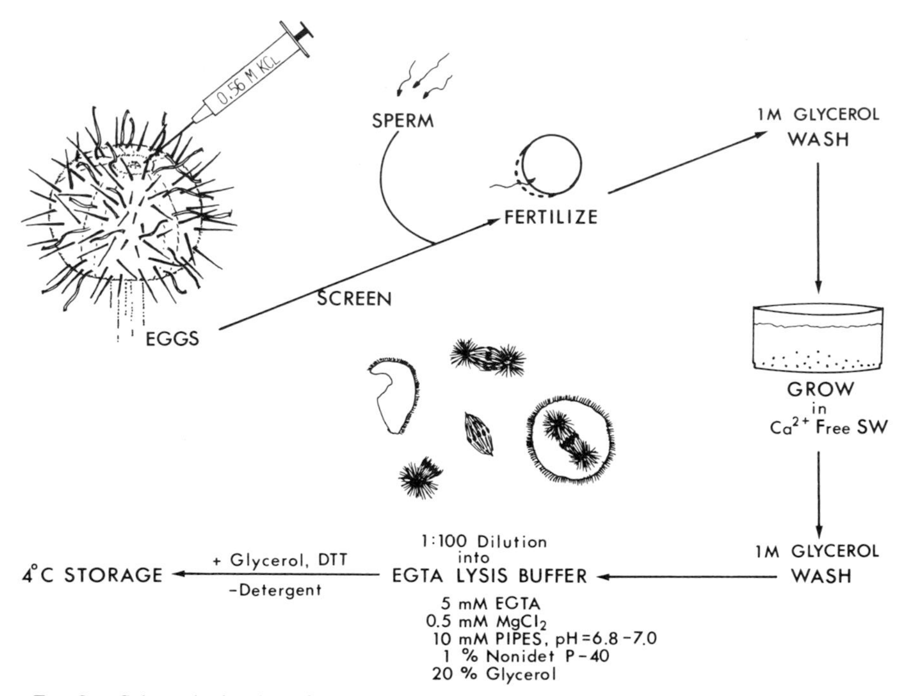

FIG. 8. Schematic drawing of the procedures for making detergent-extracted cytoskeletons and isolated spindles from the embryos of *L. variegatus*. Procedures for *S. droebachiensis* embryos are basically similar, but have several important differences in detail, as described in the text.

8. Wash the eggs into an iso-osmotic medium free of divalent cations (and low ionic strength if the eggs are not adversely affected).
9. Lyse the eggs into isolation buffer, and collect spindles by low-speed centrifugation.
10. Wash spindles in lysis buffer, then store spindles in an EGTA-glycerol buffer.

Note that sea urchins ripen on a seasonal basis in their natural habitat, hence it is best to work at a marine laboratory. Although methods for maintaining ripe sea urchins or ripening them in artificial seawater away from the marine laboratory are in general use, there is not, to my knowledge, any detailed published summary of the procedures used. Commercial sources of sea urchins include the following:

1. Marine Biological Laboratories, Woods Hole, Massachusetts 02543
   Arbacia punctulata (June–Aug.), *S. droebachiensis* (Feb.–April)
2. Gulf Specimen Company, Box 237, Panacea, Florida 32346
   L. variegatus (May–Oct.), *Arbacia punctulata* (May–Oct.)
3. Marine Specimens Unlimited, R.R. 2, Box 879, Summerland Key, Florida 33042
   L. variegatus (March–Sept.)
4. Pacific Bio-Marine Laboratories, Inc., P.O. Box 536, Venice, California 90291
   S. purpuratus (Dec.–May), *L. pictus* (May–Sept.)

I have indicated within the parentheses my experience as to the seasonal availability of ripe sea urchins from the above suppliers. Other species may also be available.

Special attention should be given to maintaining nearly the same salinity in the artificial seawater as the native salinity of the organisms. This is usually nearer to 31‰ rather than the 34–35‰ normally given for the open ocean. See Cavanaugh (1956) and Table II for formulas for artificial seawaters.

Procedures for obtaining gametes from ripe sea urchins; for fertilization; for removal of the egg jelly coat, fertilization membrane, and hyaline layer; and for development all are species dependent, yet well established. The best general manuals available are those of Costello and Henley (1971) and Harvey (1956). Harvey also gives an in-depth description of the biology of the sea urchin. Most of the current methods for culturing eggs for spindle isolation were devised by Mazia and co-workers and by Kane and co-workers (see refs. given previously).

A major factor to be considered in isolating spindles from sea urchin eggs is that the jelly coat and the tough extracellular layers that form around the egg on fertilization must be removed without damaging synchronous development (for review see Epel, 1977). Ripe eggs are covered by a thick jelly coat that hydrates

TABLE II

Composition of Buffers Used in Isolation of Mitotic Spindles from *L. variegatus* and *S. droebachiensis*

Buffer	Composition
Artificial seawater[a]	423 m*M* NaCl, 9 m*M* KCl, 23 m*M* $MgCl_2 \cdot 6H_2O$, 25 m*M* $MgSO_4 \cdot 7H_2O$, 2 m*M* $NaHCO_3$, 9 m*M* $CaCl_2 \cdot 2H_2O$, 5 m*M* Tris (pH 8.0–8.3)
Calcium-free artificial seawater[b]	436 m*M* NaCl, 9 m*M* KCl, 34 m*M* $MgCl_2 \cdot 6H_2O$, 16 m*M* $MgSO_4 \cdot 7H_2O$, 1 m*M* EGTA, 5 m*M* Tris (pH 8.0–8.3)
Glycerol buffer	1 *M* glycerol, 5 m*M* Tris (pH 8.0–8.3)
19:1 Isotonic buffer	560 m*M* NaCl, 30 m*M* KCl, 5 m*M* Tris (pH 8.0–8.3)
EGTA Buffer	0.2–5.0 m*M* EGTA, 0.5 m*M* $MgCl_2$, 10 m*M* PIPES (pH 6.8–7.0)
EGTA Lysis buffer	5 m*M* EGTA, 0.5 m*M* $MgCl_2$, 10–50 m*M* PIPES (pH 6.8–7.0); 1% Triton X-100 or 1% Nonidet P-40, 20% (v/v) glycerol (optional)
EGTA–Glycerol storage buffer	5 m*M* EGTA, 0.5 m*M* $MgCl_2$, 10 m*M* PIPES (pH 6.8); 50% (v/v) glycerol, 1 m*M* DTT
Reassembly buffer	1 m*M* EGTA, 0.5 m*M* $MgCl_2$, 0.5 m*M* GTP, 100 m*M* PIPES (pH 6.8)

[a]M.B.L. formula, salinity 31°/oo (Cavanaugh, 1956).
[b]Modified Moore calcium-free artificial seawater (Cavanaugh, 1956).

and swells on shedding into seawater. Jelly layers 100 μm thick are not uncommon. After sperm-egg membrane fusion, the cortical granules fuse with the egg plasma membrane and release their contents within the perivitelline space. This raises the vitelline membrane up off the egg surface to become the fertilization membrane. In addition, enzymes are released that cross-link and harden the fertilization membrane within 1–2 minutes. The egg jelly may be removed by using calcium-free seawater, by lowering pH to 5.25, or by mechanical shearing; the hardening of the fertilization membrane may be prevented by using either 1 *M* glycerol or 1 *M* urea, or by pretreatment of the unfertilized eggs with pronase, trypsin, or DTT (Begg and Rebhun, 1979; Epel, 1970; Kane, 1962, 1965, 1967; Mazia *et al.*, 1972; Pratt *et al.*, 1980; Salmon and Segall, 1980; Stephens, 1972b, 1973; Zimmerman *et al.*, 1977). Following fertilization, the egg secretes an extracellular layer called hyaline. In some urchins (e.g., *Arbacia, Lytechinus*) it is highly insoluble; however, it requires calcium to harden. Consequently, we grow fertilized embryos in calcium-free seawater to prevent formation of the hyaline layer. Some urchins do not secrete excessive amounts of hyaline (e.g., *S. droebachiensis* and *S. purpuratus*), so for these sea urchins regular artificial seawater may be used.

The development of ripe sea urchin eggs is generally uniform and synchronous under a constant set of conditions: temperature, handling of the eggs, composition of culture medium, and so on. Development rates are very sensitive to temperature ($Q_{10} \sim 2.5$–3.0). Costello and Henley (1971) or Harvey (1956) is a good source of information for timetables. Fry (1936) gives an excellent pictorial analysis of developmental timetables for *Arbacia* at several temperatures. Stephens (1972a) also provides a similar analysis for *S. droebachiensis*. We have found that the developmental timetables for *L. variegatus* are similar to those reported by Fry for *Arbacia*. However, because procedures and culture media differ, it is best to generate a custom timetable for development of specific sea urchins under your laboratory conditions. Controlling temperature for all media used is of the utmost importance if timetables are to be useful.

Temperature is also an important parameter because it affects the extent of spindle microtubule assembly. The optimal temperature for development and spindle assembly is highly species specific. As seen in Fig. 9, *S. droebachiensis* and *L. variegatus* must be grown at distinctly different temperatures. Based on spindle BR measurements and development, the optimal temperature for *S. droebachiensis* is 7–9°C (Stephens, 1972b, 1973), whereas the optimal temperature for *L. variegatus* is 23–25°C. *S. droebachiensis* lives and spawns in the cold northern Atlantic waters, whereas *L. variegatus* is tropical, found on the Atlantic coast usually south of Virginia.

To isolate spindles it is also important to identify accurately the mitotic stage. Polarization microscopy is the most useful method (see Fig. 2) if the egg is

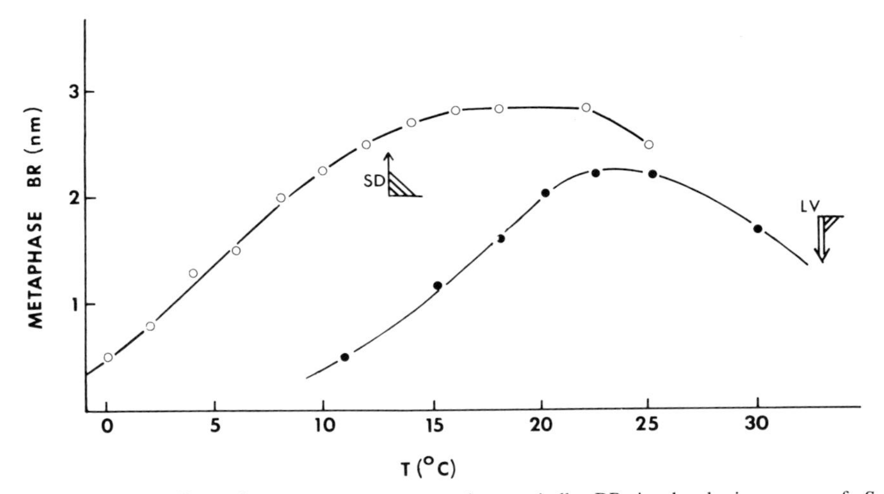

FIG. 9. The effect of temperature on metaphase spindle BR in developing eggs of *S. droebachiensis* (○, SD) and *L. variegatus* (●, LV). *L. variegatus* eggs were grown initially at 22.5°C and *S. droebachiensis* eggs were grown at 8°C. After the eggs reached prometaphase they were transferred to the experimental temperature and the spindle BR was measured (Stephens, 1973; Salmon, 1975; Salmon and Ellis, 1976). Single arrow (12–14°C) indicates the upper temperature limit of development for *S. droebachiensis*, and double arrow (33–34°C) indicates the inhibitory temperature for *L. variegatus*. The data for *S. droebachiensis* were derived from Stephens (1973); data for *L. variegatus* are from E. D. Salmon (unpublished). Data points represent average values. BR range ±15% of average values.

relatively clear like *L. variegatus* or *S. droebachiensis*. Identifying mitotic stage without a polarization microscope is a difficult problem in pattern recognition. Because the spindle actively excludes large cellular inclusions, the spindle can be viewed with a bright-field, phase-contrast, or differential-interference contrast microscope as a clear zone that changes in morphology along with the spindle as mitosis proceeds (Fig. 2). Harris (1978b) gives a series of light micrographs of fixed preparations of *S. purpuratus* that should be instructive. Another method of identifying the mitotic stage is to time the development of the embryos after fertilization; however, this is rarely sufficiently accurate because variations in development rate occur between different batches of eggs.

The procedure for handling eggs and for isolating spindles must be tailored for each species of sea urchin. Described here first are the procedures I use for isolating spindles from *L. variegatus*. The procedures for isolating spindles from *S. droebachiensis,* described second, differ principally in the methods used to remove the egg jelly, fertilization membrane, and the hyaline layer. The current spindle-isolation procedures were derived from information and results reported by Rebhun *et al.* (1974b), Kane (1962, 1965, 1967), Stephens (1972b), and Sakai and co-workers (Sakai and Kuriyama, 1974; Sakai *et al.*, 1975, 1977).

B. Methods for *L. variegatus*

1. Collecting Gametes

Extensive spawning is induced by injecting 1–3 ml of 0.56 *M* KCl into the sea urchin's body cavity. Alternately, Fuseler (1973) has described a technique for obtaining small numbers of gametes repetitively from sea urchins over a long duration. If the urchin is a male, the sperm is collected "dry" (i.e., by not allowing the urchin to shed into seawater). The concentrated sperm can be pipetted into a test tube on ice; or the whole urchin can be inverted over a watch glass sitting on ice; or the testes can be excised and stored at 4°C and the sperm will be secreted slowly around the outside of the testis. Dry sperm, free of coelomic fluid contamination, will remain viable when diluted into seawater for 1 or 2 days. Female sea urchins are inverted over a 100-ml beaker filled to the brim with seawater at physiological temperature (20–30°C for *L. variegatus*); usually room temperature is satisfactory. The sea urchin is allowed to spawn for 10–15 minutes. The last eggs spawned tend to be immature, and these should be avoided.

2. Ripeness

At this time, some decision about the quality of the eggs should be made. Unripe urchins produce few eggs and may spawn many unripe oocytes. In immature oocytes, the large germinal vesicle is very apparent. The KCl treatment will also cause overripe urchins to spawn eggs that are in a state of deterioration, readily identified by their brownish-red appearance and by the presence of many egg fragments. Ripe sea urchin eggs (about 125 μm diameter for *L. variegatus*) are arrested at the end of meiosis II, have released both polar bodies, and the female pronucleus sits up against the egg cortex at the cell surface. A ripe urchin should yield 10–30 ml of packed eggs containing their jelly coats.

3. Egg Jelly

Removal of the egg jelly will reduce the volume of the egg pellet more than 10-fold. The jelly layer is removed from *L. variegatus* eggs, immediately after shedding, by decanting the eggs in 100 ml of seawater through a 150-μm-pore Nitex screen (Tetko, Inc., Elmsford, New York). A dense sample of eggs that no longer have their jelly coats will be seen on a slide in the dissection scope to touch each other. Otherwise, they will be spaced by the thickness of the jelly layers. If the egg jelly is not released by screening in normal seawater, substitution of calcium-free seawater (Table II) during the screening should solve the problem. Complete removal of the jelly coat is critical for preservation of spindle BR during isolation procedures. After screening, the eggs are washed twice in 100 ml of artificial seawater by letting them settle and decanting the seawater. Eggs will fertilize and develop normally if used within 8 hours of the washing process.

4. Fertilization and Removal of the Fertilization Membranes

Check fertilizability and have a control for timing mitotic events by starting a small sample 10 minutes ahead of the large batch. The procedures described here must be done rapidly, on schedule, without interruption. We frequently use four 15-ml conical centrifuge tubes and a hand centrifuge to pellet the eggs. An aspirator with a Pasteur pipet tip is used to remove the supernatants quickly.

A comment is necessary about the concentration of sperm used for fertilization. If the concentration is too low, not all the eggs will be rapidly fertilized; too high, more than one sperm will enter an egg, yielding polyspermy. A polyspermic egg will generally have mutipolar spindles and may not develop at the same rate as monospermic eggs. The procedures given here include a sperm dilution we have found to be optimal, but the dilution may have to be altered for different laboratory conditions. We dilute one drop of dry sperm into 12 ml of seawater near 23–25°C (the best developmental temperature for *L. variegatus*). One milliliter of diluted sperm is used to fertilize about 1 ml of packed eggs suspended in 60 ml of seawater. After 30 seconds, the eggs are gently pelleted in the hand centrifuge (this takes about 20 seconds), and the supernatant is quickly aspirated (20–30 seconds) and replaced by 1 *M* glycerol–5 m*M* Tris, pH 8.3 (glycerol buffer) (20 seconds) to soften and remove the elevating fertilization membrane. The fertilized eggs are gently pelleted in the hand centrifuge (speed is no longer necessary), resuspended in glycerol buffer again, and after 1 minute in glycerol buffer the eggs are gently pelleted and resuspended in calcium-free artificial seawater (Table II).

5. Embryo Development

The fertilized eggs are grown in calcium-free artificial seawater in large flat-bottom finger bowls (~ 20 cm diameter) as monolayers. About 1 cm of calcium-free seawater should cover the eggs. (Calcium-free seawater is used to prevent formation of the extracellular hyaline layer.)

The development stage is established by observing an aliquot of eggs with a low-power polarization microscope or bright-field microscope (defocus condenser from Koehler illumination for bright field to increase the contrast of the yolk-free spindle region). At 23–25°C, *L. variegatus* eggs should reach metaphase 50–60 minutes after fertilization. Timing, as mentioned above, is extremely dependent on temperature.

6. Lysis

About 5 minutes before the desired stage for spindle isolation, the embryos are collected into 15-ml centrifuge tubes. Excess medium is aspirated off the monolayer of eggs in the finger bowls and the eggs are collected into four 15-ml conical centrifuge tubes by several centrifugations (gently, to avoid premature egg lysis). One minute before lysis, the eggs are suspended in isotonic glycerol buffer to remove the seawater salts. Then they are pelleted and about 0.25 ml of pelleted eggs is resuspended rapidly into 12–15 ml of lysis buffer. The EGTA lysis buffer for *L. variegatus* is given in Fig. 8 and Table II. Spindles are freed from the cortices, with difficulty, by vigorous pipetting. After 15 minutes in the lysis buffer, the isolated cytoskeletons (spindle plus cortex), spindles, and cortices are pelleted in a clinical centrifuge at 500 *g* for 10 minutes; resuspended in a glycerol storage buffer (Table II); then stored at 4°C. We have kept spindles for as long as 5–6 weeks without significant deterioration, but the maximum possible length of storage has not been determined. Spindles can be satisfactorily shipped between labs in the glycerol storage buffer on ice or, for biochemical analysis, on dry ice.

When *L. variegatus* spindles are isolated in EGTA lysis buffer *without* glycerol, initial spindle BR is nearly identical to the BR of spindles in the living cell before lysis; it decays slowly in EGTA buffer, however, reaching half its initial value in 30–45 minutes. For *L. variegatus* eggs, consequently, I add glycerol to the isolation buffer to stabilize the microtubules during the lysis, washing, and storage procedures. When ready to use the spindles for experiments, I wash out the glycerol with EGTA buffer, thus restoring the normal lability of isolated *L. variegatus* spindles.

C. Methods for *S. droebachiensis*

Whereas *L. variegatus* embryos are grown at 23–25°C, *S. droebachiensis* embryos develop best at 7–9°C. All solutions should be kept at this low temperature except for the lysis buffer.

1. Obtaining Gametes

Procedures are basically the same as for *L. variegatus*, except that the females should be shed into 100-ml beakers containing seawater on ice. Allow the urchins to shed for 20–30 minutes. The ripe egg is about 175 μm in diameter. One ripe female should yield about 30–40 ml of packed eggs with jelly coats. Eggs of *S. droebachiensis* are much more fragile than those of *L. variegatus*.

2. Fertilization, Removal of Egg Jelly and Fertilization Membrane

Twenty to thirty milliliters of eggs (including jelly coats) are suspended into an equal volume of seawater at 8°C in a 100-ml glass beaker; then they are fertilized with about 10 drops of diluted semen (3 drops of dry sperm in 12 ml of seawater). Ninety seconds after fertilization a control sample is taken to check the percentage of fertilization. Two minutes after the addition of sperm, a pH electrode is inserted into the egg suspension, which is stirred with a glass rod. At 2.5 minutes the pH of the seawater is quickly reduced to 5.25 (*not* less!) by pipetting in 0.1 *N* HCl while stirring constantly (this should be done within 30 seconds). This removes the egg jelly. The eggs are then pelleted gently by hand centrifugation (about 30 seconds) and washed twice in glycerol buffer (speed is no longer important) to remove the fertilization membranes. Finally, the eggs are resuspended in seawater near 8°C and plated out into large finger bowls as described earlier for *L. variegatus* eggs.

3. Lysis

About 5 minutes before the time of isolation, eggs are collected into four 15-ml centrifuge tubes and washed twice in an isotonic solution (19 parts NaCl to 1 part KCl) to remove divalent cations and to weaken any hyaline layer (this is not usually troublesome with *S. droebachiensis* eggs). The pelleted eggs are lysed in an EGTA lysis buffer without glycerol at 19–22°C (note the warmer temperature now) and shaken vigorously to release the mitotic spindles and to disperse the cortices. Unlike *L. variegatus* eggs, the cortex of *S. droebachiensis* rapidly solubilizes in this low-ionic-strength EGTA lysis buffer so that the spindle structures are the dominant elements in the lysate. After 30 minutes of lysis, whole eggs and fertilization membranes can be removed by filtering the preparation with a 50-μm-pore Nitex screen. Spindles are then pelleted at room temperature in a clinical centrifuge for 10 minutes at 500 *g*. The pellet is washed once in 1.5 ml of EGTA lysis buffer. For biochemical analysis, washed pellets are directly resuspended in the desired analysis medium. Since the BR of *S. droebachiensis* spindles is very stable in EGTA lysis buffer alone, no glycerol is necessary to stabilize spindles during the lysis and washing process. However, to store the spindles, they are kept in the EGTA-glycerol storage buffer developed for *L. variegatus* spindles. Stored spindles remain birefringent for several months when kept at −20°C. *S. droebachiensis* can be isolated and washed at 8°C instead of 19–22°C. Such preparations appear similar in protein composition on SDS-PAGE gels, but yield less microtubule polymer (Pratt *et al.,* 1980).

VI. Characteristics of Spindles Isolated in EGTA Lysis Buffers

The characteristics of both the *L. variegatus* and *S. droebachiensis* isolated spindles are described in detail elsewhere (Salmon and Segall, 1980; Pratt *et al.*, 1980), so will be presented only briefly here. As mentioned before, many characteristics of the isolated spindles are species dependent.

A. Structural Characteristics

The spindles from both species, when isolated in EGTA lysis buffer, are highly extracted, membrane free, and made up primarily of microtubules (Figs. 4, 5, 6, 7). With lower concentrations of detergent in the EGTA lysis buffer (0.25–0.50%), membrane sheets and vesicles are present in the isolated preparations. Actin filaments have been identified by myosin-S_1 labeling in both species (Fig. 7), but whether this actin is endogenous to the spindle or filters into the spindle during lysis is an unresolved question. Unknown ribosome-like particles and clusters of particles are distributed along and among the microtubules. The density of these particles appears highest in the centrosome complex, particularly for spindles isolated from *S. droebachiensis* eggs (Figs. 4, 5). The particles also appear to be associated with a fuzzy network of 3- to 4-nm filaments that do not decorate with myosin-S_1, but form a microtrabecular matrix in which the microtubules are embedded.

The pattern and distribution of BR is typical of mitotic spindles in living cells. In *S. droebachiensis* isolated spindles, swelling of the centrospheres was noticeable during lysis in the low-ionic-strength EGTA buffer, but the amount has not been determined. Immediately after lysis without glycerol, the magnitude of BR for *L. variegatus,* measured in the central half-spindle region, is about 2.5–3.0 nm for metaphase spindles at 23°C. The measured BR in living cells is about 25% lower. Based on form birefringence calculations, the microtubule BR of the spindles in living cells is expected to be 0.71–0.83 times the value in EGTA lysis buffer because of the difference in the refractive index of the surrounding medium (1.352–1.364 for the living cytosol versus 1.333 for EGTA buffer) (Hiramoto *et al.*, 1981; Sato *et al.*, 1975). The BR of the *S. droebachiensis* isolated spindles is somewhat higher than this ratio would predict from the BR measured in living cells (5–6 nm versus about 3 nm). This increased BR over the expected value may reflect microtubule polymerization during the lysis procedure, or—more likely—the concentration of ribosome-like particles and amorphous material that appears to be aligned along the microtubules and to coat them. The density of this material, which is higher than in *L. variegatus* spindles, appears species specific. It does not depend on the presence or absence of glycerol, but has been seen to vary in density with the concentration of Mg^{2+} in the EGTA lysis buffer (E. D. Salmon, unpublished observation).

During isolation the spindles frequently are broken in two distinct locations: at the interzone of anaphase spindles and at the junction between the central spindle poles and the aster complex (see Figs. 3, 4, 11). The mechanical weakness of the isolated spindle in the interzone is understandable because there are significantly fewer microtubules in the interzone in comparison with the half-spindles, as indicated by the distribution of BR in Fig. 2. The aster complex appears to be broken off the ends of the half-spindle easily because the half-spindle microtubules are not anchored in the centrosome as is usually described for mammalian spindles (see McIntosh, 1977, 1982). As seen in Fig. 4, the half-spindle microtubules terminate at a noticeable boundary zone at the periphery of the aster complex. This structural discontinuity in the spindles merits further attention because it suggests that detergent-solubilized components may be the mitotic centers for organizing the central-spindle microtubules.

Another interesting feature of the EGTA-isolated spindles is the spindle "remnant." As discussed below, calcium concentrations above 2 μM cause the microtubules of the isolated spindles to be completely depolymerized. However, after the BR of the spindle fibers has disappeared, a remnant of the spindle can still be seen with phase-contrast microscopy (Figs. 3, 11). We have not yet completely characterized this remnant, but it appears to be composed largely of the 3- to 4-nm fine filaments and the ribosome-like particles.

B. Biochemical Characteristics

Because the cortices of *L. variegatus* eggs are well preserved when the spindles are isolated, we have not pursued any biochemical analysis of the preparations. In contrast, only a low percentage of cortex-actin filament complexes

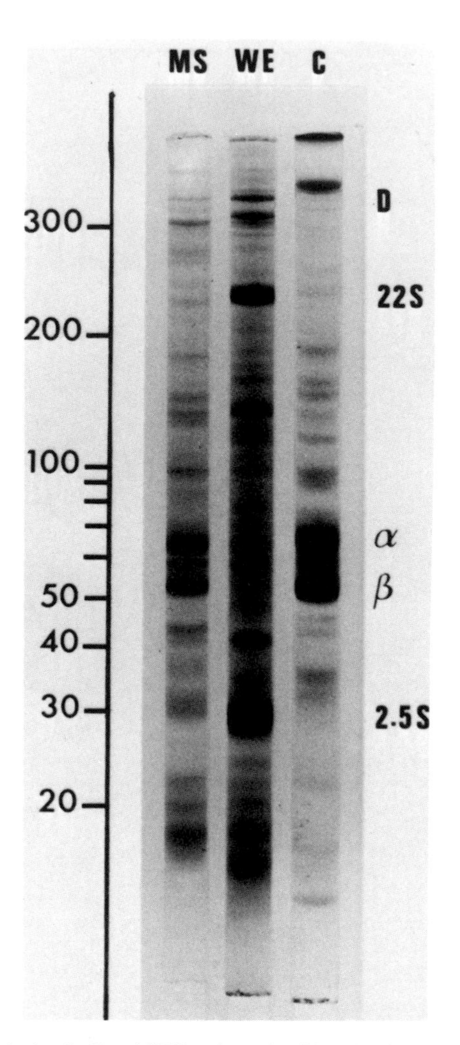

FIG. 10. 5% Tris-glycine-buffered SDS-polyacrylamide gels of *S. droebachiensis* isolated mitotic spindles (MS), whole eggs (WE), and blastula cilia (C). α- and β-Tubulin subunits are the dominant protein components of the isolated spindle, whereas they are barely detectable in the whole egg. No 22S or 2.5S cytoplasmic yolk protein and little actin (43K) from the cortex contaminate the spindle preparation. An approximate molecular weight scale is at the left ($\times$ 10^{-3} daltons). (From Pratt *et al.*, 1980, with permission.)

can be seen in electron micrographs of pelleted spindle preparations from *S. droebachiensis*. *S. droebachiensis* has proved to be an ideal source of native spindle proteins. At 20°C, lysis of 60 mg (0.6 ml) of first-division eggs yielded 0.3 mg of spindle protein, or 0.5% of the total egg protein. As seen from

SDS–PAGE analysis (Fig. 10), α- and β-tubulin make up at least 20% of the total spindle protein. Little 22S yolk protein contaminated this preparation. An unidentified 55,000-dalton protein, not tubulin, is a significant component (about 5%); it may be an element of the 3- to 4-nm filaments that enmesh the microtubules. Actin is usually only 2–8% of the total protein, indicating little cortical contamination of the spindle preparation. Two HMW bands (about 300,000–400,000 daltons) have been identified to be similar to cytoplasmic egg dynein and a dynein-like Mg-ATPase activity has been demonstrated (Pratt *et al.*, 1980). No evidence was found for myosin ATPase, oubain-sensitive Na^+/K^+ ATPase, or oligonmycin-sensitive mitochondrial ATPase or the mitotic CA^{2+}-ATPase. Since the Ca^{2+}-ATPase is believed to be bound to membranes, one would not expect to find it in detergent-extracted spindles (Mazia *et al.*, 1972; Nagle, 1979; Petzelt, 1972, 1979; Petzelt and Auel, 1978; Petzelt and von Ledebur-Villiger, 1973).

C. Physiological Characteristics

The isolated spindles of *S. droebachiensis* are unusually stable. The spindle-fiber BR remains stable for hours at room temperature in EGTA buffer without any exogenous tubulin or glycerol, and it is not significantly altered by cooling to 4°C or pressurization to 8000 psi. However, 0.4 *M* KCl rapidly abolishes the spindle BR, which may indicate that low ionic strength has a profound stabilizing effect on the *S. droebachiensis* spindles.

The isolated spindles from *L. variegatus* are labile and have been useful for studying variables that control microtubule assembly and spindle-fiber shortening. (Spindles from the sea urchin *Arbacia* have properties similar to those of *L. variegatus*.) For physiological studies on isolated spindles, I sandwich a drop of the isolated spindle preparation in glycerol storage buffer between an ethanol-cleaned slide and coverslip (previously rinsed in deionized water). The coverslip is supported by ridges of silicone grease (Dow Corning Corp., Midland, Michigan). After 10–15 minutes, many spindles will have settled and become weakly stuck to the slide. Fragments of #1 filter paper are then used to draw buffers through the slide-coverslip preparation. Experiments are done on a sensitive polarization microscope that has been modified so that phase-contrast or differential-interference contrast optics can be switched in without moving the specimen (Salmon and Segall, 1980). Most routine recording of the changes in spindle morphology is done using time-lapse video-recording techniques. When stored spindles are transferred out of EGTA-glycerol storage buffer into an EGTA or microtubule-reassembly buffer (Table II), the initial BR, rate of BR decay (half-life ~30 minutes), and calcium lability are similar to freshly isolated spindles in the absence of glycerol.

A major feature of the EGTA-isolated spindle is its calcium lability. Calcium in low micromolar concentrations has two distinct effects on the isolated spindles (Fig. 3). First, spindle BR decays as microtubules depolymerize rapidly, the rate increasing with calcium concentration above 0.2–0.5 μM, as assayed using Ca–EGTA buffers. Microtubules depolymerize almost completely in less than 6 minutes at 2 μM Ca^{2+}, and within several seconds at 10 μM Ca^{2+}. Second, concurrent with the depolymerization of the microtubules, the spindle shrinks and the spindle fibers shorten considerably, changing the spindle's morphology (Figs. 3 and 11). On the addition of 2 μM Ca^{2+}, the astral fibers curl and shorten inward toward the centrosome or centrosphere. The half-spindle fibers shorten, with the chromosome-to-pole distance frequently decreasing by 40–50% of its initial length. The spindle shape becomes more triangular. The initial rate or shortening of metaphase half-spindles is typically 10 μm/minute for 2 μM Ca^{2+} and 100 μm/min for Ca^{2+} concentrations of 10 μM and above (Salmon and Segall, 1979). If only one pole of a metaphase spindle is anchored to the slide, this calcium-induced spindle-fiber shortening produces a mass movement of the chromosomes toward the attached pole (Fig. 11). This effect is similar to the experimentally induced movement of chromosomes toward an anchored pole observed by Inoué and co-workers in living meiotic metaphase-arrested oocytes of *Chaetopterus* (Inoué and Ritter, 1975; Salmon, 1975, 1976). Although in our initial report of these observations, our calcium buffers contained ATP, we have since then been unable to demonstrate a requirement for ATP hydrolysis for the

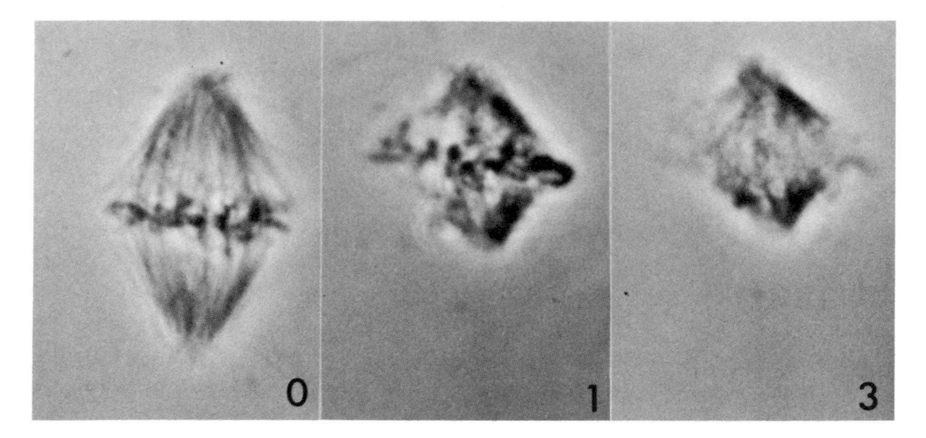

FIG. 11. Shortening of an isolated metaphase spindle from *L. variegatus* induced by addition of 2 μM Ca^{2+}. (As mentioned in the text, asters can be easily sheared off metaphase spindles during isolation.) The spindle remains anchored to the slide near the upper pole. Spindle-fiber shortening results in the chromosomes being transported toward the anchored pole, but half-spindle shortening is symmetrical. This spindle, initially in an EGTA buffer without glycerol, was perfused with an EGTA buffer containing 2 μM Ca^{2+} and 1 mM MgATP (Salmon and Segall, 1980) at time (in minutes) set = 0. The chromosomes move toward the upper anchored pole at an initial rate of 6 μm/minute. The presence of ATP induces decondensation of the chromosomes seen at 3 minutes.

FIG. 12. Isolated metaphase spindle from *L. variegatus* augmented with purified brain MTP for 30 minutes at 23°C as viewed with polarization microscopy at zero compensation. Over 30 minutes the maximum spindle BR increased from 3.2 nm to 6.8 nm, with the location of maximum BR moving from the center of each half-spindle to the metaphase plate region. The MTP, purified from porcine brain (Salmon and Segall, 1980), was at 1.35 mg/ml in reassembly buffer (Table II). The spindles were pretreated in EGTA–glycerol storage buffer with 10 m*M* EDTA at pH 6.8 for 1 hour before perfusion with the reassembly buffer containing MTP. Perfusion of other EDTA-treated spindles with reassembly buffer without tubulin resulted in the disappearance of spindle BR within about 10 minutes. Bar = 10 μm; magnification, $\times$1200.

calcium-induced microtubule depolymerization on spindle-fiber shortening (Salmon and Segall, 1979).

Note that the effects of calcium on *L. variegatus* isolated spindles are not general effects of divalent cations (Salmon and Segall, 1980), and they are distinctly different from the general ionic effects on the hydration of isolated mitotic apparatuses prepared by the hexylene glycol method (Cohen, 1968). The swelling and shrinking in low- and high-ionic-strength buffers of the mitotic apparatuses isolated with hexylene glycol may result from osmotic effects on the volume of the membrane vesicles that are preserved in mitotic apparatuses isolated with hexylene glycol from sea urchins.

Isolated spindles transferred into EGTA buffer or microtubule-reassembly buffer are relatively stable to the absence of an exogenous tubulin pool, cooling to 4°C, pressurization to 8000 psi, or 100 μM colchicine—agents that depolymerize the mitotic spindles rapidly (within 1–2 minutes) in the living cell. However, if the spindles in EGTA–glycerol storage buffer are treated for 1 hour with 10 m*M* EDTA, then the stability of the spindle microtubules when transferred to an EGTA or reassembly buffer is directly dependent on the assembly characteristics of an exogenous tubulin pool (Salmon and Segall, 1980; Salmon *et al.*, 1980). The EDTA-treated spindles are similar to spindles *in vivo* in their spindle-fiber BR, their morphology, and their lability to cold and pressure. The isolated spindles' calcium lability is not inhibited by pretreatment with EDTA.

Both EGTA- and EDTA-treated spindles will incorporate purified brain MTP (Fig. 12). The central-spindle BR increases from 3 to 6 nm with 1.5 mg/ml exogenous MTP, but little interpolar elongation occurs. Isolated spindles augmented with calcium-insensitive 6S brain tubulin lose their lability to micromolar calcium (Salmon *et al.*, 1980). Preliminary results indicate that addition of 100 μM colchicine to the MTP-reassembly medium will block augmentation of spindle BR *in vitro,* but does not promote net disassembly as colchicine does *in vivo* (E. D. Salmon, unpublished observations). The effects of colchicine on isolated spindles are similar to the effects of colchicine on the self-assembly of brain MTP *in vitro,* but not to the effects of colchicine on spindle-microtubule assembly in living cells (Inoué, 1952).

VII. Future Considerations

Current theories of mitosis are all based on proposed structural-functional properties of spindle-fiber microtubules, but no theory has yet provided a completely satisfactory explanation of the mitotic process. Microtubules are undoubtedly key functional components for moving chromosomes. The EGTA-isolated spindles have permitted us to focus attention on the structure and function of the spindle-fiber microtubules and have suggested that two additional components of the mitotic apparatus may be important: (1) the fine filamentous network that enmeshes the microtubules of EGTA-isolated spindles; and (2) the membranes of the SER, which are totally absent in EGTA-isolated spindles. Our work with EGTA-isolated spindles from *L. variegatus* suggests that spindle shortening can be induced in a manner that transmits forces sufficient to displace the chromosomes simply by adding calcium. No ATP hydrolysis appears to be necessary. Perhaps the meshwork of 3- to 4-nm fine filaments associated with and cross-linking the microtubules acts as a contractile gel, or induces spindle shortening by a mechanism similar to the action of calcium-activated contractile fine filaments in the spasmoneme (Amos, 1975). Interestingly, chromosome movement continues to a limited extent in anaphase mitotic apparatuses isolated by Sakai *et al.* (1979) with an EGTA–glycerol lysis buffer that contains no detergent. Silver *et al.* (1980), by using a low-ionic-strength EGTA lysis buffer similar to the one described here but without detergent, have recently obtained from *S. purpuratus* isolated mitotic apparatuses that retain the endogenous membrane components and that actively sequester calcium ions. It would be of interest to compare closely spindles and mitotic apparatuses isolated in buffers that are similar except for the presence or absence of detergent. However, I have not yet been successful in isolating mitotic apparatuses from *L. variegatus* or *Arbacia* with Silver's method; this also may be a problem of species specificity.

The EGTA-isolated spindles provide a simplified preparation for analyzing spindle structural and functional characteristics, including microtubule-assembly mechanisms and polarity; tubulin flux; possible microtubule-associated proteins (MAPs, calmodulin, dynein, for example); the mechanism of calcium depolymerization; the action of other possible physiological regulators of microtubule assembly such as pH, phosphorylation, sulfhydryl oxidation-reduction; and action of drugs that affect spindle-microtubule assembly *in vivo,* such as colchicine. The role of dynein or myosin in the generation of motility can be tested in "add-back" experiments. Future attention should also be given to the structure and composition of the centrosome-centrosphere complex, the ribosome-like particles, and the spindle remnant that persists after the microtubules are depolymerized.

We are currently studying the physiological properties of isolated spindles that have been returned to buffers containing estimated normal physiological concentrations of K^+, Cl^-, Mg^{2+}, glycine, glutamate, and soluble protein. Mg^{2+} concentrations above 0.5 m*M* clearly stabilize the spindle microtubules, and studies of isolated spindles in buffers with high Mg^{2+} concentrations may be misleading. Although raising the ionic strength of the EGTA buffer makes the spindle microtubules more labile and lifelike, lysis of cells in a high-ionic-strength physiological buffer without polymerizable exogenous tubulin does not yield isolated spindles. Purified brain tubulin stabilizes microtubules in isolated spindles, but can modify their calcium lability (Salmon *et al.,* 1980). It is not clear whether this effect is a result of the source of tubulin (egg tubulin might be better) or to the tubulin-purification process (a calcium-binding microtubule-associated protein may be lost during purification). Kuriyama (1977) has developed a method of purifying egg tubulin from sea urchins, but enormous quantities of eggs are needed to yield a useful quantity of tubulin. The best approach currently seems to be to use the low-ionic-strength EGTA lysis buffer for isolating and storing the spindles, and then for experiments to return the spindles to a physiological buffer that contains egg tubulin and that more nearly duplicates the composition of egg cytoplasm.

Acknowledgments

I would like to thank Tim Otter for his critical review of the manuscript and Nancy Salmon for her usual outstanding editorial assistance. I also appreciate the aid of Mike Spillane and Wilma Hanton with the electron microscopy.

This work was supported by grants from the National Institutes of Health (GM 24364) and the National Science Foundation (76-09654 and 77-07113).

References

Amos, W. B. (1975). *In* "Molecules and Cell Movement" (S. Inoué and R. E. Stephens, eds.), pp. 411-436. Raven, New York.

Andersen, B., Osborn, M., and Weber, K. (1978). *Cytobiologie* **17,** 354-364.

Bajer, A. S. (1973). *Cytobios* **8,** 139-160.

Baker, P. F., and Whitaker, M. J. (1979). *Nature (London)* **279,** 513-515.

Begg, D. A., and Rebhun, L. I. (1979). *J. Cell Biol.* **83,** 241-248.

Begg, D. A., Rodewald, R., and Rebhun, L. (1978). *J. Cell Biol.* **79,** 846-852.

Bryan, J., and Kane, R. E. (1982). *Methods Cell Biol.* **25,** 175-199.

Borisy, G. G., and Bergen, L. G. (1982). *Methods Cell Biol.* **24,** 171-187.

Borisy, G. G., and Gould, R. R. (1977). *In* "Mitosis; Facts and Questions" (M. Little, N. Paweletz, C. Petzelt, H. Ponstingl, D. Schroeter, and H.-P. Zimmermann, eds.), pp. 167-184. Springer-Verlag, Berlin and New York.

Cande, W. Z. (1982). *Methods Cell Biology* **25,** 57-68.

Cande, W. Z., Snyder, J., Smith, D., Summers, K., and McIntosh, J. R. (1974). *Proc. Natl. Acad. Sci. U.S.A.* **71,** 1559-1563.

Cavanaugh, G. M., ed. (1956). "Formulae and Methods IV of the Marine Biological Laboratory Chemical Room." Marine Biological Laboratory, Woods Hole, Massachusetts.

Cohen, W. D. (1968). *Exp. Cell Res.* **51,** 221-236.

Cohen, W. D., and Rebhun, L. I. (1970). *J. Cell Sci.* **6,** 159-176.

Costello, D. P., and Henley, C. (1971). "Methods for Obtaining and Handling Marine Eggs and Embryos," 2nd ed. Marine Biological Laboratory, Woods Hole, Massachusetts.

Epel, D. (1970). *Exp. Cell Res.* **61,** 69-70.

Epel, D. (1977). *Sci. Am.* **237,** 128-138.

Forer, A. (1969). *In* "Handbook of Molecular Cytology" (A. Lima-de-Faria, ed.), pp. 553-601. North-Holland Publ., Amsterdam.

Forer, A. (1974). *In* "Cell Cycle Controls" (G. M. Padilla, I. L. Cameron, and A. M. Zimmerman, eds.), pp. 319-336. Academic Press, New York.

Forer, A. (1976). *Cold Spring Harbor Cont. Cell Proliferation* **3** [Book C], 1273-1293.

Forer, A., and Goldman, R. D. (1972). *J. Cell Sci.* **10,** 387-418.

Fuseler, J. R. (1973). *J. Cell Biol.* **57,** 879-881.

Fry, H. J. (1936). *Biol. Bull. Woods Hole, Mass.* **70,** 89-99.

Fuge, H. (1978). *Int. Rev. Cytol., Suppl.* **6,** 1-58.

Fujiwara, K., and Pollard, T. D. (1978). *J. Cell Biol.* **77,** 182-195.

Harris, P. (1962). *J. Cell Biol.* **14,** 425–487.
Harris, P. (1975). *Exp. Cell Res.* **94,** 409–425.
Harris, P. (1978a). *In* "Cell Cycle Regulation" (E. D. Bretow, I. L. Cameron, and G. M. Padilla, eds.), pp. 75–104. Academic Press, New York.
Harris, P. (1978b). *In* "Cell Reproduction" (E. R. Dirksen, D. M. Prescott, and C. F. Fox, eds.), ICN-UCLA Symp. Mol. Cell. Biol., Vol. 12, pp. 505–514. Academic Press, New York.
Harvey, E. B. (1956). "The American *Arbacia* and Other Sea Urchins." Princeton Univ. Press, Princeton, New Jersey.
Head, J. F., Mader, S., and Kaminer, B. (1979). *J. Cell Biol.* **80,** 211–218.
Hepler, P. K. (1977). *In* "Mechanisms and Control of Cell Division" (T. L. Rost and E. M. Gifford, eds.), pp. 212–232. Dowden, Hutchinson & Ross, Stroudsburg, Pennsylvania.
Hepler, P. K. (1979). *J. Cell Biol.* **83,** 372a.
Hepler, P. K. (1980). *J. Cell Biol.* **86,** 490–499.
Herman, I. M., and Pollard, T. D. (1978). *Exp. Cell Res.* **114,** 15–25.
Herman, I. M., and Pollard, T. D. (1979). *J. Cell Biol.* **80,** 509–520.
Hiramoto, Y., Hamaguchi, Y., Shoji, Y., Schroeder, T. E., Shimoda, S., and Nakamura, S. (1981). *J. Cell Biol.* **89,** 121–130.
Inoué, S. (1952). *Exp. Cell Res., Suppl.* **2,** 305–314.
Inoué, S. (1964). *In* "Primitive Motile Systems in Cell Biology" (R. D. Allen and N. Kamiya, eds.), pp. 549–598. Academic Press, New York.
Inoué, S. (1976). *Cold Spring Harbor Conf. Cell Proliferation* **3** [Book C], 1317–1328.
Inoué, S., and Kiehart, D. P. (1978). *In* "Cell Reproduction" (E. R. Dirksen, D. M. Prescott, and C. F. Fox, eds.), pp. 433–444. Academic Press, New York.
Inoué, S., and Ritter, H., Jr. (1975). *In* "Molecules and Cell Movement" (S. Inoué and R. E. Stephens, eds.), pp. 3–30. Raven, New York.
Inoué, S., and Sato, H. (1967). *J. Gen. Physiol.* **50,** 259–292.
Inoué, S., Fuseler, J., Salmon, E. D., and Ellis, G. W. (1975). *Biophys. J.* **15,** 725–744.
Johnson, J. D., Epel, D., and Paul, M. (1976). *Nature (London)* **262,** 661–664.
Kane, R. E. (1962). *J. Cell Biol.* **12,** 47–55.
Kane, R. E. (1965). *J. Cell Biol.* **25,** 137–144.
Kane, R. E. (1967). *J. Cell Biol.* **32,** 243–254.
Kavanau, J. L. (1953). *J. Exp. Zool.* **122,** 285–337.
Kiehart, D. (1979). Ph.D. Thesis, University of Pennsylvania, Philadelphia (available from University Microfilms, Ann Arbor, Michigan).
Kiehart, D. (1982). *Methods Cell Biol.* **25,** 13–31.
Kiehart, D., and Inoué, S. (1976). *J. Cell Biol.* **70,** 230a.
Kirschner, M. W. (1979). *Int. Rev. Cytol.* **54,** 1–71.
Kuriyama, R. (1977). *J. Biochem. (Tokyo)* **81,** 1115–1125.
McIntosh, J. R. (1977). *In* "Mitosis; Facts and Questions" (M. Little, N. Paweletz, C. Petzelt, H. Ponstingl, D. Schroeter, and H.-P. Zimmermann, eds.), pp. 167–184. Springer-Verlag, Berlin and New York.
McIntosh, J. R. (1982). *Methods Cell Biol.* **25,** 33–56.
McIntosh, J. R., Hepler, P. K., and Van Wie, D. G. (1969). *Nature (London)* **224,** 659–663.
Margolis, R. L. (1978). *In* "Cell Reproduction" (E. R. Dirksen, D. M. Prescott, C. F. Fox, eds.), pp. 445–446. Academic Press, New York.
Margolis, R. L., Wilson, L., and Kiefer, B. I. (1978). *Nature (London)* **272,** 450–452.
Mazia, D., and Dan, K. (1952). *Proc. Natl. Acad. Sci. U.S.A.* **38,** 826–838.
Mazia, D., Petzelt, C., Williams, R. O., and Meza, I. (1972). *Exp. Cell Res.* **70,** 325–332.
Murphy, D. B. (1980). *J. Cell Biol.* **84,** 235–245.
Nagle, B. (1979). *J. Cell Biol.* **83,** 378a.

Nicklas, R. B. (1971). *Adv. Cell Biol.* **2,** 225-297.
Nicklas, R. B. (1975). *In* "Molecules and Cell Movement" (S. Inoué and R. E. Stephens, eds.), pp. 97-118. Raven, New York.
Nicklas, R. B. (1977). *In* "Mitosis; Facts and Questions" (M. Little, N. Paweletz, C. Petzelt, H. Ponstingl, D. Schroeter, and H.-P. Zimmermann, eds.), pp. 150-155. Springer-Verlag, Berlin and New York.
Petzelt, C. (1972). *Exp. Cell Res.* **70,** 333-339.
Petzelt, C. (1979). *Int. Rev. Cytol.* **60,** 53-92.
Petzelt, C., and Auel, D. (1978). *In* "Cell Reproduction" (E. R. Dirksen, D. M. Prescott, and C. F. Fox, eds.), pp. 487-494. Academic Press, New York.
Petzelt, C., and von Ledebur-Villiger, M. (1973). *Exp. Cell Res.* **81,** 87-94.
Pratt, M., Otter, T., and Salmon, E. D. (1980). *J. Cell Biol.* **86,** 738-745.
Raff, E. C. (1979). *Int. Rev. Cytol.* **59,** 2-96.
Rappaport, R. (1975). *In* "Molecules and Cell Movement" (S. Inoué and R. E. Stephens, eds.), pp. 287-304. Raven, New York.
Rebhun, L. I. (1976). *Am. Zool.* **16,** 469-482.
Rebhun, L. I. (1977). *Int. Rev. Cytol.* **49,** 1-54.
Rebhun, L. I., Mellon, M., Jemiolo, D., Nath, J., and Ivy, N. (1974a). *J. Supramol. Struct.* **2,** 466-485.
Rebhun, L. I., Rosenbaum, J. L., Lefebvre, P., and Smith, D. W. (1974b). *Nature (London)* **249,** 113-115.
Rebhun, L. I., Jemiolo, D., Ivy, N., Mellon, M., and Nath, J. (1975). *Ann. N.Y. Acad. Sci.* **253,** 362-377.
Robbins, E., and Jentzsch, G. (1969). *J. Cell Biol.* **40,** 678-691.
Rothchild, L., and Barnes, H. (1953). *J. Exp. Biol.* **30,** 534-544.
Sakai, H. (1978a). *Int. Rev. Cytol.* **55,** 23-48.
Sakai, H. (1978b). *In* "Cell Reproduction" (E. R. Dirksen, D. M. Prescott, and C. F. Fox, eds.), pp. 425-431. Academic Press, New York.
Sakai, H., and Kuriyama, R. (1974). *Dev., Growth Differ.* **16,** 123-133.
Sakai, H., Hiramoto, Y., and Kuriyama, R. (1975). *Dev., Growth Differ.* **17,** 265-274.
Sakai, H., Shimoda, S., and Hiramoto, Y. (1977). *Exp. Cell Res.* **104,** 457-461.
Sakai, H., Hamaguchi, M., Kimura, I., and Hiramoto, Y. (1979). *In* "Cell Motility: Molecules and Organization" (S. Hatano, H. Ishikawa, and H. Sato, eds.), pp. 609-620. Univ. of Tokyo Press, Tokyo.
Salmon, E. D. (1975). *Ann. N.Y. Acad. Sci.* **253,** 383-406.
Salmon, E. E. (1976). *Cold Spring Harbor Conf. Cell Proliferation* **3** [Book C], 1329-1342.
Salmon, E. D., and Begg, D. A. (1980). *J. Cell Biol.* **85,** 853-865.
Salmon, E. D., and Ellis, G. W. (1976). *J. Microsc. (Oxford)* **106,** 63-69.
Salmon, E. D., and Jenkins, R. (1977). *J. Cell Biol.* **75,** 295a.
Salmon, E. D., and Segall, R. R. (1979). *J. Cell Biol.* **83,** 377a.
Salmon, E. D., and Segall, R. R. (1980). *J. Cell Biol.* **86,** 355-365.
Salmon, E. D., Pape, G., and Segall, R. R. (1980). *J. Cell Biol.* **87,** 239a.
Sanger, J. W. (1977). *In* "Mitosis; Facts and Questions" (M. Little, N. Paweletz, C. Petzelt, H. Ponstingl, D. Schroeter, and H.-P. Zimmermann, eds.), pp. 98-120. Springer-Verlag, Berlin and New York.
Sanger, J. W., and Sanger, J. M. (1976). *Cold Spring Harbor Conf. Cell Proliferation* **3,** [Book C], 1295-1316.
Sato, H., Ellis, G. W., and Inoué, S. (1975). *J. Cell Biol.* **67,** 501.
Schatten, G., and Mazia, D. (1976). *Exp. Cell Res.* **98,** 325-337.

Schroeder, T. E. (1975). *In* "Molecules and Cell Movement" (S. Inoué and R. E. Stephens, eds.), pp. 305-332. Raven, New York.

Schroeder, T. E. (1978). *Dev. Biol.* **64,** 342-346.

Schroeder, T. E. (1979). *Dev. Biol.* **70,** 306-326.

Shen, S., and Steinhart, R. (1978). *Nature (London)* **272,** 253-254.

Silver, R. B., Cole, R. D., and Cande, W. Z. (1980). *Cell* **19,** 505-516.

Steinhardt, R. A., Shen, S. S., and Zucker, R. S. (1978). *In* "Cell Reproduction" (E. R. Dirksen, D. M. Prescott, and C. F. Fox, eds.), pp. 415-424. Academic Press, New York.

Stephens, R. E. (1972a). *Biol. Bull. (Woods Hole, Mass.)* **142,** 132-144.

Stephens, R. E. (1972b). *Biol. Bull. (Woods Hole, Mass.)* **142,** 145-149.

Stephens, R. E. (1973). *J. Cell Biol.* **57,** 133-147.

Stephens, R. E., and Edds, K. T. (1976). *Phys. Rev.* **56,** 709-777.

Turner, J. L., and McIntosh, J. R. (1977). *Methods Cell Biol.* **16,** 373-379.

Weisenberg, R. C. (1972). *Science* **177,** 1104.

Weisenberg, R. C. (1978). *In* "Cell Reproduction" (E. R. Dirksen, D. M. Prescott, and C. F. Fox, eds.), pp. 359-366. Academic Press, New York.

Welsh, M. J., Dedman, J. R., Brinkley, B. R., and Means, A. R. (1979). *J. Cell Biol.* **81,** 624-634.

Zimmermann, A. M., Zimmerman, S., and Forer, A. (1977). *Methods Cell Biol.* **16,** 361-372.

Chapter 6

Molecular and Genetic Methods for Studying Mitosis and Spindle Proteins in Aspergillus nidulans

N. RONALD MORRIS, DONALD R. KIRSCH, AND BERL R. OAKLEY

Department of Pharmacology
College of Medicine and Dentistry of New Jersey
Rutgers Medical School
Piscataway, New Jersey

ISBN 0-12-564125-7

I. Introduction

The fungus *Aspergillus nidulans* is one of the most promising organisms being used to study the molecular biology of mitosis. The special utility of *Aspergillus* is that it has a sophisticated and powerful genetic system (Pontecorvo *et al.*, 1953; Clutterbuck, 1974) that can be applied to problems in the biology of mitosis that hitherto have seemed inaccessible (see review by Morris, 1980). The morphological events of mitosis in *A. nidulans* are well known and have been thoroughly described by light and electron microscopy (Robinow and Caten, 1969; Oakley and Morris, 1981). A large number of temperature-sensitive conditional lethal mutants blocked in mitosis or unable to enter mitosis have been identified and characterized (Orr and Rosenberger, 1976a,b; Morris, 1976), as well as mutants that fail to septate (Morris, 1976; Trinci and Morris, 1979) and mutants that are blocked with respect to nuclear movement (Morris, 1976; Oakley and Morris, 1980). Mutations in structural genes for the α- and β-subunits of tubulin and in other genes that impinge on tubulin function have also been identified and characterized (Van Tuyl, 1977; Sheir-Neiss *et al.*, 1978; Morris *et al.*, 1979; Morris, 1980). The drug-resistant and temperature-sensitive tubulin mutants of *A. nidulans* are unique in being the only such mutants known in an organism with a sophisticated genetic system.

A. nidulans is easy to handle using conventional microbiological techniques (Clutterbuck, 1974), and it grows on simple, inexpensive media to high densities (up to 30 gm/liter), to provide ample material for biochemical analysis. High-molecular-weight DNA suitable for cloning experiments can be prepared from spores or from protoplasts (Morris, 1978) and, since *A. nidulans* has a genome of low complexity (Timberlake, 1978), identification of specific cloned genes is relatively straightforward. In this chapter we provide an outline of the general methods used in working with *A. nidulans* and a summary of the genetic system based on the detailed descriptions of Pontecorvo *et al.* (1953) and Clutterbuck (1974). A comprehensive description of the physiology and genetics of *Aspergillus* was the topic of a recent symposium (edited by Smith and Pateman, 1977), and the text by Fincham *et al.* (1979) provides an excellent general summary of fungal genetics.

II. General Methods

A. Strains

All strains in common laboratory use are derived from the Glasgow collection (Pontecorvo *et al.*, 1953) and all are closely related. The Fungal Genetics Stock Center, Humboldt State University Foundation (Arcata, California 95521), is the major source of *A. nidulans* strains in North America. A large number of genetically marked strains, including multiply marked strains that are especially useful for mapping, are available from this source. The first step that should be taken by anyone starting to work with *A. nidulans* is to write to Dr. William Ogata, curator, for the strain catalog. There are also many strains that have not yet been included in this collection; these are usually obtainable from the investigators who isolated them. The mitotic and tubulin mutants of *A. nidulans* may be obtained by writing to Dr. N. Ronald Morris, Dept. of Pharmacology, CMDNJ-Rutgers Medical School (Piscataway, New Jersey 08854).

B. Preparation of Conidiospores

Cultures of *A. nidulans* are generally inoculated with conidiospores (asexual spores, which are also called conidia) and conidiospores are used as the starting material for mutagenesis. Conidiospores are produced by most strains grown as surface cultures on nutritionally complete medium. They are harvested by scraping the surface of the culture with a sterile wire or by shaking a culture with sterile 2- to 3-mm glass beads to dislodge the spores, and then washing the surface vigorously with a solution of 0.1% Tween 80 detergent in isotonic (0.85%) saline. Conidiospores that are not to be used immediately should be washed once or twice with sterile saline without Tween 80 before storage at 4°C. Spores from most strains can be stored in saline at 4°C for weeks without significant loss of viability.

C. Growth in Liquid Media

Wild-type strains of *A. nidulans* grow well over a wide temperature range, from about 25° to 45°C, with an optimum at 37°C. At 37°C, *A. nidulans* grows with a nuclear-doubling time of about 90 minutes on a variety of simple media. For general purposes YG medium [yeast extract (Difco) 5 gm/liter; glucose, 20 gm/liter] gives rapid growth and good yields. Vigorous aeration is needed to achieve optimal growth rates in liquid medium. Two hundred milliliters of liquid medium in a 1-liter siliconized, indented Erlenmeyer shake flask (e.g., type 2542, Bellco Glass Inc., scaled up or down according to need), shaken at 200 rpm in a New Brunswick shaking incubator, gives good growth. From an inoculum of 10^6 spores/ml, yields of 5–10 gm of wet pressed log-phase mycelium per liter can be achieved overnight, and up to 30 gm/liter in 24 hours.

D. Growth on Solid Media

Addition of 20 gm/liter of agar to YG gives a good solid medium (YAG) for use in petri plates and in small screw-top tubes. YAG is the medium best suited for growing stock cultures of most strains, although some strains produce better growth and higher yields of conidiospores on other media, for example, MAG (malt extract, 20 gm/liter; peptone, 1 gm/liter; glucose, 20 gm/liter; agar, 20 gm/liter). Since YAG medium is nutritionally complete, it is not useful for genetic-mapping studies using nutritional markers. For this purpose Czapek-Dox minimal medium (Difco) can be used, or, alternatively, minimal medium can be prepared from simple ingredients (see Pontecorvo *et al.*, 1953). Like YG, these media can be solidified by addition of 20 gm/liter of agar. These defined media are also useful for biochemical studies when cellular constituents must be labeled with radioisotopes at known specific activities (see below).

E. Storage of Strains

For relatively short-term storage, cultures can be grown on YAG in slant tubes and kept in the refrigerator. Most strains are viable for at least 3 months under these conditions. For permanent storage, conidiospores are stored dry on silica gel according to the method of Ogata (1962). Conidiospores are harvested (by scraping with a sterile, wetted wire loop) from the surface of a well-sporulated culture, mixed with 0.5 ml of ice-cold Carnation instant nonfat dry milk solution (7.5 gm/100 ml, sterilized by autoclaving 10 minutes) and added to a 1-dram screw-cap tube approximately half-full of 6–12 mesh silica gel presterilized with dry heat at 160°C for 2 hours. The tubes are precooled and kept in ice to prevent overheating, which may occur as a consequence of wetting the silica gel. The tubes are allowed to stand loosely capped for 1 week at room temperature, and are then stored in the refrigerator. Strains can be recovered from storage by sprinkling a few grains of silica gel onto the surface of a petri dish containing YAG and incubating at a suitable temperature (usually 37°C) for several days.

III. Light Microscopy

Light microscopy is an extremely valuable tool for studying mitosis in *Aspergillus* and, in particular, for determining nuclear number, nuclear locations, and mitotic indices. Since hyphae of *A. nidulans* are usually only 2–3 μm in diameter and nuclei are smaller still, many nuclear structures are so minute that they approach the limit of resolution of the light microscope. Thus good optics and care in aligning the microscope are necessary. The problem of size can be partially solved by using diploid strains in which the nuclei are significantly larger than in haploid strains (see below). Despite the small size of *Aspergillus* nuclei, nuclear numbers and mitotic indices can be determined easily, and Robinow and Caten (1969) have shown that, with care in staining and observation, different mitotic morphologies can be distinguished.

In general, more information can be obtained from fixed and stained material than from living material. Chromatin and chromosomes, which are not visible in living material, can be observed easily in stained material. We have found the two most generally useful chromatin stains to be mithramycin (Oakley and Morris, 1980) and aceto-orcein (Robinow and Caten, 1969), but *Aspergillus* chromatin can also be stained by the Feulgen procedure or by the Giemsa stain (Robinow and Caten, 1969). The acid-fuchsin stain is especially useful because it is the only available spindle stain.

A. Mithramycin Staining

Mithramycin is an antibiotic that binds specifically to double-stranded DNA and fluoresces in direct proportion to the amount of DNA present (Crissman and Tobey, 1974; Ward *et al.*, 1965; Behr *et al.*, 1969). The chief advantages of mithramycin staining are that it is specific, simple, and easy to use. This technique is particularly valuable for determining numbers and positions of nuclei. It is also possible to distinguish mitotic from interphase nuclei. Nucleoli are generally visible in interphase nuclei as nonstaining regions (Fig. 1a). When nuclei enter mitosis, chromosomes condense, giving a characteristic "lumpy" appearance, and the nucleolus is no longer visible (Fig. 1b). A problem with the mithramycin technique is that the stain fades rapidly during observation, so one cannot observe a nucleus indefinitely to determine whether it is in mitosis. With practice, however, the distinctions between mitotic and interphase nuclei become apparent and one can determine mitotic indices with accuracy. The mithramycin-staining procedure is as follows.

Samples grown in liquid medium are mixed with an equal volume of a fixative solution containing 1% glutaraldehyde and 0.1 *M* cacodylate buffer, pH 7.0. After 10 minutes the sample is sedimented by centrifugation, the fixative decanted, and the sample resuspended in distilled water. After 5 minutes the sample is sedimented again, the distilled water replaced, and the sample resuspended. For staining, 10 μl of a stock solution containing 200 μg/ml mithramycin (now available commercially from Sigma) in 300 m*M* $MgCl_2$ (Hill and Whatley, 1975) and 10 μl of the detergent Nonidet P-40 are added to 200 μl of sample. Nuclei

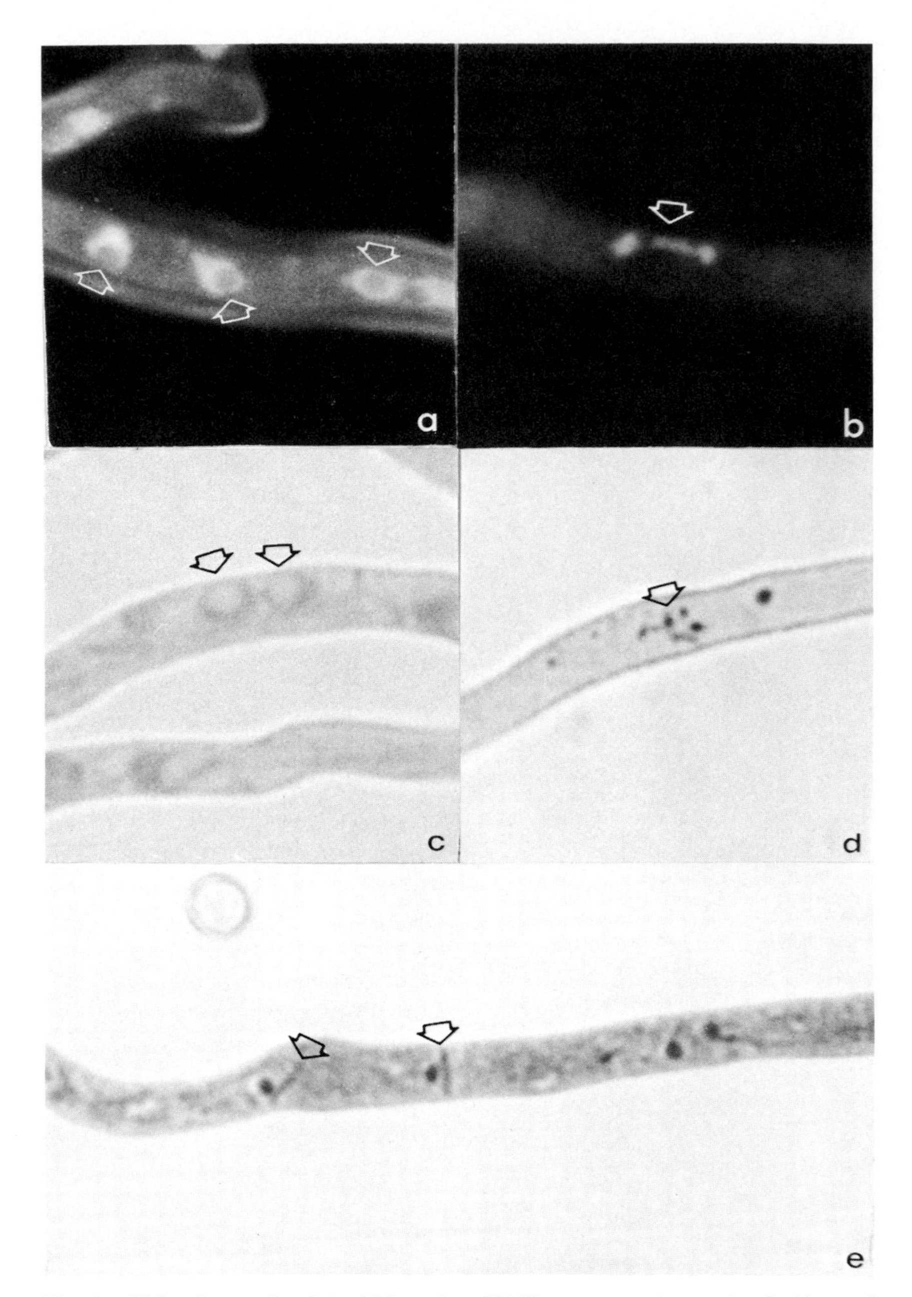

FIG. 1. Light micrographs of *A. nidulans*. (a and b) Fluorescence micrographs of mithramycin-stained material. In (a) the nuclei (brightly staining areas) are in interphase and nucleoli are visible as unstained areas (arrows) within the nuclei. The nucleus (arrow) in (b) is in mitosis. The chromo-

begin to stain after a few minutes, and the level of staining increases for several hours. We generally view the stained material with Zeiss epifluorescent illumination using a BP 390–440 excitation filter, an FT 460 beam splitter, and an LP 470 barrier filter. Mithramycin-stained material is usually suitable for observation for 1 day after staining, but hyphal background fluorescence increases and/or nuclear fluorescence decreases, rendering the nuclei difficult to see.

Williamson and Fennell (1975) have used another fluorescent DNA-binding agent, 4′,6-diamidino-2-phenylindole (DAPI) to stain nuclear and mitochondrial DNA of Saccharomyces. DAPI also stains *Aspergillus* nuclei and does not fade as rapidly as mithramycin (C. F. Roberts, personal communication).

B. Aceto-Orcein Staining

Although its use is somewhat laborious, aceto-orcein is the stain of choice for obtaining accurate mitotic indices because it is easy to distinguish between mitotic and interphase nuclei with this technique. Aceto-orcein-stained mitotic nuclei not only look "lumpy" due to chromosomal condensation (Fig. 1c), but also stain more densely than interphase nuclei (Fig. 1d). Because aceto-orcein does not fade, it is the method of choice if one is trying to distinguish between mitotic stages at the light-microscopic level. It is also easier to photograph aceto-orcein-stained material than mithramycin-stained material, again because it does not fade. Aceto-orcein-stained material may be stored for several weeks or more at 4°C. The aceto-orcein-staining procedure is as follows.

Conidiospores are spread onto a single thickness of sterile dialysis tubing lying on the surface of solid medium (usually YAG) in a petri dish. (This method of growth is useful for experiments that require rapid changes in temperature, nutrition, or other growth conditions because one can change the conditions merely by transferring the dialysis tubing from one plate to another.) For fixation and staining, the tubing is removed from the plate and submerged in modified Helly's fixative (Robinow, 1961). This is made by adding 6 parts of a 3% formaldehyde solution to 94 parts of a 5% mercuric chloride–3% potassium dichromate solution immediately before use. The staining sequence is outlined in Fig. 2. Specimens are examined in the stain under a coverslip, which must be sealed to the slide to prevent drying (nail polish is a suitable sealant). Orcein

---

somes have condensed, giving rise to a distinctive "lumpy" appearance. The lumps are individual or small groups of chromosomes. Although electron microscopy and acid-fuchsin staining reveal that the nucleoli are still present, they are no longer apparent. This nucleus is elongate, signifying that it is in a late-mitotic stage. (c and d) Micrographs of aceto-orcein-stained material, (c) two interphase nuclei (arrows) with typical weak staining. In the mitotic nucleus (arrow) in (d), the condensed chromatin has stained much more strongly and individual condensed chromosomes are visible. (e) An acid-fuchsin-stained hypha in mitosis. The mitotic spindles are indicated by arrows and the dark regions beside the spindles are nucleoli. All micrographs ×2500.

Both stains

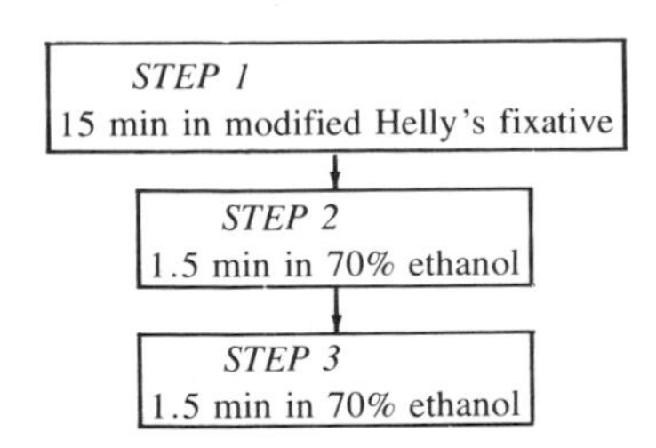

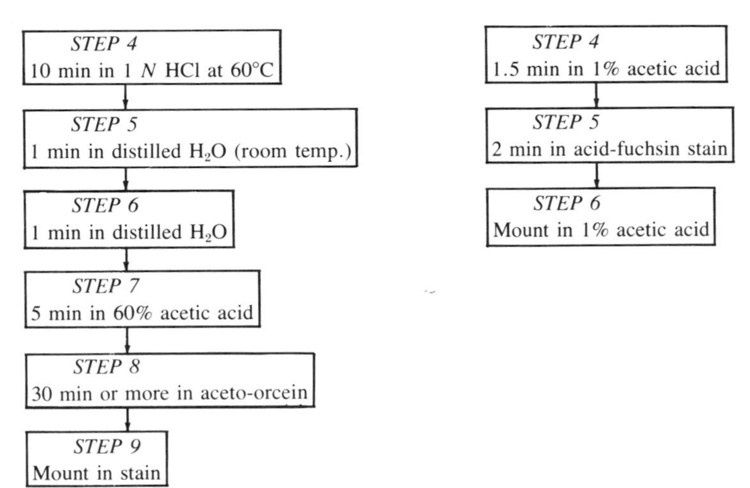

FIG. 2. Flowchart showing procedures for staining mitotic spindles using aceto-orcein and acid-fuchsin modified stains. Helly's fixative: 6 ml of 3% formaldehyde added to 94 ml of 5% mercuric chloride–3% potassium dichromate in water. Aceto-orcein, 1% orcein in 60% acetic acid. Acid fuchsin, 1 part acid fuchsin to 60,000 parts of 1% acetic acid.

obtained from various suppliers differs in its effectiveness as a stain. Orcein supplied by Fisher works well. We have also found that staining is enhanced if the aceto-orcein is warmed and filtered through filter paper immediately before use.

C. Acid-Fuchsin Staining

Acid fuchsin is uniquely useful because it stains mitotic spindles in *A. nidulans*. Specimens are grown and fixed as for aceto-orcein staining. The acid-fuchsin-staining procedure is given in Fig. 2. Specimens are examined under a coverslip in 1% acetic acid (Robinow and Marak, 1966). As with aceto-orcein staining, care must be taken to ensure that the edges of the coverslip are sealed. This stain gives variable results, but under optimal conditions, spindles (and nucleoli) are clear (Fig. 1e). Since *Aspergillus* spindles are small, excellent optics are required. Diploid spindles are easier to see than haploid spindles because they are larger.

D. Determination of Mitotic Index and Nuclear-Division Rate

Mitotic index is usually defined as the percentage of nuclei in mitosis, but *A. nidulans* is coenocytic and nearly all nuclei within a hyphal segment enter mitosis at the same time (Rosenberger and Kessel, 1967). Thus the mitotic index as normally defined would be affected by variation in the number of nuclei per hyphal segment. For *Aspergillus* it is more sensible to define the mitotic index as the percentage of hyphal tips containing mitotic nuclei (Morris, 1976). In practice, the fact that all the nuclei within a single hyphal segment are in mitosis at the same time is an aid in determining the mitotic index, because if one has difficulty in determining if a particular nucleus is in mitosis, one need only examine other nuclei within the same hyphal segment.

Determining nuclear-division rates in *Aspergillus* is easiest in germinating conidiospores. Since conidiospores are uninucleate, division rates can be determined simply by counting nuclei in young germlings as a function of time. This is a particularly useful screen for detecting nuclear-division mutants (see below).

E. Microscopy of Living Material

The major problem in examining nuclei in living material of *A. nidulans* is obtaining enough contrast to distinguish nuclear morphology. *Aspergillus* hyphae are highly refractile so phase-contrast microscopy of wet-mount material reveals little structure. Although Nomarski optics reveal more subcellular structure than phase contrast, it is still impossible to identify nuclei. If care is taken to match the refractility of the medium to the refractility of the hyphae, however, subcellular structures—including nuclei, nucleoli, and even spindles—can be seen (Robinow and Caten, 1969). In principle, a number of techniques could be used to match the refractility of the medium with the refractility of the hyphae, and a technique that has been used successfully by Robinow and Caten (1969) is as follows.

Conidiospores are inoculated onto a microscope slide covered by a thin film of 2% glucose, 0.5% yeast extract, and 18% gelatin. The conidiospores are allowed to germinate and grow suspended over a film of water in a petri dish to prevent drying of the thin gelatin film. Excess gelatin is then trimmed away and the remaining gelatin is covered by a coverslip, which is then sealed to the slide with wax to prevent evaporation.

IV. Electron Microscopy

The small size of *Aspergillus* nuclei is a great advantage for electron microscopy and for studying mitotic spindles in particular. Only a few sections are needed to pass through an entire *Aspergillus* spindle, as opposed to hundreds of sections from most higher plant and animal spindles. This makes the task of determining spindle structure much easier in *Aspergillus* than in higher organisms.

In an effort to obtain optimal fixation, we have systematically varied primary and post fixatives and block stains, and have found that the best general fixation scheme is that of Aist and Williams (1972). Specimens are fixed for 1 hour in 3% glutaraldehyde in 0.1 *M* cacodylate buffer (pH 7.0), washed twice in buffer (about 15 minutes for each wash), postfixed in 1% aqueous osmium tetroxide for 5 minutes, and then washed twice with distilled water. After washing, the specimens are placed in 0.5% aqueous uranyl acetate for 2 hours, then dehydrated in an acetone series (15 minutes each in 25%, 50%, 75%, 100%, 100%, 100%) and embedded in Spurr's resin (Spurr, 1969) (at least 3 hours in a mixture of 50% acetone and 50% Spurr's resin, followed by three changes of Spurr's resin allowing at least 8 hours between changes). The low viscosity of Spurr's resin is an advantage for infiltration, which can be a problem with fungi. Since Spurr's resin is somewhat more difficult to stain than other resins, we routinely stain for 45 minutes in 1% aqueous uranyl acetate followed by 7 minutes in Reynolds' lead citrate solution (Reynolds, 1963). Material grown in liquid culture can be harvested rapidly by filtration through Miracloth (Calbiochem) and handled in centrifuge tubes. Material may also be grown on dialysis tubing, which can simply be transferred from solution to solution. It can then be flat-embedded, and particular areas may be selected for sectioning. This procedure may cause some sectioning difficulty because sections tend to split at the tubing.

The fixation procedure just outlined gives good general fixation of cytoplasmic and nuclear structures and of microtubules; however, as with many fungi, mitotic chromosomes are not identifiable in section even though chromosomal condensation is evident during mitosis at the light-microscopic level. Small clumps of fibrous material resembling heterochromatin, however, are visible at the end of some microtubules. These have been suggested to be kinetochores in other fungi, and our evidence (Oakley and Morris, 1978, 1981) suggests that this is the case in *Aspergillus*.

V. Genetics

Aspergillus nidulans has eight well-marked linkage groups. The genetic map and a list of well-characterized mutations with appropriate literature references can be found in the Chemical Rubber Company's *Handbook of Microbiology* (Clutterbuck, 1974). Most strains of *A. nidulans* are haploid, but aneuploids exist (Kafer, 1974), and heterokaryons and diploids can be produced relatively easily (see below). Diploids are used to determine whether mutations with similar pheonotypes are in the same gene (complementation analysis) and for mapping mutations to linkage group.

A. Mutagenesis

In general, the strains chosen as parental strains for mutant isolation should have as few markers as possible, and these markers should be chosen with regard to subsequent genetic analysis of the mutants (see below). Mutations are induced by treating conidiospores with a mutagen (e.g., ultraviolet light). To mutagenize *A. nidulans* with UV light, conidiospores are washed several times by centrifugation in saline, resuspended in saline at a concentration of 1×10^6 per ml, and irradiated with UV light from a germicidal lamp at a distance of 24 cm with gentle agitation in a open petri dish. The duration of irradiation is determined individually for each source of UV light and for the spores of each strain by plotting the rate of spore kill as a function of the time of irradiation (and/or the distance of the UV lamp from the spores). This is done by plating appropriate spore dilutions onto YAG plates as a function of time to determine the number of spores able to form colonies (the viable count). Optimal mutant yields are usually obtained when the viable count is reduced by not much more than 50% (Van Tuyl, 1977). The duration and intensity of mutagenesis should be minimized to avoid producing multiply mutant strains carrying unwanted mutations.

Ultraviolet light is a good general mutagen and gives all types of mutations; however other mutagens can be used including X-rays, nitrous acid, alkylating agents, and others. A special mutagen that has been reported to give a high proportion of deletions is 1,2,7,8-diepoxyoctane (Aldrich Chemical Co.) (Ong and de Serres, 1972).

B. Complementation Analysis

The first step in classifying a set of mutants with similar or identical phenotypes is to ask whether the mutations are in the same gene or in different genes. The classical method for doing this is by complementation analysis. Complementation analysis is based on the fact that most mutations are recessive in heterozygous diploids. If a diploid is made between two mutant strains with similar recessive mutant phenotypes, the expression of this phenotype will depend on whether or not the mutations are in the same or different genes. If the mutations are in different genes, the phenotype of the diploid will be wild type because for each gene there will be a functioning dominant wild-type allele. If they are in the same gene, there will be no functioning wild-type allele and the mutant phenotype will be expressed. Mutations are said to be complementary if the mutant phenotype is extinguished in diploid strains containing both mutations on opposing chromosomes. If the mutant phenotype is expressed, the mutations are said to be noncomplementary. Failure to complement indicates that two mutations are alleles of the same gene. Complementation, however, does not necessarily mean that mutations are in different genes, since complementation between different alleles of the same gene can occasionally occur. [See *Fungal Genetics* by Fincham *et al.* (1979) for a discussion of intragenic complementation.] Complementation analysis obviously can only be used to compare mutations that are known to be recessive.

Complementation analysis requires that two (haploid) mutant strains be induced to form a diploid. Most strains will form diploids when they are grown together under conditions designed to select against the parental haploids and for the diploid. In order to save time and effort, a useful procedure is to isolate any new set of mutants in two different parental strains with complementary, recessive spore-color and nutritional markers. This allows independently isolated mutations to be tested directly against each other. We now routinely use as our parental strains R21 and R153 (obtained from Dr. Clive Roberts of the Univ. of Leicester, England). R21 carries the mutations pabaA1 (a nutritional requirement for *p*-aminobenzoic acid) and yA2 (a yellow-spore-color marker), whereas R153 carries the mutations pyroA4 (a nutritional requirement for pyridoxine) and wA3 (a white spore-color marker). All of these mutations are recessive. Heterokaryons (which are haploids with two different types of nuclei, one type from each of the parents) and diploids heterozygous for these genes are able to grow on minimal medium (e.g., Czapek-Dox) lacking *p*-aminobenzoic acid and pyridoxine.

To make diploids, spores from two complementary parentral strains are mixed on the surface of a YAG plate and incubated overnight. Small pieces of agar with germinating spores are then transferred to a minimal agar plate and incubation is continued until heterokaryons appear, usually after several days. Heterokaryons contain nuclei from both parents and are recognizable by virtue of the fact that they produce both yellow and white spores. Once obtained, the heterokaryons are then cultured on minimal medium until a diploid appears. This may take several weeks and may require repeated subculturing of small blocks of agar containing vigorously growing portions of the heterokaryon to fresh minimal plates. Diploids eventually appear as vigorously growing sectors and are additionally recognizable by the fact that they are wild-type in color, that is, green.

To determine by complementation analysis whether two newly isolated mutant strains with similar or identical phenotypes have mutations in the same gene, a diploid is made between the two mutant strains (e.g., one in the R21 background, the other in the R153 background) and is tested for restoration of wild-type phenotype. To test two (known recessive) temperature-sensitive mutations for complementation, the diploid would be tested at restrictive temperature. Failure to grow would be strong evidence that the two mutations were in the same gene.

C. Mapping

Genetic mapping in *A. nidulans* is usually done in two stages. Mutations are first mapped to linkage group by the parasexual method of Pontecorvo *et al.* (1953), and then mapped to locus by (sexual) genetic crosses.

1. Mapping to Linkage Group

To map a new mutation to linkage group, a heterozygous diploid is made (as described above for complementation analysis) between the strain carrying the mutation and another strain that has an easily scorable genetic marker on each chromosome. Such multiply marked strains are available from the Fungal Genetics Stock Center in Arcata, California. The diploid, after it has been purified and characterized, is then streaked onto a petri dish containing YAG and a drug that causes chromosome loss [e.g., *p*-fluorophenylalanine (0.013%), benomyl (1.4 μg/ml), or chloral hydrate (0.02 *M*)]. After 5 days the streak is cut out and transferred to a drug-free YAG plate, which is reincubated. Chromosome loss under these conditions appears to be random and the diploids generate haploid segregants. Haploid segregants are identified by virtue of the fact that they express recessive color and nutritional markers. If, for example, the diploid was constructed to be heterozygous for white (wA) or yellow (yA) spore-color markers, haploid segregants will radiate as white and/or yellow sectors from a green diploid parent. Under these conditions little recombination occurs between genes on the same chromosome. Therefore, by analyzing the linkage of a new mutation with respect to each of the chromosome markers, the new mutation can be assigned to linkage group. The mutation should segregate in opposition to one of the mutations used to mark the chromosomes and should segregate randomly with respect to the others. For a more complete discussion of parasexual mapping consult Fincham *et al.* (1979).

2. Mapping to Locus

After a mutation has been assigned to linkage group, it can be mapped to locus by crossing it to a mapping strain carrying several markers on the chromosome of interest. For convenience, the mapping strain should also carry a spore-color marker different from that of the strain to be mapped. Conidiospores from each strain are mixed together or streaked across each other on the surface of a petri dish about two-thirds full of YAG. The plate is wrapped in plastic wrap and incubated at a suitable temperature (usually 37°C but lower for heat-sensitive mutations, higher for cold-sensitive mutations, see below) for 20 to 30 days. At the end of this time the culture is inspected under a dissecting microscope for the presence of cleistothecia. Cleistothecia are dark-purple shiny spheres, several tenths of a millimeter in diameter, which are filled with ascospores (the sexual spores of *A. nidulans*). Several cleistochecia from each cross are transferred aseptically to the surface of a petri dish filled with 4% agar. The cleistothecia are cleaned by rolling them around the surface of the agar while inspecting them for cleanliness (absence of adhering conidiospores) under a dissecting microscope. Each cleaned cleistothecium is then removed to a small vial containing 1.0 ml of sterile 0.1% Tween 80 detergent and crushed with a glass rod to release the spores. *A. nidulans* does not have mating types, therefore most healthy strains will mate with each other. A consequence of this lack of mating types is that most strains will also mate with themselves, and crosses result in selfed cleistothecia (strain A × strain A and strain B × strain B) as well as hybrid cleistothecia (strain A × strain B). Because selfed cleistothecia are not useful for mapping, it is necessary to distinguish between hybrid and selfed cleistothecia. To find the hybrid cleistothecia, spores from three to five crushed cleistothecia are streaked separately onto YAG and incubated at 37°C for 3–4 days. Hybrid cleistothecia will show both parental spore-color markers, whereas selfed cleistothecia will show only one or the other marker. Hybrid cleistothecia are also generally larger than selfed cleistothecia.

The genetic map distance between two genes is classically defined as the proportion of nonparental types (recombinants) observed with respect to the two parental types. For example, if there is 10% recombination between a new mutation and known mutation (marker), the mutation lies 10 map units from the marker. For accurate mapping a new mutation must be located with respect to more than one closely linked marker. To determine recombination frequency, spores from hybrid cleistothecia are plated on YAG to give about 50 colonies per petri dish. These are incubated until conidiated colonies form and then are gridded (see below) onto a master plate and a series of plates designed to test for each marker and for the new mutation.

VI. Biochemial Methods

A. Labeling of Proteins and Nucleic Acids with Radioisotopes

Proteins and nucleic acids can be readily lableled *in vivo* by germinating conidiospores in a liquid minimal medium supplemented for nutritional requirements and containing a radioactively-labeled precurser. For radiolabeling, we generally use a modification of Czapek-Dox minimal medium (Difco). Standard Czapek-Dox minimal medium contains 30 gm saccharose, 3 gm $NaNO_3$, 1 gm K_2HPO_4, 0.5 gm $MgSO_4$, 0.5 gm KCl, and 0.01 gm $FeSO_4$ per liter of solution. For labeling with radioactive sodium sulfate (^{35}S), $MgCl_2$ is substituted for $MgSO_4$. Tritium or ^{14}C-labeled amino acids may be added directly to standard Czapek-Dox broth. Cultures are inoculated at 1×10^6 conidiospores per ml of medium and germinated for 12 to 14 hours at 32°C with shaking. In a typical experiment in which 30 μCi of a ^{14}C-amino acid mixture with a specific activity of 10 mCi/mmol was added to 50 ml of labeling medium, approximately 50% of the radioactivity was converted into a high-molecular-weight form. Nucleic acids can be labeled with radioactive isotopes of adenosine, cytosine, or phosphate. Adenosine can be converted to histidine by *A. nidulans*. Therefore, when adenosine is used to label nucleic acids, 0.2 mg/ml of histidine should be added to the growth medium to reduce the incorporation of label into protein. Selective labeling of DNA with thymidine is not possible because *A. nidulans* does not incorporate this nucleotide. However, DNA labeling can be determined by labeling with adenosine or cytosine, followed by the degradation of RNA with NaOH (Schmidt, 1957).

B. Copolymerization of *Aspergillus* Tubulin with Brain Tubulin

Tubulin from *A. nidulans* may be copolymerized with porcine or bovine brain tubulin (Sheir-Neiss *et al.*, 1976, 1978). A crude homogenate of radioactively labeled *A. nidulans* protein is prepared by grinding a weighed amount of labeled mycelium at 4°C with one-half weight of homogenization buffer consisting of 100 m*M* PIPES (pH 6.95)–1 m*M* EGTA–1 m*M* GTP–1 m*M* PMSF (phenylmethylsulfonylfluoride). The homogenate is then centrifuged at 35,000 *g* for 30 minutes at 4°C and the supernatant is collected. *Aspergillus* tubulin is then sedimented from this supernatant. Freshly depolymerized and repolymerized brain microtubule protein is added to the supernatant in the ratio of 2–4 mg of brain protein per mg of *Aspergillus* extract protein. GTP is added to 1 m*M* and the mixture of brain and *Aspergillus* protein is incubated at 37°C for 30 minutes. Because *Aspergillus* extracts contain an inhibitor of microtubule polymerization, only a fraction of the brain and *Aspergillus* tubulin will polymerize. This is collected by centrifugation at 35,000 *g* for 30 minutes at 37°C, homogenized with a very small volume of homogenization buffer (200–400 μl per mg of *Aspergillus* protein in the original homogenate), and depolymerized at 4°C for 30 minutes. The homogenate is clarified by a 4°C centrifugation at 35,000 *g* for 30 minutes. GTP is again added to 1 m*M* and the tubulin in the supernatant is repolymerized at 37°C for 30 minutes. The microtubules are collected with a 37°C centrifugation at 35,000 *g* for 30 minutes, and redisolved in an appropriate buffer for further analysis. The protocol described above is for use with brain microtubule protein purified in the absence of glycerol (Borisy *et al.*, 1974). If glycerol-purified microtubule protein (Shelanski *et al.*, 1973) is used, the following modifications must be made: Glycerol should be present in the polymerization mixtures at a concentration of 4 *M* and centrifugation speed and time should be increased in all steps to 100,000 *g* and 1 hour.

C. Gel Electrophoresis of Protein Extracts

Labeled mycelium is harvested by filtration through Miracloth (Calbiochem), rinsed with a large volume of ice-cold water, pressed dry with paper towels, and weighed. The mycelium is suspended in three times its weight of 1 m*M* Tris (pH 7.4)–0.2% SDS–2 m*M* PMSF and disrupted at 4°C in a Teflon homogenizer. The extract is then boiled for 2 minutes, cooled, and centrifuged for 1 minute in a Beckman Microfuge. For analysis by standard two-dimensional gel electrophoresis (O'Farrell, 1975), the supernatant is adjusted to contain 10 *M* ultrapure urea (Schwartz/Mann)–2% NP-40–15% 2-mercaptoethanol–1% Ampholine (pH 5–7). A sample containing 250,000 cpm allows the detection of several hundred protein spots after a 1-week exposure to X-ray film. For the analysis of basic proteins, a nonequilibrium-electrophoresis system (O'Farrell *et al.*, 1977) may be used. Sample preparation is the same as for standard two-dimensional gels except that the sample is adjusted after centrifugation to contain 0.4 mg/ml protamine–0.22 *M* NaCl–0.2% NP-40. Cytochrome C (which serves as a visible marker during nonequilibrium pH-gradient electrophoresis) is added to a final concentration of 100 μg/ml before the samples are loaded onto isoelectric-focusing gels.

D. Preparation of *Trichoderma* Lytic Enzyme

Trichoderma lytic enzyme, used to prepare *Aspergillus* protoplasts, is obtained from growing cultures of *Trichoderma harzianum* (de Vries, 1974). *Trichoderma* growth medium is prepared by adding, to 1 liter of water, 3.0 gm of glucose, 1.0 gm of bactopeptone, 0.3 gm of urea, 2.0 gm of KH_2PO_4, 1.4 gm of $(NH_4)_2SO_4$, 0.3 gm of $MgSO_4$, 0.2 gm of $CaCl_2$, and in 1 ml of trace elements solution made by adding 0.1 gm $FeSO_4$, 0.88 gm $ZnSO_4$, 0.04 gm $CuSO_4$, 0.015 gm $MnSO_4$, 0.01 gm $Na_2B_4O_7$, and 0.005 gm $(NH_4)_6Mo_7O_{24}$ to 100 ml of water). A cell wall preparation is made by grinding lyophilized *A. nidulans* mycelium in a mortar and pestle and washing the ground mycelium with a large volume of 0.85% saline. Twenty grams of cell walls is added to each liter of *Trichoderma* growth medium. The medium is inoculated with *Trichoderma* spores to a final concentration of 1×10^6 spores/ml and incubated at 28°–30°C for 4 days with agitation and constant incandescent light. The medium should be cleared by filtration through Miracloth and a brief low-speed centrifugation at 1000 *g* for 10 minutes. The enzyme is precipitated from the medium with ammonium sulfate (767 gm/liter) in the cold, collected by centrifugation, dialyzed against distilled water, and lyophilized. The enzyme may be stored for 6 months at −60°C and reused several times.

E. Preparation of Protoplasts

Aspergillus nidulans, like most filamentous fungi, has a tough cell wall that poses an obstacle to preparation of organelles. The wall can be removed by any one of several different lytic enzymes, the best of which appears to be the *Trichoderma harzianum* lytic enzyme preparation described above. However, even with the *Trichoderma* enzyme, protoplasts do not start to appear until after 45 minutes to 1 hour of digestion, and digestion is far from complete even after 10 hours of incubation, since many individual mycelia appear to be resistant to degradation. Mycelia older than about 18 hours are very resistant to protoplasting with *Trichoderma* enzyme.

We have tested combinations of lytic enzymes to determine whether they possess complementary activities that might lead to faster and more efficient protoplasting of *Aspergillus* mycelia. The most useful combination is a mixture of *Trichoderma* lytic enzyme, snail gut lytic enzyme (Glusulase), and Driselase. Glusulase and Driselase are available from Endo Labs and Kyowa Hakko Kogyo Ltd., respectively.

A. nidulans (3 × 10^6 conidia/ml) is grown in Czapek-Dox (Difco) medium for 15 hours at 32°C in an incubator shaker at 150 rpm. The harvested mycelium is suspended in 32°C buffer containing 0.5 *M* $MgSO_4$, 0.05 *M* Na maleate (pH 5.8), 2 mg/ml *Trichoderma* enzyme, 10 mg/ml Driselase, and 2% Glusulase (exhaustively dialyzed against Mg-maleate buffer) at a concentration of about 25 mg of pressed wet mycelium per ml of protoplasting mixture, and the suspension incubated with gentle shaking at 32°C. This mixture of enzymes starts to generate protoplasts in 15 minutes, and gives virtually complete conversion of mycelia to protoplasts in 30–45 minutes.

As described by de Vries (1974), fungal protoplasts form vacuoles on extended incubation in Mg-maleate buffer. This allows protoplasts to be purified away from residual mycelial debris, as the protoplasts float and the mycelial fragments sink when centrifuged at 16,000 *g* for 10 minutes. Protoplasts prepared with *Trichoderma* lytic enzyme, Driselase, and Glusulase, or any two of these enzymes, are suitably well vacuolated at 5 hours and are easily purified by floatation.

As little as 0.5 mg/ml of *Trichoderma* enzyme gives satisfactory digestion and conserves enzyme. If on the other hand, rapid release of protoplasts is needed, then the three-enzyme combination is best. Old mycelia (21 hours) are digested less rapidly and completely than young (13 hours) mycelia, and conidiospores, conidiophores, and Hulle cells are resistant to digestion.

A variation of these standard procedures that gives a high yield of viable protoplasts that undergo nuclear division at the same rate as the mycelium is as follows. A 25-ml suspension of 10^6–10^7 spores/liter is germinated in a 250-ml flask at 32°C with shaking for 9 hours in minimal medium with 0.2% Tween 80, then washed in warm medium without Tween 80. This gives a uniform culture of very young germlings with only one to two nuclei per germling. To this culture is added 25 ml of sterile prewarmed 1 *M* $MgSO_4$ containing 8 mg/ml *Trichoderma* enzyme and 20 mg/ml Driselase, and incubation is continued as above. Essentially all germlings are converted to protoplasts within 1 hour and these protoplasts continue to grow and to undergo nuclear division with a doubling time of about 140 minutes, which is identical to the nuclear-doubling time of mycelium under these conditions. Protoplasts made in this way will continue rapid nuclear division for at least five to six nuclear divisions. At later times they tend to regrow their cell walls.

F. Preparation of DNA

High-molecular-weight DNA can be prepared either from conidiospores or from protoplasts (Morris, 1978). Conidiospores are harvested into a siliconized 30-ml glass centrifuge tube and carefully washed with cold lysis buffer [5 m*M* spermidine–10 m*M* KCl–10 m*M* EDTA–10 m*M* Tris (pH 7.4)–0.25 *M* sucrose]. The conidiospores are ruptured and the nuclei are released by vortexing 2 gm of conidiospores with 10 gm of acid-washed 0.17-mm Braunwell (Braunwell Industries) glass beads and 2.5 ml of cold lysis buffer. The glass beads, conidiospores, and buffer are mixed with a glass rod and then vortexed on a Vortex Genie mixer at a setting of 10 for a total of 1 to 2 minutes. Vortexing should be continued until about 90% of the spores are ruptured as determined by phase-contrast microscopy. The ruptured spores and nuclei are separated from the glass beads by adding 3 to 4 ml of cold lysis buffer, stirring with a glass rod, allowing the glass beads to settle, and pipetting off the supernatant fluid. The nuclei are then lysed by the addition of sodium dodecyl sulfate (SDS) to a final concentration of 1%. The lysate is heated to 60°C for 20 minutes and then incubated with the addition of proteinase K to 200 μg/ml for 15 hours at 42°C. The lysate is deproteinized by gentle mixing with an equal volume of freshly distilled water-saturated phenol. The phenol–water emulsion is broken by centrifugation and the aqueous phase is reextracted with an equal volume of phenol. The phenol is removed from the aqueous phase by dialysis against 50 m*M* Tris (pH 8.0)–10 m*M* EDTA–100 m*M* KCl. RNAse A (preheated to 80°C for 50 minutes) is added to the dialyzed aqueous phase to a final concentration of 200 μg/ml and the solution is incubated for 1 hour at 42°C. The RNAase is then removed by phenol extraction and the phenol is removed by dialysis. The DNA may be concentrated by burying the dialysis bag in polyethylene glycol for 30 to 60 minutes. This will reduce the volume by 50–75%. A final dialysis is used to put the DNA into a desired storage buffer. We routinely use 0.1 × SSC (0.015 *M* NaCl–0.0015 *M* Na citrate). If carbohydrate contamination is high at this point, the DNA solution should be centrifuged in a type 50 rotor at 48,000 rpm for 1 hour. This will pellet most of the high-molecular-weight carbohydrate, which is then discarded. As a final purification step, the DNA can be banded in a preparative CsCl gradient. Up to 150 μg of DNA can be centrifuged in a 4.5-ml CsCl gradient using a Spinco 50 rotor at 42,000 rpm for 18 to 22 hours at 18°C. The starting density of the solution, 1.709 gm/ml, is obtained by adding 1.05 gm of DNA solution to 6.62 gm of saturated aqueous CsCl. After reaching equilibrium the gradients are fractionated from the bottom and the DNA-containing fractions (identified by absorbance at 260 nm) are pooled and NaCl is added to 0.2 *M*. The DNA is precipitated by the addition of 3 volumes of cold 70% ethanol, collected by centrifugation (10,000 *g* for 10 minutes), and gently redissolved in 0.1 × SSC. We have found that DNA prepared in this way serves as a good substrate for a variety of restriction endonucleases. When DNA is extracted from protoplasts, the protoplasts are carefully washed and resuspended in 10 to 15 volumes of 50 m*M* Tris (pH 8.0)–50 m*M* EDTA, and SDS is added to a final concentration of 1%. The remainder of the procedure is the same as for obtaining DNA from spores. Recovery of DNA is approximately 250 μg of DNA per gram of spores, and 450 μg of DNA per gm of protoplasts.

Although the above method is suitable for the preparation of significant quantities of highly purified DNA, it is laborious. A much simpler, modified protocol that allows the rapid preparation of small quantities of restriction-endonuclease-digestible DNA is as follows. Nuclei are prepared from conidiospores (as described above), lysed by the addition of Sarkosyl to 1%, and heated to 60°C for 20 minutes. The lysate is then added to a saturated solution of CsCl to achieve a starting density of 1.709 gm/ml and centrifuged at 100,000 *g* to equilibrium (about 60 hours).

The gradient may be fractionated very crudely. We generally collect eight 1-ml fractions and discard the top two fractions, which contain the bulk of the protein, and the bottom two fractions, which contain the bulk of the RNA. The middle fractions are pooled, and the DNA is precipitated by the addition of 3 volumes of cold 70% ethanol. The DNA is collected by centrifugation (12,000 *g* for 10 minutes), and the pellets are rinsed with 70% ethanol, dried, and redissolved in a small volume of 0.1 × SSC. The nuclei from 1 gm of wet-weight conidia may be processed in a single 8-ml gradient and will yield at least 100 μg of DNA.

Although the DNA is at this point contaminated with protein, RNA, and carbohydrate, we have been able to digest such samples with restriction endonuclease by adding a severalfold excess of enzyme activity.

VII. Mitotic Mutants

Mutations that block mitosis or otherwise interrupt the cell cycle are usually lethal. They therefore must be isolated as conditional lethal mutations—that is, mutations that exert their effects only under some special condition. For example, heat-sensitive (hs) mutations are lethal at high temperatures, but not at lower temperatures. Conversely, cold-sensitive (cs) mutations are lethal at low but are normal at high temperatures. Most of the mitotic mutants of *A. nidulans* were isolated as hs mutants (Morris, 1976), but cs mutants have also been isolated (C. F. Roberts, unpublished data).

Temperature-sensitive mutants of *A. nidulans* are identified as follows. Conidiospores are mutagenized with UV light (see preceding sections). The spores are then plated on petri dishes containing YAG to give 50–100 surviving spores per plate. The petri dishes are incubated at permissive temperature until well-sporulated colonies appear. Using sterile toothpicks, spores from each colony are transferred to two fresh petri dishes at the same position on a predetermined grid. The plates are carefully marked and one is incubated at 32°C, the other at 42°C. Temperature-sensitive colonies are identified by comparing the petri dishes incubated at the two different temperatures. Colonies that grow well at 32°C but poorly at 42°C are putative hs mutants, whereas colonies that grow well at 42°C but poorly at 32°C are putative cs mutants. Putative cs mutants are retested at 25°C, where they usually exhibit a more pronounced cs phenotype. Only tight hs and cs mutants should be collected. The putative hs and cs mutants should be purified by streaking twice to single colony on YAG (at permissive temperature) and then reverified.

The best method for identifying mitotic and cell cycle mutants of *A. nidulans* is to characterize them by microscopy. The simplest procedure is to germinate conidiospores at restrictive temperature and count nuclei. A loopful of spores from each hs or cs mutant is inoculated into a few milliliters of sterile YG. The spores are germinated overnight at restrictive temperature, stained with mithramycin or DAPI (see earlier), and examined by fluorescence microscopy. Mithramycin and DAPI stain nuclei, and it is very easy to determine whether nuclear division is blocked at restrictive temperature. *Aspergillus* conidiospores are uninucleate. If, after overnight incubation at restrictive temperature, germlings, which normally have multiple nuclei, have only one nucleus, the mutation being studied must preferentially affect nuclear division. Leaky mutants and mutants that have a gene product "dowry" may exhibit more than one nucleus per germling, but can usually be identified by unusually wide spacing between nuclei. Another method for identifying cell cycle mutants is to grow cultures overnight at permissive temperature and then switch to restrictive temperature for several hours before staining. This method, although more cumbersome that the first, has several advantages. It allows evaluation of mutants that fail to sporulate at restrictive temperature. Because the shift to restrictive temperature is relatively short, nuclear morphology is less distorted than when germlings are kept at restrictive temperature overnight, and the location of the block, whether at interphase or at metaphase, can be more easily determined. Septation-deficient mutants as well as other interesting mutants affecting cellular morphology will also be found by this method (Morris, 1976).

VIII. Tubulin Mutants

Mutants of *A. nidulans* selected for resistance to antimicrotubule drugs may have mutations in genes coding for microtubule proteins. We have shown that mutants with altered resistance to the benzimidazole antifungal agents benomyl, thiabendazole, and/or nocodazole carry mutations in *ben*A, a structural gene for β-tubulin (Sheir-Neiss *et al.*, 1978). Possibly other antimitotic drugs can be used in the same way to find other mutations in tubulin or in other mitotic proteins.

The following procedure can be used to select *ben*A mutants. Conidiospores are mutagenized and plated on YAG containing 2.4 μg/ml of benomyl (or nocodazole) or 20 μg/ml of thiabendazole to give about 50 resistant colonies per petri dish. The resistant colonies are then twice restreaked to single colony and retested on benomyl, thiabendazole, and nocodazole to determine the 50% inhibitory concentration for each drug and the pattern of cross-resistance between drugs. The mutants can also be scored for growth in the absence of drug and at 20°C and 42°C in both the presence and absence of drug to identify hs, cs, and drug-dependent *ben*A mutations. The mutants of interest are then characterized genetically by crossing them with previously identified *ben*A mutants. If all of the progeny of a cross between a new drug-resistant mutant and a known *ben*A mutant are drug resistant, then the new mutation must be linked to and may be identical with *ben*A, but if some moderate fraction of the progeny is sensitive, the new mutation cannot be in the *ben*A gene, but must be in a different gene.

Many of the *ben*A mutants studied in our laboratory have been found to have electrophoretically abnormal β-tubulins (Sheir-Neiss *et al.*, 1978). The demonstration of an abnormal β-tubulin by two-dimensional gel electrophoresis is the most convincing way to prove that a mutation is in β-tubulin.

A somewhat different strategy has been used to identify mutations in the

α-subunit of tubulin. Mutations in *tub*A, the structural gene for α-tubulin have been identified as suppressors of hs *ben*A (β-tubulin) mutants (Morris *et al.*, 1979). The rationale for this work (based on Jarvik and Botstein, 1975) was that since β-tubulin interacts with α-tubulin, it might be possible to find mutations in α-tubulin that counteract temperature-sensitive mutations of β-tubulin. *Ben*All (Van Tuyl, 1977) was used as the initial hs *ben*A mutant.

*Ben*All was stabbed in the center of several large petri dishes, which were then incubated at permissive temperature (32°C) for several days until the colonies were 3–4 cm in diameter. The plates were then shifted to restrictive temperature (44°C) and examined daily for the presence of fast-growing fans at the periphery of the colonies. These were picked, streaked twice to single colony, and retested at permissive and restrictive temperatures. Each of the hs+ revertants was then crossed to wild type. The reappearance of temperature sensitivity among the progeny of the crosses indicated the presence of an extragenic suppressor of the hs *ben*All mutation. Strains carrying extragenic suppressor mutations were then examined for electrophoretically altered α-tubulins by the two-dimensional gel-electrophoresis method of O'Farrell (1975).

IX. Concluding Remarks

Many useful characteristics of *A. nidulans* are common to other simple eukaryotes such as *Saccharomyces cerevisiae,* and indeed, mitotic and cell cycle mutants have been isolated in *S. cerevisiae* and other organisms, and *S. cerevisiae* has been the system of choice for solving many problems in cell and molecular biology. *A. nidulans*, however, has a number of advantages over *S. cerevisiae* for studying mitosis and other microtubule-mediated processes. First, chromosomes in *A. nidulans* condense during mitosis, whereas those of *S. cerevisiae* do not. This facilitates searching for mitotic mutants in *A. nidulans* and allows one to ask questions about chromosomal condensation in *A. nidulans* that are impossible to ask in *S. cerevisiae*. Second, the mitotic spindle assembles rapidly in *A. nidulans* and is only present for a small portion of the cell cycle (Oakley and Morris, 1981). In *S. cerevisiae,* spindle microtubules are present through much of the cell cycle, and the assembly of spindle microtubules is a very gradual process (Byers and Goetsch, 1973). This means that one can ask questions about the regulation of the assembly of spindle microtubules in *A. nidulans* that would be difficult to ask in *S. cerevisiae*. Another relevant point is that since chromosomes do not condense in *S. cerevisiae,* and since spindle microtubules are present through much of the cell cycle, the definition of mitosis and of mitotic mutants is somewhat arbitrary in *S. cerevisiae*. Thus an apparent mitotic mutant in *S. cerevisiae* might not be defective for a gene product that is essential for mitosis. For example, mutants of *S. cerevisiae* that are apparent mitotic mutants are actually defective for DNA-ligase activity (Johnston and Nasmyth, 1978). Third, since *A. nidulans* is filamentous and thus must engage in organellar transport, one can ask questions about the role of microtubules in organellar transport [e.g., nuclear movement (Oakley and Morris, 1980)]; this is not possible in *S. cerevisiae*. Fourth, *A. nidulans* differentiates into several cell types, so it is possible to ask questions about the role of microtubules in differentiation that are impossible in *S. cerevisiae,* because *S. cerevisiae* essentially does not differentiate.

These specific advantages coupled with general genetic, biochemical, and morphological utility make *A. nidulans* a uniquely powerful system for studying mitosis and other forms of microtubule-mediated motility.

ACKNOWLEDGMENTS

We would like to acknowledge many useful discussions with Dr. C. F. Roberts. This work was supported by grant GM 23060, from the National Institutes of Health.

REFERENCES

Aist, J. R., and Williams, P. H. (1972). *J. Cell Biol.* **55,** 368-389.

Behr, W., Honikel, K., and Hartmann, G. (1969). *Eur. J. Biochem.* **9,** 82-92.

Borisy, G. G., Olmstead, J. B., Marcum, J. M., and Allen, C. (1974). *Fed. Proc., Fed. Am. Sco. Exp. Biol.* **33,** 167-174.

Clutterbuck, J. (1974). *In* "Handbook of Genetics" (R. C. King, ed.), Vol. 1, pp. 447-510. Plenum, New York.

Crissman, H. A., and Tobey, R. A. (1974). *Science* **184,** 1297-1298.

deVries, O. M. H. (1974). Thesis, University of Gröningen, The Netherlands.

Fincham, J. R. S., Day, P. R., and Radford, A. (1979). "Fungal Genetics." Blackwell, Oxford.

Hill, B. T., and Whatley, S. (1975). *FEBS Lett.* **56,** 20-23.

Jarvik, J., and Botstein, D. (1975). *Proc. Natl. Acad. Sci. U.S.A.* **72,** 2738-2742.

Johnston, L. H., and Nasmyth, K. A. (1978). *Nature (London)* **274,** 891-893.

Kafer, E. (1974). *Genetics* **79,** 7-30.

Morris, N. R. (1976). *Genet. Res.* **26,** 237-254.

Morris, N. R. (1978). *J. Gen. Microbiol.* **106,** 387-389.

Morris, N. R. (1980). *Symp. Soc. Gen. Microbiol.* **30,** 41-75.

Morris, N. R., Lai, M. H., and Oakley, C. E. (1979). *Cell* **16,** 437-442.

Oakley, B. R., and Morris, N. R. (1978). *J. Cell Biol.* **79,** 300a.

Oakley, B. R., and Morris, N. R. (1980). *Cell* **19,** 255-262.

Oakley, B. R., and Morris, N. R. (1981). In preparation.

O'Farrell, P. H. (1975). *J. Biol. Chem.* **250,** 4007-4021.

O'Farrell, P. Z., Goodman, H. M., and O'Farrell, P. H. (1977). *Cell* **12,** 1133-1142.

Ogata, W. N. (1962). *Neurospora Newsl.* **1,** 13.

Ong, T., and de Serres, F. J. (1972). *Cancer Res.* **32,** 1890-1893.

Orr, E., and Rosenberger, R. F. (1976a). *J. Bacteriol.* **126,** 895-902.

Orr, E., and Rosenberger, R. F. (1976b). *J. Bacteriol.* **126,** 903-906.
Pontecorvo, G., Roper, J. A., Hemmons, L. M., McDonald, K. D., and Bufton, A. W. J. (1953). *Adv. Genet.* **5,** 141-238.
Reynolds, E. S. (1963). *J. Cell Biol.* **17,** 208-212.
Robinow, C. F. (1961). *J. Biophys. Biochem. Cytol.* **9,** 879-892.
Robinow, C. F., and Caten, C. E. (1969). *J. Cell Sci.* **5,** 403-431.
Robinow, C. F., and Marak, J. (1966). *J. Cell Biol.* **29,** 129-151.
Rosenberger, R. F., and Kessel, M. (1967). *J. Bacteriol.* (C. P. Colowick and N. Kaplan, eds.), Vol. **94,** 1464-1469.
Schmidt, G. (1957). *Methods Enzymol.* **3,** 671-679.
Sheir-Neiss, G., Nardi, R. V., Gealt, M. A., and Morris, N. R. (1976). *Biochem. Biophys. Res. Commun.* **69,** 285-290.
Sheir-Neiss, G., Lai, M. H., and Morris, N. R. (1978). *Cell* **15,** 638-647.
Shelanski, M. L., Gaskin, F., and Cantor, C. R. (1973). *Proc. Natl. Acad. Sci. U.S.A.* **70,** 765-768.
Smith, J. E., and Pateman, J. A., eds. (1978). "Genetics and Physiology of *Aspergillus*." Symp. British Mycological Society. Academic Press, London.
Spurr, A. R. (1969). *J. Ultrastruct. Res.* **26,** 31-43.
Timberlake, W. E. (1978). *Science* **202,** 973-975.
Trinci, A. P. J., and Morris, N. R. (1979). *J. Gen. Microbiol.* **114,** 53-59.
Van Tuyl, J. M. (1977). Thesis, Dept. of Phytopathology, Dept. of Genetics, Agricultural Univ. Wageningen, The Netherlands.
Ward, D. C., Reich, E., and Goldberg, I. H. (1965). *Science* **149,** 1259-1263.
Williamson, D. H., and Fennell, D. J. (1975). *Methods Cell Biol.* **12,** 335-351.

Chapter 7

Actin Localization within Cells by Electron Microscopy

ARTHUR FORER

Biology Department
York University
Downsview, Ontario, Canada

I. Introduction

Actin, a universal protein coded for by several genes (Vandekerckhove and Weber, 1979; Tobin *et al.*, 1980; Fyrberg *et al.*, 1980; Kaine and Spear, 1980; Durica *et al.*, 1980), functions in a variety of motile processes (for reviews, see Huxley, 1972, 1973; Forer, 1974, 1978b; Hepler and Palevitz, 1974; Pollard and Weihing, 1974; Goldman *et al.*, 1976; Clarke and Spudich, 1977; Korn, 1978; Goldman *et al.*, 1979; Taylor and Condeelis, 1979). Like other cellular components, actin can be studied chemically, genetically, using immunofluorescence, and in a variety of other ways. Some of these methods are discussed in other chapters in this volume (namely, those by Brown; Bryan and Kane; Lazarides; Pollard; Mooseker and Howe; Hartwig and Stossel; Pardee and Spudich). This chapter concentrates on electron-microscopic localization of actin, which, to date, has been solely of actin-containing filaments. [The term ''actin-containing filaments'' is used for the following reason: purified globular actin (G-actin) monomers polymerize to form filaments called F-actin (filamentous actin), so pure actin *can* form filaments. However, filaments *in vivo* in muscle and elsewhere contain in addition to actin other components such as tropomyosin and troponin. Thus,

ISBN 0-12-564125-7

since components other than actin may be present in the filaments, the term "actin filaments" may be inaccurate; hence the term "actin-containing filaments is used."]

There are several reasons for studying intracellular actin using electron microscopy rather than other techniques. One is that electron microscopy has the potential to define the spatial arrangements of the various cytoskeletal components at resolvable distances more than two orders of magnitude smaller than is possible using light microscopy. As is evident from our knowledge of striated-muscle myofibrils and of cilia and flagella, identifying the spatial arrangements of the component parts is crucial to our understanding the motile function of these organelles. Thus electron microscopy is a powerful tool in helping to identify the arrangements of individual actin-containing filaments, as well as the arrangements of actin-containing filaments relative to other components, such as myosin, or microtubules (e.g., Schloss *et al.,* 1977; Griffith and Pollard, 1978; Forer and Jackson, 1979). Another reason for studying intracellular actin using electron microscopy is to determine the polarities of individual actin-containing filaments: Polarity can be identified only by using electron microscopy. The polarity is visible because the complexes formed between heavy meromyosin (HMM) and F-actin, or between HMM and actin-containing filaments, look like a series of "arrowheads," and because this directionality reflects a polarity inherent in the actin-containing filaments (Huxley, 1963; Moore *et al.,* 1970; review in Forer, 1978b). The polarities of the actin-containing filaments and myosin-containing filaments in muscle help determine how these filaments interact (Huxley, 1963, 1972); in other systems as well, information on the polarities of actin-containing filaments is important evidence in understanding how these filaments might function (e.g., Mooseker and Tilney, 1975; Kersey *et al.,* 1976; Mooseker, 1976; Forer, 1978a; Forer *et al.,* 1979). Finally, the electron-microscopic identification of actin-containing filaments is unambiguous when one uses arrowhead formation after binding by HMM: the formation of arrowheads is specific to actin-containing filaments (review in Huxley, 1973; Forer, 1978b). On the other hand, identification of actin-containing components by light-microscopic methods using fluorescent HMM or fluorescent antibodies against actin often is ambiguous (for various reasons discussed in Forer, 1978a, especially from p. 65).

There are some obvious drawbacks to studying actin localizations using the electron microscope. One is the set of problems inherent in electron microscopy per se: cells must be embedded, sectioned, and stained. Further, if, as in most cases, one is not blessed with a structure as regular as the myofibril or the ciliary axoneme, one must rely on serial sectioning and three-dimensional reconstruction if one wants to define the spatial arrangement of components. These electron-microscopic procedures are long (and depending on one's temperament, may also be quite tedious), and are not without ambiguities (e.g., in fixation, or staining). A further ambiguity arises in the particular case of studying actin-containing filaments after reaction with HMM because one needs to make the cells permeable to the HMM: if the HMM can move throughout the cell, so can the actin! This (and related) ambiguities will be discussed after describing the methods in detail.

The descriptions in this chapter rely on the identification of actin-containing filaments using HMM to form arrowheads with these filaments. One could imagine that antibody techniques could be used to study actin ultrastructurally, but I know of no such work. Other techniques are possible, but these are not as specific as the HMM binding. For example, filaments have been identified as containing actin because they were not seen after treatment with DNase (Raju *et al.,* 1978). DNase *does* depolymerize F-actin filaments *in vivo* (Mannherz *et al.,* 1975; Hitchcock *et al.,* 1976), but it also breaks down chromatin to nucleosomes, so the reaction is not necessarily specific. This chapter thus concentrates on experimental approaches to identifying actin-containing filaments *in situ* using electron microscopy and the specific reaction of such filaments with HMM to form arrowheads or "decorated filaments." I have elsewhere presented a detailed review of previous electron-microscopic work on actin-containing filaments, as well as a review of some of the basic chemistry of muscle proteins and protocols for preparing actin, myosin, and HMM (Forer, 1978b). Other protocols for preparing these proteins are found in this volume in chapters by Spudich and by Pollard.

II. Electron-Microscopic Methods

A. First Steps

I strongly recommend that at the start those persons who have not previously looked at actin begin by looking at actin-containing filaments purified from (or isolated from) muscle before trying to localize actin-containing filaments *in situ*. In my experience, actin-containing filaments can be hard to see in sections in among other cellular components, and it is a help if one is familiar with what purified actin-containing filaments look like. Hence I suggest as a first step that one study both negatively stained actin-containing filaments and fixed, embedded, and sectioned actin-containing filaments; likewise one should study actin-containing filaments that have reacted with HMM.

Another purpose as well is served by this first step: by familiarizing oneself with the appearances of actin-containing filaments and of HMM-decorated filaments (arrowhead complexes) in negatively stained preparations, one gains experience necessary for assessing the purity of the HMM preparations. It is important to ascertain that the HMM preparations that are used for reacting with actin-containing filaments *in situ* are not contaminated with actin-containing filaments, because, obviously, one does not want to add external actin-containing filaments to cells when adding HMM. When actin-containing filaments *are* present in HMM, they can be seen easily (as arrowhead complexes) in negatively stained preparations. Indeed, using negative staining I have seen some actin-containing filaments in "electrophoretically pure" HMM: electron microscopy is a very sensitive way to detect small amounts of contaminating actin. Thus prior study of biochemically prepared actin both prepares one for looking at actin-containing filaments *in situ* and prepares one for checking on the purity of the HMM preparations. Following is a brief discussion of procedures for studying actin in negatively stained or in sectioned preparations.

General procedures for negatively staining specimens are given in Horne (1964), for example. Any such procedure can be used to study actin-containing filaments and arrowhead complexes. Some procedures give better results than others, however, and I have elsewhere reviewed the literature on the various parameters that have been studied (Forer, 1978b). Two points seem particularly relevant. One point is that arrowheads are best seen when the filaments in question are embedded in a pool of stain that fills holes in the formvar (or colloidion) film. This situation is achieved as follows. Grids are covered with colloidion or formvar films that contain holes; the actin–HMM mixture is placed on the film-covered grid; the filaments are rinsed with a spreading agent (such as bacitracin or cytochrome oxidase); and the filaments are rinsed with uranyl acetate (or other stain) and blotted dry. [Detailed description of this procedure is given in Forer (1978b).] That the filaments have reacted with HMM can be seen whether filaments are on the colloidion (or formvar) film or are embedded in stain that fills the holes in the film, but arrowheads and directionality (e.g., Fig. 1) are seen more clearly and regularly when the decorated filaments are in holes in the film than when they are not (e.g., Huxley and Zubay, 1960; Huxley, 1963; reviewed in Forer, 1978b). On the other hand, actin-containing filaments are distorted least—and hence are preferred—when they are on the colloidion or formvar film rather than in holes in the film (reviewed in Forer, 1978b). The second point is that the effects of various fixations can be tested using negatively stained preparations: one can treat the filaments with the fixative (or series of fixatives) in question to see if the fixatives cause alteration in repeat distances, lengths of filaments, or other parameters. It is easier and quicker to check the effects of fixatives in this way than to wait until specimens have been embedded and sectioned. Some work along these lines has been published (review in Forer, 1978b); for example, it has been found that osmium tetroxide causes shrinkage and distortion of actin-containing filaments isolated from myofibrils (e.g., Page and Huxley, 1963; Hanson, 1967).

Procedures for embedment and sectioning of actin-containing filaments (and of arrowhead complexes) are those used routinely for tissues (e.g., Glauert,

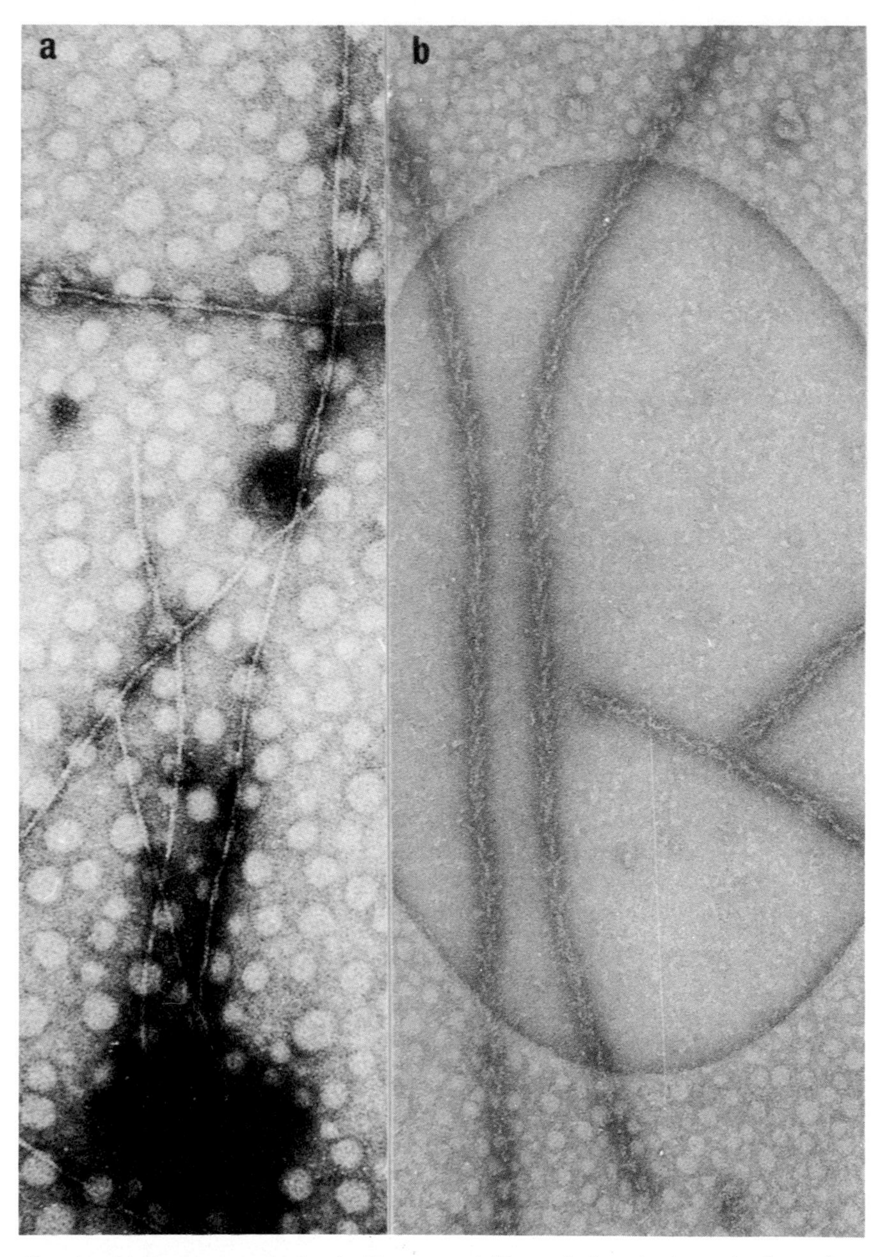

FIG. 1. (a) Negatively stained actin filaments and (b) negatively stained arrowhead complexes, both at the same magnification, ×124,500. The actin filaments are on the colloidion film; the arrowhead complexes are in a pool of uranyl acetate in a hole in the film. The procedures are as described in Forer (1978b).

1975). One can either fix pellets of filaments (or arrowhead complexes), or fix the filaments in solution and then centrifuge them into a pellet, depending on which is most convenient. Study of sections of *in vitro* actin-containing filaments not only familiarizes one with the appearances of these filaments but also gives some assessment of the fixation procedure: not all fixatives preserve filaments equally well (review in Forer, 1978b). For example, it has been known since 1963 (Page and Huxley, 1963; Page, 1964; Hanson, 1967) that osmium tetroxide can have deleterious effects on actin-containing filaments in muscle. Not all actin-containing filaments are necessarily the same, though (e.g., Forer, 1978b), and actin-containing filaments in tissue need not necessarily respond as do actin-containing filaments in muscle. Nonetheless, as a preliminary assessment of a particular fixation procedure one can study the effect of that particular fixation (and embedment) procedure on the *in vitro* filaments. I discuss below various fixatives for studying actin *in situ*.

To sum up, I recommend that as the first steps in studying actin-containing filaments *in situ,* one study purified actin-containing filaments and arrowhead complexes, using both negative-staining and sectioning techniques.

B. Studying Actin-Containing Filaments *in Situ*

An overview of the procedures is first given, followed by a discussion of details of the various aspects.

The main criterion for studying HMM-labeled actin-containing filaments *in situ* is to get HMM to the sites of the actin-containing filaments under conditions in which the HMM will bind to the filaments—that is, in the absence of ATP or pyrophosphate and under conditions in which the filaments are stable. Thus one needs to make the cells permeable to the HMM (so that the HMM can diffuse to the filaments). ATP usually leaves the cell at the same time; if it does not, the HMM is an ATPase and the ATPase activity will eventually remove ATP and allow the HMM to bind to the actin-containing filaments (if an ATP-regenerating system is not present). Ideally, then, one would see 5- to 6-nm diameter filaments *in situ,* in glutaraldehyde-fixed cells, and after adding HMM one would see exactly the same number of filaments in exactly the same positions. One would only need to identify those filaments that have and have not been decorated by the HMM. This ideal is often not satisfied, for reasons discussed below. Details of the procedures will now be given, from fixation of filaments *in situ,* to making cells permeable, to identifying polarity *in situ.*

Actin-containing filaments are not always preserved using the usual fixation-embedment procedures (review in Forer, 1978b; also Maupin-Szamier and Pollard, 1978; Seagull and Heath, 1979). One deduces this from differences in numbers of actin-containing filaments present after negatively staining preparations from living cells compared to after embedment and sectioning (e.g.,

Forer and Behnke, 1972; Palevitz and Hepler, 1975; Kersey and Wessells, 1976), or one deduces this from differences in numbers of actin-containing filaments present after different fixation procedures (e.g., Seagull and Heath, 1979). By studying the effect of various fixatives on purified actin-containing filaments, one deduces that osmium tetroxide can be destructive to the filaments (Pollard, 1976; Maupin-Szamier and Pollard, 1978; Gicquaud *et al.*, 1980; Gicquaud and Loranger, 1981), in agreement with earlier work on actin-containing filaments from muscle, described previously. Alternative fixation and embedment procedures may be necessary, then, in order to preserve actin-containing filaments *in situ*. Alternative procedures that have been used with some success include fixation under very specific conditions (Tilney, 1975), fixation with a glutaraldehyde–spermidine mixture (Hauser, 1978), and fixation with a glutaraldehyde–tannic acid mixture (Seagull and Heath, 1979). The method of dehydration may also be important, because acetone dehydration seems to preserve actin-containing filaments in muscle better than ethanol dehydration does (review in Forer, 1978b). Thus absence of 5- to 6-nm filaments *in situ* after usual embedment and sectioning procedures does not necessarily mean absence of actin-containing filaments from the cells.

Cells can be made permeable to HMM in various ways. In the first application of HMM-binding to nonmuscle cells, the cells were glycerinated with cold 50% glycerol for 24 hours, followed by gradual dilution of the glycerol at 4- to 14-hour intervals (Ishikawa *et al.*, 1969). In later work the glycerination was at room temperature, for greatly reduced times (review in Forer, 1978b)—even minutes (e.g., Schloss *et al.*, 1977). Detergents such as saponin, Triton X-100, and sodium dodecyl sulfate (SDS) have also been used, and these do not seem to interfere with arrowhead formation (review in Forer, 1978b). Some cells have been treated for short times with single solutions (containing Triton X, glycerol, *and* HMM) that acted both as an agent to cause the cells to be permeable and as a source of HMM to bind to intracellular actin-containing filaments (e.g., Forer and Jackson, 1979). The general conclusion seems to me to be that the cells in question should be made permeable using the agent that is least disruptive to the morphology of that particular cell, and that is most convenient to use (e.g., shorter rather than longer time periods).

How does one test whether the agent in question indeed makes the cell permeable? An obvious test, of course, is to look at sections of treated cells to see if putative actin-containing filaments are decorated. There are alternatives that may be more convenient, however. One is to use negatively stained preparations: treat the cells to make them permeable and to allow HMM to get inside the cell, and then wash away the HMM and break open the cells on an electron-microscope grid to see if arrowhead complexes are visible. (An obvious control, for example, would be to break open living cells to see if filaments are visible that look like actin-containing filaments, and then to determine whether filaments on the grid react with HMM.) Another possibility is to use fluorescently labeled HMM. One can fluorescently label the HMM in such a way that no unbound label is present in the HMM preparation (e.g., Schloss *et al.*, 1977; Herman and Pollard, 1978); one can then run through the test procedure using fluorescently labeled HMM (using fluorescence microscopy) to see whether fluorescently labeled HMM entered the cells.

Cellular components might move around after the cells are made permeable; if the HMM can reach the actin-containing filaments, then those filaments and/or other cellular components can move around. How does one decide whether relocation has occurred, and, if so, how to avoid relocation of cellular components? One way to decide whether relocation might have occurred is to use phase-contrast microscopy and/or other light-microscopic methods to study cells during glycerination, to see if there are changes in cell shape or in positions of light-microscopically visible organelles; indeed, there are such changes when some cells are glycerinated (e.g., Forer and Jackson, 1979). Another way is to use electron microscopy directly, to compare the arrangements of organelles (especially of actin-containing filaments) before and after reaction with HMM [and with the agent(s) used to make the cells permeable]. Yet another way is to use fluorescence microscopy and fluorescently labeled actin (and/or other proteins). Fluorescently labeled actin can be injected into cells and can be followed during the cell cycle (e.g., Taylor and Wang, 1978; Wang and Taylor, 1979). If one has a cell that contains fluorescent actin, one can follow that cell using fluorescence microscopy while that cell is made permeable and is reacted with HMM, to see if the fluorescent actin relocates during the procedure.

Pre-fixation is one way to avoid relocation during the procedure for making the cells permeable. The idea is to pre-fix the cells (before making them permeable) using a fixation procedure that does not prevent HMM from forming arrowheads with actin-containing filaments. Then the filaments are held in place and do not move during the treatment (with detergent or other agent) that is used to make the cells permeable. It is also important that the fixation does not cross-link the cell components so much that HMM is prevented from reaching the filaments. Glutaraldehyde does not prevent HMM from reacting with actin-containing filaments *in vitro* (Gadasi *et al.*, 1974; Poo and Hartshorne, 1976; Prochniewicz, 1979), and there are published methods in which cells are pre-fixed with glutaraldehyde and then treated with detergent, after which the actin-containing filaments indeed were able to react with HMM (Ohtsuki *et al.*, 1978) or with antibody (Willingham and Yamaha, 1979). However, these methods are not necessarily transferable to other cell types: we have been unable to get them to work with *Haemanthus* endosperm—in which we know there is considerable actin (Forer *et al.*, 1979)—without extensive modification (A. Forer, W. T. Jackson, and B. Doyle, unpublished data). To start experiments along these lines, then, one can follow the general procedures in the published work, but one should not expect them necessarily to work without modification. An initial step might be to determine whether the fixation in question prevents arrowhead formation. This is easily achieved by putting actin-containing filaments on an electron-microscope grid, adding fixative, rinsing away the fixative, and then adding HMM and negatively staining, to see if arrowheads are found (e.g., Fig. 1).

Another potential artifact (in addition to relocation) is that HMM can cause polymerization of monomeric G-actin to form decorated filaments (review in Forer, 1978b). In the ideal case, in which the same numbers of filaments are seen *in situ* before and after treatment with HMM, there is no problem. If, on the other hand, decorated filaments are seen after treatment with HMM, whereas none (or far fewer) are seen prior to HMM treatment, two possibilities exist: (1) The HMM caused polymerization of filaments; or (2) the HMM stabilized filaments against the deleterious effects of the fixation-embedment procedures. Both interpretations are possible, and, in different cases, one or the other has been shown to be the case (review in Forer, 1978b). One way to minimize the ambiguity is to use the digest of HMM called subfragment 1, or S_1: S_1 does *not* cause polymerization of G-actin (review in Forer, 1978b). Another way is to use pre-fixation, as previously described.

At this point the discussion has covered fixation of filaments *in situ,* and the possible need for alternate fixations to preserve actin-containing filaments. Various ways to test whether the agent in question makes the cells permeable have been discussed, as have potential artifacts of the procedure (namely, relocation of filaments, and polymerization induced by the HMM) and possible ways to overcome them. Given that all these steps are done or considered, how does one preserve and stain the decorated filaments?

In general, decorated actin-containing filaments are not seen with clear arrowheads in sectioned material when the usual fixation (glutaraldehyde) and postfixation (osmium tetroxide) procedures are followed, with rare exceptions (e.g., Mooseker and Tilney, 1975). Arrowheads (and polarity) are very clearly seen in sections, however, when decorated filaments are fixed with a glutaraldehyde–tannic acid mixture (e.g., Flock and Cheung, 1977; Begg *et al.*, 1978; Seagull and Heath, 1979), as illustrated in Fig. 2. Hence the recommended procedure is to fix the cells with a mixture of glutaraldehyde and tannic acid rather than glutaraldehyde alone.

A final point of the procedure to be considered is this: do negative results (i.e., no detectable decorated filaments seen in the sections) mean that actin-containing filaments are absent? My answer is "no," for the reason that actin-containing filaments may not necessarily bind HMM, even if present, and even if the cells are made permeable and are free from ATP. One reason is that the filaments may be packed in such a way that the HMM is sterically hindered from reaching the binding sites: crane fly spermatids contain tightly packed bundles of actin-containing filaments that do not react with HMM *in situ* in permeable cells,

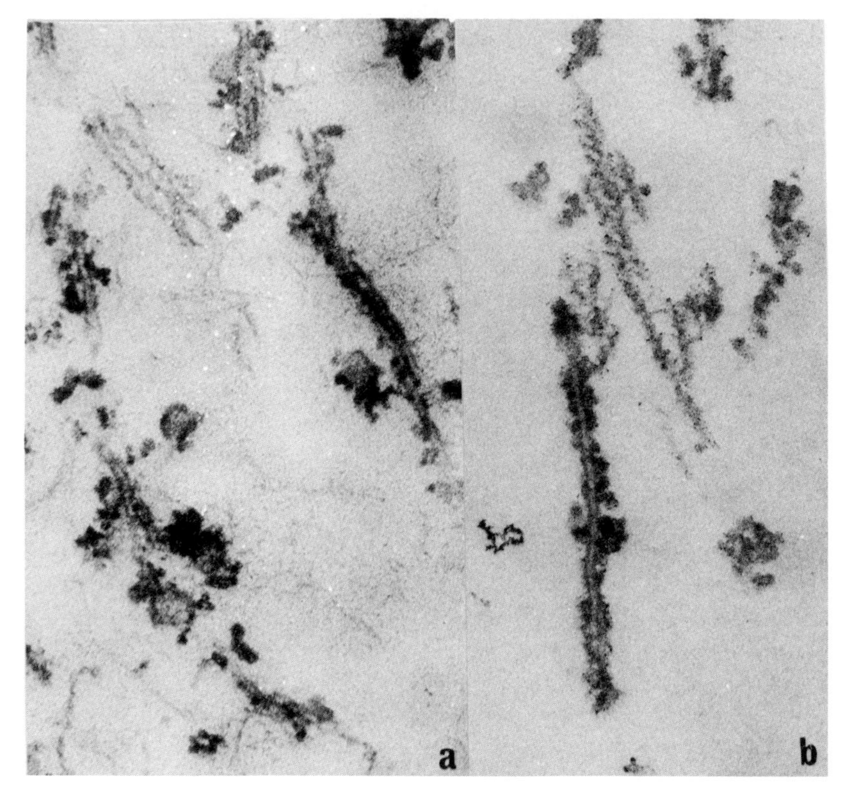

FIG. 2. Actin-containing filaments in *Haemanthus* endosperm spindles, after sectioning. HMM-treated cells were fixed using standard glutaraldehyde fixation procedures (a) described in Forer and Jackson (1979), or using glutaraldehyde–tannic acid (b), as in Begg *et al.* (1978). The polarities of the arrowheads can be seen after the tannic-acid fixation, but are only rarely seen after the glutaraldehyde fixation (see Forer *et al.*, 1979). (a) ×67,000; (b) ×81,500.

but will react with HMM after the bundles are flattened on a grid so that the filaments become separated (Forer and Behnke, 1972). Another reason is that the binding sites on the actin are already bound to myosin: in skeletal-muscle myofibrils in rigor exogenous HMM binds only to actin-containing filaments in the I band, because the myosin in the A band remains tightly bound to the actin-containing filaments and ties up all the sites to which HMM would bind. Thus negative results must be interpreted with caution.

In summary, this chapter has discussed various aspects of the procedure, from initial steps of looking at purified proteins to fixation of filaments *in situ,* in untreated cells, to observation of arrowheads after cells are treated with HMM.[1]

[1]The concentrations of HMM to use and other details of the procedures can be found in the articles cited, or in Forer (1978b).

The reaction with HMM to form arrowheads is specific for actin-containing filaments, and in theory one can determine the polarities of actin-containing filaments *in situ* to get some clues about how these filaments might be functioning. The advantage of using electron microscopy is the resolution, the identification of polarity, and the lack of ambiguity from a positive result (arrowhead formation, or decoration). There are disadvantages, however, and ambiguities in the localizations of the filaments. These were described, followed by an indication as to the kinds of control experiments that one can do to minimize these ambiguities. Perhaps the most disquieting aspect of the method is that one looks at those sites on the filaments that are *not* interacting with myosin, and, because actin in nonmuscle cells is present in large excess compared with myosin (Pollard, 1975), one may therefore be looking at actin-containing filaments that do not function in motility in that particular system.

References

Begg, D. A., Rodewald, R., and Rebhun, L. I. (1978). *J. Cell Biol.* **79,** 846-852.

Clarke, M., and Spudich, J. A. (1977). *Annu. Rev. Biochem.* **46,** 797-822.

Durica, D. S., Schloss, J. A., and Crain, W. R., Jr. (1980). *Proc. Natl. Acad. Sci. U.S.A.* **77,** 5683-5687.

Flock, Å., and Cheung, H. C. (1977). *J. Cell Biol.* **75,** 339-343.

Forer, A. (1974). *In* "Cell Cycle Controls" (G. M. Padilla, I. L. Cameron, and A. M. Zimmerman, eds.), pp. 319-336. Academic Press, New York.

Forer, A. (1978a). *In* "Nuclear Division in the Fungi" (I. B. Heath, ed.), pp. 21-88. Academic Press, New York.

Forer, A. (1978b). *In* "Principles and Techniques of Electron Microscopy" (M. A. Hayat, ed.), Vol. 9, pp. 126-174. Van Nostrand-Reinhold, Princeton, New Jersey.

Forer, A., and Behnke, O. (1972). *J. Cell Sci.* **11,** 491-519.

Forer, A., and Jackson, W. T. (1979). *J. Cell Sci.* **37,** 323-347.

Forer, A., Jackson, W. T., and Engberg, A. (1979). *J. Cell Sci.* **37,** 349-371.

Fyrberg, E. A., Kindle, K. L., Davidson, N., and Sodja, A. (1980). *Cell* **19,** 365-378.

Gadasi, H., Oplatka, A., Lamed, R., Hochberg, A., and Low, W. (1974). *Biochim. Biophys. Acta* **333,** 161-168.

Gicquaud, C., and Loranger, A. (1981). *Eur. J. Cell Biol.* **24,** 320-323.

Gicquaud, C., Druda, J., and Pollender, J. M. (1980). *Eur. J. Cell Biol.* **20,** 234-239.

Glauert, A. M. (1975). "Fixation, Dehydration and Embedding of Biological Specimens." North-Holland Publ., Amsterdam.

Goldman, R., Pollard, T., and Rosenbaum, J., eds. (1976). "Cell Motility," Vols. A, B, and C. Cold Spring Harbor Lab. Press, Cold Spring Harbor, New York.

Goldman, R. D., Milsted, A., Schloss, J. A., Starger, J., and Yerna, M.-J. (1979). *Annu. Rev. Physiol.* **41,** 703-722.

Griffith, L. M., and Pollard, T. D. (1978). *J. Cell Biol.* **78,** 958-965.

Hanson, J. (1967). *Nature (London)* **213,** 353-356.

Hauser, M. (1978). *Cytobiologie* **18,** 95-106.

Hepler, P. K., and Palevitz, B. A. (1974). *Annu. Rev. Plant Physiol.* **25,** 309-362.

Herman, I. M., and Pollard, T. D. (1978). *Exp. Cell Res.* **114,** 15-25.

Hitchcock, S. E., Carlsson, L., and Lindberg, U. (1976). *Cell* **7,** 531–542.
Horne, R. W. (1964). *In* "Techniques for Electron Microscopy" (D. Kay, ed.), pp. 328–355. Blackwell, Oxford.
Huxley, H. E. (1963). *J. Mol. Biol.* **7,** 281–308.
Huxley, H. E. (1972). *In* "The Structure and Function of Muscle" (G. H. Bourne, ed.), Vol. 1, Part 1, pp. 301–387. Academic Press, New York.
Huxley, H. E. (1973). *Nature (London)* **243,** 445–449.
Huxley, H. E., and Zubay, G. (1960). *J. Mol. Biol.* **2,** 10–18.
Ishikawa, H., Bischoff, R., and Holtzer, H. (1969). *J. Cell Biol.* **43,** 312–328.
Kaine, B. P., and Spear, B. B. (1980). *Proc. Natl. Acad. Sci. U.S.A.* **77,** 5336–5340.
Kersey, Y. M., Hepler, P. K., Palevitz, B. A., and Wessells, N. K. (1976). *Proc. Natl. Acad. Sci. U.S.A.* **73,** 165–167.
Kersey, Y. M., and Wessells, N. K. (1976). *J. Cell Biol.* **68,** 264–275.
Korn, E. D. (1978). *Proc. Natl. Acad. Sci. U.S.A.* **75,** 588–599.
Mannherz, H. G., Leigh, J. B., Leberman, R., and Pfrang, H. (1975). *FEBS Lett.* **60,** 34–38.
Maupin-Szamier, P., and Pollard, T. D. (1978). *J. Cell Biol.* **77,** 837–852.
Moore, P. B., Huxley, H. E., and DeRosier, D. J. (1970). *J. Mol. Biol.* **50,** 279–295.
Mooseker, M. S. (1976). *Cold Spring Harbor Conf. Cell Proliferation* **3** [Book B], 631–650.
Mooseker, M. S., and Tilney, L. G. (1975). *J. Cell Biol.* **67,** 725–743.
Ohtsuki, I., Manzi, R. M., Palade, G. E., and Jamieson, J. D. (1978). *Biol. Cell.* **31,** 119–126.
Page, S. (1964). *Proc. R. Soc. London, Ser. B* **160,** 460–466.
Page, S. G., and Huxley, H. E. (1963). *J. Cell Biol.* **19,** 369–390.
Palevitz, B. A., and Hepler, P. K. (1975). *J. Cell Biol.* **65,** 29–38.
Pollard, T. D. (1975). *In* "Molecules and Cell Movement" (S. Inoué and R. E. Stephens, eds.), pp. 259–285. Raven, New York.
Pollard, T. D. (1976). *J. Cell Biol.* **68,** 579–601.
Pollard, T. D., and Weihing, R. R. (1974). *CRC Crit. Rev. Biochem.* **2,** 1–65.
Poo, W.-J., and Hartshorne, D. J. (1976). *Biochem. Biophys. Res. Commun.* **70,** 406–412.
Próchniewicz, E. (1979). *Biochim. Biophys. Acta* **579,** 346–358.
Raju, T. R., Stewart, M., and Buckley, I. K. (1978). *Cytobiologie* **17,** 307–311.
Schloss, J. A., Milsted, A., and Goldman, R. D. (1977). *J. Cell Biol.* **74,** 795–815.
Seagull, R. W., and Heath, I. B. (1979). *Eur. J. Cell Biol.* **20,** 184–188.
Taylor, D. L., and Condeelis, J. S. (1979). *Int. Rev. Cytol.* **56,** 57–144.
Taylor, D. L., and Wang, Y.-L. (1978). *Proc. Natl. Acad. Sci. U.S.A.* **75,** 857–861.
Tilney, L. G. (1975). *J. Cell Biol.* **64,** 289–310.
Tobin, S. L., Zulauf, E., Sánchez, F., Craig, E. A., and McCarthy, B. J. (1980). *Cell* **19,** 121–131.
Vandekerckhove, J., and Weber, K. (1979). *Differentiation* **14,** 123–133.
Wang, Y.-L., and Taylor, D. L. (1979). *J. Cell Biol.* **82,** 672–679.
Willingham, M. C., and Yamada, S. S. (1979). *J. Histochem. Cytochem.* **27,** 947–960.

Chapter 8

The Brush Border of Intestinal Epithelium: A Model System for Analysis of Cell-Surface Architecture and Motility

MARK S. MOOSEKER AND CHRISTINE L. HOWE

Department of Biology
Yale University
New Haven, Connecticut

ISBN 0-12-564125-7

I. Introduction

The apical surface of absorptive epithelial cells, such as those lining the lumen of the intestine or proximal tubule of the kidney, consists of a tightly packed array of microvilli variously referred to as the striated, microvillous, or brush border. This distinctive cell surface was first described by Henle (1841) as a "free border" on the apical surface of intestinal epithelial cells made up of a clear layer without structure. Subsequent studies revealed the striated nature of this surface, and by the 1940s there were at least two theories regarding the structure and function of the brush border based on high-resolution light-microscopic investigations. The first viewpoint suggested that the striated border consisted of rodlets or cilia-like structures; the second notion was that this surface contained "pore-canals" through which nutrients were absorbed [see Baker (1942) for a review of the early literature]. The advent of electron microscopy in the late 1940s settled the controversy in favor of the first alternative (Granger and Baker, 1949, 1950), but functionally, at least, the second alternative is also correct with regard to the numerous absorptive functions of the brush border. It has taken over 25 years of further investigation, facilitated by numerous advances in fixation procedures, to elucidate the structural organization of this highly complex and ordered cell surface. Some of the ultrastructural studies that contributed to an understanding of the structural organization of the brush border include the following: Dalton (1951), Zetterqvist (1956), Palay and Karlin (1959), Clark (1959), Brown (1962), Millington and Finean (1962), Overton and Shoup (1964), McNabb and Sandborn (1964), Overton (1965), Overton *et al.* (1965), Mukherjee and Wynn Williams (1967), Boyd and Parsons (1969), Brunser and Luft (1970), and Mukherjee and Staehelin (1971). The main features of this organization for the brush border of intestinal epithelial cells are depicted in Fig. 1. Although this diagram and the discussion that follows specifically refer to the brush border of epithelial cells in the vertebrate small intestine, many of the structural features, particularly with respect to cytoskeletal contractile elements, are present in other brush-border surfaces as well.

The outer surface of the brush-border membrane is covered by a polysaccharide coat, termed the glycocalyx (Bennett, 1963), which is a secretion product of the underlying cell (Ito, 1961, 1965, 1969, 1974; Ito and Revel, 1964; Rambourg and Leblonde, 1967). The extent and structural complexity of this surface coat is not readily apparent in standard EM preparations (for example, see Fig. 2), but can be readily observed in tissues stained specifically for carbohydrate (see Ito, 1969; Ito and Revel, 1964), or in freeze-etch preparations (Swift and Mukherjee, 1976). Various functions for the glycocalyx have been suggested. It may provide mechanical protection and/or lubricate the lumenal surface of the intestine, and perhaps serve as a filter or diffusion barrier to gut contents of a

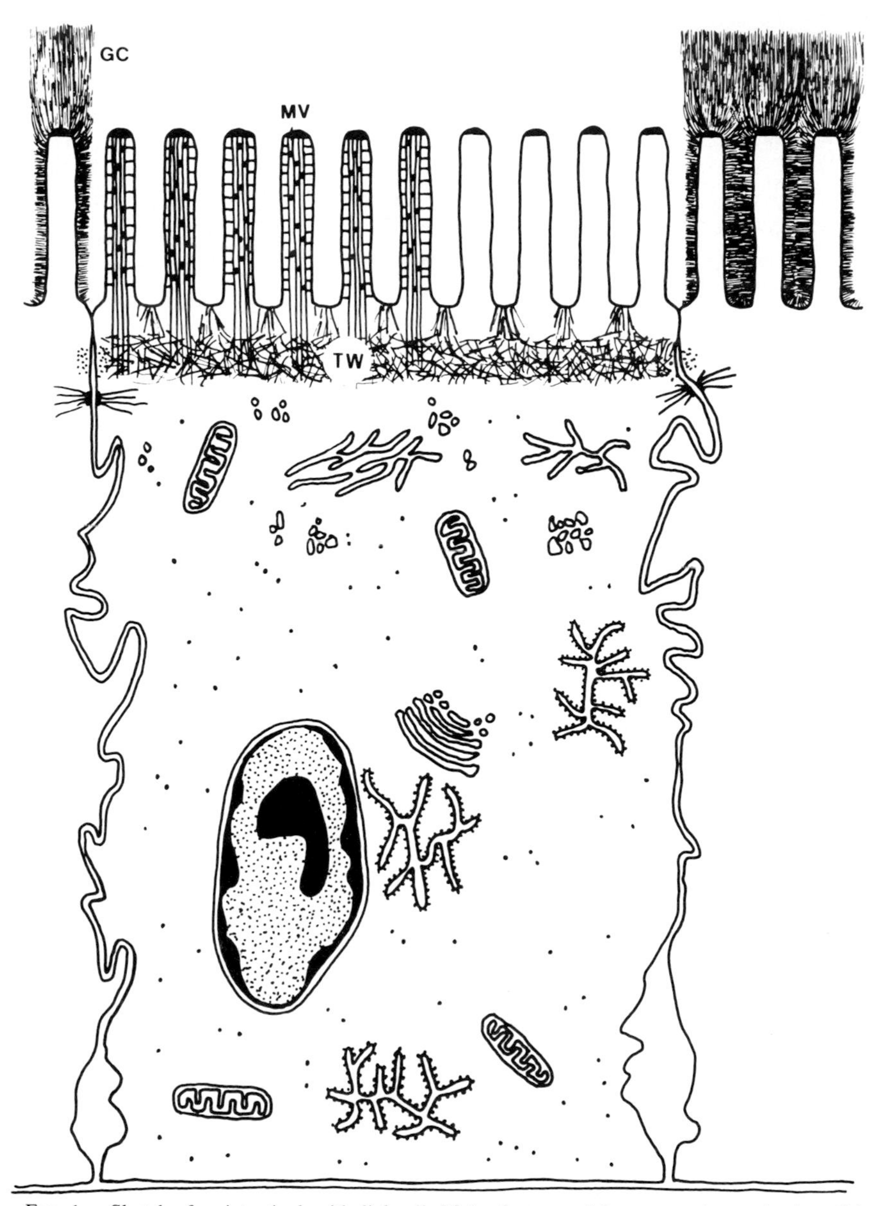

FIG. 1. Sketch of an intestinal epithelial cell. Major features of the structural organization of the brush border are depicted. GC, Glycocalyx; MV, microvillus; TW, terminal web.

certain size or type. A number of hydrolytic enzymes are adsorbed to the glycocalyx (Rambourg, 1971), thus localizing the action of these enzymes close to the site of product absorption by the brush-border membrane.

The brush-border membrane contains a variety of membrane proteins, most of which are enzymes involved in the processing and transport of the nutrients absorbed by the small intestine. Examples include disaccharidases, peptidases, lipases, and nucleotidases (for reviews, see Sacktor, 1977a,b). These enzymes are restricted for the most part to the apical, microvillous portion of the plasma membrane, and the basal-lateral membranes are characterized by their own set of enzymes (e.g., Na-K ATPase). Presumably this lateral asymmetry of the plasma membrane is maintained by the band of intercellular junctions at the apical end of the cell (see Pisam and Ripoche, 1976). First described in detail by Farquhar and Palade (1963), the junctional complex of the intestinal epithelial cell consists of the *zonula occludens* (tight junction), *zonula adherens,* and desmosomes or *macula adherens* (for recent reviews on the junctional complex, see Staehelin, 1974; Staehelin and Hull, 1978; Hull and Staehelin, 1979). In addition to the well-characterized roles of the junctional complex in providing a permeability barrier and maintaining mechanical integrity of the epithelium, the junctions of the intestinal epithelial cell may also serve as important organizing centers for the cytoskeletal/contractile elements of the brush border (see later).

The most striking ultrastructural feature of the brush border is, of course, the extensive and highly ordered array of ''fibrous'' elements that comprise the so-called cytoskeleton of this cell surface. The microvilli of the brush border contain, and presumably are supported by, a bundle of thin filaments that have been identified as actin containing[1] (Ishikawa *et al.*, 1969; Tilney and Mooseker, 1971). The distal tip of the microvillar-filament bundle is embedded in a dense, osmophilic plaque that may serve to attach the bundle to the cytoplasmic surface of the plasma membrane (see Fig. 2). The microvillus core is also connected to the membrane along its length by ''cross-filaments'' that span the distance between the filament bundle and the membrane. These cross-filaments were first detected by Millington and Finean (1962), and more convincing evidence for their existence was presented by Mukherjee and Staehelin (1971), who observed these structures in freeze-etch preparations of intestinal epithelium. By treating isolated brush borders with high Mg^{2+} concentrations (15 m*M*), we were able to demonstrate that these cross-filaments are quite numerous, and are bound to the core at periodic intervals (33 nm) along its length (Fig. 2b; Mooseker and Tilney, 1975). Subsequent work by Matsudaira and Burgess (1979) using the tannic-

[1]Although it has been convincingly shown that alpha-actinin is present in the terminal-web region of intact cells (Bretscher and Weber, 1978b; Craig and Pardo, 1979; Geiger *et al.*, 1979), this protein is usually lost from the brush border during isolation (Craig and Lancashire, 1980; Mooseker and Stephens, 1980).

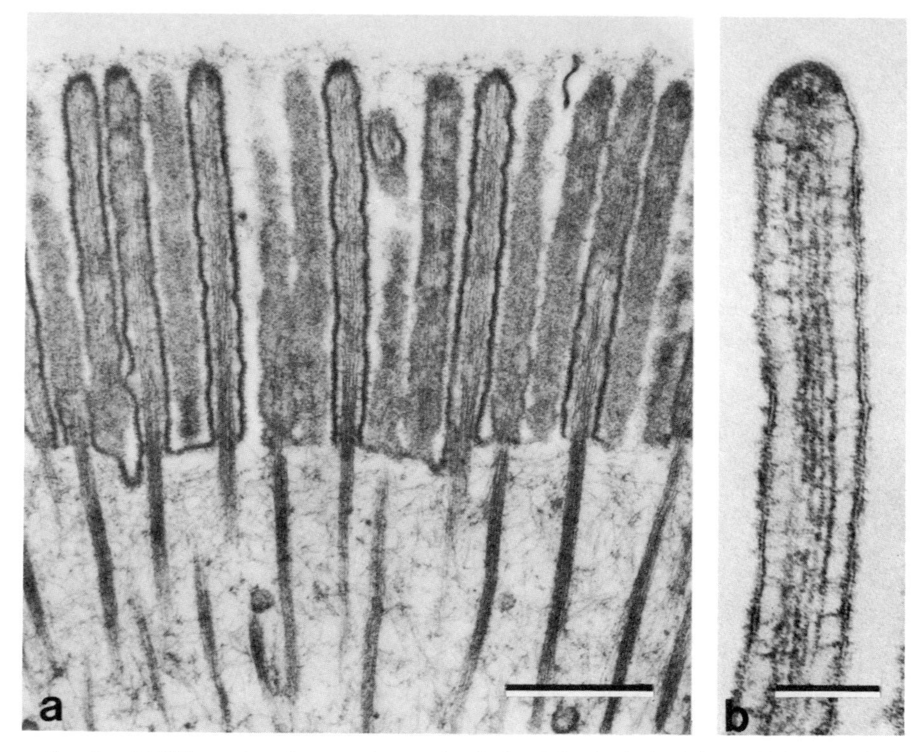

FIG. 2. (a) Thin-section electron micrograph of a brush border isolated from chicken intestinal epithelium. Bar = 0.5 μm. From Mooseker and Tilney (1975). ×38,000. (b) Microvillus of an isolated brush border treated with buffer containing 15 mM Mg^{2+}. Note the periodic cross-filaments attaching the microvillar actin-filament bundle to the plasma membrane. Bar = 0.1 μm. (From Mooseker and Tilney, 1975; reproduced with permission of the *Journal of Cell Biology*.) ×132,000.

acid-fixation procedure of Begg *et al.* (1978) has revealed that the cross-filaments are attached in a helical fashion along the length of the microvillar-filament bundle. Bridges of similar dimensions have been observed within the bundle connecting the core filaments to each other (Mukherjee and Staehelin, 1971; Mooseker and Tilney, 1975).

The basal ends of the microvillus cores descend into the apical cytoplasm of the cell, a region usually referred to as the terminal web. It is of interest to note that the lateral projections connecting the core to the microvillar membrane are not present on this "rootlet" portion of microvillus core. The terminal web is a complicated meshwork of filaments of several different types; unfortunately, we do not yet have a detailed understanding of its organization and, in particular, the chemical composition of its constituent filaments. Nevertheless, the recent studies of Hull and Staehelin (1979) using the high-voltage EM have helped a great deal in defining the basic structural, if not chemical, organization of the terminal web. It is clear from their work and that of many other investigators, that the junctional complexes associated with the lateral margins of the cell play important roles as organizing centers for many of the filaments that make up the terminal-web meshwork. In at least some types of absorptive epithelial cells, there are microfilaments associated with the zonula occludens (Hull and Staehelin, 1979). The zonula adherens has associated with it a thick bundle of actin filaments that runs circumferentially around the cell, much like the contractile ring of filaments in dividing cells. In addition to this belt of filaments, there are thin filaments that extend away from the zonula adherens into the cell interior (Rodewald *et al.*, 1976; Hull and Staehelin, 1979). These filaments presumably are among those that interlace with the basal ends of the microvillar filaments. In some cell types, another source of these interdigitating filaments is those emanating from the apical membrane at points between adjacent microvilli (see Fig. 1). Below the zonula adherens lies the macula adherens, and the desmosomes of this junctional region serve as nucleating centers for numerous intermediate filaments, which extend throughout the basal region of the terminal web, forming a dense mat of filaments just below the ends of the microvillus cores. (Most of these filaments are lost during brush-border isolation.)

In addition to the filament types discussed above, the terminal-web region contains a high concentration of myosin, which may or may not be in the form of bipolar, thick filaments *in vivo* (Mooseker and Tilney, 1975; Mooseker, 1976b; Mooseker *et al.*, 1978; Bretscher and Weber, 1978b; Drenckhahn and Gröschel-Stewart, 1980). We hedge on this point because even though we have observed short bipolar filaments interdigitating with the basal ends of microvillus cores (Mooseker and Tilney, 1975), and also in even greater numbers at the lateral margins of the cell (M. S. Mooseker, unpublished observations), such filaments have only been observed in isolated brush borders incubated with relatively high concentrations of Mg^{2+} (>5 mM). Such conditions may favor the preservation of thick filaments, but they also might induce the formation of such filaments. Thus we cannot assign an organizational state to the large amounts of myosin found in the terminal web of the brush border.

Begg *et al.* (1978) have identified a fourth filament type in the terminal web: very thin filaments (2–3 nm) that, like the myosin filaments, appear to be associated with the basal ends of the microvillus cores. Once again, the existence of such filaments *in vivo* has not been established, as they have only been observed in isolated brush borders treated with the myosin fragment S_1. Although the chemical identity of these filaments is not known, they are not actin, since they were not decorated with S_1.

Although it is clear from the above description of brush-border morphology that the interaction of the cytoskeletal components with the plasma membrane is a key organizational feature of this cell surface, it is also clear that the structural integrity of cytoskeletal/contractile apparatus is not dependent on these interac-

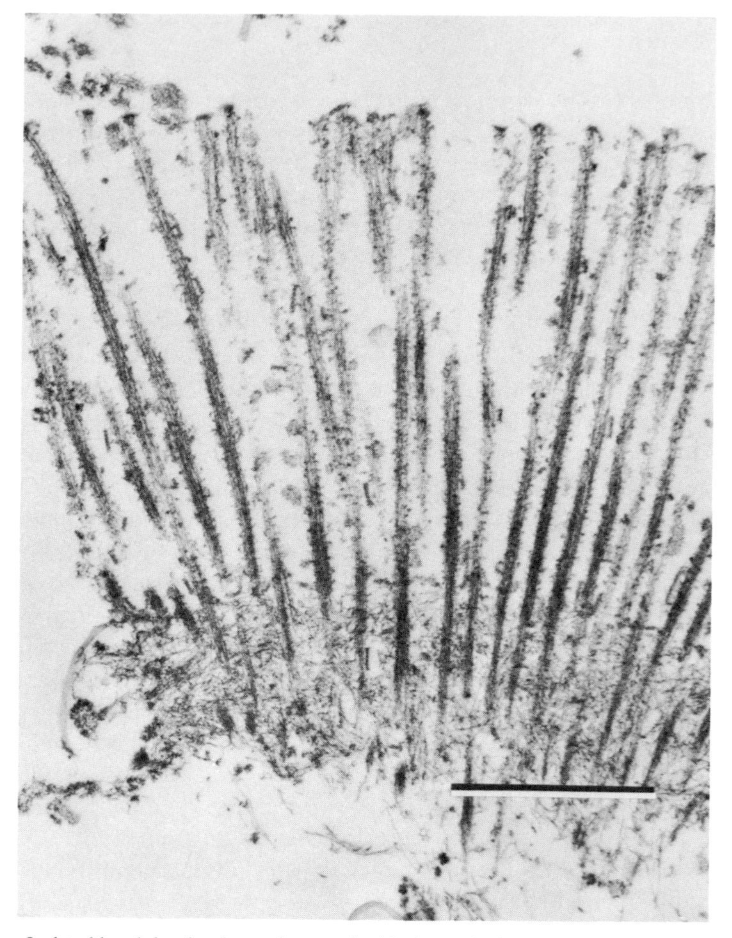

FIG. 3. Isolated brush border demembranated with the nonionic detergent, Triton X-100. Bar = 1.0 μm. (From Mooseker, 1976a; reproduced with permission of the *Journal of Cell Biology*.) $\times$27,000.

tions. Treatment of isolated brush borders with nonionic detergents can be used to remove the plasma membrane without perturbing this remarkable array of filaments (Fig. 3; Mooseker and Tilney, 1975).

The brush border has been a useful model for investigating the organization and function of contractile proteins in nonmuscle cells. This is due in part to its exquisite structural design, but more importantly, because it can be isolated in a functionally intact form in quantities sufficient for coordinated biochemical and cytological analysis. The rest of this chapter is devoted to the elucidation of techniques and approaches we and other laboratories have used for probing the organization, chemistry, and motility of the brush border.

II. Brush-Border Isolation

Procedures for the isolation of brush borders were first developed by Miller and Crane (1961) almost 20 years ago, and the subsequent modification of their methods by Forstner *et al.* (1968) has provided the basis for isolation procedures currently in use in most laboratories. The methods used in our laboratory for the isolation of brush borders from chicken intestine are outlined in Fig. 4, and are discussed in detail below.

For most of the cytological and biochemical studies conducted in our laboratory, the chicken has been the organism of choice. Chickens are inexpensive, the microvilli of their brush borders are quite long (up to 4 μm), and preparations of sufficient size for biochemical procedures can be obtained. Nevertheless, we have used the procedures outlined below to isolate intestinal brush borders from a variety of organisms including monkey, dog, rabbit, mouse, rat, hamster, guinea pig, cow, pig, frog, and sea robin (a teleost fish). Slightly modified procedures have been used for isolation of brush borders from the spiral valve of the elasmobranchs, dogfish and skate.

In brief, brush borders are isolated from dissociated intestinal epithelial cells by homogenization and differentical centrifugation. For cytological experiments, one chicken is used, and for biochemical studies, 5–10 animals are used. All the procedures outlined below are conducted at 0–4°C. Chickens are sacrificed by decapitation, and the proximal one third to one half of the small intestine is removed, cut into 20- to 30-cm segments, and immersed in cold saline (0.15 *M* NaCl–4 m*M* NaN_3–10 m*M* imidazole Cl, pH 7.4). The segments are flushed with cold saline using sufficient pressure to ensure thorough discharge of gut contents. For convenience, we use saline dispensed from a 10-liter bottle at a pressure head of about 1.5 m. Epithelial cells are dissociated from the intestines by a modification of the method of Evans *et al.* (1971). The cleaned segments are filled, sausage-fashion, using a 50-cc syringe, with cell-dissociation medium consisting of 0.20 *M* sucrose–0.076 *M* Na_2HPO_4–0.019 *M* KH_2PO_4–12 m*M* ethlenediamine tetraacetic acid (EDTA). Soft-fiber packing string is used to tie off the ends. The filled segments are massaged vigorously between the thumb and two fingers to facilitate cell dissociation. The dissociation process is monitored by holding the segments up to a bright light; the segments become almost transparent when dissociation is complete. The time required for this process is quite variable (10–30 minutes), but in general, the older the bird, the longer the time required. It is wise to avoid mature roosters, as they are difficult to handle, and their intestinal epithelium is resistant to dissociation. When the dissociation process is complete the segments are drained and flushed with 30–50 ml of dissociation medium containing 0.1 m*M* phenylmethylsulfonyl fluoride (PMSF). The effluents are pooled and the epithelial cells collected by low-speed

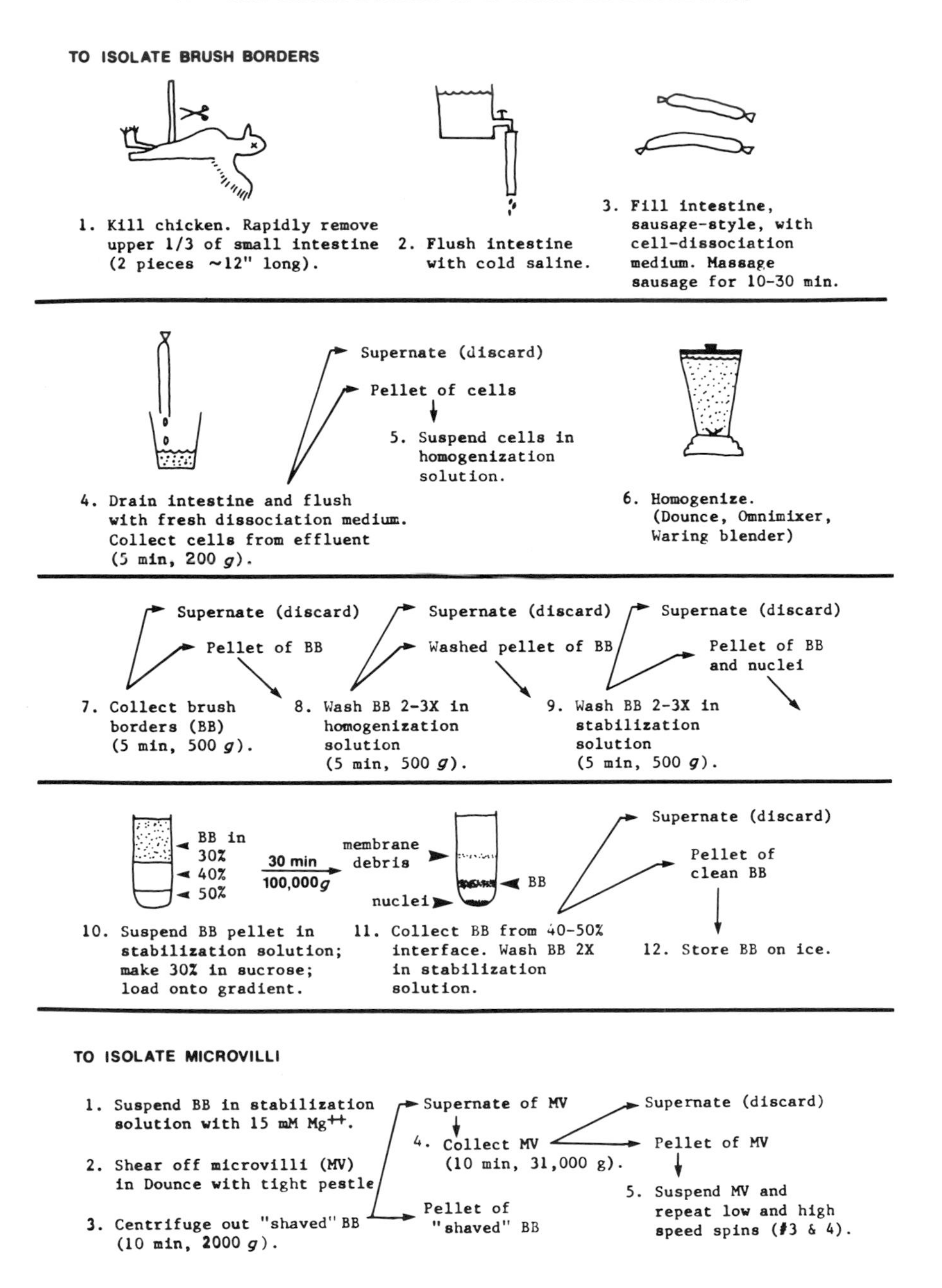

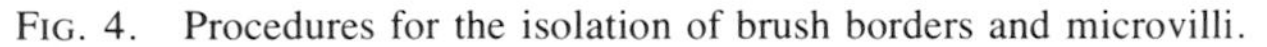

FIG. 4. Procedures for the isolation of brush borders and microvilli.

centrifugation (200 *g* for 5 minutes). The pellets are suspended in 15–30 volumes of homogenization buffer consisting of 4 m*M* EDTA–1 m*M* (EGTA)–10 m*M* imidazole Cl (pH 7.0)–0.1 m*M* dithiothreitol (DTT)–0.1 m*M* PMSF—4 m*M* NaN_3. A variety of homogenization procedures can be used depending on the size of the preparation. For relatively small volumes, a Dounce homogenizer fitted with a tight pestle works well (20–30 strokes), and for larger preparations a blender or other mechanical homogenizer such as an Omnimixer (Dupont Instr., Norwalk, Connecticut) can be used. Generally, 10–20 seconds at three-quarter speed, delivered in 1- to 2-second bursts to avoid heating, in either a blender or Omnimixer will be sufficient to produce free brush borders such as those de-

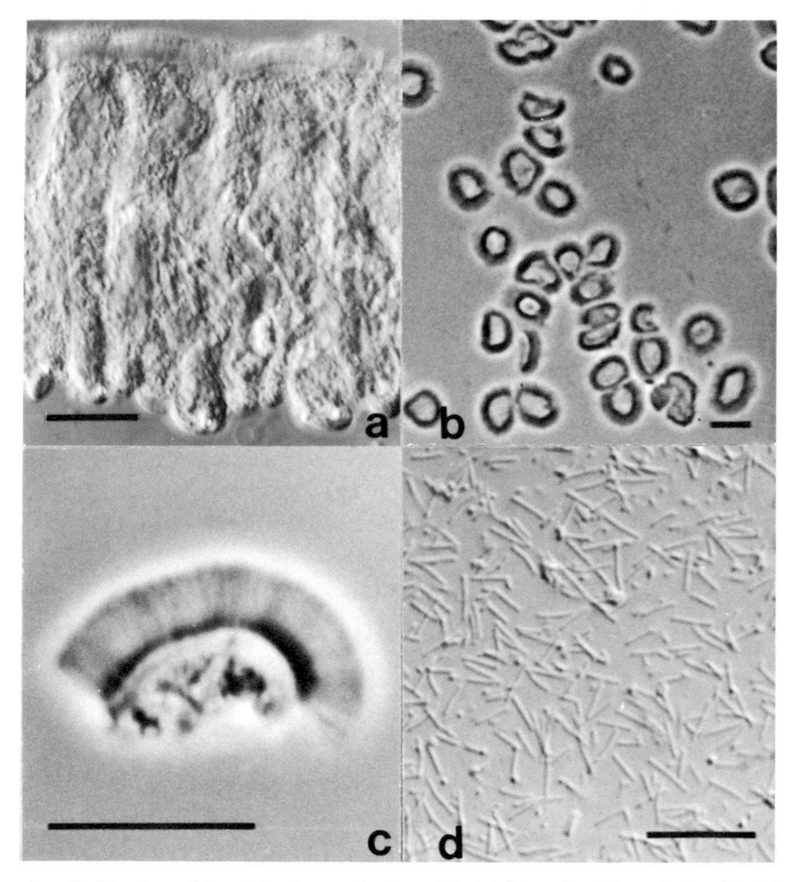

FIG. 5. Purification of brush borders and microvilli. (a) Dissociated intestinal epithelial cells. Differential-interference contrast light micrograph (LM). ×1165. (b) Isolated brush borders after sucrose-gradient purification. Phase-contrast LM. (From Mooseker *et al.*, 1978; reproduced with permission of the *Journal of Cell Biology*.) ×500; (c) High-magnification LM of an isolated brush border. ×2700; (d) Isolated microvilli. Bars (a–d) = 10 μm.

picted in Fig. 5. However, it is imperative that the progress of homogenization be monitored by light microscopy to avoid overhomogenization—a process that results in "frazzled" brush borders that are disheveled in appearance because many of their microvilli have been sheared away during homogenization. The brush borders and contaminating nuclei are collected by centrifugation at 500 *g* for 5 minutes. Often a layer of froth will float to the surface of the supernate. This froth layer contains many brush borders, thus care should be taken not to lose it when the supernate is decanted (and discarded). The pellets (and accompanying froth layer, if present) are resuspended in homogenization buffer and once again collected by centrifugation at low speed. This washing procedure is repeated two or three times and then the brush borders are suspended in stabilization buffer (BBSB) consisting of 75 m*M* KCl-5 m*M* $MgSO_4$-1 m*M* EGTA-4 m*M* NaN_3-0.1 m*M* DTT-0.1 m*M* PMSF-10 m*M* imidazole Cl (pH 7.0). The brush borders are washed twice in BBSB using the same centrifugation procedure outlined above. Contaminating nuclei are removed from the preparation by centrifugation on a sucrose step gradient consisting of 50% (w/w), 40%, and 30% sucrose. The sucrose solutions are prepared by mixing the appropriate amount of sugar (60%—772 gm/liter; 50%—614 gm/liter; 40%—470 gm/liter) with 500 ml/liter double-strength BBSB, and then adding H_2O to 1 liter. The brush borders are suspended in BBSB and an equal volume of 60% sucrose added. This suspension is layered onto the 50%-40% step gradient. The gradients (6 × 34 ml) are spun for 30-45 minutes at 26,000 rpm in an SW-27 rotor (Beckman Instruments, Inc., Palo Alto, California 94304). A vertical rotor can be used to reduce centrifugation time (Matsudaira and Burgess, 1979). The purified brush borders collect at the 40%/50% interface and the nuclei form a pellet at the bottom of the tube. The brush borders are collected from the gradient using a syringe fitted with a long, 18 g pipetting needle. The brush borders are diluted with three volumes of BBSB, collected by centrifugation (5000 *g* for 10 minutes), then washed one or two additional times (1000 *g* for 5 minutes) with BBSB to remove residual sucrose. At this point the purified brush borders are stored as dry pellets on ice until use. Although proteolysis of constituent proteins does occur, most brush-border preparations can be used for several days after isolation before proteolysis becomes a severe problem. If one works quickly, the above procedure can be completed in 4 hours, with an average yield of 60-100 mg total brush-border protein per chicken.

III. Microvillus Isolation

Microvilli can be separated from the brush border by vigorous homogenization, as first shown by Booth and Kenny (1974) for kidney brush borders. The procedures used in our laboratory for the isolation of microvilli (Howe *et al.*, 1980) are modified from those of Bretscher and Weber (1978a). Purified brush borders are suspended in five volumes of BBSB and the Mg^{2+} concentration is increased to 15 m*M* by addition of 1 *M* $MgSO_4$. The increased Mg^{2+} concentration somehow improves the yield of microvilli, probably by increasing the rigidity of the microvillus core and/or membrane. Microvilli are freed from the brush borders by homogenization in a Dounce homogenizer, fitted with a tight pestle. Homogenization is continued until the brush borders begin to fragment (30–50 strokes). As is the case for brush-border isolation, it is important to monitor the progress of homogenization by light microscopy; the freed microvilli can be observed by either phase or differential-interference contrast using a 100× oil-immersion objective (Fig. 5). The "shaved" brush borders are removed by centrifugation at 3000 *g* for 10 minutes and the microvilli are collected from the supernate at 35,000 *g* for 15 minutes. The microvillus pellets are resuspended in BBSB and the above centrifugation procedure is repeated to remove contaminating brush-border fragments.

IV. Preparation of Demembranated Brush Borders and Microvillus Cores

The plasma membrane can be removed from the underlying cytoskeletal elements of the brush border and microvillus by treatment with nonionic detergents. Isolated brush borders or microvilli are suspended in five volumes of BBSB containing 1–2% Triton X-100 or Nonidet P-40 (Mooseker and Tilney, 1975; Howe *et al.*, 1980). Demembranated brush borders are collected by centrifugation at 2000 *g* for 10 minutes, microvillus cores are pelleted at 40,000 *g* for 15 minutes. To ensure complete membrane removal, two or three washes with detergent solution are required, as determined by both biochemical and morphological criteria.

V. Characterization of Brush-Border Motility

A. *In Vivo* Studies of Microvillar Movements

Although the brush border may be one of the best systems available to study the organization of contractile proteins in nonmuscle cells, the system does possess a rather large (at least a size 13) Achilles' heel. That is, we do not know how microvilli move *in vivo,* and in fact, the evidence that any such movements actually occur is somewhat slim. There are only two reports in the literature describing observations of microvillar motility, and unfortunately neither set of observations provides adequate documentation for the movement of microvilli *in vivo*. In an abstract submitted over a decade ago, Thuneberg and Rostgaard (1969) outline observations of microvillar motility in epithelial cells of both the small intestine and kidney proximal tubule, using cell preparations derived from mucosal scrapings of the intestine and minced kidney tissue, respectively. The experimental conditions were not defined in these abstracts, but the authors did report that movements were optimally maintained in a medium buffered to pH 7.1–7.3, containing high K^+ and low Na^{2+}, 2 m*M* cysteine, 10 m*M* glucose, and 1 m*M* EDTA or EGTA. Although the exact nature of the movements was not discernible, there was a microvillar "beat" frequency of 10–20 hertz, as determined by stroboscopic illumination; in addition, cooling of the preparation resulted in a cessation of movment.

A second set of observations has been reported by Sandström (1971), in which he has *detected,* but not resolved movements of microvilli on intestinal epithelia in culture. Unfortunately, the documentation provided is not convincing, but admittedly such observations are much easier to make visually than they are to record photographically, particularly under the suboptimal optical conditions of standard tissue culture preparations. Nevertheless, neither we nor any other laboratory have been able to observe or record examples of microvillar movements on intact epithelial cells, despite considerable effort to that end. Thus the cloud of doubt regarding the existence of such movements remains.

B. Brush-Border Contraction *in Vitro*

Although the evidence for microvillar movements *in vivo* is scant, studies on the contractility of isolated brush borders certainly establish the potential for movements of some kind, and also provide insight into the chemical and structural basis for such movements. Rodewald *et al.* (1976) have studied the effects of "contraction" solutions on brush borders isolated from the small intestines of neonatal rats. They observed that the addition of ATP in the presence of either Mg^{2+} or Ca^{2+} resulted in a contraction of the terminal web (see Fig. 6), but neither movements nor detectable shortening of microvilli was observed. The authors proposed that the contraction of the terminal web was caused by sarcomere-like myosin interactions with the basal ends of the microvillar core filaments and the actin filaments extending from the zonula adherens. It is also possible that this contraction is due to a constriction of the circumferential bundle of actin filaments associated with the zonula adherens (Hull and Staehelin, 1979).

We have observed a second form of contractility is isolated brush borders, which, unlike the terminal-web contraction described above, is Ca^{2+} dependent

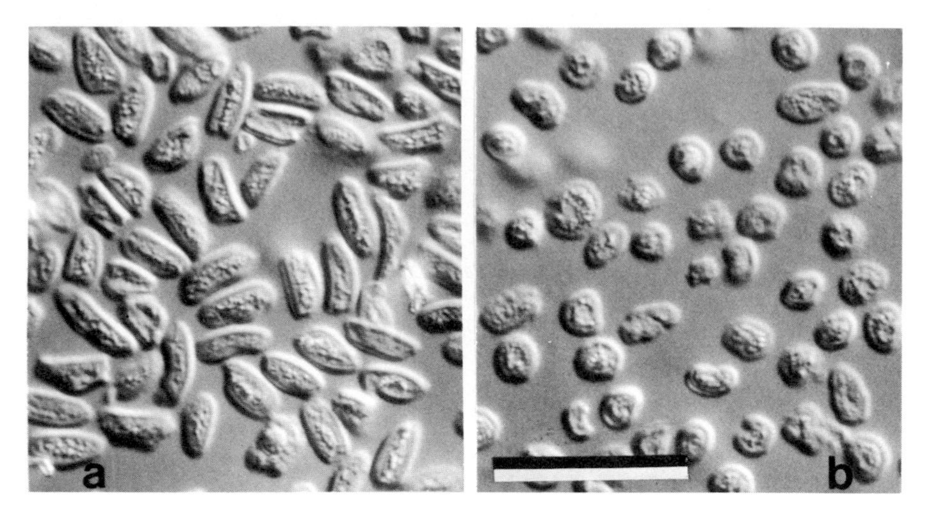

FIG. 6. Terminal-web contraction in brush borders isolated from intestinal epithelium of neonatal rats. Brush borders before (a) and after (b) addition of ATP. In the presence of divalent cations, ATP causes a contraction of the terminal-web region, but no movements or shortening of microvilli are observed. Micrographs courtesy of R. Rodewald. Bar = 50 μm. (From Rodewald *et al.*, 1976; reproduced with permission of the *Journal of Cell Biology*.) $\times$500.

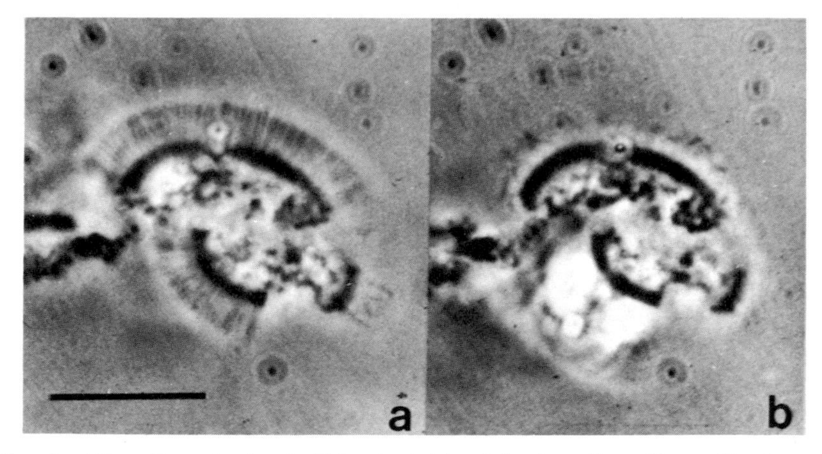

FIG. 7. Microvillar retraction in Triton-treated brush borders. From 16-mm film frames of detergent-treated brush borders before (a) and after (b) perfusion with contraction solution containing ATP and Ca^{2+}. Microvilli have contracted into the terminal-web region. Time elapsed between (a) and (b) is 3 seconds. Bar = 10 μm. (From Mooseker, 1976b). $\times$2000.

FIG. 8. (*Opposite*) Reversible effects of Ca^{2+} on microvillus core structure. Thin-section electron micrographs of 100,000-*g* pellets of isolated microvillus cores (a) in the absence of Ca^{2+}; (b) 5×10^{-6} *M* Ca^{2+}; and (c) after removal of Ca^{2+} by addition of EGTA before centrifugation. Ca^{2+} treatment causes a complete disruption of actin bundle and filament structure that is partially reversed after removal of Ca^{2+}. (From Mooseker *et al.*, 1980; reproduced with permission of the *Journal of Cell Biology*.) $\times$65,000.

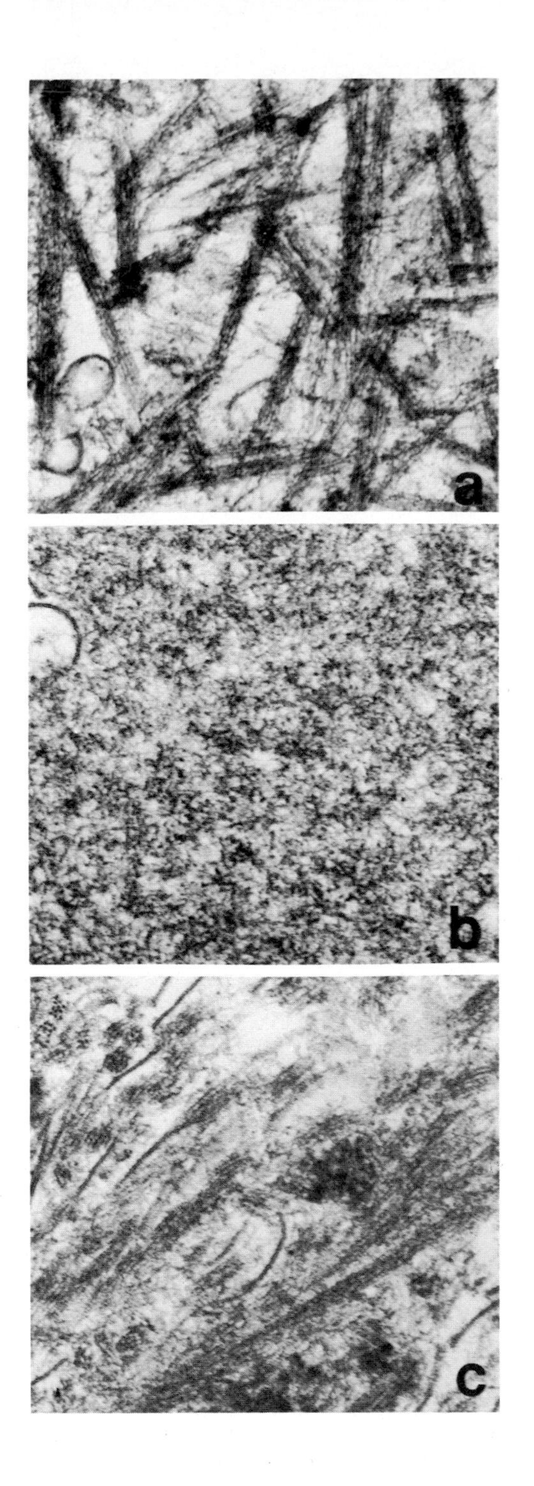
a
b
c

and does involve the microvilli. Addition of ATP and Ca^{2+} ($> 10^{-6}M$) to brush borders treated with Triton X-100 (0.1–1.0% in BBSB) causes a rapid irreversible retraction of the microvillus cores into the terminal-web region (Fig. 7). Ultrastructural studies of contracted brush borders indicate that there is a drastic loss of structural organization, which we had previously assumed was due to the unrestricted interaction of brush-border myosin with the actin filaments of the microvilli and terminal web, analogous to the loss of sarcomere organization in the contraction of glycerinated myofibrils of skeletal muscle (see Bendall, 1969). However, recent studies in our laboratory indicate that Ca^{2+} also causes a disruption of microvillus core structure (Howe *et al.*, 1980; Mooseker *et al.*, 1980; see below), which occurs independently of core retraction. Consequently, we are currently reexamining brush-border contraction *in vitro* in order to investigate the possible interrelationships between these two Ca^{2+}-dependent effects on brush-border structure and contractility. One approach we are pursuing is the study of brush-border contractility in phalloidin-treated brush borders. This actin-stabilizing drug (see Wieland, 1977) blocks the Ca^{2+}-dependent solation of the microvillus core (Howe *et al.*, 1980), but does not inhibit brush-border contraction (M. S. Mooseker and C. L. Howe, unpublished observations).

C. Ca^{2+}-Dependent Solation of the Microvillus Core

The addition of Ca^{2+} ($> 10^{-6}M$) to isolated, demembranated microvillus cores causes a rapid disruption of the microvillar filament bundle that is at least partially reversible if Ca^{2+} is then removed by chelation with EGTA (Howe *et al.*, 1980; Mooseker *et al.*, 1980; Fig. 8). This disruption of the microvillus core results from a solation of core filaments that is mediated by one of the major proteins of the microvillus core, the 95,000-dalton subunit (see later). The significance of this *in vitro* solation of microvillar actin filaments with respect to brush-border function *in vivo* is currently under investigation.

VI. Ultrastructural Characterization of the Brush-Border Contractile Apparatus

A. Fixation Conditions for Preservation of Actin Filaments

There has been considerable debate over the structural organization of actin-containing filaments in nonmuscle cells based on images derived from fixed and embedded tissues. The problem, stated simply, is that actin filaments are difficult to fix, even using conditions that generally satisfy the morphologist's gestalt impression of what constitutes good preservation of cytoplasmic structure. The term *microfilamentous networks* has often been used to describe arrays of filaments, generally observed in the cortical cytoplasm of cells. Do such networks exist or are they the result of the destructive effects of chemical fixatives on what had been long, straight actin filaments in, perhaps, more highly organized arrays? On the side of artifact are studies such as those of Maupin-Szamier and Pollard (1978), which provide convincing and quantitative evidence for the disruptive effects of OsO_4 on actin-filament structure. In addition, there are many examples, including the brush border, in which one can observe either microfilamentous networks or highly organized bundles of actin filaments, depending on the fixation conditions used. On the other hand, there are actin-binding proteins such as the 95,000-dalton subunit of the microvillus core (see later), which when added to pure F-actin can convert these filaments into networks closely resembling the arrays observed in thin sections of intact cells (see later). These results support the notion that filamentous networks do represent a functional state of actin in the cell.

The conditions we have found satisfactory for preservation of actin filaments in the brush border are based on a protocol developed by Gibbons and Gibbons (see Gibbons, 1975) for the fixation of so-called rigor waves in reactivated axonemes of sea urchin sperm. Recently we have included the modification of Begg *et al.* (1978), which calls for the addition of tannic acid in the glutaraldehyde-fixation step. The protocol is outlined here.

Pellets of brush borders or microvilli are suspended in 5–10 volumes of cold 2.0% glutaraldehyde (Electron Microscopy Sciences, Fort Washington, Pennsylvania 19034), 0.2% tannic acid, and 0.1 *M* sodium phosphate buffer (pH 7.0), and then incubated on ice for 45 minutes. The brush borders or microvilli are collected by centrifugation, washed three times with 0.1 *M* sodium phosphate buffer (pH 7.0), and then fixed as pellets in 4–5 volumes of 1.0% OsO_4–0.1 *M* sodium phosphate buffer, pH 6.0. After 45 minutes' incubation on ice, the samples are washed two or three times with cold water, then *en bloc* stained with 0.5% aqueous uranyl acetate. Standard dehydration and embedding procedures are followed.

B. Determination of Actin-Filament Polarity in the Microvilli of the Brush Border

The above procedure is also excellent for determination of actin-filament polarity by "decoration" with the myosin head fragments heavy meromyosin (HMM) or subfragment 1 (S_1; see Begg *et al.*, 1978; and Fig. 9). Fixation using the tannic-acid procedure results in preservation of the arrowhead structures generated by HMM of S_1 binding to actin, which is far superior to the results we obtained earlier (Mooseker and Tilney, 1975), in which it was demonstrated that all the actin filaments of the microvillus core have the same polarity: The ar-

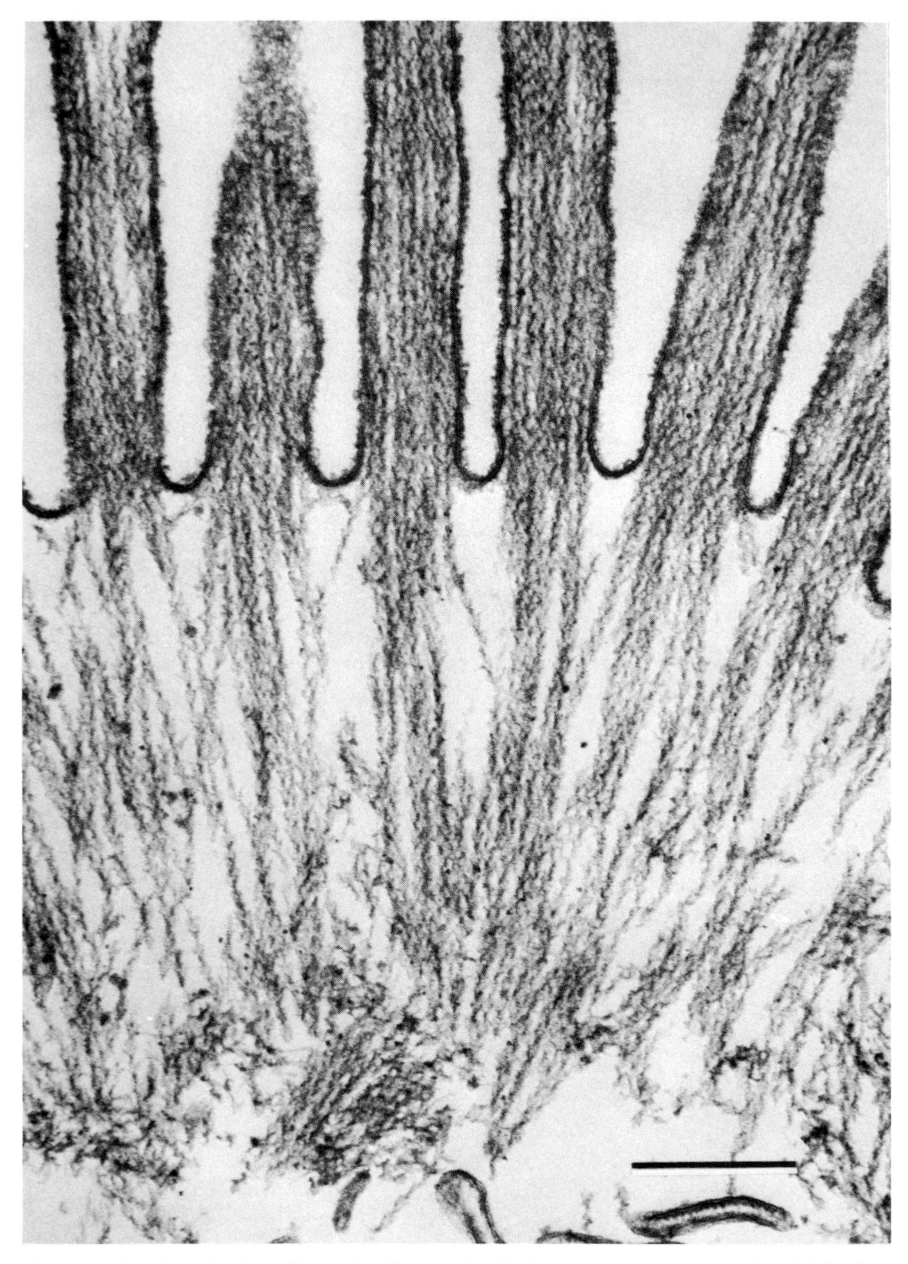

FIG. 9. Polarity of microvillar actin filaments by S_1 decoration using the tannic-acid-fixation procedure of Begg *et al.* (1978). Note that all the core filaments have uniform polarity—the S_1 arrowheads point basally, away from the tip of the microvillus. Micrograph courtesy of D. Begg. Bar = 0.25 μm. ×88,000.

rowheads point away from the point of membrane attachment at the tip of the microvillus. For detailed discussions regarding the functional significance of the sarcomere-like organization of actin (and myosin) in the brush border, the reader is referred to the following: Mooseker and Tilney (1975), Rodewald *et al.* (1976), Mooseker (1976a,b), and Mooseker *et al.* (1978). We are currently reinvestigating actin-filament polarity in the brush border using intact epithelial cells in order to establish the organization of nonmicrovillar actin filaments in the terminal web.

VII. Characterization of Brush-Border Contractile/Cytoskeletal Proteins

Two very useful fractionation schemes for the analysis and localization of brush-border proteins are the separation of membrane and cytoskeletal components by detergent treatment and the isolation and subsequent detergent extraction of microvilli. A comparison by SDS-gel electrophoresis of the fractions obtained by these procedures is shown in Figs. 10 and 11. The detergent-solubilized brush-border membrane fraction contains several prominent glycoproteins (Fig. 10), all of which are enriched in the microvillus-membrane fraction (Fig. 11). A comparison of the cytoskeletal fractions derived from intact brush borders and isolated microvilli indicates that there is a nonuniform distribution of proteins. The proteins of the microvillus core comprise a subset of the total cytoskeletal proteins of the brush border. In addition to actin, major proteins of the core include calmodulin (Howe *et al.,* 1980), and subunits of 105,000 (105K), 95,000 (95K), and 70,000 (70K) daltons (see Bretscher and Weber, 1978a, 1979; Mooseker and Stephens, 1980; Howe *et al.,* 1980; Matsudaira and Burgess, 1979). A number of other brush-border proteins, including myosin (Mooseker *et al.,* 1978; Bretscher and Weber, 1978a,b), tropomyosin (Mooseker, 1976a; Bretscher and Weber, 1978a), and alpha-actinin (see footnote 1) (Bretscher and Weber, 1978b; Craig and Pardo, 1979; Craig and Lancashire, 1980; Geiger *et al.,* 1979; Mooseker and Stephens, 1980), are found only in the terminal-web region of the brush border. With the exception of actin, it is not known whether the proteins of the microvillus core are exclusively localized there, but several of these proteins, including calmodulin (Howe *et al.,* 1980) and the 105K- and 95K-dalton subunits are enriched in the microvillus fractions as compared to the intact brush border. Conversely, all these proteins are depleted in brush borders from which most of the microvilli have been removed by shear (M. S. Mooseker and C. L. Howe, unpublished observation).

We outline next procedures used for the isolation and characterization of cytoskeletal/contractile proteins of the brush border, including actin,

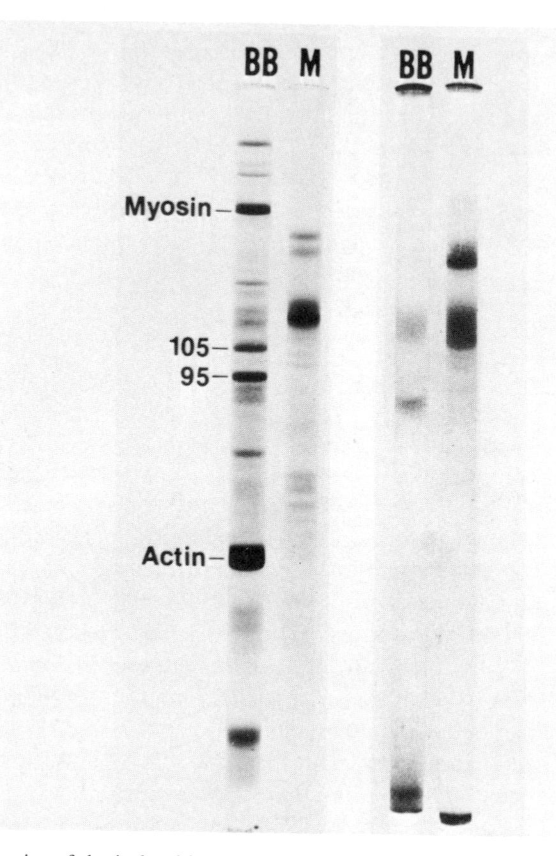

FIG. 10. Proteins of the isolated brush border. SDS-polyacrylamide gel electrophoresis (SDS-PAGE) of demembranated brush borders (BB) and the detergent-solubilized membrane (M). Gels on left are stained with Coomassie Blue, and on the right, identical gels stained for carbohydrate with periodic acid-Schiff reagent. The major components of the brush-border membrane are glycoproteins. None of the major cytoskeletal proteins of the brush border are PAS positive. (From Mooseker and Stephens, 1980.)

tropomyosin, myosin, calmodulin, and the 95K-dalton (95K) protein of the microvillus core [referred to as villin by Bretscher and Weber (1979) and flaccin (from flaccid) in our laboratory (see Mooseker *et al.*, 1980)].

A. Brush-Border Actin

Actin was first purified from acetone powders of brush borders using procedures essentially identical to those for the isolation of actin from skeletal muscle (Tilney and Mooseker, 1971). However, the yields from this approach are quite low, probably because of the destructive effects of acetone on actin in the absence of bound tropomyosin (L. G. Tilney, Department of Biology, University of Pennsylvania, personal communication). Much better yields of brush-border actin can be obtained by gel filtration of either low- or high-ionic-strength extracts of demembranated brush borders (M. S. Mooseker, unpublished observations). Brush-border actin, like other vertebrate nonmuscle actins, consists of the two isotypes, β-and γ-actin, which are present in roughly equimolar amounts (Matsudaira and Mooseker, 1977; Bretscher and Weber, 1978a). As has been the case in other systems, no functional significance for these two isotypes of actin can be assigned based on differential distribution into distinct populations of actin filaments in the brush border. The proportion of β- and γ-actin in the filaments of the microvillus core is the same as that for actin in the terminal-web

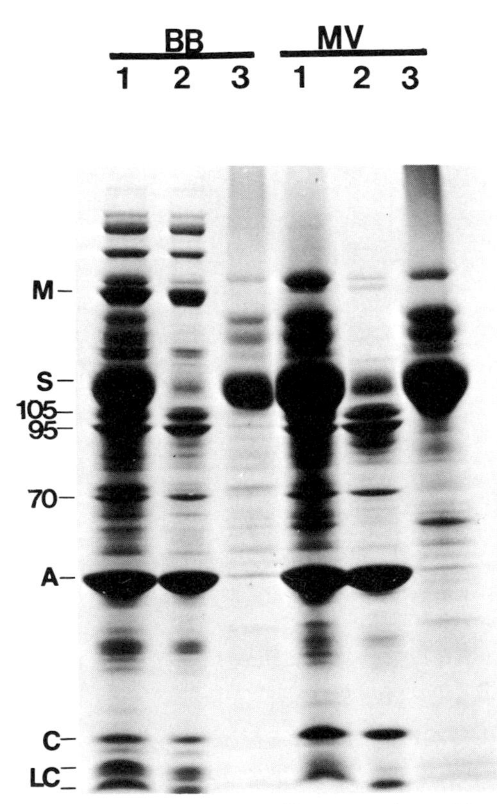

FIG. 11. Proteins of the brush border and microvillus: comparison of the cytoskeletal and membrane components of the brush border (BB) and microvillus (MV). The proteins of the intact organelle (1), cytoskeletal fraction after detergent treatment (2), and solubilized membrane (3), are shown for both the brush border and microvillus by 4–16% SDS-PAGE. M, Myosin heavy chain; LC, myosin light chains; S, sucrase-isomaltase; A, actin; C, calmodulin; 105, 95, 70, molecular weight $\times 10^3$. See text for details. (From Mooseker *et al.*, 1980; reproduced with permission of the *Journal of Cell Biology*.)

fraction of the brush border [the fraction that remains after most of the microvilli have been removed by shear; Bretscher and Weber (1978a); M. S. Mooseker and C. L. Howe, unpublished observations].

B. Brush-Border Tropomyosin

There is only preliminary biochemical evidence for the presence of tropomyosin in the brush border. Using the techniques of Cohen and Cohen (1972) and Fine *et al.* (1973) for the isolation of tropomyosin from nonmuscle cells, a protein with a subunit molecular weight of 30,000 daltons was purified from preparations of porcine brush borders (Mooseker, 1976a). In brief, pellets of brush borders are extracted with ethanol and ether, the dried powder is extracted with 1 *M* KCl, and the supernate fraction is heated in a boiling-water bath for 10 minutes. Heat-denatured protein is removed from the extract by centrifugation, and "presumptive" brush-border tropomyosin is purified from the supernate by ammonium sulfate fractionation (70% saturation cut). Although this procedure is quite selective for highly alpha-helical proteins such as tropomyosin, the characterization of brush-border tropomyosin has not been conducted with respect to important aspects of tropomyosin structure and actin-binding properties. The identification of a tropomyosin in the brush border is probably correct, however, because antibodies prepared against tropomyosin from smooth muscle and bovine brain have been used to stain brush borders at both the light- and electron-microscope levels of resolution (Bretscher and Weber, 1978b; Drenckhahn and Gröschel-Stewart, 1980). These studies indicate that brush-border tropomyosin is restricted in its location to the terminal-web region, primarily in association with the rootlet portion of the microvillus core.

C. Brush-Border Myosin

The brush border is an excellent source of nonmuscle myosin, since this organelle contains a much greater concentration of this protein than is typically seen in other nonmuscle cell types. Myosin is purified from KI extracts of demembranated brush borders using the gel-filtration method of Pollard *et al.* (1974). In brief, demembranated brush borders are extracted with 0.6 *M* KI and the supernate fraction is chromatographed in the presence of 0.6 *M* KCl by agarose-gel filtration (Biorad A 15m; exclusion limit, 15 million daltons). Fractions highly enriched in myosin elute just after the void volume, and final purification and concentration of brush-border myosin is achieved by low-ionic-strength precipitation of the thick filaments that form when the salt concentration is lowered by dialysis against low-ionic-strength buffer (Mooseker *et al.*, 1978). The physical and enzymatic properties of brush-border myosin are similar to those of other vertebrate, nonmuscle myosins that have been analyzed (Mooseker *et al.* 1978). It consists of heavy chains (200,000 daltons) and two light chains (16,000 and 19,000 daltons) in a 1:1:1 molar ratio. The Mg-ATPase activity of brush-border myosin is not activated by actin, presumably because this myosin has become dephosphorylated during purification. Studies are now in progress to determine the role of myosin light- and heavy-chain kinases in the regulation of brush-border motility [see Dedman *et al.* (1979) for a brief review of myosin phosphorylation in nonmuscle systems].

D. Brush-Border Calmodulin

As a first step in deciphering the basis for Ca^{2+} regulation of brush-border contraction *in vitro* (Fig. 7), and presumably microvillar movement *in vivo,* we conducted a series of experiments to determine whether the ubiquitous Ca^{2+} regulatory protein, calmodulin, was present in the brush border in association with either the membrane or cytoskeletal fractions of this organelle (see Wolff and Brostrom, 1979, for a recent review of calmodulin). We were surprised to find that not only is calmodulin present in the brush border, it is one of the major constituent proteins of the isolated microvillus core (scc Howe *et al.,* 1980). Calmodulin can be purified from either isolated demembranated brush borders or microvillus cores using a simple boiling procedure. Demembranated brush borders or microvilli are suspended in 10 m*M* imidazole buffer (pH 7.3), and heated in a boiling-water bath for 5 minutes. Heat-denatured protein is removed by centrifugation at 100,000 *g* for 1 hour, and the supernate fraction contains partially purified calmodulin. The supernate from boiled brush borders contains, in addition to calmodulin, a 30,000-dalton protein that is probably tropomyosin, another heat-stable protein (see above, and Mooseker, 1976a). The supernate fraction prepared from microvillus cores contains only one major polypeptide, calmodulin (Howe *et al.,* 1980; Fig. 12). No calmodulin is present in the detergent-solubilized membrane fraction from either intact brush borders or isolated microvilli. There is a substantial enrichment of calmodulin in the isolated microvillus (relative to both total protein and actin), as compared with the intact brush border, but we cannot exclude the possibility that calmodulin is also present in the terminal-web region.

One unusual aspect of brush-border calmodulin is its binding interaction with the microvillar actin-filament bundle. Calmodulin remains tightly bound to the microvillus core in the absence rather than presence of Ca^{2+}; in most systems that have been characterized thus far, calmodulin binding is activated by the presence of Ca^{2+}. The molecular basis for the association of calmodulin with the microvillus core is currently under investigation. Preliminary experiments indicate that approximately 30–40% of the calmodulin in the microvillus core dissociates in the presence of Ca^{2+} (Howe *et al.,* 1980; Fig 13). However, as we have already noted above, Ca^{2+} also causes a drastic disruption of the microvillus core, result-

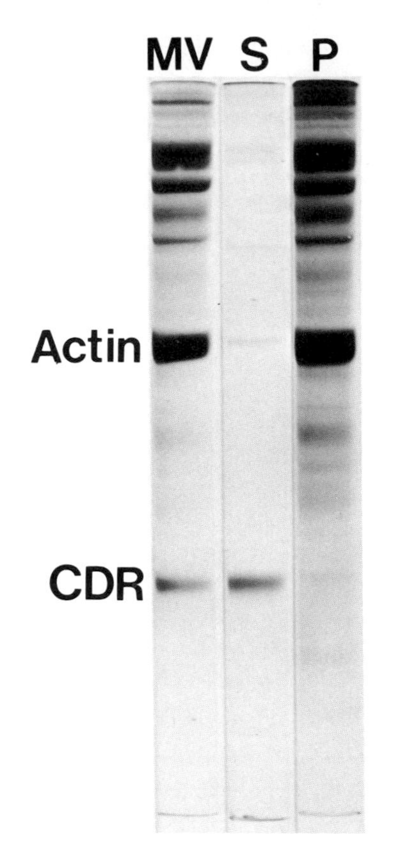

FIG. 12. Isolation of calmodulin from microvillus cores. Supernate (S) and pellet (P) after heat treatment and centrifugation of isolated demembranated microvilli (MV). The supernate contains one prominent polypeptide identified as calmodulin (CDR). (From Howe *et al.*, 1980; reproduced with permission of the *Journal of Cell Biology*.)

ing in a solubilization of other core proteins as well, including 50% of the actin and most of the 95K and 70K (Fig. 13). Gel-filtration chromatography of such Ca^{2+} extracts indicates that calmodulin is in the free-subunit state (Mooseker *et al.*, 1980), suggesting that calmodulin is not bound to any of the other proteins solubilized from the core by Ca^{2+} treatment. Cosedimentation experiments with purified calmodulin and F-actin from skeletal muscle indicate that calmodulin does not bind directly to actin in the presence or absence of Ca^{2+} (Howe *et al.*, 1980). Similarly, calmodulin does not cosediment with 95K–F-actin complexes (these experiments could only be done in the absence of Ca^{2+} because the 95K converts actin into a nonpelletable form in the presence of Ca^{2+}; see later). Presumably, calmodulin binds to the microvillar actin-filament bundle through its association with one of the other core proteins (e.g., the 105K or 70K subunits). The function(s) of calmodulin in the brush border has not been determined.

E. The 95K and Ca^{2+}-Dependent Solation and Cross-Linking of Actin Filaments of the Microvillus Core

Some time ago, we had speculated that the 95K subunit of the brush border was an alpha-actinin-like protein, perhaps involved in the attachment of microvillar actin filaments to the plasma membrane (Mooseker and Tilney, 1975). We and several other laboratories have shown this speculation to be incorrect Mooseker and Stephens, 1978, 1980; Bretscher and Weber, 1979; Craig and Pardo, 1978; Craig and Lancashire, 1980; Geiger *et al.*, 1979; Bretscher and

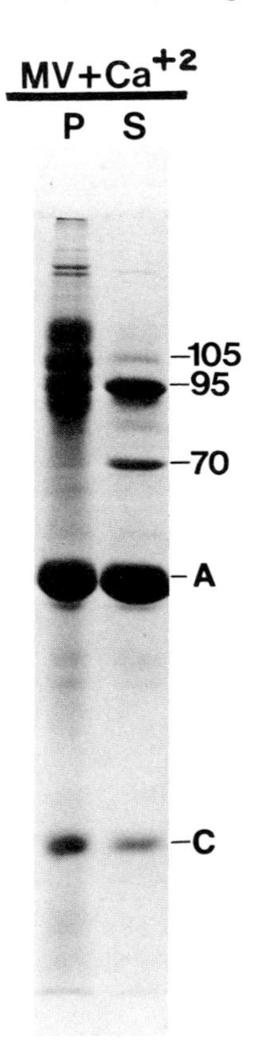

FIG. 13. Effect of Ca^{2+} on microvillus cores: 10% SDS-PAGE of pellet (P) and supernate (S) fractions after 100,000-*g* spin of microvillus cores treated with $10^{-6}M$ Ca^{2+}. Much of the core actin has been "solated." The supernate fraction also contains about half of the calmodulin (C) and most of the 95K and 70K subunits. Other abbreviations as in Fig. 11. (From Mooseker *et al.*, 1980; reproduced with permission of the *Journal of Cell Biology*.)

Weber, 1978b; see footnote 1). The 95K has now been purified and characterized by several laboratories including our own, and the results of these studies indicate that this protein of the microvillus core is one of a growing family of proteins that have been identified in the Ca^{2+}-dependent regulation of actin-filament structure and filament–filament interaction (Mooseker *et al.*, 1980; Bretscher and Weber, 1980b; S. W. Craig and L. D. Powell, Department of Physiological Chemistry, Johns Hopkins Medical School, personal communication; P. T. Matsudaria and D. R. Burgess, Dept. of Biology, Dartmouth College, personal communication). Further investigation of the Ca^{2+}-dependent solation of the microvillus core (see preceding; Figs. 8 and 13; Howe *et al.*, 1980) led us to the finding that "solated" actin (i.e., actin that is not pelletable) released from the microvillus core in the presence of Ca^{2+} is an oligomeric state, bound to the 95K-dalton subunit (Mooseker *et al.*, 1980). This result suggested that the 95K is somehow involved in the Ca^{2+}-dependent solation of the core. We also found that removal of Ca^{2+} from these Ca^{2+} extracts of microvilli results in the reformation of actin filaments and that the 95K still remains bound to the newly assembled filaments. We have purified 95K from Ca^{2+} extracts of microvillus cores by ion-exchange chromatography in DEAE–Sephadex. In brief, microvillus cores are extracted with 0.3 *M* KCl–0.2 m*M* $CaCl_2$–0.1 m*M* ATP–0.1 m*M* DTT–0.1 m*M* PMSF–10 m*M* imidazole Cl (pH 7.2), and the 100,000 *g* supernate is dialyzed against DEAE starting buffer consisting of the same solution as above, except that the KCl is 20 m*M* rather than 0.3 *M*. The dialyzed supernate, which contains actin, calmodulin, 70K, and 95K, is applied to a DEAE–Sephadex column (~10 ml total volume) and eluted with increasing steps of KCl (50, 100, 150, and 500 m*M*). Purified 95K elutes in the 50 m*M* KCl step (see Mooseker *et al.*, 1980).

Bretscher and Weber (1980b) have purified 95K (they refer to this protein as villin) from KI extracts of brush borders using a DNAase I affinity column. In the presence of Ca^{2+}, the 95K, together with actin, is retained by the column. Presumably, the 95K is bound to the actin, which in turn is bound to the DNAase. The 95K is eluted from the column by application of an EGTA wash.

In our laboratory, experiments with purified 95K and muscle actin indicate that 95K will inhibit the assembly of actin in the presence but not absence of Ca^{2+} (Mooseker *et al.*, 1980). Furthermore, addition of 95K to actin filaments in the presence of Ca^{2+} results in a fragmentation of those filaments, and the degree of fragmentation depends on the concentration of 95K added, as determined both by viscometry and by sedimentation (Figs. 14 and 15). Similar results have been reported by Bretscher and Weber (1980b). In the absence of Ca^{2+}, 95K cross-links actin filaments into bundles, suggesting that this core protein is one of the bundling proteins of the microvillus core (Fig. 15; Bretscher and Weber, 1980b; P. T. Matsudaira and D. R. Burgess, personal communication; Mooseker *et al.*, 1980). The solation of actin filaments by 95K is at least partially reversible.

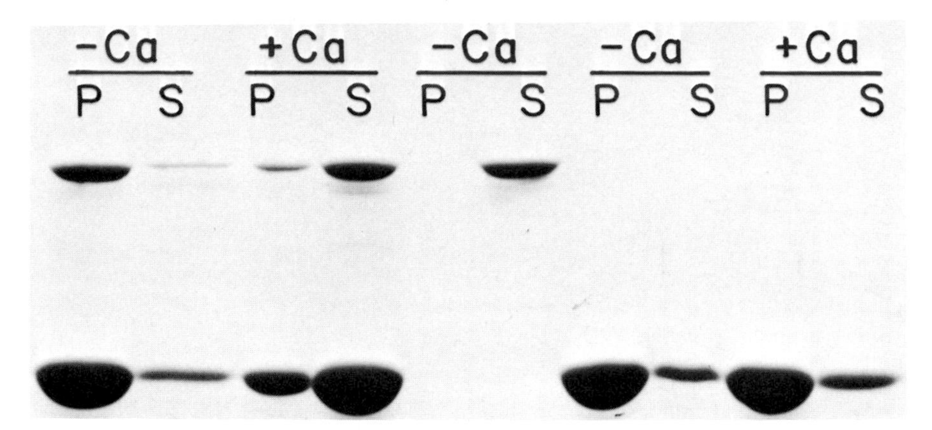

FIG. 14. Cosedimentation of purified 95K and F-actin in the presence and absence of Ca^{2+}. SDS-PAGE of 100,000-*g* supernates (S) and pellets (P) of 95K-F-actin mixtures (1 μM 95K: 6 μM F-actin) in the presence and absence of 0.1 m*M* $CaCl_2$. In the absence of Ca^{2+} (first two lanes), 95K and actin cosediment; under these conditions 95K alone (fifth and sixth lanes) remains in the supernate. In the presence of Ca^{2+} (third and fourth lanes), most of the actin and 95K remain in the supernate. No difference in the sedimentation of F-actin alone in the presence and absence of Ca^{2+} is observed (last four lanes). (From Mooseker *et al.*, 1980; reproduced with permission of the *Journal of Cell Biology*.)

Removal of Ca^{2+} from actin–95K complexes formed in the presence of Ca^{2+} results in a re-formation of F-actin filaments, but these filaments tend to be somewhat shorter than those observed in control preparations of F-actin that contain no 95K (Mooseker *et al.*, 1980).

Actin-binding proteins with properties similar to the 95K have been identified in several other cell types including macrophages (Yin and Stossel, 1979), Ehrlich's tumor cells (Mimura and Asano, 1979), and *Dictyostelium* (Hellewell and Taylor, 1979), but the direct comparison of these various actin-binding proteins will be required to determined what, if any, chemical homologies exist.

Presumably the 95K is involved in Ca^{2+}-dependent structural changes in the core, but the role of such changes, and the extent to which "sol-gel" transformations of the microvillar-filament bundle occur *in vivo* have not yet been determined.

F. Other Brush-Border Contractile/Cytoskeletal Proteins

There are a number of other major proteins of the brush border whose properties and function are not known. These proteins include several high-molecular-weight subunits associated with the terminal-web region (i.e., they are absent from preparations of isolated microvilli). A second set of high-molecular-weight subunits is present in the microvillus core fraction (Fig. 11).

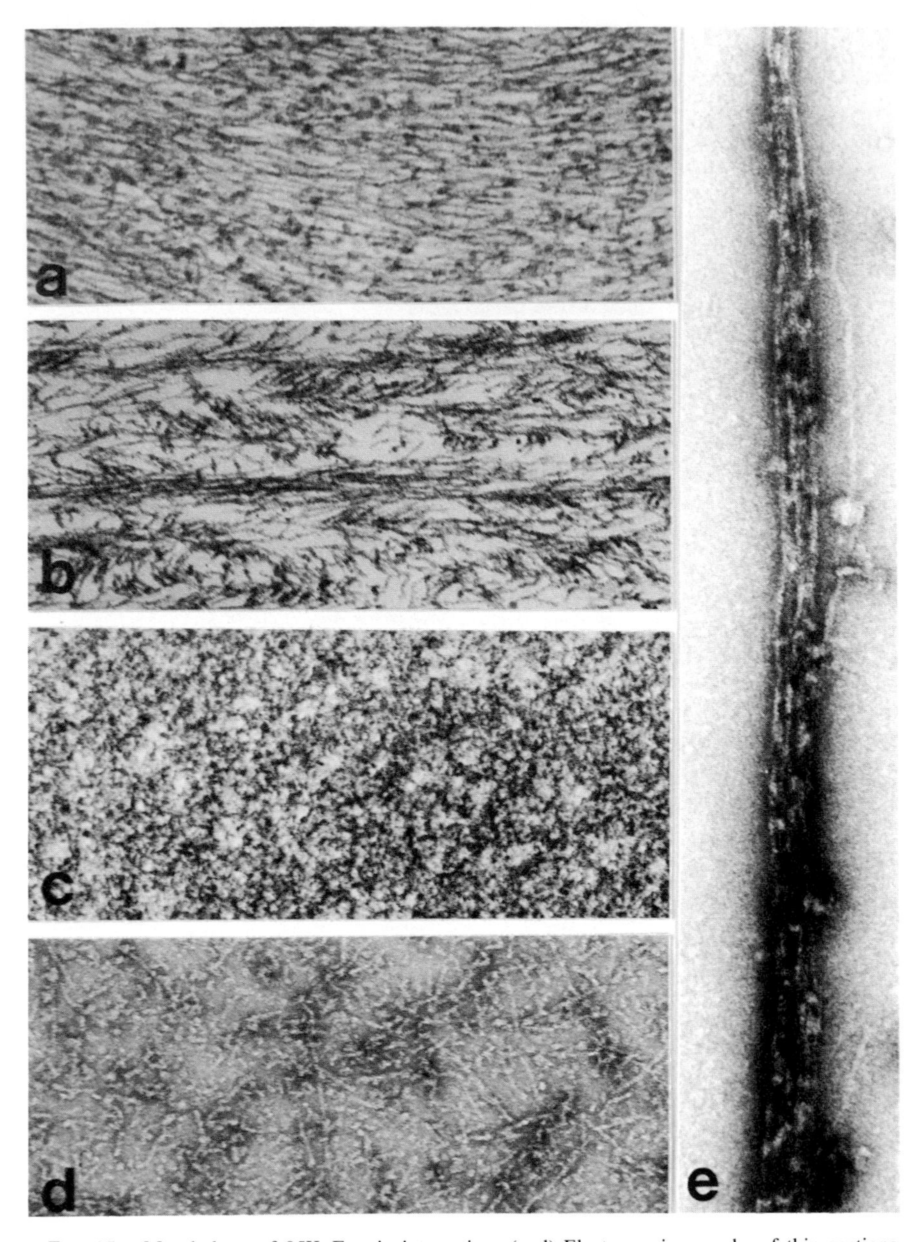

FIG. 15. Morphology of 95K–F-actin interaction. (a–d) Electron micrographs of thin sections (a–c) and negatively stained (d) samples of F-actin (a) and 95K plus F-actin (1:6 molar ratio) in the absence (b) and presence (c,d) of Ca^{2+}. In the absence of Ca^{2+}, the 95K induces the aggregation of F-actin into loosely packed bundles. In the presence of Ca^{2+} (c,d), 95K causes the disruption of F-actin into small filamentous fragments best resolved in negatively stained preparations (d). Mag-

One of the high-molecular-weight subunits in the terminal web may be filamin, since antibodies prepared against this smooth-muscle protein (the function of which remains unknown, except that it binds to and cross-links actin filaments *in vitro*) cross-react with the brush border, and stain the terminal-web region (Bretscher and Weber, 1978b). Among the most prominent structural features of the microvillus core are the lateral bridges that attach the core to the brush-border membrane (see Fig. 2b). Based on a series of selective-extraction experiments, Matsudaira and Burgess (1979) have tentatively identified the 105K-dalton subunit (110K daltons on their gel system) as the lateral-bridge protein. They observed that extraction of isolated microvillus cores with solutions containing ATP and low Mg^{2+} concentrations resulted in a loss of bridges from the core, along with a reduction in the 105K protein as determined by SDS–PAGE of the cores. The appearance of 105K in the supernate fraction was not reported, however. In a second series of experiments, Matsudaira and Burgess (1979) extracted cores with sodium deoxycholate, which results in a splaying of the bundles into individual filaments, sometimes tethered together at one end. Attached to the filaments are lobules resembling the lateral bridges, and on gels these preparations contain actin and the 105K-dalton protein. The 95K and 70K are released into the supernate fraction. These experiments suggest that the 95K and/or the 70K are involved in the cross-linking of microvillar actin filaments into bundles. As noted above, the cross-linking function of 95K has already been established in *in vitro* experiments. The functional properties of the 70K have not yet been determined, but this protein may be a common constituent of "membrane-associated" actin filaments. Bretscher and Weber (1980a) have prepared an antibody against the 70K and have observed that this antibody cross-reacts with a wide variety of tissue culture cells. Much of the observed staining was associated with surface ruffles (lamellapodia) and microvilli.

VIII. Concluding Remarks

Over a decade has passed since the possible analogy between the structural organization of the brush border and striated muscle was first suggested by Tilney and Cardell (1970). Since then, the unique usefulness of the brush border as a model system for analysis of structure-function relationships among contractile proteins in a nonmuscle cell has been established. As a result, the brush border has become a system of choice for investigating questions regarding contractile/cytoskeletal protein function in a dozen laboratories throughout the

---

nification (a–d), ×82,000. (e) At higher ratios of 95K–F-actin (1:2), the cross-linking of actin by 95K into tightly packed bundles is observed. (From Mooseker *et al.*, 1980; reproduced with permission of the *Journal of Cell Biology*.) ×144,000.

world. This active interest in the brush border will undoubtedly lead to a rapid unraveling of many of the questions we have raised or left unanswered in this brief review. Among those questions that stimulate the most interest in our own laboratory include the determination of the *in vivo* function of the brush-border contractile apparatus, and in particular, its role in the membrane-mediated transport activities of the intestinal epithelium. A second focus of our laboratory will be the analysis of brush-border assembly and how the polymerization of actin is controlled during this process.

Acknowledgments

We would like to acknowledge the important contribution of Lewis Tilney in his recognition of the brush border as a useful organelle for the study of contractile-protein structure and function. This is just one of many examples of his "biological foresight" in exploiting systems in which many of the difficult questions have already been answered by nature. We also thank Kristi Wharton for excellent technical assistance, and Kristine Hall Mooseker for her help in manuscript preparation. The most recent experiments reported in this review from our laboratory were supported by NIH Grant #AM 2538.

References

Baker, J. R. (1942). *Q. J. Microsc. Sci.* [N.S.] **84,** 73-104.
Begg, D. A., Rodewald, R., and Rebhun, L. I. (1978). *J. Cell Biol.* **79,** 846-852.
Bendall, J. R. (1969). "Muscles, Molecules and Movement." Am. Elsevier, New York.
Bennett, H. S. (1963). *J. Histochem. Cytochem.* **11,** 14-23.
Booth, A. G., and Kenny, A. J. (1974). *Biochem. J.* **142,** 575-581.
Boyd, C. A. R., and Parsons, D. S. (1969). *J. Cell Biol.* **41,** 646-651.
Bretscher, A., and Weber, K. (1978a). *Exp. Cell Res.* **116,** 397-407.
Bretscher, A., and Weber, K. (1978b). *J. Cell Biol.* **79,** 839-845.
Bretscher, A., and Weber, K. (1979). *Proc. Natl. Acad. Sci. U.S.A.* **76,** 2321-2325.
Bretscher, A., and Weber, K. (1980a). *J. Cell Biol.* **86,** 335-340.
Bretscher, A., and Weber, K. (1980b). *Cell* **20,** 839-847.
Brown, A. L. (1962). *J. Cell Biol.* **12,** 623-627.
Brunser, O., and Luft, J. H. (1970). *J. Ultrastruct. Res.* **31,** 291-311.
Clark, S. L. (1959). *J. Biophys. Biochem. Cytol.* **5,** 41-50.
Cohen, I., and Cohen, C. (1972). *J. Mol. Biol.* **68,** 382-383.
Craig, S. W., and Lancashire, C. L. (1980). *J. Cell Biol.* **84,** 655-667.
Craig, S. W., and Pardo, J. V. (1979). *J. Cell Biol.* **80,** 203-210.
Dalton, A. J. (1951). *Am. J. Anat.* **89,** 109-133.
Dedman, J. R., Brinkley, B. R., and Means, A. R. (1979). *Adv. Cyclic Nucleotide Res.* **11,** 131-174.
Drenckhahn, D., and Gröschel-Stewart, U. (1980). *J. Cell Biol.* **86,** 475-582.
Evans, E. M., Wrigglesworth, J. M., Burdett, K., and Pover, W. F. R. (1971). *J. Cell Biol.* **51,** 452-464.
Farquhar, M. G., and Palade, G. E. (1963). *J. Cell Biol.* **17,** 375-409.

Fine, R. E., Blitz, A. L., Hitchcock, S. E., and Kaminer, B. (1973). *Nature (London) New Biol.* **245,** 182-186.
Forstner, G. G., Sabesin, S. M., and Isselbacher, K. J. (1968). *Biochem. J.* **106,** 381-390.
Geiger, B., Tokuyasu, K. T., and Singer, S. J. (1979). *Proc. Natl. Acad. Sci. U.S.A.* **76,** 2833-2837.
Gibbons, I. R. (1975). *In* "Molecules and Cell Movement" (S. Inoué and R. E. Stephens, eds.), pp. 207-231. Raven, New York.
Granger, B., and Baker, R. F. (1949). *Anat. Rec.* **103,** 459.
Granger, B., and Baker, R. F. (1950). *Anat. Rec.* **107,** 423-441.
Hellewell, S. B., and Taylor, D. L. (1979). *J. Cell Biol.* **83,** 633-648.
Howe, C. L., Mooseker, M. S., and Graves, T. A. (1980). *J. Cell Biol.* **85,** 916-923.
Hull, B. E., and Staehelin, L. A. (1979). *J. Cell Biol.* **81,** 67-82.
Ishikawa, H., Bischoff, R., and Holtzer, H. (1969). *J. Cell Biol.* **43,** 312-328.
Ito, S. (1961). *J. Biophys. Biochem. Cytol.* **11,** 333-347.
Ito, S. (1965). *J. Cell Biol.* **27,** 475-491.
Ito, S. (1969). *Fed. Proc., Fed. Am. Soc. Exp. Biol.* **28,** 12-25.
Ito, S. (1974). *Philos. Trans. R. Soc. London, Ser. B* **268,** 55-66.
Ito, S., and Revel, J. P. (1964). *J. Cell Biol.* **23,** 44a.
McNabb, J. D., and Sandborn, E. (1964). *J. Cell Biol.* **22,** 701-704.
Matsudaira, P. T., and Burgess, D. R. (1979). *J. Cell Biol.* **83,** 667-673.
Matsudaira, P. T., and Mooseker, M. S. (1977). *Biol. Bull. (Woods Hole, Mass.)* **153,** 439.
Maupin-Szamier, P., and Pollard, T. D. (1978). *J. Cell Biol.* **77,** 837-852.
Miller, D., and Crane, R. K. (1961). *Biochim. Biophys. Acta* **52,** 293-298.
Millington, P. F., and Finean, J. B. (1962). *J. Cell Biol.* **14,** 125-139.
Mimura, N., and Asano, A. (1979). *Nature (London)* **282,** 44-48.
Mooseker, M. S. (1976a). *J. Cell Biol.* **71,** 417-432.
Mooseker, M. S. (1976b). *Cold Springs Harbor Conf. Cell Proliferation* **3** [Book A], 631-650.
Mooseker, M. S., and Stephens, R. E. (1978). *Biophys. J.* **21,** 24a.
Mooseker, M. S., and Stephens, R. E. (1980). *J. Cell Biol.* **86,** 466-474.
Mooseker, M. S., and Tilney, L. G. (1975). *J. Cell Biol.* **67,** 725-743.
Mooseker, M. S., Pollard, T. D., and Fujiwara, K. (1978). *J. Cell Biol.* **79,** 444-453.
Mooseker, M. S., Graves, T. A., Wharton, K. A., Falco, N. A., and Howe, C. L. (1980). *J. Cell Biol.* **87,** 809-822.
Mukherjee, T. M., and Staehelin, L. A. (1971). *J. Cell Sci.* **8,** 573-599.
Mukherjee, T. M., and Wynn Williams, A. (1967). *J. Cell Biol.* **34,** 447-461.
Overton, J. (1965). *J. Exp. Zool.* **159,** 195-202.
Overton, J., and Shoup, J. (1964). *J. Cell Biol.* **21,** 75-85.
Overton, J., Eichholz, A., and Crane, R. K. (1965). *J. Cell Biol.* **26,** 693-706.
Palay, S. L., and Karlin, L. J. (1959). *J. Biophys. Biochem. Cytol.* **5,** 363-372.
Pisam, M., and Ripoche, P. (1976). *J. Cell Biol.* **71,** 907-920.
Pollard, T. D., Thomas, S. M., and Niederman, R. (1974). *Anal. Biochem.* **60,** 258-266.
Rambourg, A. (1971). *Int. Rev. Cytol.* **31,** 57-114.
Rambourg, A., and Leblond, C. P. (1967). *J. Cell Biol.* **32,** 27-53.
Rodewald, R., Newman, S. B., and Karnovsky, M. J. (1976). *J. Cell Biol.* **70,** 541-554.
Sacktor, B. (1977a). *Curr. Top. Bioenerg.* **6,** 39-81.
Sacktor, B. (1977b). *In* "Mammalian Cell Membranes" (G. A. Jamieson and D. M. Robinson, eds.), Vol. 4, pp. 221-254. Butterworth, London.
Sandström, B. (1971). *Cytobiologie* **3,** 293-297.
Staehelin, L. A. (1974). *Int. Rev. Cytol.* **39,** 191-283.
Staehelin, L. A., and Hull, B. E. (1978). *Sci. Am.* **238,** 140-152.

Swift, J. G., and Mukherjee, T. M. (1976). *J. Cell Biol.* **69,** 491–494.
Thuneberg, L., and Rostgaard, J. (1969). *J. Ultrastruct. Res.* **29,** 578a.
Tilney, L. G., and Cardell, R. R., Jr. (1970). *J. Cell Biol.* **47,** 408–422.
Tilney, L. G., and Mooseker, M. S. (1971). *Proc. Natl. Acad. Sci. U.S.A.* **68,** 2611–2615.
Wieland, T. (1977). *Naturwissenschaften* **64,** 303–309.
Wolff, D. J., and Brostrom, C. O. (1979). *Adv. Cyclic Nucleotide Res.* **11,** 27–88.
Yin, H. L., and Stossel, T. P. (1979). *Nature (London)* **281,** 583–586.
Zetterqvist, H. (1956). ''Aktiebolaget Godvil.'' Karolinska Institutet, Stockholm.

Chapter 9

Actin Gelation in Sea Urchin Egg Extracts

JOSEPH BRYAN

Department of Cell Biology
Baylor College of Medicine
Houston, Texas

ROBERT E. KANE

Pacific Biomedical Research Center
University of Hawaii
Honolulu, Hawaii

ISBN 0-12-564125-7

I. Introduction

The purpose of this chapter is to review our work on actin gelation in sea urchin egg extracts and to integrate these data with recent studies on the postfertilization changes in the egg cortical cytoplasm. Whereas sea urchin eggs are not usually considered to be highly motile cells, they do display a rather striking extension of microvilli following fertilization and are a classical material for studying cell division. In addition, egg-cytoplasmic extracts contain relatively large amounts of actin (Hatano *et al.*, 1969; Mabuchi, 1979; Otto *et al.*, 1980), myosin (Mabuchi, 1973, 1979; Kane, 1980), and other proteins (Ishimoda-Takagi, 1979) associated with nonmuscle motility, although it is not clear to what extent these molecules are involved in fertilization and division as opposed to storage for use in the motile events of later embryogenesis.

The ease of obtaining large numbers of gametes and the synchronous nature of the events of fertilization and first division make this material particularly attractive for studying the function of contractile proteins during early embryogenesis. The approach taken to exploit the system has been to obtain active cytoplasmic extracts that will assemble actin filaments from a soluble monomeric pool and organize these filaments into either structured or contractile actin-filament networks or gels. In order to determine what cytoplasmic components are involved in the regulation of filament assembly, the further organization of actin filaments into gels, and their contractile activity, the gels have been isolated and fractionated into their components. Although the story is far from complete, progress has been made in determining the function of several of these components in the *in vitro* gelation reaction by doing reconstitution experiments with the isolated proteins. In addition, the use of specific antibodies has made it possible to identify the function of at least one component in both eggs and sea urchin coelomocytes.

II. Sea Urchin Egg Procedures

A. Sea Urchin Egg Cytoplasmic Preparations

Eggs were obtained from the Hawaiian sea urchin *Tripneustes gratilla* by injection of isotonic KCl into the body cavity. The jelly coat was removed by rapidly titrating a 10% egg suspension to pH 5 with HCl, then washing the eggs several times with Millipore-filtered seawater. The eggs were sedimented by hand centrifugation and the packed egg volume estimated. To remove the seawater divalent cations, eggs were washed at 25°C with 10 volumes of a 19 to 1 mixture of isotonic sodium and potassium chlorides containing 2 m*M* EDTA.

The eggs were then washed once with 2 volumes of 0.9 *M* glycerol, 0.005 *M* EGTA, and 0.1 *M* PIPES at pH 7.15. The volume of packed eggs was measured and resuspended in 1.5 volumes of this isotonic glycerol medium and homogenized by 10–15 passes in a tight-fitting glass Dounce homogenizer. The homogenate was checked using phase-contrast microscopy and found to be free of unbroken eggs, then sedimented in the SW 50.1 rotor at 40,000 rpm for 1 hour at 2–4°C. The clear supernatant can be recovered by gently puncturing the lipid layer with a needle and withdrawing fluid by syringe. Alternatively, the lipid layer can be removed with a water aspirator and the clear supernatant recovered with a Pasteur pipet. The supernatant is then dialyzed overnight against approximately 20 volumes of 0.005 *M* PIPES (pH 7.15), containing 0.1 m*M* EGTA and 0.1 m*M* ATP. The dialysate is sedimented at 25,000 *g* for 15 minutes at 4°C to remove particulate material and any residual lipid. Sodium azide is added to the final extract at a concentration of 1 m*M* to minimize bacterial growth. The final pH of the extracts varies somewhat, but normally is 6.9–7.0.

B. Storage of Extracts

Extracts can be stored on ice for several days with little deterioration. We have been able to store the extracts for an indefinite period of time in an active form using the following procedure. Three to five milliliters of extract are dispensed into 5-ml cryotubes (Vanguard International, Inc.) and rapidly frozen in liquid nitrogen. These samples are stored at −80°C or in a liquid-nitrogen refrigerator, and recovered by rapid thawing with constant agitation in a 25° or 35°C water bath. The contraction and gelation behavior of this material is qualitatively similar to unfrozen samples. The actin content of frozen samples, as measured by the DNase I-inhibition assay (Blikstad *et al.*, 1978), is equivalent to fresh controls. Myosin isolated from frozen extracts retains full ATPase activity.

III. Isolation and Characterization of Structured Gels and Cytoskeletal Proteins

A. Structural Gels from Sea Urchin Egg Extracts

The formation of a structured actin gel consisting of linear aggregates of actin filaments can be induced by warming the cytoplasmic extracts as first described by Kane (1975, 1976). ATP and EGTA are added to a final concentration of 1 m*M*, and KCl to 10–20 m*M*. Gelation proceeds slowly (6–8 hours) at 25°C, but is complete in less than 1 hour at 40°C. After gelation at 40°C, samples are cooled to room temperature and the gel collected by sedimentation at 25,000 *g* for 15 minutes at 25°C. The structured gel is stable in 0.1 *M* KCl–0.001 *M* ATP–0.01 *M* PIPES (pH 6.9), and can be resuspended in this buffer, gently disrupted with a loose-fitting glass homogenizer, then resedimented at 25,000 *g* for 15 minutes to remove contaminating cytoplasmic proteins. The washed gel pellets have been used for isolation of the major gelation proteins including actin, a 58,000-dalton protein termed fascin, and a 220,000-dalton protein.

B. Isolation of Sea Urchin Actin

Actin has been isolated from acetone powders of sea urchin eggs (Hatano *et al.*, 1969) and partially characterized (Miki-Noumura and Oosawa, 1969; Miki-Noumura and Kondo, 1970). The procedure described here depends on the ability of this invertebrate actin to form large bundles in the presence of high salt and appears to be generally applicable to echinoderms.

The gel is resuspended in three to five volumes of 1 *M* KCl–1 m*M* EGTA–10 m*M* PIPES (pH 6.9) and gently disrupted in a tight-fitting glass homogenizer. The solution clarifies on standing at 0–4°C for 20–30 minutes, and no structured gel should be visible in the phase microscope. Undissolved material is removed by centrifugation at 25,000 *g* for 15 minutes, the clear supernatant is recovered, and after addition of ATP to 1 m*M*, is stored on ice. After 10–12 hours, the solution is turbid and bundles of actin filaments are visible in the phase microscope. The filament bundles can be collected by centrifugation (25,000 *g* for 15 minutes) and will disaggregate in 0.1 *M* KCl–1 m*M* ATP–10 m*M* PIPES (pH 6.9). The actin that results is essentially as clean as material prepared from rabbit muscle using the procedures of Spudich and Watt (1971). Residual contaminants can be reduced by an additional high-salt bundling cycle. The ability to form actin bundles in high salt appears to be intrinsic to this actin. Chromatography on phospho- or DEAE-cellulose does not change the critical concentration, which is approximately 1.5 mg/ml in 0.6 *M* KCl–1 m*M* ATP–1 m*M* EGTA, using 0.01 *M* PIPES at pH 6.9 at a buffer.

A similar procedure can be used to recover large quantities of actin from either sea urchin lantern muscles or the smooth muscles of sea cucumbers (*Parastichopus californicus*). We discuss this work here to illustrate some of the properties of the bundling reaction. Muscles are isolated and washed in 19:1 isotonic NaCl : KCl, then homogenized in a Waring blender in three volumes of 1.5 *M* KCl–5 m*M* EGTA–10 m*M* PIPES (pH 6.9). The homogenate is centrifuged at 100,000 *g* for 60 minutes and the clear supernatant recovered. Addition of ATP (1 m*M*) to these concentrated actin solutions results in the rapid formation of microscopic bundles that continue to grow in size for several hours (Fig. 1). The efficiency of this bundling of actin is dependent on the salt concentration, as shown in Fig. 2. The process appears to be a critical concentration (C_c) phenom-

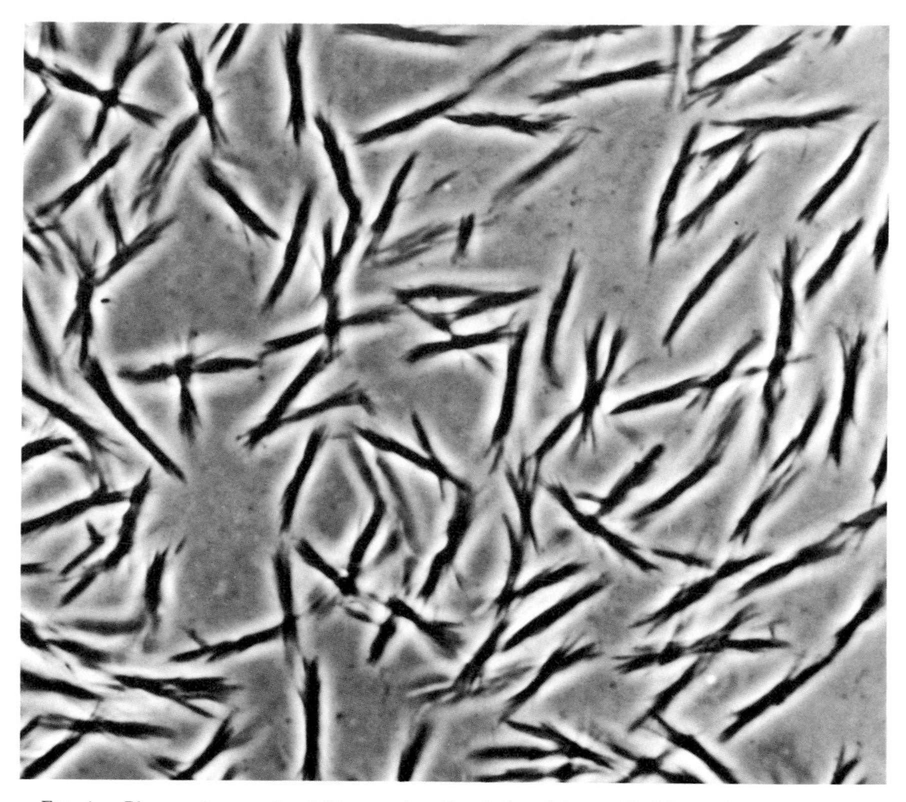

FIG. 1. Phase micrograph of filament bundles induced in purified invertebrate actin by the addition of KCl. The actin was purified as described in the text from the five longitudinal body-wall muscles of the sea cucumber, *Parastichopus californicus*. Bundle formation was induced by addition of KCl to 0.6 *M*. The actin concentration was 4 mg/ml in 0.001 *M* EGTA-0.001 *M* ATP-0.01 *M* PIPES (pH 6.9). Magnification, ×585.

enon, with a C_c of about 0.45 mg/ml in 1 *M* KCl and 2 m*M* ATP (Fig. 3). Both KCl and NaCl appear to be equally efficient at causing bundle formation.

C. Isolation of Fascin

Fascin, the 58,000-dalton actin-cross-linking protein, can be isolated from the structured gel using methods we have described in detail elsewhere (Bryan and Kane, 1978). In brief, 20–30 mg of the structured gel is dissolved in an equal volume of 1.5 *M* KI-10 m*M* ATP-0.001 *M* EGTA-0.01 *M* PIPES (pH 6.9) to depolymerize the actin filaments. The gel is solubilized by homogenization and incubation on ice for 30 minutes, then clarified at 100,000 *g* for 60 minutes.

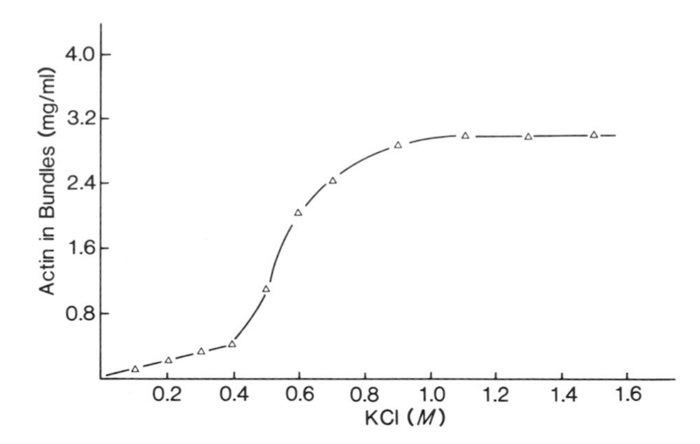

FIG. 2. The effect of salt concentration on the formation of actin-filament bundles. The actin was prepared from *Parastichopus* as described in the text. The final actin concentration was 3.5 mg/ml in 0.001 *M* EGTA-0.003 *M* ATP-0.01 *M* PIPES (pH 6.9). After addition of KCl to the indicated final concentration, the solutions were incubated for 18 hours on ice, then the bundles were collected by centrifugation at 12,000 *g* for 5 minutes in an Eppendorf microfuge. Suitable aliquots were then removed for protein determinations. The addition of NaCl to these actin solutions has a similar effect.

The supernatant is fractionated on a small agarose A-5m column that has been equilibrated with 0.6 *M* KI-0.01 *M* PIPES-0.001 *M* EGTA. The fractionation is monitored using SDS-slab gels, and the fractions containing actin and fascin are pooled, then dialyzed versus two changes of 0.6 *M* KI-0.001 *M* EGTA-0.01 *M* PIPES (pH 6.9) to reduce the ATP concentration. The sample can then be dialyzed against 0.01 *M* PIPES (pH 6.9) to reduce the ionic strength without assembly of actin filaments. Fascin can be purified from the actin and other proteins by chromatography on DEAE-cellulose. The sample is loaded onto a small column of well-washed DEAE-cellulose and eluted either with a KCl gradient (0-0.5 *M*) or with an 0.09 *M* KCl step. The yield of fascin is approximately 20-30% of the total. The major loss occurs in the ion-exchange step where a fraction of the fascin appears to bind very tightly to the DEAE-cellulose.

D. Isolation of the 220,000-Dalton Protein

Relatively little work has been done with this higher-molecular-weight protein, which we originally obtained in a partially purified form from the first agarose-chromatography step described in the fascin purification. Using 10% Laemmli gels, this fraction appeared as a singlet with an apparent molecular weight somewhat greater than myosin. Our more recent data from lower-porosity gels indicates that there are two proteins in this fraction that migrate as a closely spaced doublet. These polypeptides are distinct from the myosin heavy chains and migrate faster than chicken gizzard filamin (Wang, 1977).

E. Reconstitution of Structured Gels

Figure 4 shows the three purified fractions used in the recombination experiments. Addition of fascin to purified actin produces small needle-like structures visible in the light microscope. These two proteins do not form a three-dimensional gel. The extent of this needle formation, which is a function of the concentration of actin and fascin, can be quantitated by simple centrifugation assays (Fig. 5) and the results used to determine the stoichiometry of actin to fascin in the needles. The ratio at saturation is 4.6:1 (Bryan and Kane, 1978). Essentially identical results are obtained using sea urchin egg actin or purified rabbit skeletal-muscle actin.

The addition of the 220,000-dalton (220K) protein to purified actin has no obvious effect. We have looked for direct cross-linking or gelation of actin by this protein using both centrifugation assays and negative staining, but can detect no direct interaction. When this protein is added either to the actin–fascin needles or to F-actin with fascin, a three-dimensional network is formed. The mechanism by which the 220K protein links actin-fascin needles is not yet clear.

Preliminary experiments with vertebrate filamin isolated from chicken giz-

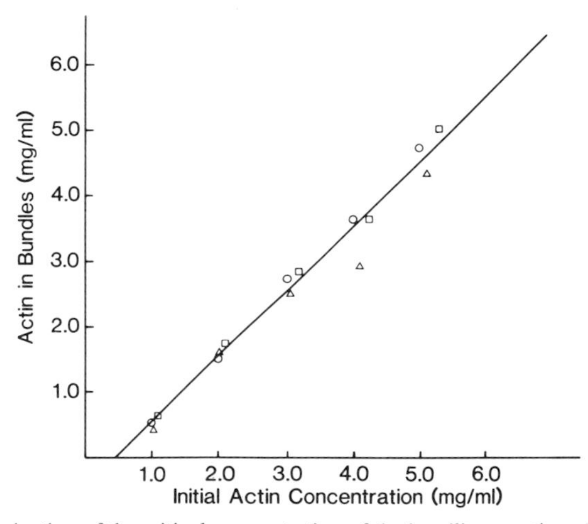

FIG. 3. Determination of the critical concentration of the bundling reaction. *Parastichopus* actin at the indicated concentrations was incubated on ice for 15 hours in 1 *M* KCl–0.001 *M* EGTA–0.002 *M* ATP–0.01 *M* PIPES (pH 6.9), to induce bundle formation. The filament bundles were then collected by centrifugation at 12,000 *g* for 5 minutes in an Eppendorf microfuge. The solid line is the linear regression on the grouped data from three preparations. The slope is 0.98, the C_c is 0.45 mg/ml. There is no major difference in the data from actin purified by one (circles), two (squares), or three (triangles) cycles of high-salt bundling. The individual critical concentrations, by linear regression, are 0.46, 0.45, and 0.43 mg/ml for one, two, and three cycles, indicating low levels of contaminating proteins.

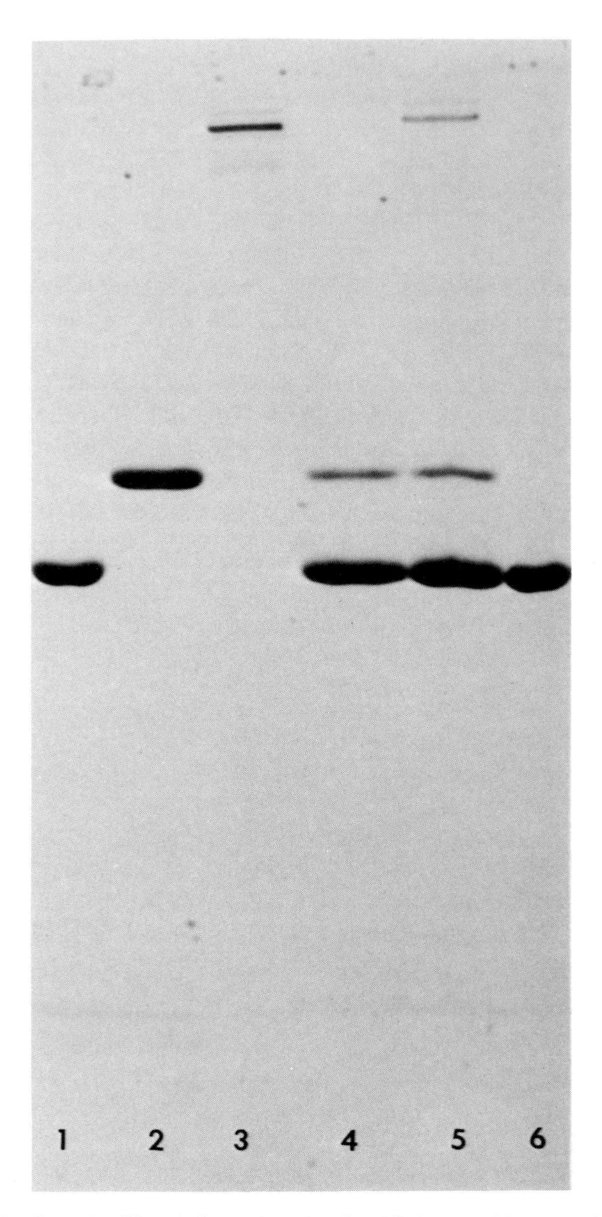

FIG. 4. SDS-polyacrylamide gel electrophoresis of purified sea urchin egg actin gel fractions. Samples: Lane 1, purified sea urchin egg actin; lane 2, purified fascin; lane 3, partially purified 220,000-dalton protein; lane 4, a sample from a 15,000 *g*-20 minute pellet from a mixture of actin and fascin; lane 5, a sample from a 15,000 *g*-20 minute pellet from a mixture of actin, fascin, and the 220,000-dalton protein after 6 hours' incubation on ice; lane 6, actin isolated from *Parastichopus californicus* using the high-salt bundling procedure described in the text. The gels are 10% Laemmli SDS-slab gels run according to Bryan and Kane (1978).

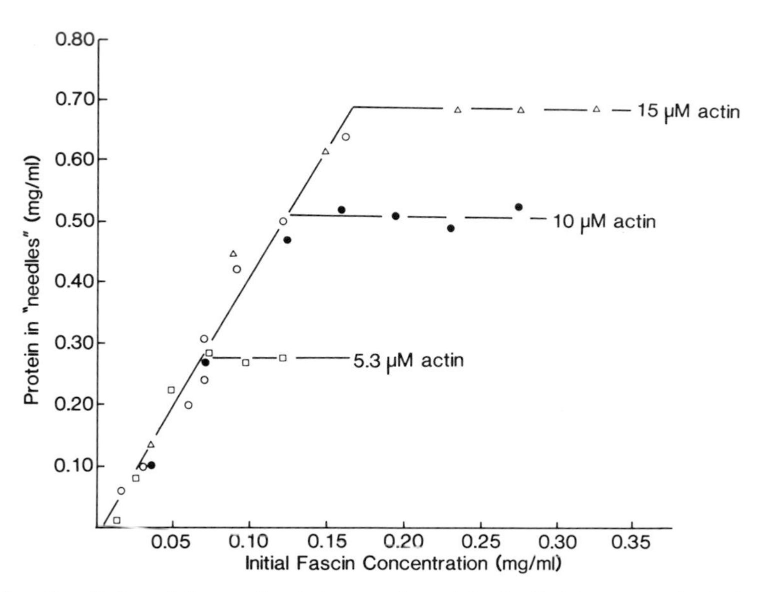

FIG. 5. Quantitation of the reaction between actin and fascin. Proteins were mixed at various initial concentrations and incubated for 18 hours on ice. Actin–fascin needles were collected by centrifugation at 20,000 *g* for 15 minutes. Samples were taken and analyzed as described in Bryan and Kane (1978). The initial concentrations of F-actin are given for each plateau value. The slope of the regression line is 4.3 mg of actin needles per mg of added fascin, which gives a minimum estimate of the actin/fascin molar ratio in the needles of 4.6. The buffer is 0.1 *M* KCl–0.001 *M* Mg-ATP–0.01 *M* PIPES at pH 6.8.

zards show that the actin–fascin interaction is not blocked by filamin and that filamin will cross-link actin–fascin needles into a three-dimensional network.

F. Structure of Actin–Fascin Gels

The ultrastructure of the actin–fascin needles was originally described by Kane (1975, 1976) and has been studied by DeRosier *et al.* (1977) using optical diffraction and image-reconstruction methods. Each needle consists of a bundle or fascicle of approximately parallel actin filaments arranged in a hexagonal lattice having an 8.3-nm spacing. Consistent with this, two characteristic views that depend on the planar projection of the hexagonally packed filaments are seen in negatively stained specimens in the electron microscope. In one view (Fig. 6), individual, filaments are clearly visible in the bundle and 33- to 35-nm transverse banding is quite apparent. In the second view (Fig. 6), filaments are not so easy to discern, but the 11- to 13-nm transverse periodicity is now prominent. Optical-diffraction patterns from these images consist of a set of strong reflections that fall on layer lines. From these patterns, DeRosier *et al.* (1977) con-

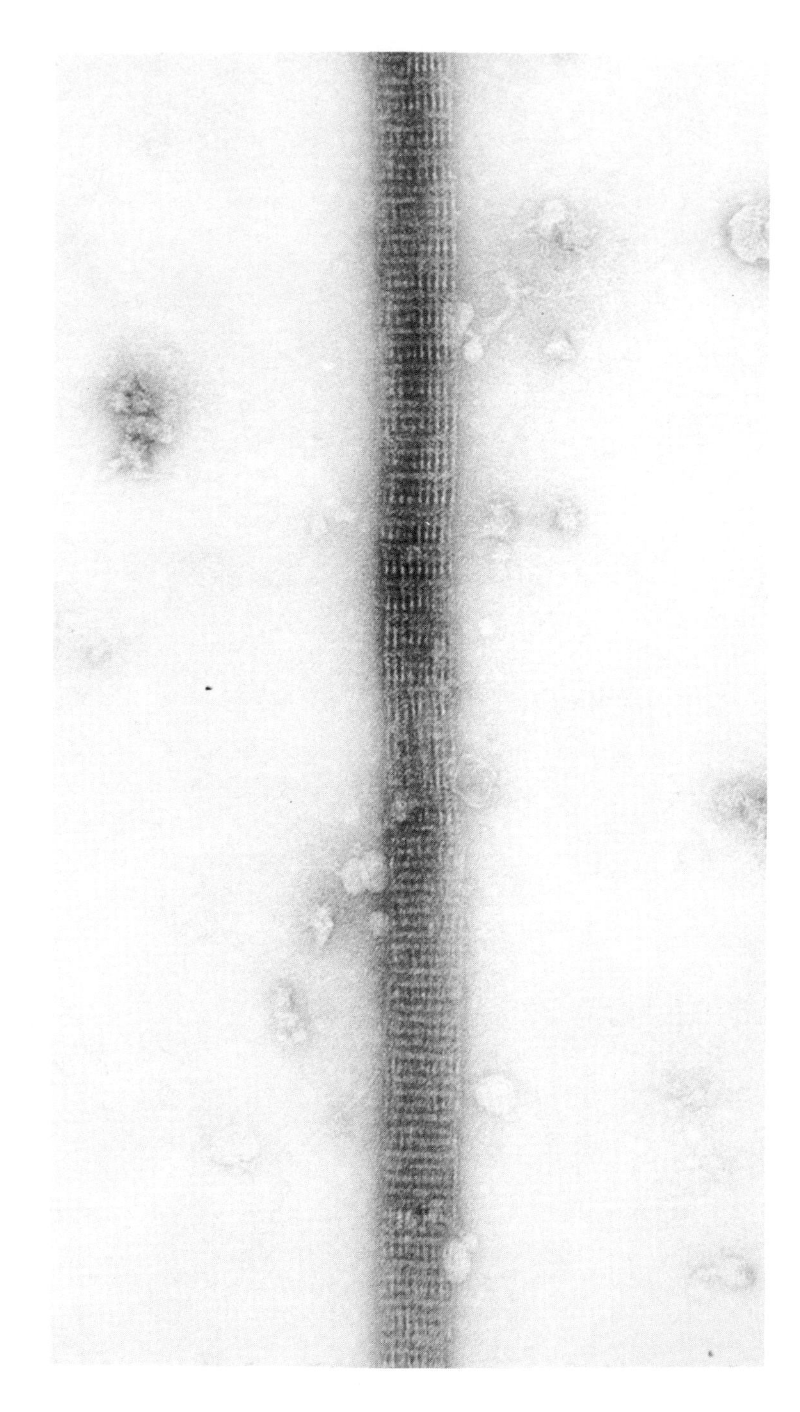

clude that there are 41 actin monomers in 19 turns of the shallow left-hand genetic helix that connects all the subunits. Based on their analysis of the acrosomal actin-filament bundle of *Limulus* sperm, DeRosier *et al.* (1977) argued that it should be possible to form a specified number of quasi-equivalent interfilament links in the actin–fascin needles. In the acrosomal actin bundle, the actin filaments have 28 monomers in 13 turns of the basic helix. Six sets of bonds are formed between these hexagonally packed filaments along this 28-subunit repeat. Extension of this bonding rule to the 41/19 geometry of the needles gives approximately 9 links per 41 monomers. The links in the actin–fascin needles are not evenly spaced along the monomers but can be described by the series 5, 4, 5, 4, 5, 4, 5, 4, 5. The predicted stoichiometry of monomers to links is $41/9 = 4.56$, which agrees with the chemical determination if each link is one fascin molecule (Bryan and Kane, 1978).

G. Biological Role for Fascin

All the currently available evidence suggests that fascin functions in sea urchin eggs and other cells to organize actin filaments into the bundles that form the cores of filopodia and microvilli. We have used one type of phagocytic coelomocyte from sea urchins to study the role of this actin-cross-linking protein. Edds (1977) has demonstrated that these cells have an actin-filament mat that is organized into filament bundles as these cells are transformed from a petaloid to a filopodial form. This sequence is shown in Fig. 7. Indirect immunofluorescent staining of these cells with antifascin indicates that fascin is localized on the developing bundles and could be responsible for the filament reorganization (Otto *et al.*, 1979). In addition, our more recent work and that of DeRosier and Edds (1980) shows that the actin bundles in the filopodia have the characteristic ultrastructure (Fig. 8). Finally, direct chemical analysis (Edds, 1979; Otto *et al.*, 1980) demonstrates that fascin is present in the filopodial cores.

Three similar lines of evidence indicate that fascin is important in the cortical changes that occur in these eggs after fertilization. A number of classical studies on sea urchin eggs have established that the region of cytoplasm immediately under the plasma membrane, called the *cortex,* has special physical properties and is actively involved in fertilization and cell division. This work has been reviewed by others (Just, 1939; Chambers and Chambers, 1961; Hiramoto,

---

FIG. 6. Negatively stained actin–fascin needle prepared by reconstitution of purified sea urchin egg actin with the "S" solution described by Kane (1975, 1976). The upper half of the needle shows the 33- to 35-nm repeat quite clearly. In this view the projected spacing of the individual filaments, which are well resolved, is $d\sqrt{3/2}$, where d is the center-to-center spacing between hexagonally packed filaments. About midway down the needle the bundle becomes slightly twisted and the projected spacing of the poorly resolved individual filaments is now $d/2$. In this view, the 11- to 13-nm repeat is enhanced and becomes clearly visible. The needle was stained with 1% uranyl acetate. Magnification, ×137,000.

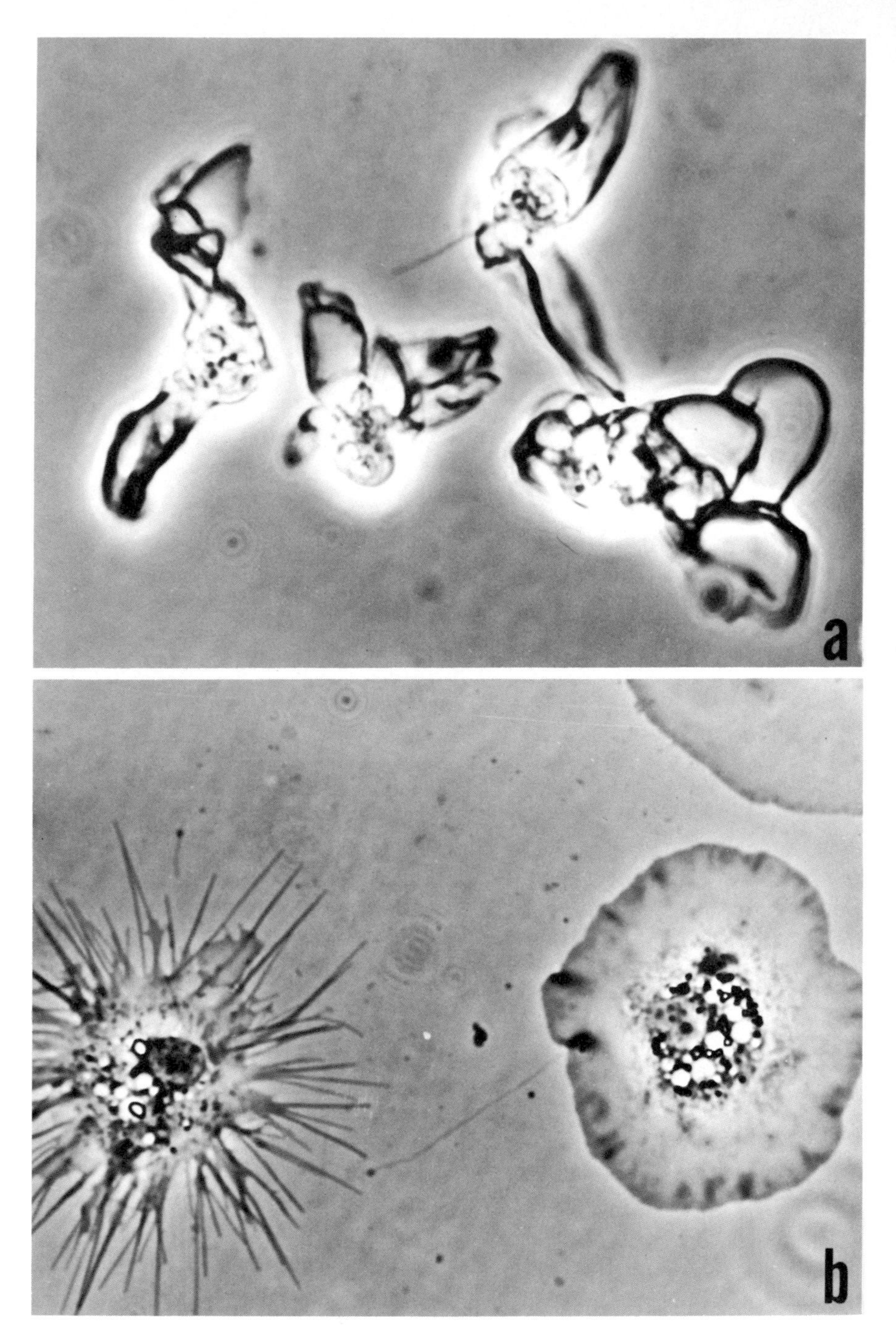

FIG. 7. Transformation of phagocytic coelomocytes from the petaloid to the filopodial form. Coelomyocytes from the sand dollar, *Dendraster eccentricus,* were purified as described in Otto *et*

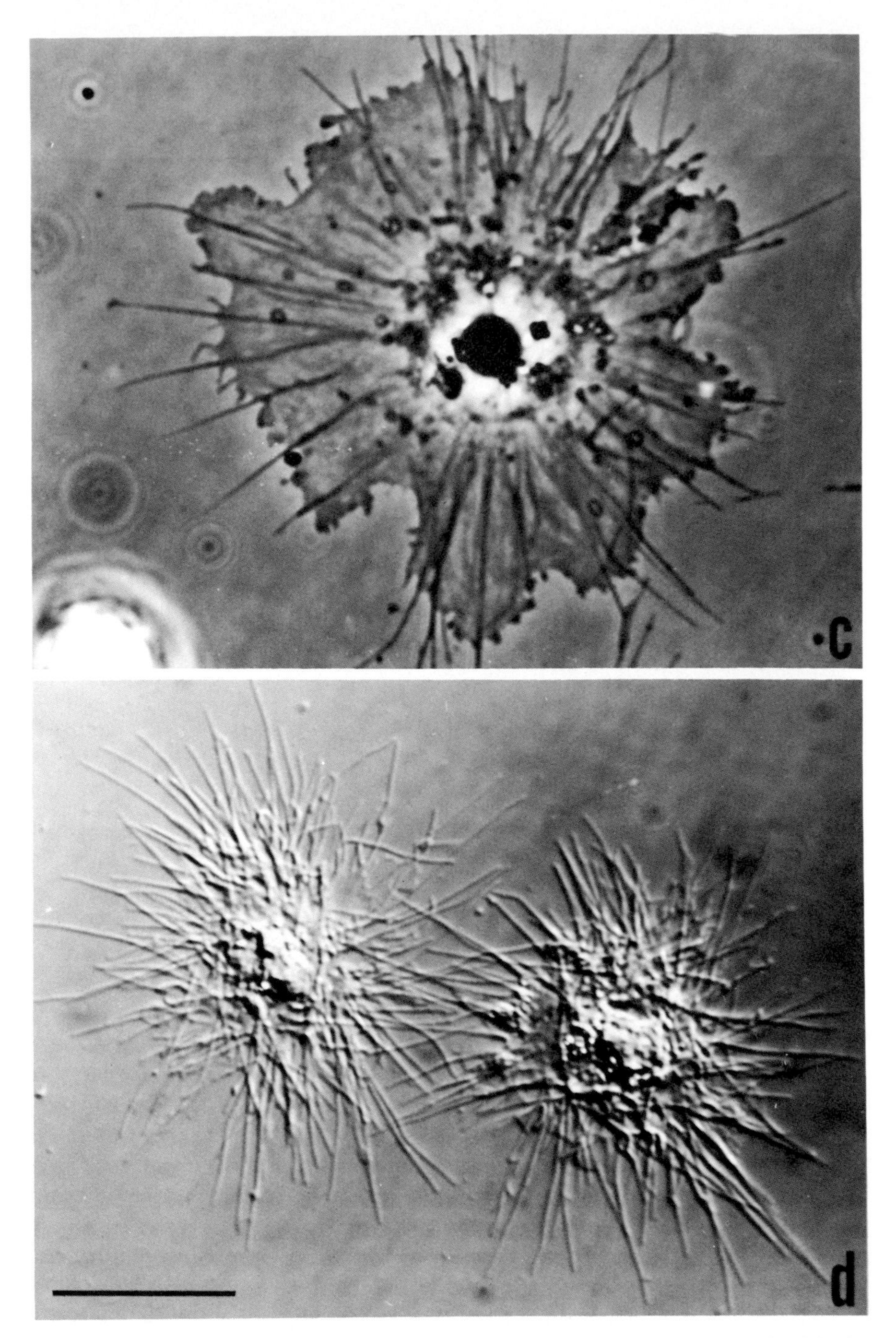

al. (1979). (a) Petaloid cells before attachment to the glass substrate. (b) One cell, on the right, has attached, is well spread, and has formed a large circumferential lamellipodia. The cell on the left has

1970), and we will not discuss it in depth. Although the molecular organization of the cortical cytoplasm is still poorly understood, the general conclusion from this earlier work using a variety of different techniques—such as compression (Cole, 1932; Hiramoto, 1963), centrifugation and pigment-granule displacement (Brown, 1934; Marsland, 1939; Zimmerman *et al.*, 1957), controlled suction in a cell elastometer (Mitchison and Swann, 1954a,b), or metal-particle displacement (Hiramoto, 1974)—is that the cortical cytoplasm undergoes a marked increase in "stiffness" shortly after fertilization and again at first division.

Studies of the outer surface of the egg with the scanning electron microscope show numerous short papillae about 0.2 μm in length (Hagstrom and Lonning, 1976; Eddy and Shapiro, 1976; Schatten and Mazia, 1976), which we presume are precursors of microvilli. These papillae can be seen in thin section as short extensions of cortical cytoplasm, but few if any microfilaments are detectable (Begg and Rebhun, 1979). These bundles project into the cortical cytoplasm, where additional filamentogenesis forms a mat of fibers resembling a terminal web. This initial elongation of microvilli, what Schroeder (1978) refers to as the "first burst," is closely coupled with the ionic events at fertilization, with the exocytosis of cortical granules, and with the stiffness increase. Begg and Rebhun (1979) have provided data suggesting that filament formation in these eggs may be triggered by the pH rise that follows fertilization (Johnson *et al.*, 1976; Shen and Steinhardt, 1978). Schroeder (1979) has demonstrated that the exocytosis of the cortical granules transiently increases the egg surface area over twofold within a minute after fertilization. This membrane is then resorbed over a 15-minute period, and Chandler and Heuser (1979) suggest that the resorption may occur through an endocytic mechanism because they observed a large, but transient, increase in coated vesicles shortly after insemination.

This cortical activity is followed by a "second burst" of elongation of microvilli approximately midway between fertilization and first division. This extension, which takes about 30 minutes in *Strongylocentrotus drobachiensis*, results in microvilli with an average length of 1.8 μm. Burgess and Schroeder (1977) have shown that these microvilli have a bundle of actin filaments in their core

---

begun to transform spontaneously. (c and d) Effect of hypotonic shock on the untransformed cells. Transformation was initiated by immersion in 0.3 *M* NaCl–0.001 *M* EGTA–0.01 *M* Tris HCl at pH 8. (c) Approximately 2–3 minutes after immersion, (d) about 10 minutes later. (a–c) Phase micrographs; (d) Nomarski micrograph of two fully transformed cells. The actin-filament bundles are clearly visible in (c) as phase-dense rods that extend into the deeper cytoplasm from the forming filopodia. Bar = 5 μm.

FIG. 8. Negatively stained isolated filopodial after treatment from Triton X-100. The fragment on the left shows the 11- to 13-nm transverse periodicity along the long axis of the filament bundle. The other three fragments show a similar periodicity and, in addition, appear to have the actin filaments inserted into an electron-lucent cap structure. Bar = 0.2 μm.

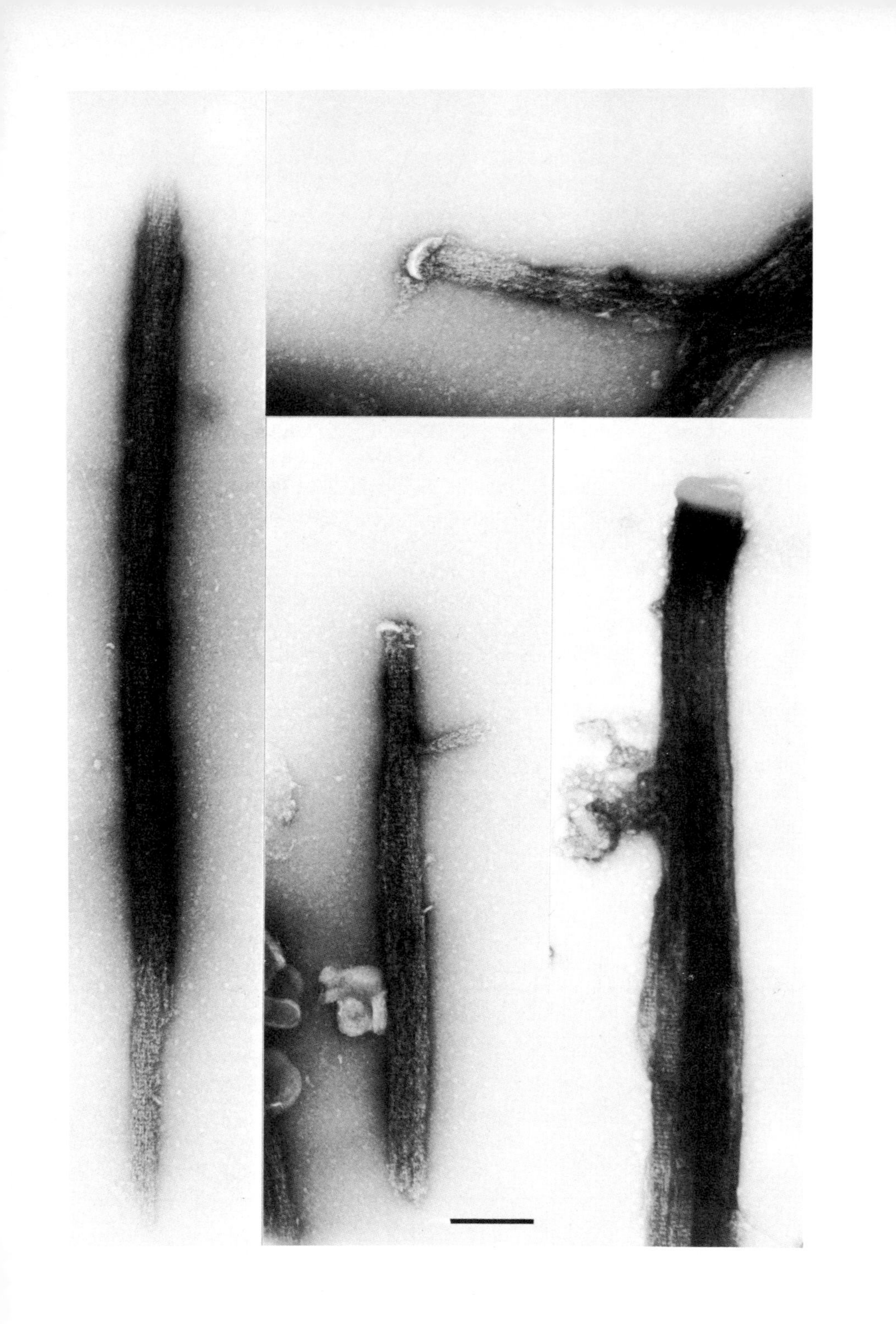

and that this bundle has an 11- to 13-nm transverse-banding pattern similar to that seen in egg actin gels (Kane, 1975, 1976). More recently, Spudich and Amos (1979) have used optical diffraction and image-reconstruction techniques to examine the cores in detail. Their findings are similar to those of DeRosier *et al.* (1977). Spudich and Amos find that their preparations give layer line reflections close to that expected for the 41/19 helix proposed by DeRosier *et al.* Their data give a value for the position of the links that is intermediate between the 41/19 helix and the 28/13 helix. They further suggest that the actin filaments within the *in situ* bundle are supercoiled with a long axial repeat of 150–200 nm. In the detailed model proposed by these workers, the heads of the fascin cross-links appear to bind in the grooves of the actin helices and may be associated with adjacent pairs of actin monomers.

Using eggs from *T. gratilla,* we have recently examined the redistribution of actin and fascin following fertilization (Otto *et al.*, 1980). The content of actin and fascin in cytoplasmic extracts prepared as described earlier was measured using the DNase I-inhibition assay for actin (Blikstad *et al.*, 1978) and a radioimmunoassay for fascin (Otto and Bryan, 1980). The results are given in Table I. Actin comprises approximately 1.4% of the total egg protein and almost 7% of the soluble cytoplasmic protein in the extracts from unfertilized eggs. The cortical actin not solubilized during preparation of the cytoplasmic extracts accounts for approximately 10% of the total actin in unfertilized *Tripneustes* eggs. Fascin is about 0.3% of the total egg protein and is essentially all soluble in unfertilized eggs under the present extraction conditions. Fascin accounts for about 1.5% of the soluble extract protein. These values are markedly reduced in extracts from fertilized eggs, which also show a reduced ability to form structural gels. The data demonstrate a depletion of about 30–35% of the soluble actin and

TABLE I

QUANTITATION OF ACTIN AND FASCIN IN SEA URCHIN EGGS[a]

	Total actin	Soluble actin	Cortical actin	Soluble fascin	Cortical fascin
Unfertilized					
ng/egg	0.64	0.58	0.026–0.06[b]	0.12	ND
molecules/egg ($\times 10^9$)	9.2	8.3	0.37–0.88	1.25	
Fertilized					
ng/egg	0.64	0.40	0.24	0.08	0.04
molecules/egg ($\times 10^9$)	9.2	5.7	2.6	0.83	0.41

Reproduced with permision of the *Journal of Cell Biology*.

[a] Total egg protein = 46 ng/egg ± 2.5 ng/egg (SD). Total soluble protein (100,000g – 60 minutes) = 8.3 ng/egg ± 0.5 ng/egg (SD).

[b] Values are either direct measurements or difference measurements; see text for details.

fascin by 45 minutes postfertilization. The EM evidence indicates microvilli are elongating at this time. Antibody-localization studies on isolated egg cortices with antifascin provide direct evidence that some or all of the insoluble fascin is incorporated into microvilli. It is not possible for the moment to decide what fraction of the redistributed actin and fascin is localized in the cortex, or whether there is an extensive subcortical actin-containing cytoskeleton in these eggs. This work and that of Spudich and Spudich (1979) show, however, that there can be a substantial incorporation of actin into the cortex after fertilization. Although detailed correlations have not been done, it seems reasonable to associate the initial increase in cortical stiffness with the actin filamentogenesis and cross-linking that occurs after fertilization. It is not yet clear if this first activity requires active contraction or the participation of egg myosin. The second increase in stiffness, however, can be directly correlated with cell cleavage (for a review, see Hiramoto, 1970), a contractile process, which can be blocked by microinjection of egg antimyosin.

H. Isolation of Myosin from Cytoplasmic Extracts

The failure to observe contraction in the structured sea urchin cytoplasmic gels was initially attributed to a lack of myosin, since Mabuchi (1976) had shown that egg myosin was insoluble at the low salt concentrations used to make cytoplasmic extracts. It is now clear, however, that the cytoplasmic extracts contain myosin, which is soluble at low ionic strength but which is inactivated on warming to 40°C (Kane, 1980). This myosin precipitates spontaneously during storage of the extracts at 0°C for long periods. The precipitation can be greatly accelerated by the addition of excess ATP (8–10 m*M*). With 8 m*M* ATP at 0°C, myosin precipitation is essentially complete in 3 hours. The extracts develop a slight turbidity during this time, which, in the phase microscope, appears to be due to the formation of large numbers of small phase-lucent spheres. These can be collected by sedimentation at 5000 *g* for 10 minutes, washed once on 0.01 *M* PIPES (pH 6.9), and finally dissolved in a small volume of 0.6–1.0 *M* KCl. On 10% Laemmli gels, this material consists of a high-molecular-weight protein, which comigrates with the rabbit-muscle myosin heavy chain (Fig. 9). On 7.5% gels, two lower-molecular-weight proteins are present, which we presume are sea urchin myosin light chains. The identification of this protein as myosin rests partially on the subunit molecular weights, the ATPase activity of this fraction, and the observation that the demyosinated extracts will not undergo the contraction phenomena we describe later (Kane, 1980). The demyosinated extracts can be returned to the original solution conditions by dialysis against 0.01 *M* PIPES (pH 6.9)–0.0001 *M* EGTA–0.0001 *M* ATP. Sodium azide is then added to 1 m*M*. There appears to be essentially no difference between the original and the demyosinated extracts in their ability to form structured gel. After addition of

FIG. 9. Co-electrophoresis of rabbit skeletal-muscle myosin with sea urchin myosin. Lane 1, rabbit skeletal-muscle myosin heavy chains. Lane 2, sea urchin egg myosin precipitated from egg cytoplasmic extract by addition of 8 m*M* ATP and incubation on ice. Lane 3, mixture of two myosin heavy chains. Electrophoresis is on 10% Laemmli slab gels.

high ATP, dialysis of the extract with myosin against low ATP does not resolubilize the myosin. It is not clear whether this ATP-induced change in solubility occurs *in vivo* or plays a physiological role.

I. Contraction

A nonstructured contractile gel can be induced in the cytoplasmic extracts by a 10-minute incubation at 35°C to accelerate the polymerization of actin without inactivating the myosin. On return to room temperature (25°C), contraction occurs in less than 1 hour and produces a contraction pellet that is about 1% of the original gel volume. The high ATP concentrations necessary for myosin precipitation are not required for contraction, provided the KCl concentration is 25 m*M* or higher. In low-concentration KCl, the high ATP concentration is necessary for contraction; at 25 m*M* KCl or above, 2 m*M* ATP is sufficient. The incubation of the demyosinated extract under equivalent conditions produces a nonstructured gel that does not contract. Full contraction, judged by time after incubation at 35°C, can be restored by addition of approximately 25% of the initial amount of myosin present in the extract.

The composition of the contraction pellet is markedly different from the structured gel (Fig. 10). Actin, myosin, and the 220K protein are present, as well as two additional proteins of 250K and 110K. The function of the 110K component is unknown. The 250,000-dalton protein will comigrate with vertebrate filamin

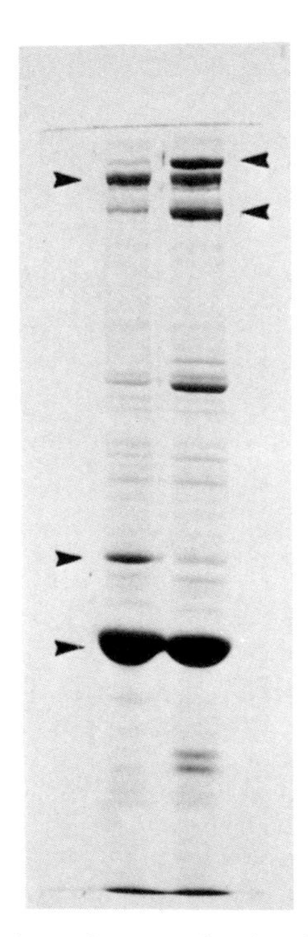

FIG. 10. Electropnoretic comparison of structured actin gel and unstructured contractile gel. The structured gel (left-hand lane) was obtained by a modification of the procedure described by Kane (1975, 1976). Egg-cytoplasmic extract with 1 m*M* ATP and 15 m*M* KCl added was warmed to 40°C for 10 minutes, then incubated at 25°C for an additional 60 minutes. The resulting actin gel was collected by centrifugation at 20,000 *g* for 15 minutes. The contractile gel was obtained by adding 4 m*M* ATP–1 m*M* $MgCl_2$, and 50 m*M* KCl to the egg-cytoplasmic extract and warming to 35°C for 10 minutes. Under these conditions the higher-MW proteins are not denatured and the resulting unstructured gel will contract into a small, dense "contraction" pellet that can be removed from the extract. The gel profile of this contraction pellet is shown in the right-hand lane. Actin, fascin, and the 220,000-dalton protein are marked by arrows in the left lane. Myosin and sea urchin filamin are marked by arrows in the right lane. The identity of the 110,000-dalton protein in the contraction pellet is not known. The separation conditions are 10% Laemmli SDS gels.

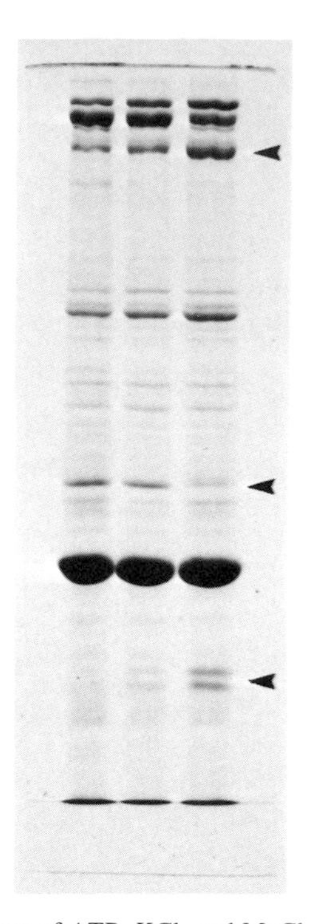

FIG. 11. Comparison of the effects of ATP, KCl, and $MgCl_2$ on contraction and composition of the contraction pellet. The left-hand lane is a noncontractile structured gel formed after addition of 1 m*M* ATP and 15 m*M* KCl followed by a 10-minute incubation at 35°C to initiate actin polymerization. The solution was then incubated at 25°C for 60 minutes and the resulting gel collected by centrifugation at 20,000 *g* for 15 minutes. The center lane shows a slowly contracting gel formed after addition of 2 m*M* ATP and 25 m*M* KCl followed by a 10-minute 35°C incubation to initiate actin assembly. The right-hand lane shows a rapidly contracted gel that was formed in 4 m*M* ATP, 1 m*M* $MgCl_2$, and 50 m or KCl followed by a 10-minute 35°C incubation. There is a marked decrease in fascin and some reduction in the 220,000-dalton protein as the rate of contraction increases. In addition there is an increase in the myosin content and two lower-MW polypeptides, which are tentatively identified as tropomyosin. The positions of myosin, fascin, and the tropomyosins are indicated by arrows. The electrophoresis is on 10% Laemmli slab gels.

on SDS gels and we presume, but have not proved, that this is the analogous sea urchin protein. The 58K polypeptide is absent from the contraction pellet under optimal contraction conditions.

J. Relation of Structured Gelation and Contraction

These observations suggest that there is an antagonistic relation between the formation of structured gelation and contraction, and that this antagonism involves fascin. A noncontractile gel is formed at 1 m*M* ATP and 15 m*M* KCl; at 2 m*M* ATP and 25 m*M* KCl, slow contraction occurs and the amount of fascin in the gel pellet is reduced. An increase in ATP and KCl causes a rapid contraction and a complete loss of fascin from the contraction pellet (Fig. 11; see also Kane, 1980). In general, the presence of the 58K protein produces a rigid, noncontractile gel. Increasing the temperature or time of incubation at 35°C progressively inactivates the myosin and reduces the amount of 250K protein in the gel, eventually eliminating the contraction and increasing the fascin content. The heat lability of myosin can also be demonstrated by heating purified myosin, then testing it for its ability to induce contraction in demyosinated extracts.

IV. Summary and Further Problems

We have described the methods used to prepare egg-cytoplasmic extracts that will both gel and contract, and have outlined the procedures for purification of some of the proteins involved. Most of the biochemical work has been done with the Hawaiian urchin, *Tripneustes gratilla,* but structured gelation, actin and myosin purification, and localization of fascin with antibodies have also been done with *Strongylocentrotus purpuratus, Dendraster eccentricus,* and *Parastichopus californicus.* With some variation, the methods outlined have general usefulness for echinoderms.

The gelation and contraction observed in cytoplasmic extracts can serve as *in vitro* models for some of the early cortical changes seen in eggs and for cytokinesis. The evidence indicates that actin and fascin form bundles *in vitro* that closely resemble those observed *in situ* in microvilli, and that the structural gelation is a massive *in vitro* assembly of microvillar cores. Our evidence demonstrates the incorporation of fascin into the extending cortical microvilli, but falls short of actually proving that fascin is the causal agent in filament-bundle formation and microvillus extension. Some preliminary work with the extracts, however, shows that antibodies against fascin will block bundle formation. This suggests that we should be able to study the role of fascin in the early cortical changes more directly by microinjection of antifascin into eggs. We are currently pursuing this more direct approach.

Another reasonably well-defined problem concerns the distribution and localization of fascin in fertilized eggs. Antibody localization and electron-microscopic data place fascin in the microvilli. The RIA measurements show that this is about 30–35% of the total egg fascin. The filaments in the terminal-weblike region, however, do not appear to be tightly bundled even though there is excess fascin available. This argues for modification of fascin or for some type of blocking molecule. There are at least two candidates for the blocking function.

As we discussed before, gel contraction is slowed or inhibited by fascin. Although we have not measured displacement of fascin from actin filaments by myosin, it is possible that myosin could attenuate actin–fascin interactions in the cortex. The second potential regulatory molecule is tropomyosin. In attempting to make the actin–fascin interaction sensitive to calcium ions, we observed that rabbit-muscle tropomyosin, but not troponin, would quantitatively displace fascin from either rabbit muscle or sea urchin actin filaments. Sea urchin tropomyosin has been localized in the cortical region (Ishimoda-Takagi, 1979) and could reduce actin–fascin interactions in the web.

One unsatisfactory aspect of extract behavior, in a physiological sense, is the necessity of using elevated temperatures to induce actin polymerization. Although 90% or more of the unfertilized egg actin is soluble in the extracts and the actin concentration is near 1 mg/ml, warming to ambient temperature (25°C) does not induce rapid filament assembly. Warming to 35–40°C is distinctly nonphysiological and can result in complete or partial denaturation of egg myosin, the 250K filamin-like protein, and presumably, other regulatory factors. The mechanism of the block and its temperature sensitivity are not yet clear. Begg and Rebhun (1979) have shown that the assembly of actin in cortical preparations from eggs can be triggered by a pH shift similar to that observed *in vivo* (Johnson *et al.*, 1976; Shen and Steinhardt, 1978). Titration of the egg-cytoplasmic extracts over a similar pH range, however, does not induce filamentogenesis. This suggests that in the egg either the nonfilamentous actin in the cortex is uniquely pH sensitive, or perhaps that the slight alkalinization of the cytoplasm releases a blocking factor from the cortical actin. It is necessary to argue that this initial assembly then triggers or nucleates additional polymerization, since straightforward quantitative considerations indicate there is insufficient cortical actin to construct all the microvilli, and we can directly measure an increase in actin in the cortex after fertilization (Otto *et al.*, 1980; Spudich and Spudich, 1979). We have partially tested the idea that the extracts might need to be nucleated. The addition of exogeneous F-actin to the extracts will produce gelation even at 4°C, demonstrating that a temperature-dependent activation of the gelation factors is not required. We have not, however, determined whether the exogeneous F-actin is further elongated or simply serves as a substrate for the gelation factors.

The presence of profilin in urchin eggs has been questioned (Mabuchi, 1979), although our own preliminary work using the methods outlined for isolation of profilin from vertebrate platelets (Markey *et al.*, 1978) indicates that a molecule that comigrates with platelet profilin does copurify with the urchin actin. We are currently trying to develop extract conditions that will allow initiation of filamentogenesis by a physiological signal.

Another problem of actin assembly concerns the effect of calcium ions on gelation. Kane originally observed that the extract gelation was sensitive to calcium. The presumption was that this ion blocked the interaction between actin filaments and the gelation factors. However, we have since found that the addition of F-actin to calcium-inhibited extracts will produce either structural or contractile gels, depending on the solution conditions. Furthermore, the recovery of sedimentable actin from extracts incubated at either 35 or 40°C is greatly reduced when calcium ion is buffered at micromolar levels. These two observations suggest that the inhibitory effect of calcium ions on gelation is directly through an inhibition of actin assembly. We have observed a similar effect in human platelet extracts (Wang and Bryan, 1980) and have partially purified a specific protein that modulates G-actin assembly in a Ca^{2+}-dependent manner. We are currently looking for similar factors in the urchin extracts.

The discovery of conditions for generating contractile gels in these extracts now provides a means to study contraction and also egg-myosin mobilization. Mabuchi (1976) has shown that egg myosin is necessary for cytokinesis, and we can infer from the work of Fujiwara and Pollard (1978) on vertebrate cells that myosin becomes localized in the contractile ring during cytokinesis. The different solubility properties of egg myosin in the extracts may be related to this localization, with the low-salt soluble form being the storage or transport state. The *in vitro* data show that fascin will block the myosin contraction of actin gels. It is not clear whether fascin is involved in the formation of the contractile ring, or the egg ATP and KCl levels are sufficiently high so that the antagonism seen *in vitro* is eliminated.

A final set of problems is concerned with the functions of the other polypeptides of the structural gels in contraction. For example, the 220K protein(s) are the other major components of the structural gel. The reconstitution experiments (Bryan and Kane, 1978) show that the function of these molecules *in vitro* is to cross-link actin–fascin needles into a three-dimensional network. The situation is not so clear *in vivo*. Attempts to colocalize the 220K polypeptides with fascin on the actin bundles of coelomocytes have not given straightforward results (J. J. Otto and J. Bryan, unpublished), but in general we do not observe colocalization. In addition, it has been surprising to find these molecules enriched in the contraction pellets that are devoid of fascin. In short, the function of these molecules is unknown.

We infer, mainly from the vertebrate gelation work (Stossel and Hartwig, 1976; Wang, 1977), that actin cross-linking in the nonstructured contractile gel is mediated by the 250K filamin-like protein. However, the location of this molecule and its functions during fertilization also remain unknown.

ACKNOWLEDGMENTS

This work was supported by NIH Grant GM26091 to Joseph Bryan, GM14363 to Robert E. Kane, and by an MDA postdoctoral fellowship and Purdue Cancer Center Grant to Dr. Joann Otto. The authors want to thank Dr. Otto for her careful reading and criticism of this review, and express their appreciation to Ms. Joyce Wu for her excellent technical assistance during the work and to Ms. Dotty Florip for her constant help in preparing the manuscript. Part of this work was done at the Friday Harbor Laboratories and we express thanks to Dr. Dennis Willows for the opportunity to work there.

REFERENCES

Begg, D. A., and Rebhun, L. I. (1979). *J. Cell Biol.* **83,** 241-248.
Blikstad, I., Markey, F., Carlsson, L., Persson, T., and Lindberg, U. (1978). *Cell* **15,** 935-943.
Brown, D. E. S. (1934). *J. Cell. Comp. Physiol.* **5,** 335-346.
Bryan, J., and Kane, R. E. (1978). *J. Mol. Biol.* **125,** 207-224.
Burgess, D. R., and Schroeder, T. E. (1977). *J. Cell Biol.* **74,** 1032-1037.
Chambers, R., and Chambers, E. E. (1961). "Explorations into the Nature of the Living Cell." Harvard Univ. Press, Cambridge, Massachusetts.
Chandler, D. E., and Heuser, J. (1979). *J. Cell Biol.* **83,** 91-108.
Cole, K. S. (1932). *J. Cell Comp. Physiol.* **2,** 1-19.
DeRosier, D., and Edds, K. (1980). *Exp. Cell Res.* **126,** 490-493.
DeRosier, D., Mandelkow, E., Silliman, A., Tilney, L., and Kane, R. E. (1977). *J. Mol. Biol.* **113,** 679-695.
Edds, K. T. (1977). *J. Cell Biol.* **73,** 479-491.
Edds, K. T. (1979). *J. Cell Biol.* **83,** 109-115.
Eddy, E. M., and Shapiro, B. M. (1976). *J. Cell Biol.* **71,** 35-48.
Fujuwara, K., and Pollard, T. D. (1978). *J. Cell Biol.* **77,** 182-195.
Hagstrom, B. E., and Lonning, S. (1976). *Protoplasma* **87,** 281-290.
Hatano, S., Kondo, H., and Miki-Noumura, T. (1969). *Exp. Cell Res.* **55,** 275-277.
Hiramoto, Y. (1963). *Exp. Cell Res.* **32,** 59-75.
Hiramoto, Y. (1970). *Biorhelogy* **6,** 201-234.
Hiramoto, Y. (1974). *Exp. Cell Res.* **89,** 320-326.
Ishimoda-Takagi, T. (1979). *Exp. Cell Res.* **119,** 423-428.
Johnson, J. D., Epel, D., and Paul, M. (1976). *Nature (London)* **262,** 661-664.
Just, E. E. (1939). "The Biology of the Cell Surface." McGraw-Hill (Blakiston), New York.
Kane, R. E. (1975). *J. Cell Biol.* **66,** 305-315.
Kane, R. E. (1976). *J. Cell Biol.* **71,** 704-714.
Kane, R. E. (1980). *J. Cell Biol.* **86,** 803-809.
Mabuchi, I. (1973). *J. Cell Biol.* **59,** 542-547.
Mabuchi, I. (1976). *J. Mol. Biol.* **100,** 569-582.
Mabuchi, I. (1979). *In* "Cell Motility" (S. Hatano, H. Ishikawa, and H. Sato, eds.), pp. 147-164. University Park Press, Baltimore, Maryland.
Markey, F., Lindberg, U., and Eriksson, L. (1978). *FEBS Lett.* **88,** 75-79.
Marsland, D. A. (1939). *J. Cell. Comp. Physiol.* **13,** 15-22.
Miki-Noumura, T., and Kondo, H. (1970). *Exp. Cell Res.* **61,** 31-41.
Miki-Noumura, T., and Oosawa, F. (1969). *Exp. Cell Res.* **56,** 224-232.
Mitchison, J. M., and Swann, M. M. (1954a). *J. Exp. Biol.* **31,** 443-460.
Mitchison, J. M., and Swann, M. M. (1954b). *J. Exp. Biol.* **31,** 461-472.
Otto, J. J., and Bryan, J. (1981). *Cell Motil.* **1,** 179-192.
Otto, J. J., Kane, R. E., and Bryan, J. (1979). *Cell* **17,** 285-293.
Otto, J. J., Kane, R. E., and Bryan, J. (1980). *Cell Motil.* **1,** 31-40.
Schatten, G., and Mazia, D. (1976). ECR **98,** 325.
Schroeder, T. E. (1978). *Dev. Biol.* **64,** 342-346.

Schroeder, T. E. (1979). *Dev. Biol.* **70,** 306–326.
Shen, S., and Steinhardt, R. (1978). *Nature (London)* **272,** 253–254.
Spudich, J. A., and Watt, S. (1971). *J. Biol. Chem.* **246,** 4866–4871.
Spudich, A., and Spudich, J. A. (1979). *J. Cell Biol.* **82,** 212–226.
Spudich, J. A., and Amos, L. A. (1979). *J. Mol. Biol.* **129,** 319–331.
Stossel, T. P., and Hartwig, J. H. (1976). *J. Cell Biol.* **68,** 602–619.
Wang, K. (1977). *Biochemistry* **16,** 1854–1865.
Wang, L. L., and Bryan, J. (1981). *Cell* **25,** 637–649.
Zimmerman, A. M., Landau, T. V., and Marsland, D. A. (1957). *J. Cell. Comp. Physiol.* **49,** 395–435.

Chapter 10

Macrophages: Their Use in Elucidation of the Cytoskeletal Roles of Actin

JOHN H. HARTWIG AND THOMAS P. STOSSEL

Hematology–Oncology Unit, Massachusetts General Hospital
and
Department of Medicine, Harvard Medical School
Boston, Massachusetts

I. Introduction

Macrophages have a specialized layer of cytoplasm located just beneath their plasma membrane. This cortical cytoplasm is dynamic, undergoing changes in size and shape during cell movement. In macrophages it forms a variety of cortical protrusions or extrusions, including pseudopodia, membrane ruffles, and

ISBN 0-12-564125-7

microvilli. It has a transparent appearance in the light microscope because it excludes granules and other organelles.

Actin filaments predominate in electron micrographs of the macrophage cortex and seem to be organized into a dense network of short, highly branched filaments (Reaven and Axline, 1973; Trotter, 1979). The presence of actin, a major muscle protein in cortical cytoplasm, suggested that it was responsible in part for the movement of this region and, because of its location, for the maintenance of cell shape and for the stability of the membrane. The cortical cytoplasm of the macrophage is similar morphologically to the cortex of many eukaryotic cells. Unlike some cells, such as fertilized oocytes, activated blood platelets, or intestinal brush-border epithelial cells, the macrophage does not usually form surface protrusions that contain parallel bundles of actin filaments.

Extracts of macrophages prepared with 0.34 *M* sucrose contain high concentrations of actin (15–20% of the total protein) and a number of proteins that interact with actin. Some of the proteins are definitely present and possibly concentrated in the cortical cytoplasm, as determined by immunofluorescence and subcellular-fractionation studies (Stendahl *et al.*, 1980; Davies and Stossel, 1977; Hartwig *et al.*, 1977). We have systematically purified and characterized these proteins and analyzed their interactions with actin. These proteins can be divided into four categories based on their effects of actin.

1. Proteins that form complexes with actin monomers or oligomers, preventing their assembly into long filaments.
2. Proteins that promote the formation of actin-filament networks, increasing the rigidity of F-actin solutions.
3. Proteins that shorten actin-filament length without affecting the monomer–polymer equilibrium.
4. Proteins that move actin filaments.

In this chapter, we summarize the purification, methods of assay, and biochemical properties and interactions of the macrophage proteins of these four categories.

II. Macrophages: Properties of the Cells and Their Cytoskeletal Components

A. State of Actin Assembly in Macrophages: Monomer–Polymer Equilibrium

Actin is the most abundant protein of macrophages, constituting 12% of the total cell protein. Macrophage actin, as is the case for all other nonmuscle actins studied thus far, is a 42,000-dalton globular protein (G-actin) indistinguishable from muscle actin in all important aspects. This finding is expected because of the high degree of conservation in the primary structure of actins (Vanderkerckhove and Weber, 1978).

Actin molecules can assemble into double-helical polymers (F-actin). The equilibrium distribution between monomers and polymers is determined by the ambient ionic conditions, temperature, and presence of exogenous proteins (Oosawa and Kasai, 1971). The assembly state of actin in macrophages is not precisely known. Only one half of the actin present in macrophage extracts will assemble into long polymers in the presence of 0.1 *M* KCl, a condition in which at least 95% of an equivalent concentration of purified macrophage actin would exist as long filaments (Stossel and Hartwig, 1976).

The inability of actin to polymerize into long filaments in extracts may be due to the presence of protein(s) that interfere with this process. Rabbit lung macrophages and human granulocytes contain a protein that can inhibit the assembly of actin monomers into long filaments (Figs. 1 and 2; Tatsumi *et al.*, 1981; Southwick and Stossel, 1981). By both electrophoresis in sodium dodecyl sulfate (SDS) and hydrodynamic measurement, this actin inhibitor protein is a globular molecule having a molecular weight (M_r) of 65,000 daltons (Table I). Macrophage actin-inhibitor protein blocks actin monomers from being incorporated into long filaments, suggesting that this protein complexes with actin oligomers

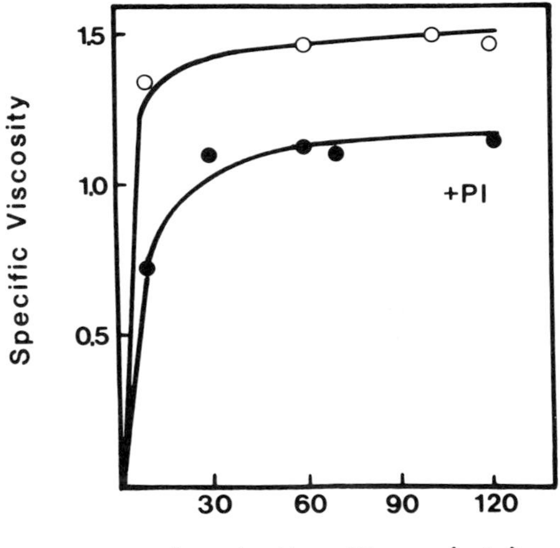

FIG. 1. Time course showing the effect of the purified macrophage actin-inhibitor protein (PI) on the specific viscosity of actin. Actin 2.1 mg/ml in the absence (○) or presence of 0.34 mg/ml inhibitor protein (●).

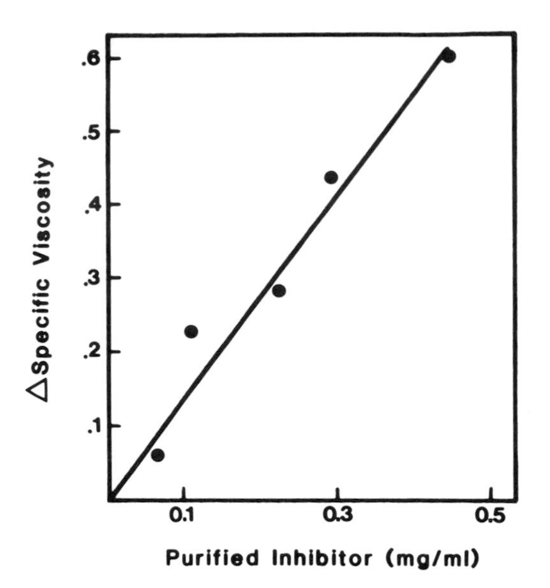

FIG. 2. Effect of increasing macrophage inhibitor protein concentration on the final viscosity of actin. G-actin (1.1 mg/ml), in the presence or absence of inhibitor protein, was made 0.1 *M* with KCl and incubated for 2 hours at 25°C. The specific viscosities of these solutions are plotted compared to actin polymerized without inhibitor—that is, Δ decrease in specific viscosity.

or forms short filaments. The concentration of the actin-inhibitor protein in macrophage extracts is sufficient to account for all of the unsedimentable actin in these extracts. Factors that might control the interaction of the 65K inhibitor protein with actin have not been identified. To be effective, the actin-inhibitor protein must be added to G-actin before or during polymerization, and it does not depolymerize preformed actin filaments. Preliminary studies indicate that the actin filament binding site of this protein is different from gelsolin (see the following). Therefore, it may interfere with actin assembly by binding to the "unpreferred" end (the pointed end of F-actin as determined by heavy meromyosin binding) of F-actin or by binding to G-actin and removing it from the filament assembly pool. Its activity is not changed by the presence or absence of calcium.

B. Network Formation

Because actin filaments are extremely long, they interact to form weak networks at physiological ionic strength and pH. These networks are reversibly disrupted by shearing forces, which cause the filaments to slide past one another and acquire parallel alignment. When the shearing force is removed, the filaments again become randomly oriented in the solution and form gels. This weak

TABLE I

MACROPHAGE PROTEINS THAT REGULATE ACTIN STRUCTURE

Protein	Cellular function	M_r	S^{0a}	a^b	Other properties
Actin-binding protein	Network formation	540,000 2 × 270,000	9.4	13.5 nm	160 × 3 nm strands containing two binding sites for actin filaments. K_a for F-actin, 2 μM^{-1}. Stimulates the rate of actin polymerization.
Gelsolin	Regulation of network structure	91,000	4.9	4.4	Reversibly regulates actin-filament lengths. Calcium dependent, K_a for calcium, 1 μm^{-1}. Binds both G- and F-actin in calcium. Binds to "preferred assembly end" of actin filaments.
65,000-MW actin-polymerization inhibitor	Regulates monomer–polymer actin equilibrium	65,000	4.0	3.3	Binds actin.
Myosin	Movement of F-actin	470,000 2 × 200,000 2 × 20,000 2 × 15,000		19.0	300-nm bipolar filaments in 0.1 M KCl. Actin-activated Mg^{2+}-ATPase superprecipitates actin filaments.

thixotropic network is sustained by entanglements of filaments and formation of weak bonds between filaments at points of intersection (Kasai *et al.*, 1960).

We would emphasize that the apparent viscosity of actin is highly shear dependent, and that at low shear rates it is proportional to an exponential function of the filament length (Graessley, 1974). Therefore, seemingly marked alterations that have been reported in the viscosity at low shear rates of actin prepared by different methods are due to small alterations in the filament-length distribution and not because of the presence or absence of "cross-linking factors." Furthermore, the biological importance of "viscosity" at very low shear rates is likely to be minimal.

The formation of a true gel, likely to be of importance to the macrophage, requires a union of the actin filaments by cross-links having sufficient bond energies to resist the forces acting on the cell cortex. The resistance of the resulting network to deformation—the rigidity—is proportional to the number of cross-links.

Actin-binding protein is the major actin-cross-linking protein of rabbit lung macrophages (Hartwig and Stossel, 1975, 1981; Stossel and Hartwig, 1975, 1976; Brotschi *et al.*, 1978; Table I). It binds to actin filaments with high affinity ($K_a = 2\ \mu M^{-1}$) and a capacity of 1 molecule of actin-binding protein per every 14 actin monomers in filaments. Actin-binding protein is a large, asymmetrical, and flexible molecule (Fig. 3A). In 0.1 *M* KCl solutions, it is an asymmetrical dimer composed of two 270K subunits that are joined together in a head-to-head fashion (Hartwig and Stossel, 1981). The free tails, each of which possesses an actin-filament binding site, are therefore available to cross-link the actin filaments (Fig. 3B) into a rigid lattice.

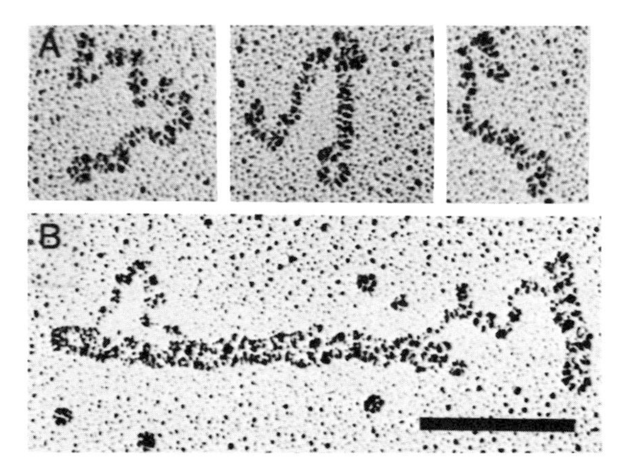

FIG. 3. (A) Electron micrographs of macrophage actin-binding protein dimers after spraying onto mica and rotary shadowing. (B) Electron micrograph showing cross-linking of actin filaments by an actin-binding protein dimer. Bar = 0.1 μM.

The formation of a true gel network is influenced by a number of factors: (1) the concentration and length of the polymers; (2) the affinity and functionality of the cross-linker for the polymer; (3) the distribution of the cross-linker in solution; and (4) the orientation of the polymers in solution. From the theory for network formation of random polymers (Flory, 1946), the physical properties of these actin and actin-binding protein networks can be related to cross-linker specificity. Network formation of a polymer solution is characterized by abrupt rises in the solution's rigidity as the cross-linker concentration approaches the critical gel point, according to the following equation.

$$V_c = \frac{C}{M_\omega} \tag{1}$$

where V_c is the cross-linker concentration at which a network first forms in moles/liter, C is the concentration of the polymer in grams/liter, and $\bar{M}_\omega$ is the weight average molecular weight of the polymer. Because these sol-gel transformations are abrupt, small alterations in polymer-to-cross-linker ratio or in the length of the polymers can provide efficient regulatory mechanisms.

The rigidity of actin filaments of known length distribution in solution is markedly increased by actin-binding protein at concentrations near those predicted by the network theory (Hartwig and Stossel, 1979, 1981; Yin *et al.*, 1980; Fig. 4). The rigidity of the networks is close to that predicted by Flory (1942) for isotropic networks, providing further evidence that actin-network behavior is not different from that of other polymer systems (Zaner, unpublished data). The agreement with theory also suggests that actin-binding protein molecules are very efficient actin-gelling agents. Since the theory for gelation assumes a random binding of the cross-linker to the polymer, the high efficiency of actin-binding protein in gelling actin requires that cross-linking of the actin filaments occur randomly at points of filament intersection. The large size and flexibility of actin-binding protein molecules allow them to bind to specific sites on filaments overlapping at random.

The isotropy of actin filaments in solution is not altered when amounts of actin-binding protein close to the critical gelling concentrations are added to the filaments. However, when actin is polymerized in the presence of actin-binding protein, the resulting gel is composed of short filaments that branch perpendicularly off one another (Hartwig and Stossel, 1980). This occurs because actin-binding protein acts as a nucleus for actin-filament assembly (Fig. 5). It stimulates the onset of actin polymerization, resulting in an increased number of shorter filaments (Oosawa and Kasai, 1971). Because actin-binding protein molecules are bipolar, filaments assemble off each of the free tails, forming branching structures observed in electron micrographs.

In theory, the cross-linking of preformed actin filaments in solution by actin-binding protein molecules, or a condensation polymerization in which cross-

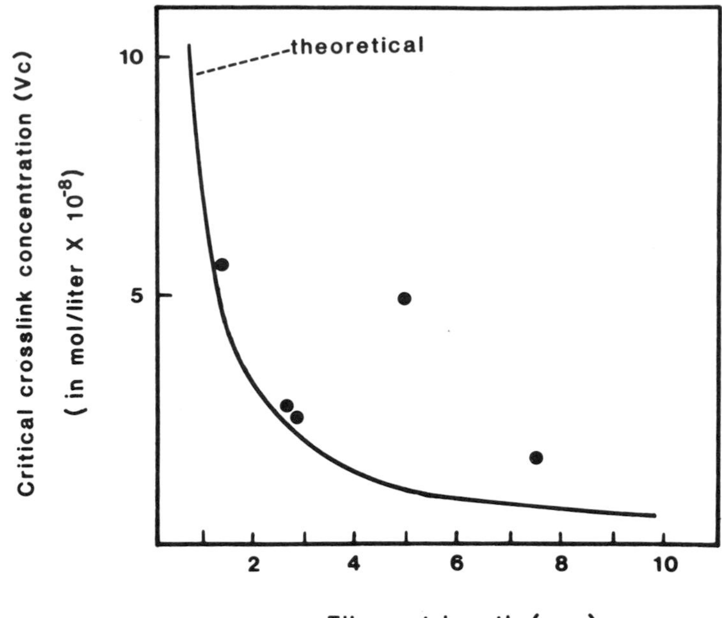

FIG. 4. Plot of the theoretical critical gel point for actin filaments of known length (—) versus the experimentally determined critical gel points with actin-binding protein dimers (●). The length distribution of actin filaments for gel-point determinations were measured in electron micrographs.

linking and polymer growth occur simultaneously (i.e., monomers assemble into filaments off the ends of actin-binding protein dimers), can generate a network structure (Flory, 1946). In practice, however, it is likely that a different network is created in each case. By nucleating filament assembly, shorter filaments are formed, creating a network composed of short actin struts. These networks will be more rigid than gels composed of long filaments cross-linked by the addition of actin-binding protein. The morphology of networks prepared by copolymerization of actin-binding protein with actin resembles that of the cortical cytoplasm.

C. Control of Filament Length

Gelsolin, a globular calcium-binding protein with a M_r value of 91,000 (Table I), regulates network structure in macrophages by decreasing actin-filament lengths (Yin and Stossel, 1979, 1980; Yin *et al.*, 1980). Gelsolin binds 2 moles of calcium with a high (1 μM^{-1}) affinity. In the presence of calcium it binds to actin filaments, shortening them. Its shortening effect on actin filaments is very rapid and reversible and is demonstrated by the decreased flow birefringence and viscosities of F-actin. Filament shortening is also demonstrated in the electron microscope (Fig. 6). The shortening of filaments by gelsolin to solate actin network occurs without increases in the actin-monomer concentration, as verified by sedimentation, viscosity, and turbidity measurements.

The gelsolin–Ca^{2+} complex increases the V_c (Eq. 1) of actin-binding protein for the gelation of actin filaments. This effect is dependent on the gelsolin concentration and, when the actin-filament lengths are determined in the electron microscope, the experimental shift in V_c agrees well with that predicted from the shift to a lower actin-filament $\bar{M}_{\omega}$. These experiments directly demonstrate the relevance of filament-length control as a regulatory mechanism for the control of sol-gel transformation.

The shortening of filaments by gelsolin could occur by two different mechanisms. The random binding of gelsolin to filaments could lead to severing and shortening of the filaments. The second mechanism is more complicated. Gelsolin molecules in solution could bind actin monomers. If binding and nucleation of filament assembly occurred rapidly, the overall length distribution of actin filaments in solution would decrease—that is, lead to a shortening of the filament-length distribution and an increased number of filaments. In the latter case, the monomer pool would be replaced by depolymerization of the filaments. Not enough information is available at present to decide between these two possibilities. The rapidity (seconds) of the shortening seems to argue against the latter mechanism. However, gelsolin binds both G- and F-actins, and can stimulate the onset of actin polymerization, properties important in the latter mechanism. Although its precise shortening mechanism is unknown, gelsolin, after filaments have become shortened, is bound on the ends of filaments. Biochemical and morphological studies demonstrate it is bound on

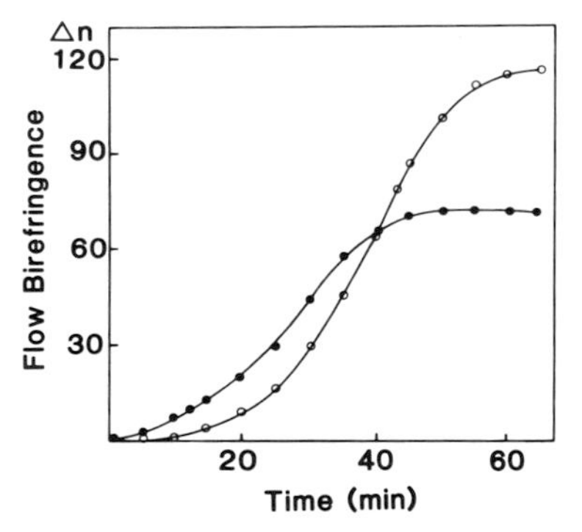

FIG. 5. Effect of actin-binding protein on the rate of actin polymerization in 0.1 M KCl measured by flow birefringence. Actin (0.33 mg/ml) polymerized in the absence of actin-binding protein (○). Actin (0.33 mg/ml) polymerized in the presence of 30 μg/ml of actin-binding protein (●).

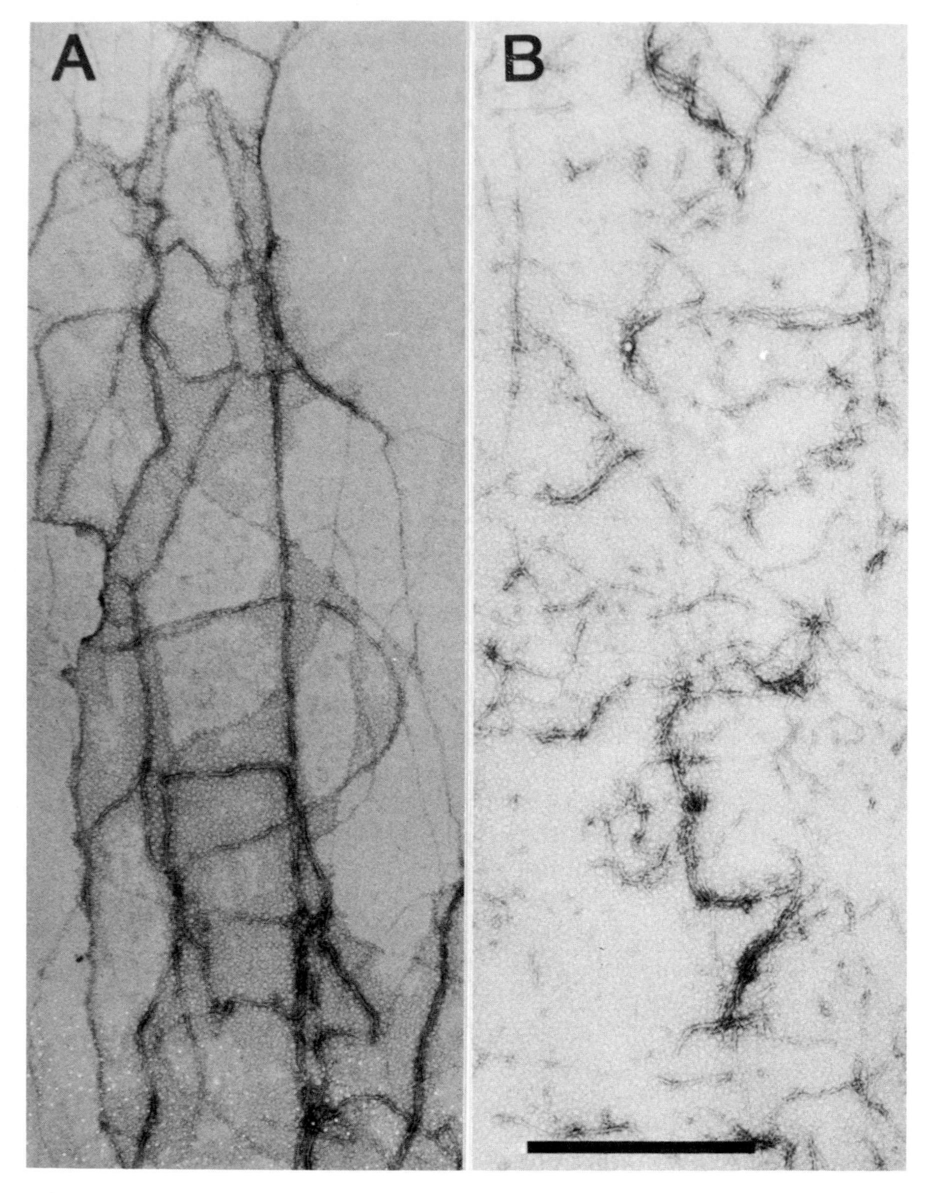

FIG. 6. Negatively stained electron micrographs demonstrating actin-filament shortening by gelsolin. (A) F-actin. (B) F-actin incubated with gelsolin at a molar ratio of 21 actin monomers in filaments per gelsolin molecule. Bar = 1 μm.

the "preferred end" for the addition of actin monomers to filaments during assembly (Yin *et al.*, 1981).

It would seem likely that gelsolin is but one of a class of calcium-dependent proteins that control the network state of actin filaments in cortical cytoplasm (Bretscher and Weber, 1979, 1980; Hasegawa *et al.*, 1980). Its widespread distribution in many mammalian cell types suggests that gelsolin may be the prototype for this type of network regulation (Yin *et al.*, 1981).

D. Movement of Actin Filaments

Macrophage myosin, in the presence of Mg^{2+} and ATP, generates a contraction of F-actin. Like other nonmuscle myosins, macrophage myosin is a hexameric molecule composed of two 200,000-dalton heavy chains, two 20,000-dalton and two 15,000-dalton light chains (Table I; Hartwig and Stossel, 1975). Purified macrophage myosin forms filaments at physiological ionic strength and pH that are 300 nm long. These filaments are composed of myosin molecules aggregated along helical portions of their heavy chains and extending heads, globular domains of the heavy chains from both ends of the filaments. The filaments bind tightly to actin filaments in the absence of ATP. Only limited information exists concerning either the state of macrophage myosin-filament assembly or the regulation of its ATP-splitting activity in the presence of F-actin and $MgCl_2$, a reaction associated with repetitive binding of the myosin heads to actin monomers in the filaments and the generation of tension in muscle. Purified macrophage myosin has a low actin-activated Mg^{2+}-ATPase activity and will slowly cause actin to contract. These activities are markedly increased independently of the calcium concentration by the addition of a crude macrophage protein fraction (Stossel and Hartwig, 1975, 1976). This crude fraction may contain a kinase that can phosphorylate the 20,000-dalton light chains of myosin to activate its actin-dependent Mg^{2+}-ATPase activity (Trotter and Adelstein, 1979).

We have utilized the macrophage as a cell from which to identify and purify proteins found in cortical cytoplasm that may be important for the structure and movements of this region of the cell. As cell biologists, we wish to know how these proteins are arranged in the cortex and how the cortical structure works to cause movements in response to various stimuli acting within and on the cell. The effects of the various proteins we have isolated from the cortex of macrophages on actin suggest some ideas concerning this arrangement and function. These ideas may have to be modified as more information becomes available and will have to be subjected to various tests, when possible, to establish or negate their validity.

E. Assembly of the Cortical Actin Lattice

G-actin synthesized on cellular polyribosomes would be expected to polymerize rapidly into filaments under the ionic conditions in the cell (Fig. 7, step 1). This assembly would very likely be detrimental because it would be unregulated and would impair the transport of actin molecules to the cortex. Such spontaneous filament assembly could be suppressed by molecules of the actin-inhibitor (PI) protein or by other proteins of similar function. The inhibitor complexes actin monomers or oligomers to block long filament growth (Fig. 7, step 2). Because the actin–PI complexes are small, relative to actin filaments, they could diffuse to the cortex. Molecules of actin-binding protein are present in the cortex and bind to actin with high affinity. Therefore, actin might be expected to dissociate from the PI and assemble in a branching polymerization onto the dimers of actin-binding protein (Fig. 7, step 3). Newly synthesized actin molecules could also assemble directly onto molecules of actin-binding protein near sites of synthesis. In the presence of calcium, assuming that calcium concentrations are greatest in the cell center and diminished outward toward the cortex, the growth of actin filaments from actin-binding protein would be prevented by molecules of gelsolin. These complexes could also diffuse to the cortex where gelsolin would dissociate in the low calcium concentrations and permit the incorporation of the actin-binding protein–actin segments into the present lattice.

The cortical actin lattice is perceived here to be in a dynamic state of constant remodeling. The supply of material for this process is not limited to synthetic origin. Remodeling can also occur in the cortex in response to changes in the concentration of calcium. Increases in the local calcium concentration would activate gelsolin, resulting in local severing of actin filaments (Fig. 7, step 4). After cutting the filaments, gelsolin molecules would cap the filaments. Since capping occurs at the end of the actin filament that has the highest affinity for

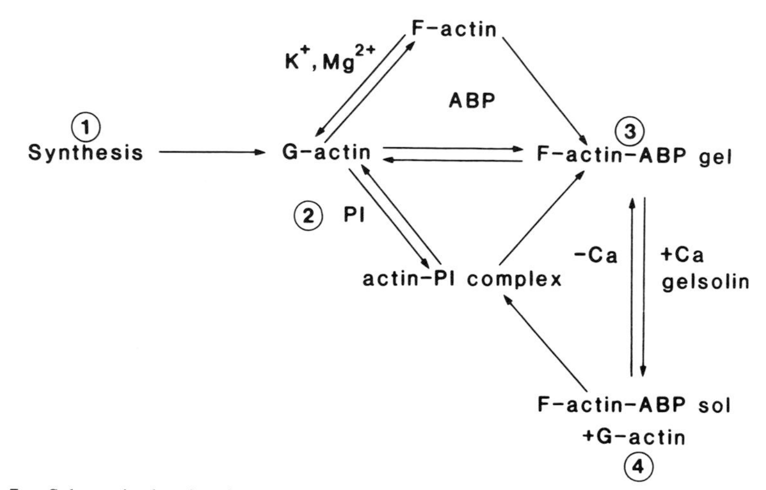

FIG. 7. Schematic showing the possible interactions of actin, actin-binding protein, gelsolin, and the actin-inhibitor protein. This figure is discussed in the text.

actin monomers, the concentration of monomeric actin would increase. These monomers could either be bound by the actin–PI, or else lead to nucleation and growth of new filaments. Decrease in the calcium concentration would reverse this process, and monomers would be available for assembly into filaments.

F. Movement of the Cortical Lattice

Directional movements of the cortical actin-filament lattice requires the ability to vary regionally the rigidity of this lattice. Lattice structure can be regulated by regional alteration in cytoplasm calcium concentration. High ($\geq 1\ \mu M$) calcium concentrations activate gelsolin and diminish network structure. Myosin filaments, dispersed uniformly within the lattice, generate movements by splitting ATP to ADP and Pi, via a sliding-filament mechanism. The principle of the hypothesis is that if actin filaments can slip past one another, no tension can develop. In an actomyosin gel, entanglements between actin filaments prevent this slippage to some extent (Hayashi and Maruyama, 1975). The cross-linking of actin filaments by actin-binding protein completely blocks slippage and therefore is expected to enhance tension generation. To some extent, this idea has been verified experimentally (Ebashi *et al.*, 1964; Stendahl and Stossel, 1980). The direction of movement of the actin filaments in response to regional changes in lattice structure is slightly controversial. We have suggested that there will be movement out of areas of decreased rigidity toward regions of greater lattice structure or rigidity (gelation-contraction coupling; Stossel, 1980). In support of this hypothesis, networks composed of F-actin and actin-binding protein gelled in capillary tubes in the presence of gelsolin and myosin move in the direction predicted above. Movement is always from the side of the capillary exposed to the higher calcium concentration (Stendahl and Stossel, 1980). It follows that membranes to which actin filaments might be attached in the cell would move in the same direction. Hellewell and Taylor (1979) have suggested that movement will be in the opposite direction (solation and contraction coupling), based on the intuitive notion that a rigid gel cannot move. "Contraction" is therefore expected only to occur in regions where the gel is dissolved.

The mechanism by which the cell controls the local cytoplasmic free-calcium concentration remains to be established. Macrophages have Mg^{2+}-ATP-dependent, calmodulin-stimulated pumps in their plasma membrane, which transport calcium ions out of the cell (Lew and Stossel, 1980). This pump may be involved in the control of directional movement, provided its activity can be coupled to environmental stimuli.

The rapid polymerization of actin in the cortex of cells can also lead to movement of cortical cytoplasm. Strong evidence for this mechanism exists in the acrosomal-extension reaction of certain starfish sperm (Tilney *et al.*, 1973; Tilney, 1976, 1978), and nonmuscle cells have large pools of unpolymerized actin (Bray and Thomas, 1975; Gordon *et al.*, 1977) that could be used in rapid filament-assembly reactions.

A polymerization mechanism in nonmuscle cells would seem to require structures in the cortex to direct the growth of the filaments. The branching polymerization of actin observed in the presence of actin-binding protein provides one such possible mechanism to organizing growth. Actin monomers could assemble onto the actin-binding sites on actin-binding protein molecules, creating a branching polymerization expanding in all directions. Structures in the cell center such as the nucleus, organelles, and intermediate filaments, would act as a buttress against inward growth, allowing only outward pseudopod expansion.

In summary, the actin-associated proteins of macrophages could function together with actin to stabilize, pull and push the plasma membrane, and account for the dynamic structure of cortical cytoplasm. Similar or identical proteins exist in polymorphonuclear leukocytes, platelets, oocytes, and epithelial cells, suggesting that the findings and ideas generated by work with the macrophage may be applicable to other cell types.

III. Purification of Actin-Binding Protein, Gelsolin, and the Actin-Polymerization Inhibitor

A. Reagents

1. Solutions

For Actin-Binding Protein and Myosin.

A. 0.6 *M* KCl–0.5 m*M* ATP–0.5 m*M* dithiothreitol–1 m*M* EGTA–10 m*M* imidazole HCl (pH 7.5)

B. 1.2 *M* KI–5 m*M* ATP–5 m*M* dithiothreitol–1 m*M* EGTA–20 m*M* imidazole HCl (pH 7.5)

C. 20 m*M* KPO_4–0.5 m*M* dithiothreitol–0.5 m*M* EGTA (pH 7.5)

D. Homogenizing buffer: 0.34 *M* sucrose–5 m*M* EGTA–5 m*M* ATP–5 m*M* dithiothreitol–5 m*M* phenylmethylsulfonyl fluoride–20 m*M* imidazole HCl (pH 7.5), containing 0.25 mg/ml soybean-trypsin inhibitor

E. Saturated solution of ammonium sulfate at 4°C in 10 m*M* EDTA (pH 7.0)

For Gelsolin.

F. 1 m*M* EGTA–1 m*M* dithiothreitol–0.5 m*M* ATP–10 m*M* imidazole HCl (pH 7.8)

G. 0.8 *M* KCl–1 m*M* EGTA–1 m*M* dithiothreitol–10 m*M* imidazole HCl (pH 7.5)

For Actin-Inhibitor Protein

H. 0.5 m*M* ATP, 0.1 m*M* $MgCl_2$, 0.75 m*M* 2-mercaptoethanol and 10 m*M* imidazole HCl, 50 m*M* KPO_4 (pH 7.5)

2. Columns

For Actin-Binding Protein and Myosin.

A. 4% agarose, 2.5 × 95 cm (A15, 200–400 mesh, Bio-gel, Bio-Rad Laboratories). The bottom 9/10 was equilibrated with solution A. The top 1/10 was equilibrated with solution B

B. DEAE–Sepharose (Pharmacia Fine Chemicals), 1.5 × 20 cm equilibrated with solution C

For Gelsolin

C. DEAE–Sepharose, 2.5 × 20 cm, equilibrated with solution F

D. Sephadex G-200, 1.6 × 90 cm, equilibrated with solution G

For Actin-Inhibitor Protein

F. Hydroxyapatite, 19 × 5 cm, equilibrated in solution H containing 50 m*M* KPO_4

A flow diagram for the purification of actin-binding protein, gelsolin, and the 65K actin-inhibitor protein from rabbit lung macrophages is shown in Fig. 8.

Macrophages are obtained from the lungs of New Zealand white rabbits injected with 1.25 ml of complete Freund's adjuvant (Cappel Laboratories, Cochranville, Pennsylvania) into the marginal ear vein 2–3 weeks prior to being sacrificed. The cells are washed from the lungs by intratracheal lavage with 0.15 *M* NaCl (6 × 50 ml) and collected by centrifugation at 280 *g* for 10 minutes at 4°C. The pelleted cells are resuspended in 0.15 NaCl, passed through cheesecloth, and collected a second time by centrifugation at 280 *g*. The average yield is 3 ml of packed cells per 5 kg rabbit.

All of the following purification steps are done at 4°C unless otherwise indicated.

B. Treatment of Macrophages with Diisofluorophosphate (DFP)

The treatment of intact macrophages with DFP prior to lysis of the cells increases the amount of actin-sedimentation activity in the extracts by 50%, presumably by inhibiting the activity of serine proteases prior to their release into the cytoplasm. Since DFP is a volatile and potent inhibitor of acetylcholinesterase, it is extremely toxic. Utmost care is therefore required in handling this chemical. Precautions include wearing gloves, working in a fume hood, and having sodium hydroxide or sodium bicarbonate nearby to inactivate the DFP in case of spillage. A supply of atropine and hypodermic syringe should be available.

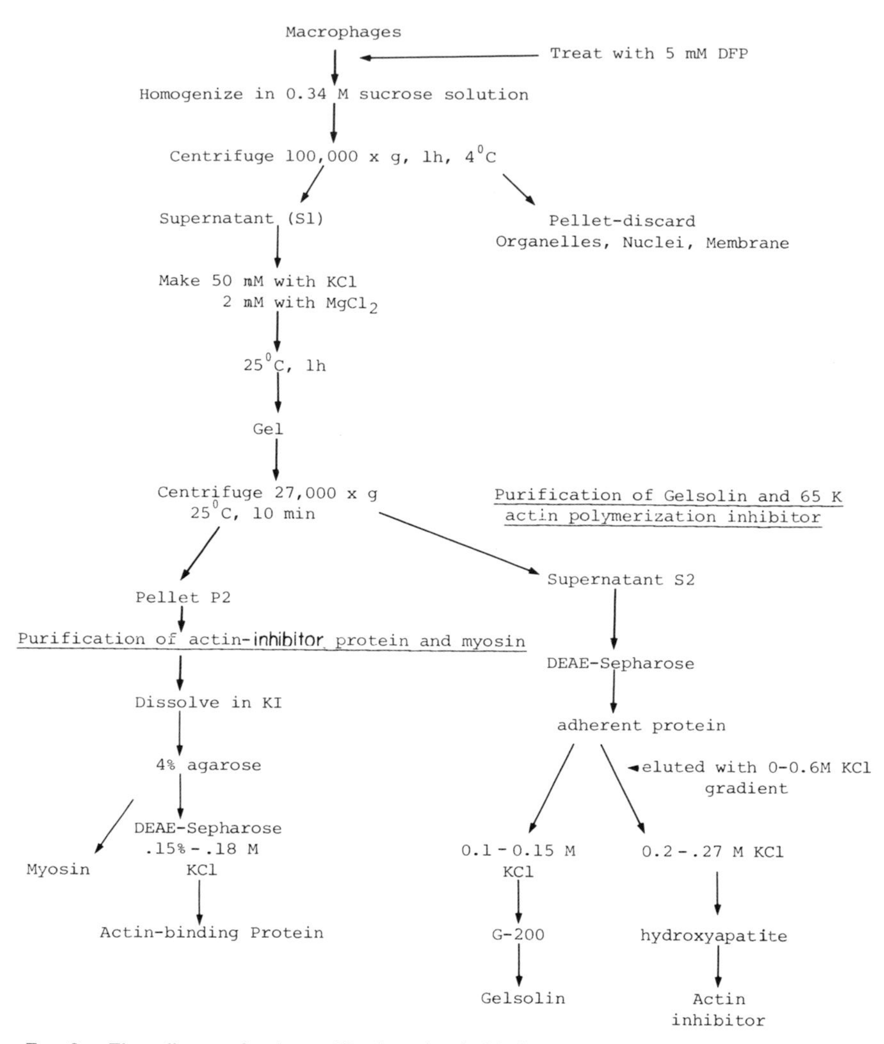

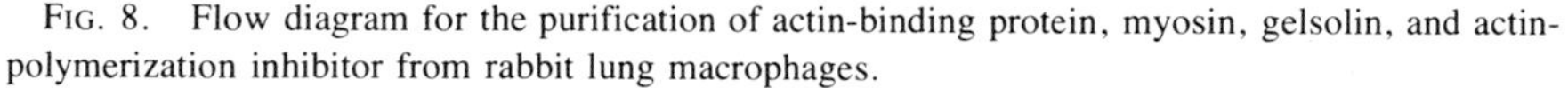

FIG. 8. Flow diagram for the purification of actin-binding protein, myosin, gelsolin, and actin-polymerization inhibitor from rabbit lung macrophages.

The cells are suspended with one volume of 0.15 *M* NaCl and this suspension is placed in 50-ml plastic tubes (20 ml per tube) having caps that can be securely tightened. To each tube, 1 ml of 0.1 *M* DFP in propylene glycol is added in the fume hood. The DFP-treated cells are incubated at ice-bath temperature for 5 minutes and the tubes filled with ice-cold deionized water, tightly capped, and centrifuged at 990 *g* for 10 minutes. The supernatant is decanted into 5 *M* NaOH to inactivate residual unreacted DFP. The cell pellets are then resuspended in 0.075 *M* NaCl in the hood, mixed, and centrifuged again at 990 *g*. Once again, the supernatant fluid is carefully decanted and inactivated in NaOH.

When chymotrypsin is added to broken cells at this point, its activity is not inhibited, indicating that there is no unreacted DFP present.

C. Homogenization of the Cells

The DFP-treated cell pellets are suspended with one volume of homogenization solution (solution D), placed in a Dounce glass tissue grinder and broken, using 50 strokes with a type B pestle. Cell rupture is monitored by phase-contrast microscopy. Nuclei, membranes, and organelles are removed from the homogenate by centrifugation at 100,000 *g* for 1 hour.

D. Extract Gelation

The supernatant liquid (S1) obtained by centrifugation is removed, made 50 m*M* with 3 *M* KCl and 2 m*M* with 1 *M* $MgCl_2$, and incubated for 1 hour in a 25°C water bath. During this time, the extract solidifies because under these conditions 50% of the actin in the extract assembles into filaments, which are then cross-linked by molecules of actin-binding protein into a rigid gel lattice. The gel is then centrifuged at 27,000 *g* for 10 minutes at 25°C, yielding a compressed pellet of gelled material (P2) and a clear supernatant fluid (S2). Actin-binding protein comprises 5.6% of the protein in the compressed gel pellet, a 3.5-fold enrichment over its S1 concentration, where it is 1.6% of the total protein. Gelsolin and the 65K actin-inhibitor protein remain in the S2 fluid expressed from the gel by centrifugation.

E. Actin-Binding Protein and Myosin

1. Chromatography on 4% Agarose

The P2 pellet is dissolved in five volumes of solution B and this solution is clarified by centrifugation at 27,000 *g* for 10 minutes. The supernatant liquid is decanted, placed on a magnetic stirrer, and diluted slowly while being stirred with an equal volume of saturated solution of ammonium sulfate. The precipitate that forms is collected by centrifugation at 15,000 *g* for 20 minutes and dissolved in 5 ml of solution B. This solution is centrifuged at 27,000 *g* for 10 minutes and the liquid phase removed and applied to a 4% agarose column equilibrated with solution A after 50 ml of solution B has been layered on the agarose beads and

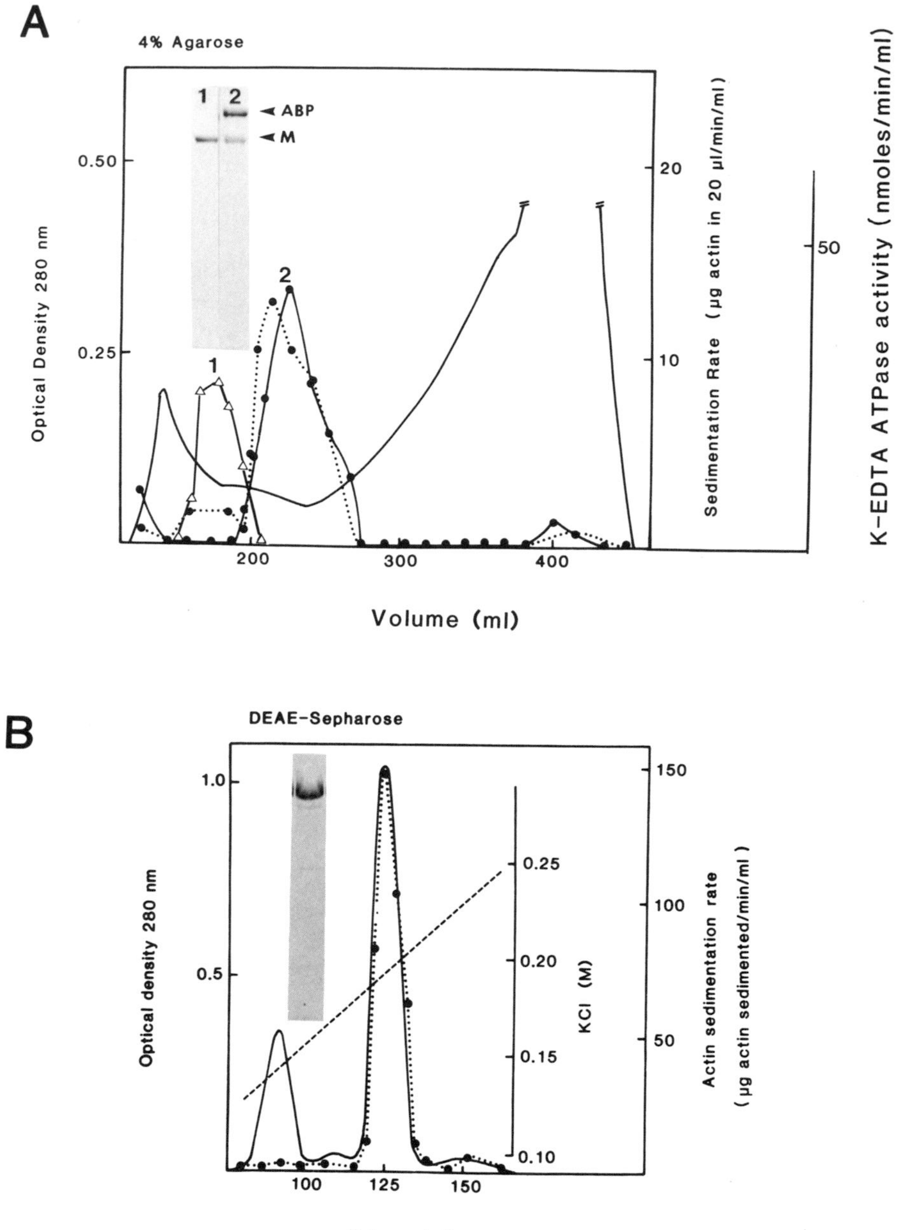

FIG. 9. (A) Gel filtration of the P2 pellet on a 4% agarose column. The absorbance at 280 nm (—), the actin-sedimentation activity in the presence (● · · · · ●) and the absence of 1 mM $CaCl_2$ (●—●), and K^+-EDTA–ATPase activity (Δ), by the eluted fractions are indicated. The insert shows

allowed to run into the gel. The sample is eluted with solution A. Figure 9A shows the elution profile of a macrophage P2 pellet on a 4% agarose column. The column fractions were monitored for both optical density, actin-sedimentation activity, and EDTA–ATPase activity in 0.6 *M* KCl (myosin). Myosin elutes first at a K_{av} of 0.24, followed by a single actin-binding protein-dependent calcium-insensitive peak of actin-sedimenting activity that elutes at a K_{av} of 0.31. The insert shows 5–15% gradient gels of the two activity peaks. Fractions with the subunit of actin-binding protein, determined by electrophoresis on polyacrylamide gels in the presence of SDS, or with actin-sedimenting activity, are pooled and dialyzed into solution C. Fractions with myosin-dependent EDTA–ATPase activity are pooled at this point.

2. DEAE–Sepharose Chromatography

A small 1.5 × 20 cm column filled with DEAE–Sepharose is equilibrated with solution C. The dialyzed fractions containing actin-binding protein are applied to this column and the adherent protein is eluted with a linear 0–0.4 *M* KCl concentration gradient. Figure 9B shows the optical density and actin-sedimentation activity profile for this column; the insert shows Coomassie Blue-stained polyacrylamide gels of the material that elutes with actin-sedimentation activity. Actin-binding protein elutes at a KCl concentration of 0.18 *M* and is 95% pure as determined by densitometry of Coomassie Blue-stained polyacrylamide gels.

F. Purification of Gelsolin and the 65K Actin-Inhibitor Protein

Both gelsolin and the inhibitor protein can be purified from the S2 fluid expressed from the centrifugally compressed gel. Gelsolin's binding to actin filaments is calcium dependent, and therefore it remains soluble in the EGTA-containing supernatant. The actin-inhibitor protein is bound to actin oligomers not sedimentable under these relatively low g-forces.

5–15% polyacrylamide gels of the K^+–EDTA–ATPase (1) and actin-sedimentation (2) activity peaks after electrophoresis in the presence of sodium dodecyl sulfate and staining with Coomassie Blue. Myosin heavy chain (M) and the subunit of actin-binding protein (ABP) are indicated. (B) Ion-exchange chromatography of the actin-sedimentation activity peak from the 4% agarose column on a DEAE–Sepharose column. The column was equilibrated as described in the text and eluted with 200 ml of a linear 0–0.4 *M* KCl concentration gradient (- - - -). Absorbance at 280 nm (—) and actin-sedimentation activity (● · · · · ●) by the eluted fractions are indicated. The insert shows a 5–15% polyacrylamide gel of the pooled activity peak after electrophoresis in the presence of sodium dodecyl sulfate and staining with Coomassie Blue.

1. Fractionation of the S2 Supernatant

The S2 supernatant fluid is dialyzed into solution F and applied to a small, 2.5 × 20 cm DEAE–Sepharose column equilibrated with the same solution. The adherent protein is eluted with a linear 0–0.6 *M* KCl gradient. Figure 10 shows the elution profile of one such S2 on this ion-exchange column. Fractions were monitored for optical density at 280 nm, for effect on calcium-dependent F-actin (gelsolin) viscosity, actin-viscosity-inhibition activity, and polypeptide composition by electrophoresis on 5–15% polyacrylamide gels in the presence of SDS. Gelsolin elutes at a starting KCl concentration of 0.1 *M* and the inhibitor at a concentration of 0.2 *M* KCl. Fractions containing gelsolin activity are pooled and made 50% with the saturated solution of ammonium sulfate. The precipitate that forms is collected at 27,000 *g* for 10 minutes at 4°C and dissolved in 1 to 2 ml of solution G. Fractions containing actin-polymerization activity are pooled and dialyzed into solution H.

2. Gel Filtration of Gelsolin Activity on Sephadex G-200

The pooled DEAE fractions with gelsolin activity are further fractionated through a Sephadex G-200 column. As shown in Fig. 11, gelsolin activity elutes

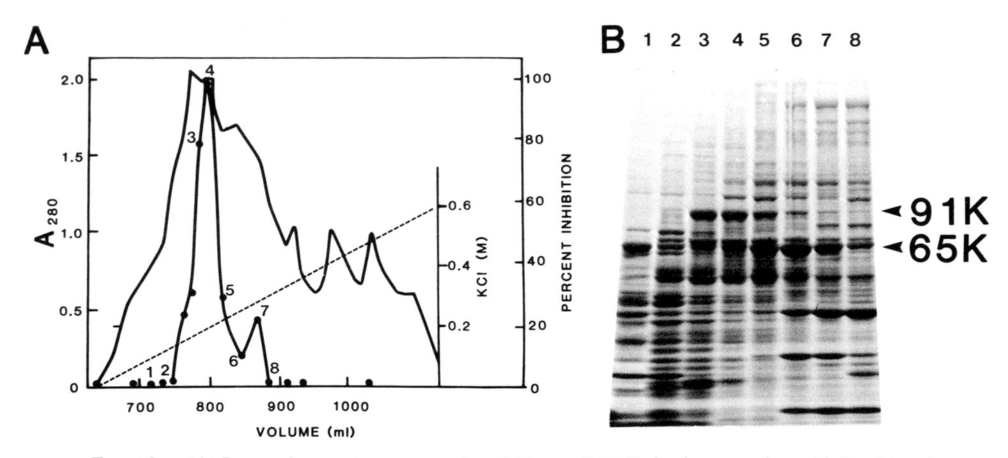

Fig. 10. (A) Ion-exchange chromatography of S2 on a DEAE–Sepharose column (1.5 × 21 cm). Adherent proteins were eluted with 500 ml of a linear 0–0.6 *M* KCl concentration gradient (- - - -) in solution containing 1 m*M* EGTA, 1 m*M* dithiothreitol, 0.5 m*M* ATP, and 0.9 m*M* imidazole HCl (pH 7.8). Absorbance at 280 nm (—) and calcium-dependent (gelsolin) inhibition of F-actin viscosity (●—●) by the eluted fractions are indicated. The numbers indicate fractions electrophoresed in B. (B) Coomassie-blue-stained 5–15% polyacrylamide gel of eluted fractions indicated in (A). Fractions 3 to 5 have the peak inhibitor activity and contain the 91K polypeptide of gelsolin. Fractions 6 to 8 contain the 65K polypeptide of the actin-polymerization inhibitor. The migration of gelsolin (91K) and the actin-polymerization inhibitor are indicated on the figure.

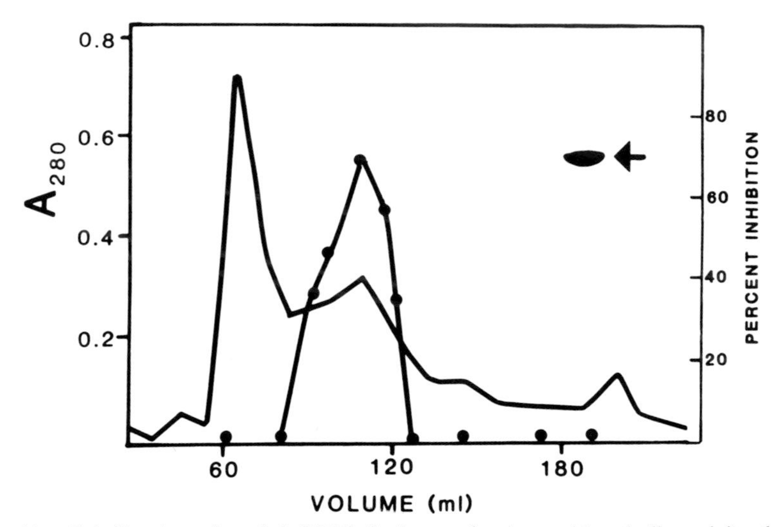

FIG. 11. Gel filtration of pooled DEAE–Sepharose fractions with gelsolin activity through Sephadex G-200 column. Fractions were eluted in a 0.8 *M* KCl solution containing 1 m*M* EGTA, 1 m*M* dithiothreitol, and 10 m*M* imidazole HCl (pH 7.5) from a 1.6 × 90 cm column. Absorbance at 280 nm (—) and calcium-dependent inhibition of F-actin viscosity (●—●) by the eluted fractions are indicated. Insert: Coomassie Blue-stained polyacrylamide gel of fractions with peak activity after electrophoresis in sodium dodecyl sulfate. The 91,000-dalton band corresponding to gelsolin is indicated.

as a single symmetrical peak. The insert shows that this peak copurifies with the 91K gelsolin polypeptide.

3. FRACTIONATION OF ACTIN-VISCOSITY-INHIBITION ACTIVITY BY HYDROXYAPATITE

Eluted fractions containing actin-viscosity-inhibition activity on the DEAE–Sepharose column are pooled and dialyzed into solution H containing 50 m*M* KPO_4. These are then applied to a small hydroxyapatite column equilibrated with the same solution. The adherent protein is eluted with a linear 0.5–0.4 *M* KPO_4 gradient (Fig. 12). Actin-viscosity-inhibition activity elutes as a sharp peak at a potassium phosphate concentration of 0.1 *M*.

G. Assays for Actin-Binding Protein, Gelsolin, and Actin-Inhibitor Protein

1. REAGENTS

Actin. Actin is prepared from the back and leg muscle of rabbits by the procedure of Spudich and Watt (1971), except that 0.8 *M* KCl is substituted for

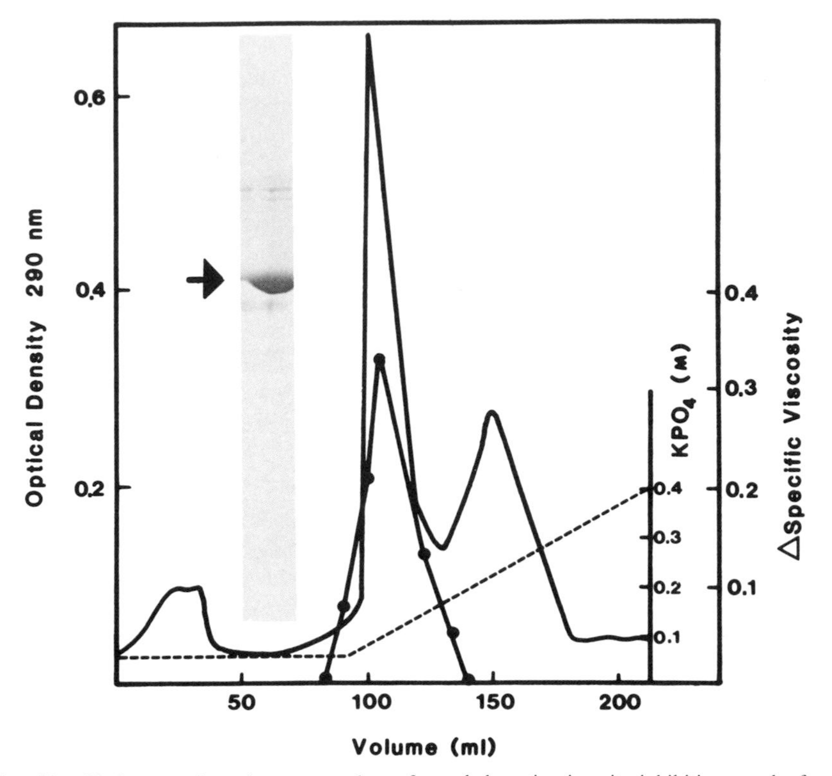

FIG. 12. Hydroxyapatite chromatography of pooled actin-viscosity-inhibition peak from DEAE–Sepharose column. The pooled material was dialyzed into solution H, applied to the hydroxyapatite column, and nonadherent protein eluted with the same buffer. Adherent protein was eluted with a linear 0.05 to 0.4 *M* potassium phosphate gradient (- - - -) in solution H. Optical density of the eluted protein at 290 nm (—) and actin-inhibitory activity (●—●) are indicated. The insert shows a 5–15% polyacrylamide gradient gel of the actin-inhibitory-activity peak fraction. The migration of the 65K polypeptide of the actin-inhibitor protein is indicated by the arrow.

0.6 *M* KCl during the first polymerization cycle. G-actin is passed through a sterile 0.45 μm Millipore filter and stored at 4°C. G-actin (10 mg/ml) is induced to polymerize by the addition of KCl to a final concentration of 0.1 *M* at 25°C for 3 hours prior to use in sedimentation and viscosity assays.

2. PRINCIPLE OF ACTIN-VISCOSITY-INHIBITION ASSAY

The effect of the inhibitor protein on the assembly of actin is assayed in Cannon-Manning microviscometers. Skeletal-muscle G-actin (0.7 mg/ml, final viscosity of 0.6 units) is induced to polymerize by the addition of KCl to a final concentration of 0.1 *M* in the presence or absence of the purified actin-inhibitor protein or less-pure fractions containing it. When actin polymerizes in the absence of the inhibitor, long filaments form and viscosity of the solution increases. Because this protein prevents the formation of long actin filaments it decreases the final actin viscosity in a dose-dependent fashion (Fig. 2).

3. Units of Specific Inhibitory-Polymerization Activity

Inhibitory activity is expressed as the change in specific viscosity of the actin solution in the absence versus the presence of the inhibitor protein per mg protein.

4. Assays for Network Formation

The point at which a polymer solution gels is characterized by an abrupt increase in the rigidity of the solution. The gel point can therefore be easily determined and the amount of cross-linking protein required to solidify the polymer quantitated. Gel points have been determined by a number of methods, including the tipping of test tubes [i.e., looking for the point where the liquid stops flowing (Maruta and Korn, 1977)], viscometry (MacLean-Fletcher and Pollard, 1980), in simple yield-point devices (Brotschi *et al.*, 1978), or in more elegant and accurate viscoelastometers (Zaner *et al.*, 1981).

We have found sedimentation to be extremely useful for the quantitation of macrophage-cross-linking activity. Sedimentation measures the aggregation state of the actin filaments. Since actin filaments are extremely long in solution, they are highly susceptible to aggregation. Under the conditions, pH, and g-forces used in our sedimentation assay, they are, however, sufficiently dispersed in the absence of cross-linking proteins to sediment slowly. Proteins capable of cross-linking actin filaments result in aggregation in the actin filaments, causing them to sediment more rapidly from solution. Since sedimentation is proportional to the amount of cross-linker protein added over a large range, it is highly suitable for the quantitation of cross-linking activity.

Procedure. Samples tested for actin-sedimenting activity are dialyzed into 0.1 *M* KCl–0.5 m*M* EGTA–10 m*M* imidazole HCl, pH 7.5 (0.1 *M* KCl solution) at 4°C. Each assay contains a final volume of 0.2 ml. 40 μl of F-actin, 10 mg/ml, and 160 μl of the sample or of the 0.1 *M* KCl solution (control) is placed in 1.5-ml plastic microcentrifuge tubes (Sarstedt Co., Princeton, New Jersey). When the sample volume is less than 160 μl, 0.1 *M* KCl solution is used to make up the difference. The protein solution is immediately mixed and a 20-μl sample is removed immediately from the solution and its protein concentration determined (Lowry *et al.*, 1951). The protein solution is then incubated at 25°C for 30 minutes and centrifuged at 10,000 *g* for 10 minutes at 25°C. After centrifugation,

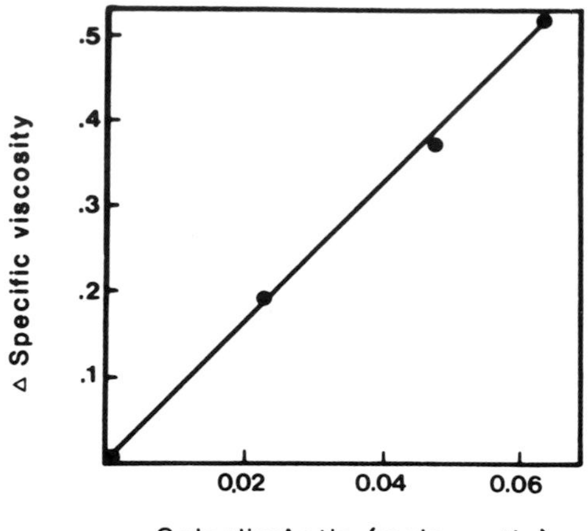

FIG. 13. Effect of increasing gelsolin concentrations on the viscosity of F-actin. The specific viscosity of F-actin in the presence of gelsolin is plotted relative to the viscosity of F-actin alone.

20 μl is carefully aspirated from the meniscus of the solution in the tube and its protein concentration determined and compared to the initial protein concentration present in the solution before centrifugation.

Actin-sedimentation activity units. Actin-sedimentation activity is expressed as micrograms of actin protein sedimented from a 20-μl volume per minute of centrifugation at 10,000 *g* per mg of protein in the test sample.

5. Assay for Gelsolin Activity

Gelsolin activity can be measured in network-formation assays or by sedimentation, by looking for a calcium-dependent inhibitor of either network formation or sedimentation (Yin and Stossel, 1979). However, since its effect on network structure occurs directly by shortening actin filaments, its activity can be determined by viscometry. F-actin solutions are incubated in the presence of gelsolin and in the presence and absence of calcium. In the presence of calcium, gelsolin is active, the filaments are shortened, and the viscosity of the solution decreased. The decrease in the viscosity of F-actin solutions is proportional to the amount of added gelsolin protein (Fig. 13).

References

Bray, D., and Thomas, C. (1975). *Biochem. J.* **147,** 221–228.

Bretscher, A., and Weber, K. (1979). *Proc. Natl. Acad. Sci. U.S.A.* **76,** 2321–2325.

Bretscher, A., and Weber, K. (1980). *Cell* **20,** 839–847.

Brotschi, E. A., Hartwig, J. H., and Stossel, T. P. (1978). *J. Biol. Chem.* **253,** 8988–8993.
Davies, W. A., and Stossel, T. P. (1977). *J. Cell Biol.* **75,** 941–955.
Ebashi, S., Ebashi, F., and Maruyama, K. (1964). *Nature (London)* **203,** 645–646.
Flory, P. J. (1942). *J. Phys. Chem.* **46,** 132–140.
Flory, P. J. (1946). *Chem. Rev.* **39,** 137–197.
Gordon, D. J., Boyer, J. L., and Korn, E. D. (1977). *J. Biol. Chem.* **252,** 8300–8309.
Graessley, W. W. (1974). *Adv. Polym. Sci.* **9,** 3–178.
Hartwig, J. H., and Stossel, T. P. (1975). *J. Biol. Chem.* **250,** 5696–5705.
Hartwig, J. H., and Stossel, T. P. (1979). *J. Mol. Biol.* **134,** 539–553.
Hartwig, J. H., and Stossel, T. P. (1981). *J. Mol. Biol.* **145,** 6–7.
Hartwig, J. H., Davies, W. A., and Stossel, T. P. (1977). *J. Cell Biol.* **75,** 956–967.
Hartwig, J. H., Tyler, J., and Stossel, T. P. (1980). *J. Cell Biol.* **87,** 841–848.
Hasegawa, T., Takahashi, S., Hayashi, H., and Hatano, S. (1980). *Biochemistry* **19,** 2677–2683.
Hayashi, T., and Maruyama, K. (1975). *J. Biochem. (Tokyo)* **78,** 1031–1038.
Hellewell, S. B., and Taylor, D. L. (1979). *J. Cell Biol.* **83,** 633–648.
Kasai, M., Kawashima, H., and Oosawa, F. (1960). *Polym. Sci.* **44,** 51–69.
Lew, P. D., and Stossel, T. P. (1980). *J. Biol. Chem.* **255,** 5841–5846.
Lowry, O. H., Rosebrough, N. J., Farr, A. L., and Randall, R. J. (1951). *J. Biol. Chem.* **193,** 265–275.
MacLean-Fletcher, S. D., and Pollard, T. D. (1980). *J. Cell Biol.* **85,** 414–428.
Maruta, H., and Korn, E. D. (1977). *J. Biol. Chem.* **252,** 399–402.
Oosawa, F., and Kasai, M. (1971). *In* "Subunits in Biological Systems, Part A" (Timasheff, S., and Fasman, G. D., eds.), pp. 261–322. Marcel Dekker, New York.
Reaven, E. P., and Axline, S. G. (1973). *J. Cell Biol.* **59,** 12–27.
Southwick, F. S., and Stossel, T. P. (1981). *J. Biol. Chem.* (in press).
Spudich, J. A., and Watt, S. (1971). *J. Biol. Chem.* **246,** 4866–4871.
Stendahl, O. I., and Stossel, T. P. (1980). *Biochem. Biophys. Res. Commun.* **92,** 675–681.
Stendahl, O. I., Hartwig, J. H., Brotschi, E. A., and Stossel, T. P. (1980). *J. Cell Biol.* **84,** 215–224.
Stossel, T. P. (1980). *In* "The Motor of Macrophages" (G. W. Bailey, ed.), pp. 589–598. Claitor's Publ. Div., Baton Rouge, Louisiana.
Stossel, T. P., and Hartwig, J. H. (1975). *J. Biol. Chem.* **250,** 5708–5712.
Stossel, T. P., and Hartwig, J. H. (1976). *J. Cell Biol.* **68,** 602–614.
Tatsumi, N., Southwick, F. S., and Stossel, T. P. (1981). In preparation.
Tilney, L. G. (1976). *J. Cell Biol.* **69,** 73–89.
Tilney, L. G. (1978). *J. Cell Biol.* **77,** 551–564.
Tilney, L. G., Hatano, S., Ishikawa, H., and Mooseker, M. S. (1973). *J. Cell Biol.* **59,** 109–126.
Trotter, J. A. (1979). *J. Cell Biol.* **83,** 231a.
Trotter, J. A., and Adelstein, R. S. (1979). *J. Biol. Chem.* **254,** 8781–8785.
Vanderkerckhove, J. and Weber, K. (1978). *FEBS Letters* **102,** 219–222.
Yin, H. L., and Stossel, T. P. (1979). *Nature (London)* **281,** 583–586.
Yin, H. L., Hartwig, J. H., Maruyama, K., and Stossel, T. P. (1979). *J. Biol. Chem.* **256,** 9693–9697.
Yin, H. L., and Stossel, T. P. (1980). *J. Biol. Chem.* **255,** 9490–9493.
Yin, H. L., Zaner, K. S., and Stossel, T. P. (1980). *J. Biol. Chem.* **255,** 9494–9500.
Yin, H. L., Albrecht, J., Fattoum, L. (1981). In preparation.
Zaner, K. S., Fotland, R., and Stossel, T. P. (1981). *Rev. Sci. Instrum.* **52,** 65–67.

Chapter 11

Crane Fly Spermatocytes and Spermatids: A System for Studying Cytoskeletal Components

ARTHUR FORER

Biology Department
York University
Downsview, Ontario, Canada

I. Introduction

This chapter describes methods for rearing crane flies in the laboratory, and for dissecting living testes to study living spermatogenic cells *in vitro*. Why study crane flies? One can rear crane flies in the laboratory and obtain cells at any time, so one is not restricted to a given season. Year-round availability is an advantage compared to seasonally available cells, such as *Haemanthus* endosperm or sea urchin zygotes; why, though, would one choose to work with cells from crane flies rather than other consistently available cells such as tissue culture cells or cells from *Drosophila* or from other animals that can be reared in the laboratory?

ISBN 0-12-564125-7

There are two major reasons for choosing to work with crane fly spermatocytes. One reason, of interest to those studying mitosis and the mitotic apparatus, is the favorable *cytology* during cell division: the spermatocytes are large (the spindle pole-to-pole distance is 20–30 μm in flattened cells), and primary spermatocytes contain only five small chromosomes (three autosomal bivalents and two sex-chromosome univalents), each of which can be followed simultaneously *in vivo,* throughout cell division (e.g., Fig. 11). Further, in each half-spindle there are only five chromosomal spindle fibers, each of which is ~ 1 μm in diameter: These large, easily studied chromosomal spindle fibers can be followed throughout cell division (e.g., Fig. 11). Before considering the second reason for studying crane fly spermatogenic cells, I give a brief overview of the kinds of experiments that have been done on the mitotic apparatus in crane fly spermatocytes.

Normal living crane fly spermatocytes have been studied using phase-contrast and polarization microscopy. Information has been obtained about chromosomal orientations (and reorientations) during various stages of meiosis; about velocities of movements during the different stages of meiosis; about the timing (and speeds) of autosomal anaphase relative to the later sex-chromosomal anaphase; about the development and birefringence of individual spindle fibers during various stages of meiosis; about spindle-shape changes during various stages; and about velocities of autosome movements and sex-chromosome movements at different temperatures (e.g., Bauer, 1931; Bauer *et al.,* 1961; Begg and Ellis, 1979b; Behnke and Forer, 1966; Dietz, 1956, 1963, 1969, 1972a,b; Forer, 1965, 1966, 1969, 1976; La Fountain, 1972, 1974a, 1976; Salmon and Begg, 1980; Schaap and Forer, 1979). Living crane fly spermatocytes have been studied after ultraviolet microbeam irradiation of single chromosomal spindle fibers; after flattening so that centrioles are found in the cytoplasm and not at the spindle poles; during pressure treatment or micromanipulation of spindle fibers or chromosomes; after colchicine treatment and irradiation with 360-nm light; and after treatment with hypotonic or hypertonic media (e.g., Begg and Ellis, 1979a,b; Dietz, 1959, 1966; Forer, 1965, 1966, 1969, 1972; Forer and Koch, 1973; Salmon, 1975; Salmon and Ellis, 1975; Sillers and Forer, 1981a,b,c). Finally, electron-microscopic data include study of kinetochore shapes and sizes during cell division; serial-section analysis of microtubule numbers and arrangements in spindles; study of microtubules in association with nonkinetochore (''lamellae'') regions of autosomes and sex chromosomes; study of spindle microtubules after tannic-acid staining or after freeze-fracture; and study of presence (or absence) of spindle microfilaments after usual fixations, after tannic-acid fixation, or after glycerination and reaction with heavy meromyosin (HMM) (e.g., Behnke and Forer, 1966; Behnke *et al.,* 1971; Forer and Behnke, 1972a; Forer and Brinkley, 1977; Fuge, 1971, 1973, 1974a,b, 1975, 1977a,b, 1980a,b, 1981; Fuge and Müller, 1972; La Fountain, 1974a,b, 1975, 1976; La Fountain and

Thomas, 1975; La Fountain *et al.*, 1977; Müller, 1972). Some of the main conclusions from these data have been summarized elsewhere (Forer, 1980). I think it is fair to say that many details of cell division have been studied quite well in crane fly spermatocytes: some progress has been made in identifying the force producers (for chromosome movement), in describing the organization and properties of spindle fibers, in describing how microtubules are distributed during cell division, in describing changes in birefringence in individual chromosomal spindle fibers during cell division, and in studying how the movements of individual chromosomes can influence the movements of other chromosomes in the same cell.

A second reason for choosing to study crane fly spermatogenic cells, perhaps of interest to those studying cytoskeletal elements other than the mitotic apparatus, is the *synchrony* of the cells inside a testis: 5000–10,000 spermatogenic cells are enclosed within the testis wall (the "wall" is a single layer of cells), and the spermatogenic cells are synchronous—that is, *all* the spermatogenic cells in a given testis are in about the same stage of spermiogenesis at the same time. For example, in *N. abbreviata,* all the cells in one testis enter meiotic S-period at about 1 day after the start of the IV-larval stage, all enter zygotene about 2 days later, all enter pachytene about a day or two later, and all take part in cell division (i.e., prometaphase-I through telophase-II) on days 8–9. (The IV-instar is the *form* of the insect; the time period during which the larvae have the IV-instar form is the IV-larval *stage,* or stadium.) Thus if one removes a testis at day 7, all cells in the testis are *pre*-prometaphase; at day 8, cells in the testis are in various stages between prophase-I and spermatid; at day 10, all cells in the testis are early spermatids; and so forth. Thus, unlike testes from *Drosophila* or mouse or grasshopper, which all contain cells at various stages of meiosis (from spermatogonia through mature sperm), each crane fly testis contains cells at one stage only. Consequently, one can easily follow developmental sequences in cytoskeletal elements (such as time course of synthesis, or of assembly, or of phosphorylation) by choosing testes that contain cells of the appropriate stage. Before describing methods for rearing the flies, I give a brief overview of the kinds of experiments that have been (and can be) done on cytoskeletal components of this system.

Electron microscopy after glycerination and HMM labeling has been used to study the changing arrangements of cortical microfilaments at different stages of meiosis in crane fly spermatocytes (Forer and Behnke, 1972b). Microtubules have been studied in young spermatids (just after meiosis), and four groups of microtubules were defined, based on their relative stabilities to different treatments (Behnke and Forer, 1967). Cytoplasmic microtubules and axoneme microtubules change during spermiogenesis. Cytoplasmic microtubules are relatively "labile" in young spermatids, but become less labile in older spermatids (Behnke and Forer, 1972; Forer and Behnke, 1972c); vinblastine treatment converts cytoplasmic microtubules in older spermatids to "helices" that remain "in place" (Behnke and Forer, 1972). The 9 + 2 (axoneme) microtubules appear hollow in young spermatids (e.g., Behnke and Forer, 1967), but those in older spermatids and in mature sperm are no longer "hollow" but are very densely staining and substructured (see Behnke and Forer, 1972; Forer and Behnke, 1972c). Actin-containing filaments seen electron-microscopically in crane fly spermatids also change their organization during maturation of the spermatids, from single filaments to tightly packed, "semicrystalline" bundles (Forer and Behnke, 1972c). Crane fly testes cells can be analyzed biochemically (e.g., Petzelt, 1970), and indeed there are biochemically detectable changes in the organization of actin during crane fly spermiogenesis (Strauch *et al.,* 1980). Crane fly spermatocytes and spermatids will live *in vitro* in Ringer's solution, and spermatocytes incorporate radioactively labeled precursors (e.g., [^{3}H]thymidine) at the appropriate stages (A. Forer, unpublished data); hence one can use the synchronous system *in vitro* to study cytoskeletal elements chemically.

II. Rearing Crane Flies in the Laboratory

This section first gives a general description of rearing crane flies in the laboratory, then details of various aspects of the procedures.

A. Procedures

Adults are placed in cages. Copulating pairs (e.g., Fig. 1) are captured in glass dishes that have moist paper (papier-mâché) in the bottom and that are covered with plastic petri dish lids. Within a few days after copulation the adult females lay eggs in the moist paper that lines the bottom of the dishes. Egg laying also can be induced by cutting the heads off the females: see Bodenheimer (1924).

After eggs are laid in the paper pad at the bottom of the dish (there are generally 150–300 eggs per female), the pad (plus eggs) is removed, placed in the bottom of a 90-mm diameter plastic petri dish, and covered with a lid. The eggs hatch about 6–8 days after they are laid; at this time dried nettle leaves are sprinkled on top, and the surface of the paper is broken in several places to allow the larvae to burrow. The nettle leaves are eaten by the larvae, and to help these early larvae eat the nettle I generally crush the leaves between my fingers, to make a fine powder that is sprinkled over the paper. For older larvae ($\gtrsim$ 2 weeks old), I use the nettle leaves as they come (i.e., as small pieces, 2–4 mm on a side).

The small pad of paper (with contained larvae) is transferred to a larger ($\sim$ 150-mm diameter) plastic petri dish at about 9–10 days after hatching: the small

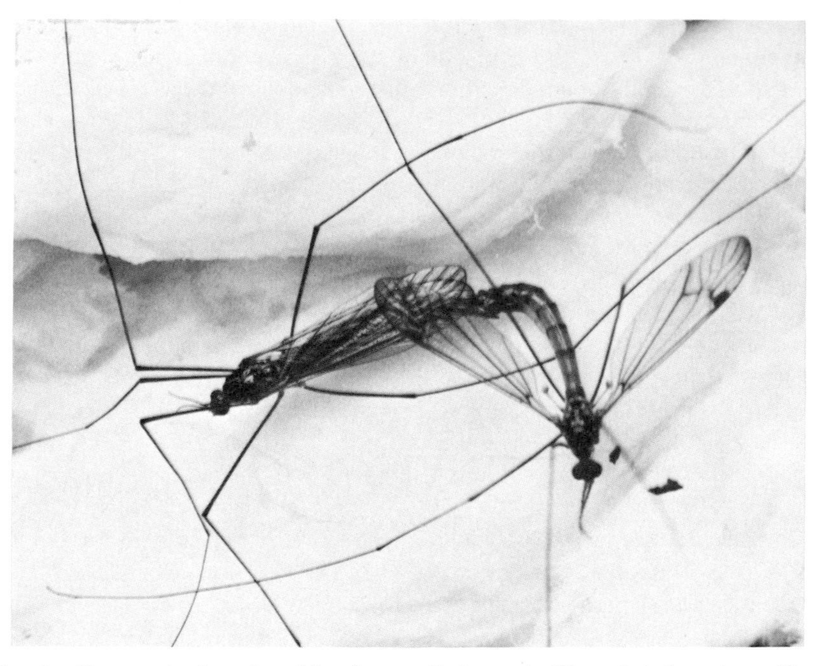

FIG. 1. Photograph of a male and female crane fly *in copulo*. The male is the partner with open wings. (This photograph and the rest of the photographs in Figs. 1–11 are of *Nephrotoma abbreviata* animals or cells.) The photograph is about 1.5× magnification.

pad of paper is torn into four or five pieces and interspersed in clean new paper that fills the larger dish. The larvae are kept in these large plastic petri dishes and are fed every 2–3 days until they become pupae. I do not powder the nettle leaves for larvae in these larger petri dishes.

Pupae begin to appear about 24–30 days after hatching, for the species I keep; pupae are removed from the petri dishes and are placed on moist paper at the bottom of a glass dish that is then covered with a plastic petri dish lid. (Pupae have external sex characteristics that enable one to separate male pupae from females, if this is required.) Adults emerge about 5–6 days later, and are placed in cages for mating.

Having described the procedure in overview, I now give some of the details.

Special cages are not required; I have used various sizes and designs. When no alternatives were available I even used cardboard boxes with arm holes and a transparent plastic sheet for the top! At present we use as cages transparent Plexiglas cubes about 50 cm on a side. One side is removable, for cleaning the insides. Holes (about 15 cm in diameter) are cut in two opposite sides and these are lined with cheesecloth or discarded panty hose and used as arm holes. The only "requirement" in designing a cage is that the arm holes be large enough for the glass dishes (which contain pupae and adults) to pass through. Our cages often contain orange peels and pieces of moist tissue containing sucrose. Fresh orange peels seem to promote mating: quiescent adults become very active when fresh orange peels are placed in the cages, and the ''frantic'' flying around often results in mating. The sucrose-containing moist tissue seems to attract the adults. Our subjective impression is that males and females that suck on the sucrose live longer than they otherwise would, and that, for some species anyway, females lay more eggs than they otherwise would.

Glass dishes used for pupae and for copulating adults are nominally 90 mm diameter × 50 mm high. I select them so that lids to plastic 90 × 15 mm plastic petri dishes fit over the top of the glass. Then, when I transfer the paper plus eggs to a plastic petri dish, I use the same lid for both the glass dish and the 15-mm-high petri dish bottom. I use 150-mm (diameter) × 25-mm plastic petri dishes for the larger larvae. The plastic petri dishes are recycled until they break; dirty ones are soaked and cleaned *without soap,* using tap water only, and are left to dry on a dish rack before the next use.

When feeding the larvae, I try to give enough food to last them for 2–3 days. That is to say, I try to add nettles so that after 2–3 days no uneaten food remains, and the paper has a mottled appearance, from being dug up by the larvae. Then I feed them again. If too much food is added, food that remains after 2–3 days tends to become moldy, requiring the animals to be removed from the paper and placed in fresh paper (as described below).

Water is added to the paper pads while waiting for adults to lay eggs or for eggs to hatch, or while feeding the larvae—in general, at all steps in the procedure. The water is added to the paper (about two or three times a week *when necessary*) in order to keep it at the original moisture level. Tap water is generally good enough: I have used tap water in Toronto, Canada; in New Hampshire; in France; and in various places in Denmark and England. Some tap water can be lethal, however: in southern California, for example, contaminants in the laboratory ''industrial water'' kill the larvae.

Nettle leaves (''cut and sifted'') can generally be obtained at health food shops in both Europe and North America. (They are said to be good for the spleen, and are used to make tea, beer, soup, and hair rinses.) The nettle leaves can be used as they come, but we generally put them in 2-liter beakers and heat them at 160°C for 1–2 hours, in case some parasites otherwise might be brought into the culture. One can also substitute lettuce for the nettle leaves (also see Section IV).

The ''pads'' of paper are made from the cheapest possible paper that readily becomes papier mâché. For this, the paper must be *without* ''wet strength.'' I previously used tissue, but we now use cheap toilet paper. The pads are made as follows. Water is added to a roll of toilet paper, in a plastic bucket. The paper is then converted to pulp (papier mâché) by strong agitation. The water is then squeezed out of the pulp, to form a round ''pad'' of paper, as follows. A plastic or glass dish is filled about three quarters full with the pulp. (I use a 150 mm diameter × 75 mm high glass crystallization dish; my present students prefer a plastic dish.) Then the water is squeezed out using a plastic disk that has holes in it. (The holey plastic disk is made from a piece of 3- to 5-mm-thick methacrylate—Plexiglas in North America or Perspex in the United Kingdom—that is cut to fit snugly on the inside of the glass dish. Many holes—about 2–3 mm diameter—are drilled in this disk. When the holey disk is pushed to the bottom of the papier mâché-filled dish the water comes to the top and the paper is pushed to the bottom, as in Fig. 2.) The water is poured out, the holey disk is removed, and the pad of paper is transferred to the bottom of a 150 × 25 mm plastic petri dish (Fig. 2). A lid is *not* placed immediately over the moist pad of paper. Rather, the paper is generally left open (e.g., overnight) to dry out somewhat: the "proper" moisture level for adding larvae is when water is not visible (the paper should not be "sopping wet"), but when water appears at the surface when the paper is pressed with a finger. (Some crane fly species do better when the paper is more moist, and others when the paper is less moist.) We generally use tap water for this procedure, since in most locations tap water is not harmful to the larvae.

Larvae sometimes have to be removed from the paper in a petri dish and placed in new paper. This is necessary when extra food has allowed too much mold to grow, when the paper has been eaten up, or to time the larvae (see below). In all cases, the larvae can be removed from the old paper quite easily and rapidly by flotation on a $MgSO_4$ solution. I buy 50-lb bags of technical-grade Epsom salts; using a graduated 1-liter beaker to measure volume of $MgSO_4$ powder, I dissolve 2200 ml of this powder in 5 gallons of tap water. A dish of larvae-plus-paper is placed in this crude $MgSO_4$ solution, and the paper is dispersed (i.e., broken up into small pieces), after which the larvae float and the paper and debris sink. The larvae are scooped up from the surface using a tea strainer or curved forceps, the larvae are rinsed by immersion in tap water (except in southern California!), and—except when we are timing the appearance of the IV-instar—the larvae are then placed in clean paper in a petri dish.

One need not worry that lengthy immersion in water or flotation on Epsom salts will harm the larvae. I have seen no effect at all after several hours in water or after several hours floating in $MgSO_4$; nor, on several accidental occasions, has overnight immersion seemed harmful to the larvae. [Ecologists do not report harmful effects either, after flotation on $MgSO_4$ (Coulson, 1962; Freeman, 1967), or after flotation on "cattle salt" (Laughlin, 1967; Meats, 1967), or after 10 days' immersion in aerated water (Hadley, 1971a).] Indeed, larvae have crawled away after a 1-hour immersion in 2% glutaraldehyde (O. Behnke and A. Forer, unpublished data), and have survived for at least a few days thereafter.

One needs to know when the IV-stage started in order to get testes that contain cells at the "proper" stage of meiosis. For timing the onset of the IV-stage, all

FIG. 2. Series of photographs illustrating how we make paper pads from papier-mâché. (a) Photograph of a 150 × 75 mm high crystallizing dish filled with papier-mâché, in front of which is the "holey disk." (b) Water layer at the top and the paper at the bottom, after the disk is pushed to the bottom. In the next step (c), the water is poured off, and then the disk is removed. Then the pad of paper is removed from the crystallization dish and placed in the bottom of a 150 × 25 mm petri dish (d).

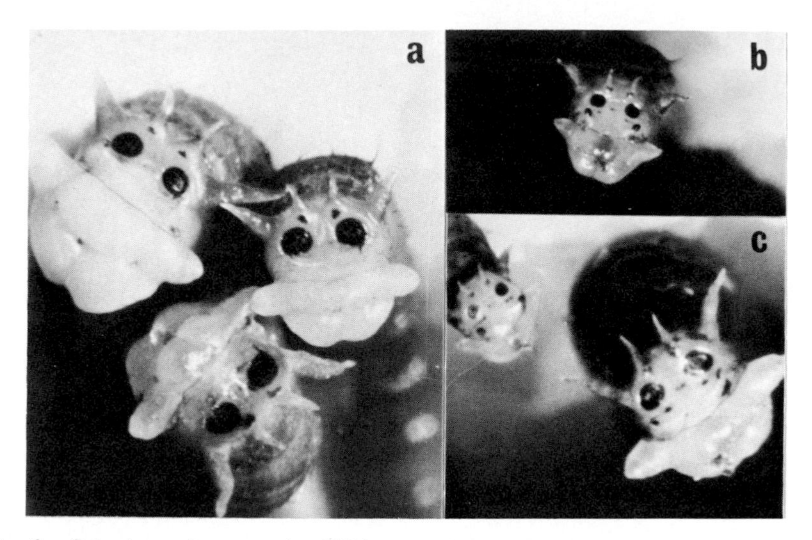

FIG. 3. Spiracles at the rear ends of III-instar and IV-instar *N. abbreviata* larvae. The larvae were killed by immersion in 70°C water; shortly thereafter they were photographed through a dissection microscope. (a) Spiracles in three IV-instar larvae; (b) spiracles in a III-instar larva. The spiracles (the more or less solid black circles) are distinctly larger in the IV-instar larvae than in the III-instar larva, as is also seen in (c), illustrating a III-instar larva and a IV-instar larva in the same photograph. All three photographs are at 11× magnification.

the animals are removed from a dish containing 14- to 18-day-old larvae, using $MgSO_4$, as described previously. Then IV-instar larvae are separated from III-instar larvae and the two groups are placed in paper in separate petri dishes. Those IV-instar larvae found the next day in the dish that previously contained only III-instar larvae must, necessarily, have molted in that 1-day period.

The IV-instar larvae are distinguishable from III-instar larvae by the size of the spiracles at the rear end of the larvae (e.g., Hemmingsen, 1965; Hadley, 1971a; Pritchard and Hall, 1971; Hofsvang, 1972). In the species we rear, the spiracles appear as black dots about 0.3 mm in diameter, whereas those in III-instar larvae are black dots about 0.15 mm in diameter, barely visible to the naked eye (Fig. 3). It is relatively simple to keep all the larvae in water in a white-bottomed dish and to sort them into two groups, based on size of spiracles. There are only two possible sources of confusion: (1) Brand-new IV-instar larvae have pinkish-grey spots rather than black spots; it takes a few hours for the black to develop. (2) Very late III-instar larvae have the old skin on top of the new ones; superficially this might look like a IV-instar animal, but close inspection shows a smaller black spot on the outside skin, with a larger pinkish spot on the inside skin. It is relevant to point out that larval size is not a good marker for IV-instar: some

III-instar larvae, with small spiracles, are larger than other IV-instar larvae, with larger spiracles.

B. Cost

We rear four different species of crane flies, and we have animals of all stages at all times, so that we can do experiments whenever we want. One or two species could be reared, easily, by an undergraduate spending 6–8 hours per week. Other than the initial expenditure for small items (cages, 5-gallon jugs, tea strainer, plastic photographic dishes for sorting out larvae, etc.), the only recurring costs are for toilet paper, $MgSO_4$, and nettle leaves. In rearing four species we use perhaps six cases of toilet paper, a dozen 50-lb bags of Epsom salts, and 25 lbs of nettle leaves per year; our total cost in 1979 was less than $550 for recurring items. Others would probably rear fewer species, so the costs would be reduced correspondingly. Thus keeping the stocks of animals is cheap and really quite easy. The most difficult part of the procedure is explaining to the purchasing department why I need cases of toilet paper together with 50-lb bags of Epsom salts!

III. Removing Testes and Making Preparations

There are two main things to learn in removing testes from crane fly larvae or pupae: (1) choosing animals of the proper age and sex; and (2) keeping the cells alive during and after dissection. I deal with each procedure in turn.

A. Choosing Animals

The various stages of spermiogenesis can be timed with respect to two occurrences, start of the last (IV) larval stage and start of the pupal period. Different species are somewhat different in their timing, but—to generalize (at least over the 4 *Nephrotoma* species I have worked with)—when larvae are reared at 22°–24°C, meiotic S-period occurs near the start of the IV-stage, and prometaphase I–telophase II occurs around 8–9 days later. From this time on, the testes contain spermatids that become progressively more and more mature until pupation (5–6 days later) and throughout the 6-day pupal period. Mature sperm are present in the adult.

Male pupae can be distinguished from female pupae using external morphological characteristics (Fig. 4), so it is a simple matter to choose male pupae with 100% accuracy. Thus in order to study stages of spermiogenesis from

FIG. 4. Male and female *N. abbreviata* pupae in bottom and in side view. (These were killed with 70° water just before photographing them.) The two male pupae (on the right) have more or less flat rear ends (uppermost), whereas the two female pupae (on the left) have "points" at the rear ends (uppermost). The scale is 1 mm between vertical lines. (Printed at about 2.6× magnification.)

pupa to adult, one would time pupae by removing them daily from dishes, then choose males of the desired age.

Male larvae cannot be distinguished from female larvae using external markers, however, so it is impossible to choose male larvae with 100% accuracy. With experience, though, one can choose male larvae reasonably reliably (say, 8 or 9 out of 10) by taking account of two factors. First, within any group of sibling larvae the male larvae tend to become IV-instar before the female larvae do. This is illustrated in a specific example (Fig. 5). In general, siblings become IV-instar over a 6- to 12-day period; in the one example in question the siblings became IV-instar over a 12-day period. For these siblings the sex of the larvae was determined at the pupal stage, where identification is unambiguous. In this particular case, which is completely unexceptional, 80% of the males had become IV-instar when only 62% of the total had become IV-instar. Thus, in general, even if there were *no* way to distinguish male from female crane fly larvae, one increases one's chances of choosing males by restricting the choice to the first 50–60% of the group to become IV-instar.

Second—and even better—there *is* a difference between males and females that one learns to recognize subjectively, and that is a difference in size (e.g., Hadley, 1971b; Pritchard and Hall, 1971). The IV-instar male larvae tend to be shorter than female larvae; in addition, even if they are near the same length, they tend to be *thinner* than female larvae.

Before describing the procedures for dissection, it is relevant to state explicitly

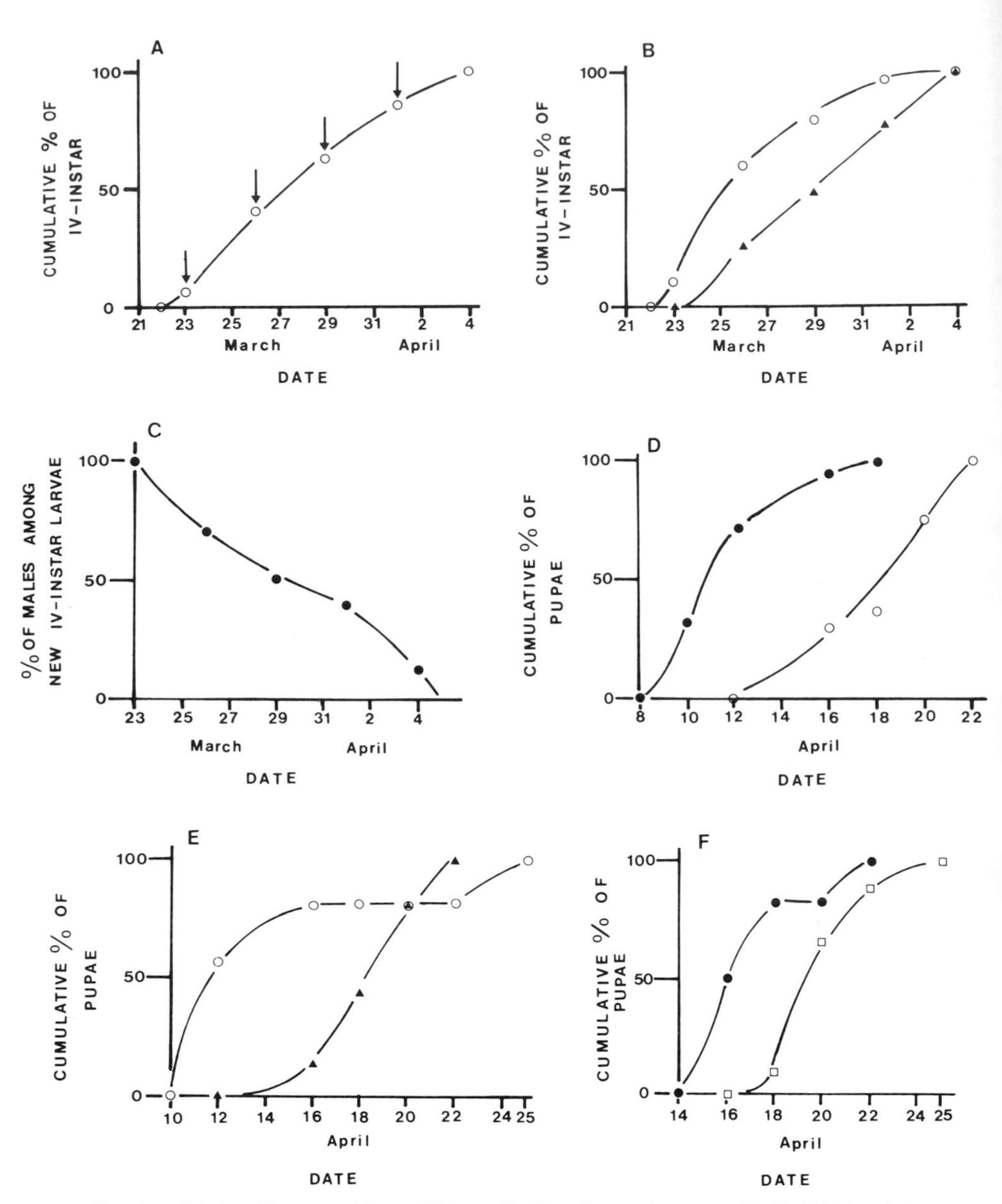

FIG. 5. Timing of larvae molting to IV-instar (A–C) and becoming pupae (D–F). (A) Cumulative percentage of larvae that are IV-instar (ordinate) versus time (abscissa), out of a total population of 86 *N. suturalis* larvae. The entire population became IV-instar in a 12-day period. (B) Same data for the same population, but with separate curves for the male and female populations (the sexes were determined at pupation). ○, Males; ▲, females. (C) Percentage of male larvae found among the new IV-instar larvae on that given day of sorting (ordinate) versus the day on which the IV-instar larvae

a point inherent in the data of Fig. 5: siblings in the same dish become less and less synchronous as they grow. Eggs are laid over a period of a few *hours*, yet IV-stage starts over a period of 6–12 *days*. Those animals that entered IV-stage in a 3-day period became pupae in a 12-day period. Even if one considers males and females separately, those that became IV-instar over a 3-day period became pupae over at least a 6-day period. Therefore, if one waits for 8–9 days after the start of the IV-stage to choose testes that contain cells undergoing cell division, the animals in the dish are not all at exactly the same stage, even if they are siblings: *some* spread has occurred. This is usually not important, but nonetheless it would help to have some additional way to recognize animals of the desired age. One way that seems to work is to look at the development of fat bodies in the larvae. Early in the IV-stage there is no internal fat, so one can see the entire gastrointestinal tract through the body wall; as the IV-stage proceeds there is more and more internal fat, which more and more obscures the gastrointestinal tract. Just before pupation the animals are almost completely white. In a relatively short time one learns to recognize the desired stage by the amount of fat in the animal.

It is relevant to point out that different species are different with respect to the synchrony between testis maturation and fat proliferation. For example, a male *N. ferruginea* larva with maximal spermatocyte cell division is much whiter (i.e., has much more fat) than a male *N. suturalis* larva with maximal spermatocyte cell division: choosing *N. suturalis* larvae using *N. ferruginea* criteria results in *N. suturalis* testes that contain only spermatids rather than cells in division!

It is conceivable that better synchrony might be obtained by keeping the animals at constant density and at constant humidity in a constant-temperature incubator. However, I have not tried to maintain constant conditions other than temperature, and I know of no work along these lines.

B. Dissections

There are two main points to keep in mind when removing testes from larvae or pupae. First evaporation must be prevented. Testes are $\leq 1\ \mu l$ in volume and evaporation occurs rapidly: cells die within seconds if testes are left uncovered on the stage of the dissection microscope. To prevent evaporation during the

---

were sorted from III-instar (abscissa). Those that molt first are male, and progressively smaller and smaller percentages of those that molt later are males. (D–F) Subsequent molting of the same larvae into pupae. (D) Animals started the IV stage between March 23 and 26—that is, between the first and second arrows in (A). ●, Males; ○, females. (E) Animals started the IV stage between March 26 and 29—that is, between the second and third arrows in (A). ○, Males; ▲, females. (F) Animals started the IV stage between March 29 and April 1—that is, between the two furthest right arrows in (A). ●, Males; □, females. These curves show that male larvae have a shorter IV-stage than do female larvae, and that synchrony is gradually lost as the animals progress through the IV-stage.

dissection, I cover larvae or pupae with nontoxic fluorocarbon oil, and the testes remain under oil throughout the dissection procedure. Further, there must be a heat filter on the lamp of the dissection microscope—even with testes under oil—or else the cells are irreparably damaged during the dissection. Second, in order to keep spermatogenic cells alive the preparations should be bacteria free. Thus I use clean dissecting tools (and rinse them in 70% ethanol between dissections, if necessary), and I make sure that the bacteria-laden gastrointestinal tract is not punctured during the dissection.

Larval testes are removed using the following procedure. A drop of Voltalef oil, on a clean microscope slide, is put on the stage of the dissection microscope, out of the field of view. Then a larva is rinsed in tap water, dried on a tissue, and placed on the stage of the dissection microscope *in* the field of view. I then press on the head capsule with one finger. This builds up the "turgor pressure" in the inside of the larva, so when I cover the rear of the animal with Halocarbon oil and then make a slit in the animal's body wall, the body contents are pushed out through the slit in the animal's body wall. I immediately add more Halocarbon oil on top of the "popped-out" internal organs (in case oil has run off from the organs), find the testes, and remove them (covered with oil) and place them in the drop of Voltalef oil on the clean slide. The procedure is illustrated in Fig. 6. Further details and rationale for some of these steps follow.

I cut the slit in the animal's body wall in the direction perpendicular to the animal's long axis, but only a small way across the animal (say one-quarter or less across the width); this is to prevent the gastrointestinal tract from being punctured, which would occur if the larva were cut in two. Similarly, if the larva defecates when I push on the head capsule, the larva is recleaned so that bacteria will not get into the preparation.

Both oils that we use—Voltalef oil 10S and Halocarbon oil 10-25—are high-viscosity polytetrafluoroethylene oils. The Halocarbon oil sometimes has a "scum" on the surface that is seen when one makes open-to-the-air preparations. Thus the Voltalef oil is preferred for some purposes. Regardless of which oil the cells finally are put into, Halocarbon oil, the less expensive of the two, is used during dissection to cover the animals; the testes are removed under Halocarbon oil and then transferred to either Halocarbon or Voltalef oil, usually the latter. (Voltalef oil is Huile 10S from Ugine Kuhlmann, Division Plastiques, 11 Boulevard Pershing, F-75017 Paris, France; and the Halocarbon oil is series 10-25 oil, from Halocarbon Products Corporation, Burlews Court, Hackensak, New Jersey 07601.)

Testes are distinguishable from ovaries throughout the IV-stage (Fig. 7). At the start of the IV-stage the testes and ovaries are round and are about the same size; however, for the species we rear, when they are under the fluorocarbon oil the testes are somewhat translucent, whereas the ovaries are transparent. (If there is doubt about which is ovary and which is testis at this early stage, the techniques described below are used to study the cells contained within.) One or two

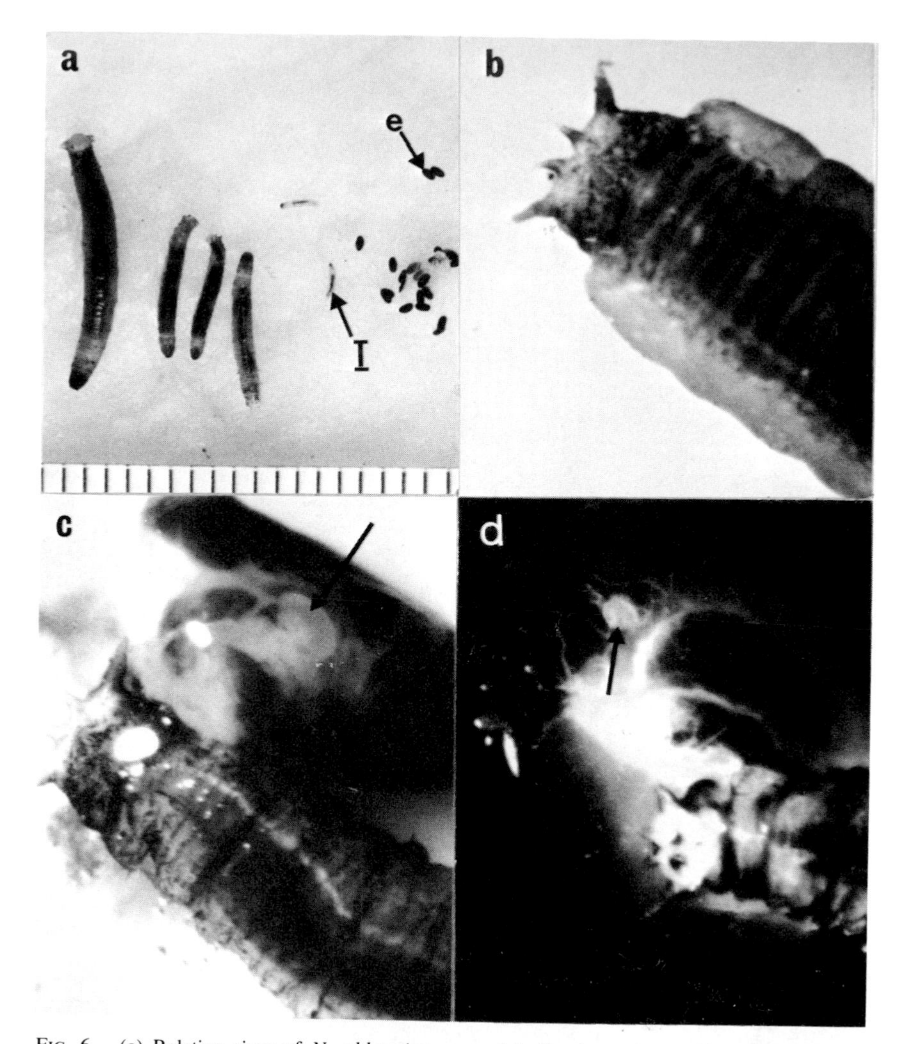

FIG. 6. (a) Relative sizes of *N. abbreviata* eggs (e), first-instar larvae (I), and other larvae, including an early IV-stage larva (at the far left). The vertical lines are 1 mm apart. (b, c) A dissection from a larva that is 8 days past the molt to the IV-stage. The rear end of the larva is illustrated in (b); this is before the larva was slit, but after I pressed on its head. (c) Same larva after I slit the body wall: the insides have "popped out" and the larva is of much smaller width. One testis (arrow) is seen on top of the GI tract. (d) Similar "popped out" insides, but this time from a larva that was only a few hours into the IV-stage. One testis (arrow) is visible, which is much smaller than that in (c). (b–d) Magnification ×11.

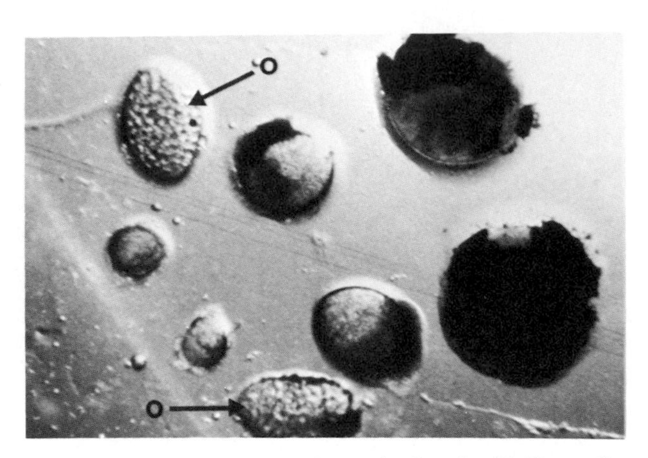

FIG. 7. Various-sized *N. abbreviata* testes plus a pair of ovaries (o). The smallest testes are from an animal just molted to IV-stage; the medium-sized testes are from an animal 8 days into IV-stage; and the largest testes are from an animal 15 days into IV-stage. The ovaries are from an animal about 15 days into IV-stage; large oocytes can be seen in the ovaries, even using a dissection microscope. Magnification ×15.

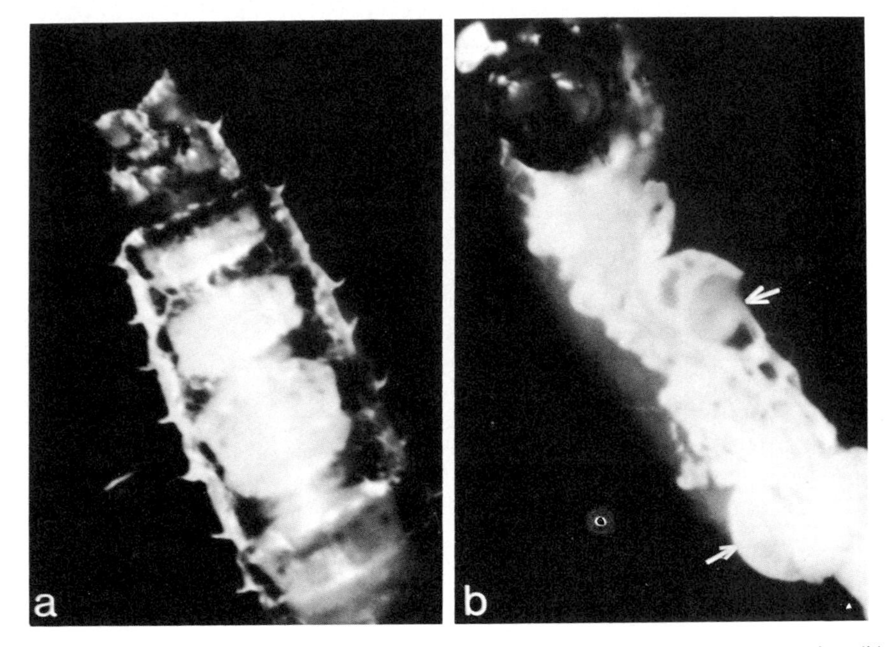

FIG. 8. Dissection of a *N. abbreviata* pupa. (a) Rear end of a pupa before the dissection; (b) appearance after the dissection, with the pupal insides spread more-or-less linearly. The two testes are indicated by arrows. Magnification ×11.3.

days after the start of the IV-stage the testes and ovaries still are round, but the translucent testes are much larger than the transparent ovaries. The testes remain round and much larger than the round ovaries until near the end of the IV-stage, about which time the ovaries grow to be as large as the testes, and even larger. The large ovary is very transparent, and contains large transparent cells, the outlines of which are visible using the dissection microscope, whereas the testes are translucent, with the spermatids identifiable in the dissection microscope as "lines." In some species the ovaries and testes are of different shapes [e.g., Pritchard and Hall, 1971; A. Forer, unpublished data, on *Limonia* (*Metalimnobia*) *annulus cinctipes* (Say)], but this is not true for the four *Nephrotoma* species that we rear.

Pupal testes are removed using procedures similar to those used in obtaining larval testes. I push on the head and cover the rear end with oil, as with larvae; but rather than cutting a slit in the body wall, as with larvae, I cut completely across the pupa. I cut at the junction to the last segment: the blade is across the whole body, and as I cut I pull the rearmost segment away from the head. I pull in this way to get the insides spread out in a more-or-less linear array. This is the easiest and most reliable way for me to find the testes (Fig. 8).

C. Living Cell Preparations

What does one do after the testes are placed in the Voltalef oil? If one wants to treat the cells with some agent (say colchicine), or to incorporate radioactive nucleotides or amino acids (for example), and if one is not interested in observing the cells at the same time, the easiest way is to transfer the testes to a Ringer's solution containing the agent (or precursor) in question—or to a glycerol or other solution. The easiest method I have found to do this is to suck up testes (and surrounding oil) into a Pasteur pipet, release them into the solution in question, and then make sure that the oil has not remained around the testes. (If oil *did* remain around the testes, the oil would seal off the testes from the solution.) To free testes from surrounding oil (should this happen), I suck the testes up and back in a Pasteur pipet until the testes are freed from the oil.

If one wants to observe the cells and is not interested in adding experimental agents while the cells are under observation, the easiest method is to transfer the testis to a drop of Voltalef or Halocarbon oil on a coverslip, and to make a smear of cells. I do this by piercing the testis with a fine needle and then moving the testis around under the oil, against the coverslip; the cells flow out into the "channels" (paths) made by the needle (Fig. 9). (I generally pierce the testis with a needle held in my left hand, and move the testis around with forceps held in my right hand.) If the cells all flow out into one big drop, they are not at all flattened, and internal details are often not easy to see; if, on the other hand, the

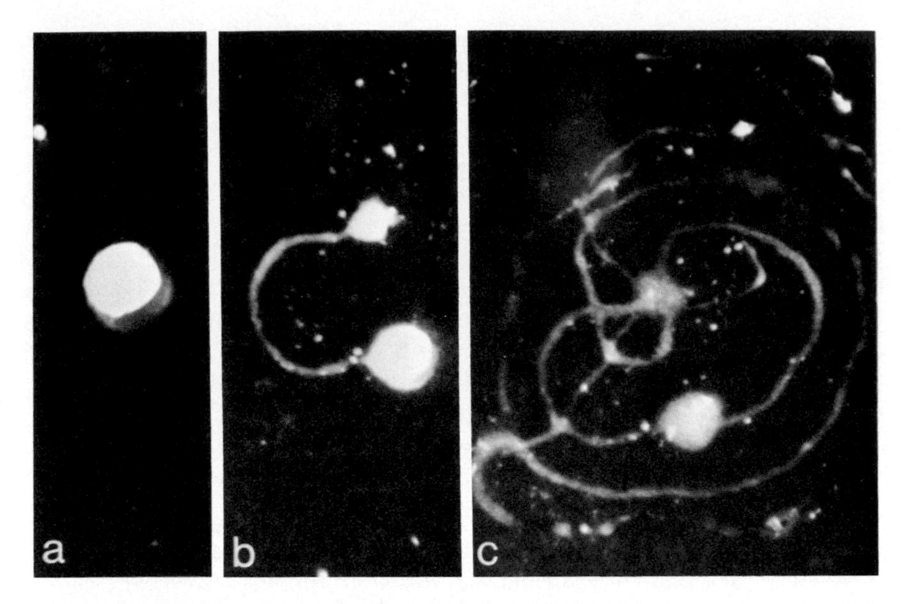

FIG. 9. Method for smearing cells. As seen in reflected light (which gives an appearance similar to that of dark-field illumination): (a) a testis, under oil; (b) the same testis after some cells have been spread into a narrow "channel"; (c) the cells from the same testis spread around in many "channels." Magnification ×11.

cells flow into channels that are only two or three cells wide, then they become flattened and internal details are quite easy to see. Bacteria-free spermatocyte smears made in this way live and divide for at least 5–8 hours. I have even seen telophase-II spermatocytes develop into spermatids when left overnight.

In order to treat cells with some experimental solution, one can transfer the testis to the experimental solution, as previously described, and then transfer the testis from the solution to a drop of Voltalef oil on a coverslip and make a smear preparation, as usual. This is not suitable, though, if one wants to observe cells while they are being perfused. Nor can one perfuse a smear preparation (say with colchicine–Ringer's), because the cells flow away when the solution is introduced under the oil. One method of observing a cell while it is being perfused is to hold the cells in place using a fibrin clot. To do this, the testis is transferred from Voltalef oil into a drop of fibrinogen (in Ringer's solution) that is on a coverslip under oil. Then one pierces the testis to release the cells, mixes the cells with fibrinogen, pulls the fibrinogen into channels, and then—when the cells are flattened—adds thrombin to cause the fibrinogen to clot (Forer, 1972). Once the cells are held in place in the clot, the oil can be removed and the cells placed in whatever aqueous medium is desired.

The Ringer's solution that I use is Ephrussi–Beadle insect Ringer's solution

(Ephrussi and Beadle, 1936), to which is added phosphate buffer at pH 6.8; the final solution contains 0.13 *M* NaCl, 0.005 *M* KCl, 0.003 *M* $CaCl_2$, and 0.01 *M* Sørensen's phosphate buffer, final pH 6.8. This Ringer's solution has been used to keep cells alive for hours in a clot (Forer, 1972), to treat testes in various ways (Behnke and Forer, 1967, 1972; Forer and Behnke, 1972c), and to keep cells alive for incorporation of tritiated thymidine during overnight incubations (A. Forer, unpublished). Begg and Ellis (1979b) have used a more complicated Ringer's solution, but in our hands the simpler Ephrussi–Beadle Ringer's solution plus buffer has been sufficient.

I have run into unexpected problems from "rubber bulbs" and from plastic tubing. We keep the Voltalef oil and the Halocarbon oil in glass bottles. The bottles have glass droppers that fit to the bottles with a ground-glass seal, and in order to get oil from the bottles we put "rubber" bulbs onto the ends of the glass droppers. If the bulbs are kept permanently on the glass droppers some vapor collects in the oil and kills the cells. This once cost me several months of work, so we now keep the bulbs on the ends of the droppers only when we need to remove oil. When the droppers are not in use we remove the rubber bulbs and cover the open ends of the droppers with Parafilm. Similarly, rubber tubing and soft plastic tubing leading from our distilled water bottle introduced into the water a contaminant that is toxic to the cells, and this plastic had to be replaced by Teflon.

All dissection tools are made from stainless steel; this is to avoid heavy-metal contamination during the dissection. We do not use special cleaning procedures for coverslips: they are simply taken new from the box, flamed, and used. For preparations using an inverted microscope we do not use wax to seal the coverslips over holes in glass slides, as we have previously done (e.g., Forer and Koch, 1973; Forer and Jackson, 1979). Rather, we tape coverslips over holes in glass or metal slides. The tape makes a leakproof seal, and is easily peeled off, especially if the end of the tape is bent over, sticky side against sticky side (Sillers and Forer, 1981a). [To affix coverslips over holes, we use "magic transparent tape" (3M Co.) for preparations kept under oil, and we use either No. 850 polyester film tape (3M Co.) or silicone vacuum grease (Dow) for preparations in which the oil is replaced by aqueous solution.]

IV. Other Relevant Data, and Hints on Rearing the Animals

Males can be distinguished from females both as adults and as pupae: the males have a cylindrical abdomen and the rear end is more or less flat, with "claspers" on the sides, whereas the female abdomen tapers to "points" at the rear end (Figs. 4 and 10). The pointed ovipositor at the rear end of females

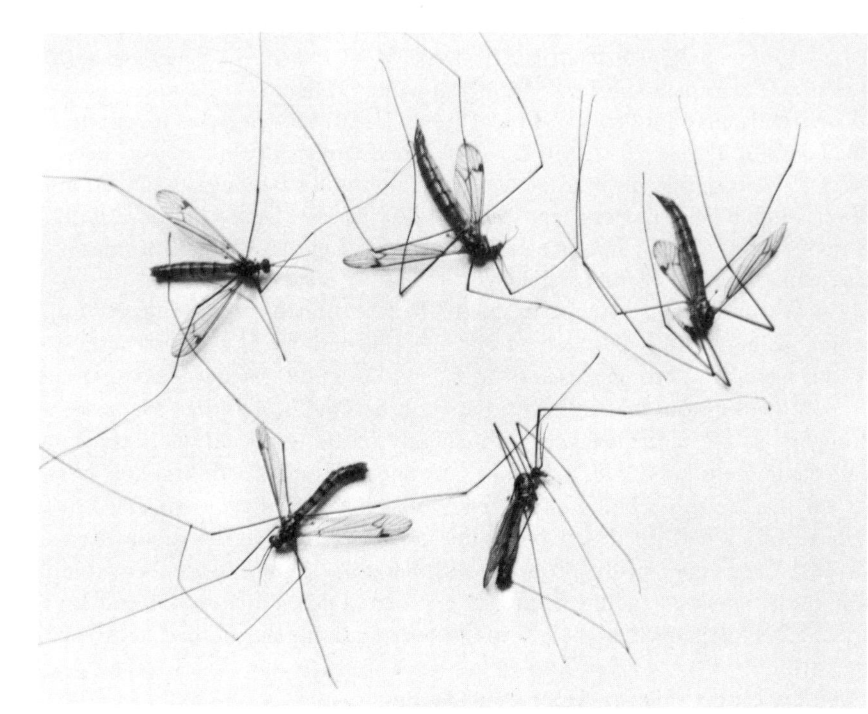

FIG. 10. Male and female *N. abbreviata* adults. The females (the two furthest right in the upper row) have pointed rear ends, whereas the males have more or less flat rear ends. Magnification about ×0.9.

appears double as viewed from the side, but usually single when viewed from the top or bottom (e.g., Fig. 4).

Males remain fertile after mating with single females; that is to say, individual males can be used to mate with at least four females and still give fertilized eggs. Females, too, can mate with several males.

Inbreeding (brother-sister mating) causes reduced numbers of progeny compared with outbreeding. I do not take precautions to avoid inbreeding, however, for the following reason. We generally have several pairs that mate every day, and we consequently need to throw away animals at about the time we transfer them into large dishes. So, rather than *pre*select against brother-sister matings, we *post*select: we discard those dishes in which the animals eat poorly and we keep those in which the animals eat well. (In the former, much of the nettle remains uneaten after 2–3 days, and the paper is not eaten; in the latter, the nettle is gone, and the paper is eaten.) This ''postselection'' method is easier for us than trying to keep track of siblings.

I keep stocks of four species: *Nephrotoma suturalis* (Loew), *N. ferruginea*

(Fabricius), *N. abbreviata* (Loew), and *N. wulpiana* (Bergroth). The methods described above have also been used to rear *N. macrocera* (Say), *N. maculata* (Meig), *N. flavipalpis* (Meig), *N. flavescens* (Linneaus), *N. breviorcornis* (Doane), *Tipula* (*Yamatotipula*) *furca* (Walker), and *Limonia* (*Metalimnobia*) *annulus cinctipes* (Say), but I no longer keep these stocks.

There are no mutant stocks of any species I rear (or of any others that I know of). Dietz has had stocks of *N. ferruginea* that contained chromosomal rearrangements (e.g., Dietz, 1969, 1972a,b; Fuge, 1975), but I do not know if he still keeps these. Before describing methods for getting chromosomal rearrangements, it is relevant to point out that Dietz and Fuge work with one of the same crane flies that I do, but they use the name *Pales ferruginea* (e.g., Dietz, 1969; Fuge, 1980a) whereas I use the name *Nephrotoma ferruginea!* According to Dr. George Byers (Entomology Department, University of Kansas), the two names arose as follows. Meigen named the genus *Pales* in 1800, and renamed it *Nephrotoma* in 1803. Macquart named it *Pachyrhina* in 1834. From 1840 to 1900 the genus was called both *Nephrotoma* and *Pachyrhina*. In the early 1900s Herdel discovered the rare 1800 paper, and hence, by "priority," the name should be *Pales*. Because of possible "taxonomic instability," however, the naming of this genus was taken to the International Commission on Zoological Nomenclature; the decision (in 1963) was that the original 1800 paper is considered "suppressed," and that the names in it (including *Pales*) should be excluded from use. [This is also described in Hemmingsen (1965), in a note on p. 104, citing *Bull. Zool. Nomencl.* **20,** 321–400 (1963).] My usage complies with this advice, and with the official decision on nomenclature.

Chromosomal rearrangements can be induced by X-irradiating adult males, single-pair mating the F1, and brother-sister mating the F2 (Rohloff, 1970). I have used this method to obtain translocations in *N. suturalis*. Adult males were X-irradiated for a total of 4000 roentgens (100 R per minute for 40 minutes), and were then mated with virgin females. Any given rearrangement in any given sperm will appear in only *one* of the F1 animals, and at that only as a heterozygote. In order to get "clones" with altered chromosomes, the F1 are single-pair mated with nonirradiated (wild-type) adults. If the F1 animal contained no chromosomal rearrangements, then all F2 animals should have no chromosomal rearrangements. If, on the other hand, the F1 animal was heterozygous for a rearrangement, then half the F2 animals will be heterozygous for the rearrangement and half will not have the rearrangement at all. To see which animals contain chromosomal rearrangements, one observes cells at metaphase-anaphase of meiosis-I and looks for tetravalents instead of bivalents (caused by a heterozygous reciprocal translocation), or for trivalents (caused by autosomal translocation onto the X- or Y-chromosome), or for chromosomal bridges (caused by heterozygous inversions). Once the desired rearrangement is found, one can obtain homozygous stocks by brother-sister matings.

Which species of crane fly should one choose to work with—and why does A. Forer keep four species instead of just one? First, note that different species have different habitats, ranging from freshwater ponds and streams (Byers, 1978; Kavaliers, 1981), to marine habitats (e.g., Toye, 1967), to pasture, to moist soil, to desert habitats (e.g., Hemmingsen, 1958, 1959), to those in the Arctic, and those that emerge in the winter and are found on the snow. Consequently the life cycles vary: those I keep in the laboratory have generation times of about 7 weeks, at room temperature, while others can be one generation per year—or, in the Arctic, for example, 2 or 4 years per generation (e.g., Hofsvang, 1972; MacLean, 1973). Some species have a diapause (e.g., Butterfield, 1976a). Some species, though, have generation times of only 3–4 weeks, and these might be preferred for genetic experiments.

We keep different species because different ones have somewhat different cytological characteristics. For example, the sex chromosomes are different sizes in different species (e.g., Henderson and Parsons, 1963). When the sex chromosomes are small, as they are in *N. suturalis,* they can be lost in among cytoplasmic granules that invade the spindle region as the sex chromosomes start to move poleward. For following sex-chromosome movements, then, *large* sex chromosomes are preferable; that is why we keep *N. abbreviata* (Fig. 11). [One can even choose crane flies that do *not* have sex chromosomes and in which $2n = 6$ rather than the more usual $2n = 8$ (e.g., White, 1949; Dietz, 1956); *Limonia* (*Metalimnobia*) *annulus cinctipes,* for example, has $2n = 6$ and has no sex chromosomes (A. Forer, unpublished data).] As another example of different cytology, the spindle birefringence can be different in different species. When studied at room temperature, *N. suturalis* spermatocytes at metaphase and anaphase have very little continuous-fiber birefringence, whereas *N. ferruginea* spermatocytes at metaphase and anaphase have considerable continuous-fiber birefringence (e.g., Forer, 1976; A. Forer and P. Shirley, unpublished data). Finally, although *N. ferruginea* has more cells per testis than *N. suturalis,* the synchrony of the cells in *N. ferruginea* testes is not as good as the synchrony of cells in *N. wulpiana* testes, or the other two species we rear.

It is relevant to point out that in some crane flies the chromosomes are achiasmatic in male meiosis (e.g., Dietz, 1956). Thus one can directly compare chiasmatic versus achiasmatic meiosis by comparing related species. It is also relevant to note that crane flies have polytene chromosomes, which are useful for some kinds of experiments.

To collect crane flies in the wild, one uses standard methods. They can be caught with a net, or by using a night-light in conjunction with a net or trap. Larvae can be collected by spraying orthodichlorobenzene onto the surface of the ground. Crane fly larvae immediately come to the surface, unharmed, and can then be collected and reared (Barnes, 1937, 1941). One should be somewhat cautious in bringing animals into the laboratory from the wild, though, because

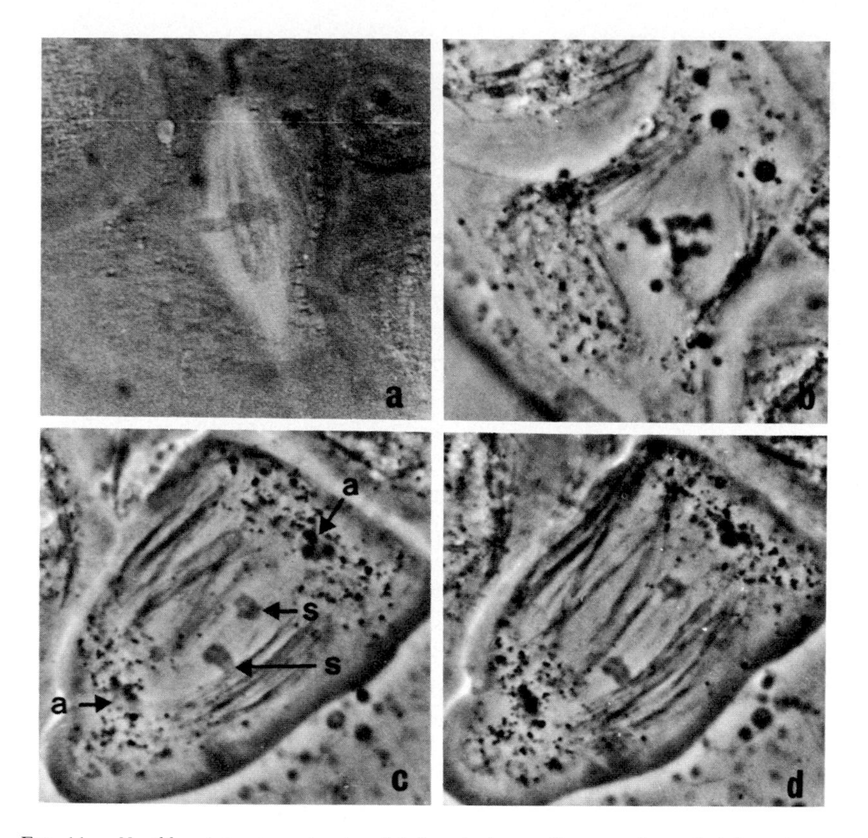

FIG. 11. *N. abbreviata* spermatocytes. (a) A metaphase cell as seen in a polarizing microscope; the spindle fibers appear bright against the gray background. (b) Same cell as (a), as seen in a phase-contrast microscope; the picture was taken shortly after the autosomal half-bivalents began to move poleward in anaphase. (c and d) Sex-chromosome movements—in a different cell than (a) and (b)—as seen in a phase-contrast microscope. The autosomal half-bivalents (a) are at the poles, and the sex chromosomes (s) then move poleward, as in (c) and (d). In this species the sex chromosomes are at least as large as, if not larger than, the autosomal half-bivalents (cf. Fig. 16c and 16d with Fig. 16b). Magnification ×980.

there are known pathogens that can kill them, such as viruses (e.g., Bird, 1961; Carter, 1973, 1974, 1975), microsporidians (e.g., Weissenberg, 1926; Carter, 1976), and coccidians (Beesley, 1977a,b). Mites also attack (and hang onto) crane flies; in fact, I lost my first laboratory stock of *N. abbreviata* because of infections by mites. (The only way I know of to fight a mite infection is to wash down table tops daily, and to remove larvae from dishes and place them in new paper daily.)

Other methods have been used for rearing crane flies in the laboratory, and these can be found in Barnes (1937), White (1951), Laughlin (1958, 1960), and

Butterfield (1976b). However, these do not seem as reliable or convenient as the method just described. Ricou (1967) has compared the nutritional value of various grains, in having crane fly larvae grow and develop; unfortunately, however, nettle leaves were not included in the list of those tried.

The temperature at which the stocks are reared affects the growth rate of the flies (A. Forer, unpublished data; Butterfield, 1976b). Our *N. suturalis* stocks do not tolerate temperatures above about 30°C, and our *N. ferruginea* stocks do not tolerate temperatures above about 27°C (see also Schaap and Forer, 1979)—but for each species the rate of development is faster at higher temperatures than at lower temperatures. [Butterfield (1976b) and Carter (1975) found the same for a different species.] For example, *N. suturalis* has a generation time of about 5.5–6 weeks at 27°C, but about 7–7.5 weeks at 22°C.

I know of no way to store animals frozen, as one does with tissue culture cells or *Dictyostelium,* for example. It would be convenient to be able just to thaw them out when they are needed, but I know no way of doing this. The growth rate does slow almost to zero when animals are kept at about 10°C, however, and I have kept larvae at 9°C for months while their siblings at room temperature went through several generations. Those at 9°C were moistened and fed when necessary (about every 1 or 2 weeks), but they stayed at about the same stage of development as when they were first put into the incubator. They developed normally when returned to room temperature. (This is the most useful way I know of to get an initial stock spread out so that one has all stages at all times: let some develop at 25°C, others at 20°C, and hold others back at 10°C.) Storing animals at 10°C is not as useful as deep-freezing tissue culture cells, but it is the only method I know of for keeping some "in reserve."

Acknowledgments

Roland Dietz first introduced me to growing crane flies, in 1961, and he gave me seven larvae of *N. suturalis,* from which my present stock derives. He and I have both modified our laboratory rearing methods, in divergent ways: My most recent methods are given here. I am grateful to Roland for introducing me to the flies and cells. I am grateful to Dr. Jim La Fountain, Jr., in Buffalo, New York, for help in replenishing our *N. suturalis* stock in our time of need, and to Dr. Donna Kubai for sending the eggs of *N. abbreviata,* from which our present laboratory stock (designated DK) derives. I continue to be grateful to Dr. George Byers (University of Kansas) for identification of stocks and for general encouragement, to Alexandra Engberg, and to F. B. T. Birkes. I thank Mrs. A. Adhihetty and Mrs. D. Gunning for excellent typing, through several versions of this chapter.

References

Barnes, H. F. (1937). *Ann. Appl. Biol.* **24,** 356–368.
Barnes, H. F. (1941). *Ann. Appl. Biol.* **28,** 23–28.

Bauer, H. (1931). *Z. Zellforsch. Mikrosk. Anat.* **14,** 138–193.
Bauer, H., Dietz, R., and Röbbelen, C. (1961). *Chromosoma* **12,** 116–189.
Beesley, J. E. (1977a). *Parasitology* **74,** 273–283.
Beesley, J. E. (1977b). *J. Invertebr. Pathol.* **30,** 249–254.
Begg, D. A., and Ellis, G. W. (1979a). *J. Cell Biol.* **82,** 528–541.
Begg, D. A., and Ellis, G. W. (1979b). *J. Cell Biol.* **82,** 542–554.
Behnke, O., and Forer, A. (1966). *C.R. Trav. Lab. Carlsberg* **35,** 437–455.
Behnke, O., and Forer, A. (1967). *J. Cell Sci.* **2,** 169–192.
Behnke, O., and Forer, A. (1972). *Exp. Cell Res.* **73,** 506–509.
Behnke, O., Forer, A., and Emmersen, J. (1971). *Nature (London)* **234,** 408–410.
Bird, F. T. (1961). *Can. J. Microbiol.* **7,** 827–830.
Bodenheimer, F. (1924). *Z. Wiss. Zool.* **121,** 393–441.
Butterfield, J. (1976a). *J. Insect Physiol.* **22,** 1443–1446.
Butterfield, J. (1976b). *Oecologia* **25,** 89–100.
Byers, G. (1978). *In* "An Introduction to the Aquatic Insects of North America" (R. W. Merritt and K. W. Cummins, eds.), pp. 285–310. Kendall/Hunt Publ. Co., Dubuque, Iowa.
Carter, J. B. (1973). *J. Invertebr. Pathol.* **21,** 123–130.
Carter, J. B. (1974). *J. Invertebr. Pathol.* **24,** 271–281.
Carter, J. B. (1975). *J. Invertebr. Pathol.* **25,** 115–124.
Carter, J. B. (1976). *J. Invertebr. Pathol.* **27,** 409–410.
Coulson, J. C. (1962). *J. Anim. Ecol.* **31,** 1–21.
Dietz, R. (1956). *Chromosoma* **8,** 183–211.
Dietz, R. (1959). *Z. Naturforsch.* **14b,** 749–752.
Dietz, R. (1963). *Zool. Anz.* **26,** Suppl., 131–138.
Dietz, R. (1966). *Chromosomes Today* **1,** 161–166.
Dietz, R. (1969). *Nátúrwissenschaften* **56,** 237–248.
Dietz, R. (1972a). *Chromosomes Today* **3,** 70–85.
Dietz, R. (1972b). *Chromosoma* **38,** 11–76.
Ephrussi, B., and Beadle, G. (1936). *Am. Nat.* **70,** 218–225.
Forer, A. (1965). *J. Cell Biol.* **25** (No. 1, Pt. 2), 95–117.
Forer, A. (1966). *Chromosoma* **19,** 44–98.
Forer, A. (1969). *In* "Handbook of Molecular Cytology" (A. Lima-de-Faria, ed.), pp. 553-601. North-Holland Publ., Amsterdam.
Forer, A. (1972). *Cytobiologie* **6,** 403–409.
Forer, A. (1976). *Cold Spring Harbor Conf. Cell Proliferation* **3** [Book C], 1273–1293.
Forer, A. (1980). *Symp. R. Entomol. Soc. London* **10,** 85–95.
Forer, A., and Behnke, O. (1972a). *Chromosoma* **39,** 145–173.
Forer, A., and Behnke, O. (1972b). *Chromosoma* **39,** 175–190.
Forer, A., and Behnke, O. (1972c). *J. Cell Sci.* **11,** 491–519.
Forer, A., and Brinkley, B. R. (1977). *Can. J. Genet. Cytol.* **19,** 503–519.
Forer, A., and Jackson, W. T. (1979). *J. Cell Sci.* **37,** 323–347.
Forer, A., and Koch, C. (1973). *Chromosoma* **40,** 417–442.
Freeman, B. E. (1967). *J. Anim. Ecol.* **36,** 123–146.
Fuge, H. (1971). *Z. Zellforsch Mikrosk. Anat.* **120,** 579–599.
Fuge, H. (1973). *Chromosoma* **43,** 109–143.
Fuge, H. (1974a). *Protoplasma* **79,** 391–393.
Fuge, H. (1974b). *Chromosoma* **45,** 245–260.
Fuge, H. (1975). *Chromosoma* **52,** 149–158.
Fuge, H. (1977a). *Int. Rev. Cytol., Suppl.* **6,** 1–58.
Fuge, H. (1977b). *In* "Mitosis: Facts and Questions" (M. Little, N. Paweletz, C. Petzelt, H.

Ponstingl, D. Schroeter, and H.-P. Zimmermann, eds.), pp. 51–68. Springer-Verlag, Berlin and New York.
Fuge, H. (1980a). *Chromosoma* **76,** 309–328.
Fuge, H. (1980b). *Eur. J. Cell Biol.* **23,** 166–170.
Fuge, H. (1981). *Eur. J. Cell Biol.* **25,** 90–94.
Fuge, H., and Muller, W. (1972). *Exp. Cell Res.* **71,** 241–245.
Hadley, M. (1971a). *J. Anim. Ecol.* **40,** 445–466.
Hadley, M. (1971b). *Oecologia* **7,** 164–169.
Hemmingsen, A. M. (1958). *Vidensk. Medd. Dan. Naturh. Foren.* **120,** 207–236.
Hemmingsen, A. M. (1959). *Entomol. Medd.* **29,** 46–64.
Hemmingsen, A. M. (1965). *Vidensk. Medd. Dan. Naturh. Foren.* **128,** 93–150.
Henderson, S. A., and Parsons, T. (1963). *Caryologia* **16,** 337–346.
Hofsvang, T. (1972). *Nor. Entomol. Tidsskr.* **19,** 43–48.
Kavaliers, M. (1981). *Can. J. Zool.* **59,** 555–558.
La Fountain, J. R., Jr. (1972). *J. Cell Sci.* **10,** 79–93.
La Fountain, J. R., Jr. (1974a). *J. Ultrastruct. Res.* **46,** 268–278.
La Fountain, J. R., Jr. (1974b). *J. Cell Biol.* **60,** 784–789.
La Fountain, J. R., Jr. (1975). *BioSystems* **7,** 363–369.
La Fountain, J. R., Jr. (1976). *J. Ultrastruct. Res.* **54,** 333–346.
La Fountain, J. R., Jr., and Thomas, H. R. (1975). *J. Ultrastruct. Res.* **51,** 340–347.
La Fountain, J. R., Jr., Zobel, C. R., Thomas, H. R., and Galbreath, C. (1977). *J. Ultrastruct. Res.* **58,** 78–86.
Laughlin, R. (1958). *Entomol. Exp. Appl.* **1,** 241–245.
Laughlin, R. (1960). *Entomol. Exp. Appl.* **3,** 185–197.
Laughlin, R. (1967). *Entomol. Exp. Appl.* **10,** 52–68.
MacLean, S. F., Jr. (1973). *Oikos* **24,** 436–443.
Meats, A. (1967). *Entomol. Exp. Appl.* **10,** 312–320.
Müller, W. (1972). *Chromosoma* **38,** 139–172.
Petzelt, C. (1970). *Chromosoma* **29,** 237–245.
Pritchard, G., and Hall, H. A. (1971). *Can. J. Zool.* **49,** 467–482.
Ricou, G. (1967). *Ann. Nutr. Aliment.* **21,** 199–215.
Rohloff, H. (1970). Ph.D. Dissertation, Eberhard-Karls-Universität zu Tübingen, Tübingen, West Germany.
Salmon, E. D. (1975). *J. Cell Biol.* **65,** 603–614.
Salmon, E. D., and Begg, D. A. (1980). *J. Cell Biol.* **85,** 853–865.
Salmon, E. D., and Ellis, G. W. (1975). *J. Cell Biol.* **65,** 587–602.
Schaap, C. J., and Forer, A. (1979). *J. Cell Sci.* **39,** 29–52.
Sillers, P. J., and Forer, A. (1981a). *J. Cell Sci.* **49,** 51–67.
Sillers, P. J., and Forer, A. (1981b). *Can. J. Biochem.* **59,** 770–776.
Sillers, P. J. and Forer, A. (1981c). *Can. J. Biochem.* **59,** 777–792.
Strauch, A. R., Luna, E. J., and La Fountain, J. R., Jr. (1980). *J. Cell Biol.* **86,** 315–325.
Toye, S. A. (1967). *Proc. R. Entomol. Soc. London* **42,** 167–170.
Weissenberg, R. (1926). *Arch. Protistenkd.* **54,** 431–467.
White, J. H. (1951). *Ann. Appl. Biol.* **38,** 847–858.
White, M. J. D. (1949). *Evolution* **3,** 252–261.

Chapter 12

Reactivation of Sperm Flagella: Properties of Microtubule-Mediated Motility

BARBARA H. GIBBONS

Pacific Biomedical Research Center
University of Hawaii
Honolulu, Hawaii

I. Introduction

Development of procedures for reactivation of cilia and sperm flagella, in which the membrane has been rendered permeable by treatment with 50% glycerol, was initiated by Hoffman-Berling (1955), who found that the motility

METHODS IN CELL BIOLOGY, VOLUME 25

ISBN 0-12-564125-7

of glycerinated grasshopper sperm could be restored by the addition of exogenous ATP. Subsequent refinements and variations of this basic approach have yielded much information about the properties of the motile mechanisms in cilia and flagella. In most cases the organelles are now demembranated by extraction with a nonionic detergent, such as Triton X-100 or Nonidet P-40, rather than with glycerol.

The most detailed studies of reactivation have utilized sea urchin sperm, and the methods given here focus on this material. However, Triton X-100 has also been used successfully to prepare reactivated cilia in *Paramecium* (Naitoh and Kaneko, 1972; Nakaoka and Toyotama, 1979), *Elliptio* (Walter and Satir, 1978), *Mytilus* gill epithelia (Tsuchiya, 1977), bracken fern spermatozoids (Wolniak and Cande, 1980), and sperm flagella of the tunicate *Ciona,* the hydroid *Tubularia* (Brokaw and Simonick, 1976), as well as the flagella of bull and human sperm (Lindemann and Gibbons, 1975). A different but related nonionic detergent, Nonidet P-40, appears to be preferable for the reactivation of flagella of *Chlamydomonas* (Hyams and Borisy, 1978; Nakamura and Kamiya, 1978; Bessen *et al.,* 1980) and of the protozoan, *Crithidia* (Holwill and McGregor, 1976), and it can replace Triton for the reactivation of sea urchin sperm (B. Gibbons, unpublished results).

II. Collection of Sea Urchin Sperm

Sperm may be collected either in seawater containing 0.1–0.2 m*M* EDTA and adjusted to pH 8.3 (Gibbons and Gibbons, 1972), or in 0.5 *M* NaCl (Brokaw, 1975).

Healthy sea urchins are induced to shed their sperm by injecting 0.5 *M* KCl through the lantern mantle into the test of the animal. The thick semen is collected under a layer of EDTA–seawater and stored undiluted. The maximum time that the semen may be stored without loss of quality varies in different species of sea urchin, with good reactivation being obtained after as long as 3–4 days at 4°C in some cases (e.g., *Colobocentrotus atratus* and *Strongylocentrotus purpuratus*), but only for 1–2 hours in others (*Tripneustes gratilla*).

Before being used, the semen is diluted 1- to 5-fold with EDTA–seawater to give a suspension with a concentration of 5–25 mg protein/ml and stored either at 0°C or at room temperature depending on the species. (For the extensively studied Hawaiian species, the sperm of *Colobocentrotus* are best stored cold, whereas those of *Tripneustes* are best stored at room temperature.) Samples of sperm are then removed from this stock suspension for demembranation. The stock suspension should be discarded after 2–4 hours.

The concentrations of the sperm suspensions in seawater may be assayed by adding a portion to 1% trichloroacetic acid, and measuring the turbidity at 550 nm, always with the same spectrophotometer for reproducibility. This turbidimetric assay may be calibrated in terms of protein concentration using the Lowry method (Lowry *et al.*, 1951).

III. Demembranation and Reactivation

A. Standard Method

The standard procedure for preparing reactivated sea urchin sperm is as follows. Twenty-five microliters of the diluted sperm suspension in seawater is added with gentle mixing to 0.3 ml of demembranating solution containing 0.04% (w/v) Triton X-100–4 m*M* $CaCl_2$–1 m*M* dithiothreitol (DTT)–0.1 m*M* EGTA–10 m*M* Tris HCl (pH 8.1). Extraction is allowed to proceed at room temperature (23°C) for 20–30 seconds, after which time 10–50 μl of the demembranated sperm is withdrawn with a capillary pipet and transferred to 2.5 ml of a reactivating solution containing 0.15 *M* KCl–2 m*M* $MgSO_4$–1 m*M* DTT–0.1 m*M* EGTA–2% polyethylene glycol (PEG) (MW 20,000)–1 m*M* ATP–10 m*M* Tris HCl buffer (pH 8.1). The swimming motion and percentage of motile sperm are observed by dark-field light microscopy in either an open-top chamber with an optically flat bottom made by gluing a glass ring onto a 2 × 3 in. microscope slide, or a well slide with coverslip. The glass surface next to which the swimming motion of the sperm is observed is precoated either with 0.03% Formvar in ethylene dichloride or with Siliclad to prevent sperm adhesion.

B. Discussion

The motility of demembranated cells is extremely sensitive to contamination by heavy-metal ions, and the quality of reactivated preparations—especially their longevity—depends on reducing such contamination as far as possible. For this reason, deionized water is used to prepare all solutions, EDTA is added to the seawater, and DTT and either 0.5 m*M* EDTA or 0.1 m*M* EGTA are added to the reactivating solution to scavenge free heavy-metal ions and minimize oxidation of sulfhydryl groups. In some cases, it has also been found desirable to purify the commercial Tris by recrystallization (Gibbons and Gibbons, 1972). The demembranating and reactivating solutions should be prepared fresh daily, mainly because of the auto-oxidation of DTT.

With these precautions, reactivated preparations can readily be obtained in which the percentage of motile sperm is close to 100%, the flagellar beating and forward progression of the sperm are homogeneous and vigorous (resembling those of the live sperm), the beat frequency is homogeneous and reproducible from one preparation to another, and bend propagation is regular and progresses all the way to the tip of the flagellum (Fig. 1). An important parameter of the flagellar beat concerns the degree of asymmetry of the waveforms, which may be assayed either by measuring the diameter of the circular paths of the sperm as they swim at the bottom surface of an observation dish or by determining the difference between the angles of the principal and reverse bends on the flagellum (Gibbons and Gibbons, 1972; Brokaw, 1979). In good preparations, the asymmetry is no greater than that of the live sperm (see Section III, D). This and the other characteristics described should prevail for at least 10 minutes, and may be expected to last up to half an hour or even longer. The first sign of deterioration is usually the appearance of incomplete propagation of waves into the distal portion of the flagellum, making this region appear stiff. The presence of 2% PEG in the reactivating solution seems to improve the longevity of reactivated-sperm preparations by delaying the appearance of this stiffness, as well as influencing the initial waveforms so that they more closely resemble those of the live sperm. Some day-to-day variability in the quality of reactivated preparations is encountered, possibly as a consequence of differences in the health of individual animals and their sperm, and it is occasionally necessary to discard batches of sperm that show abnormal or inhomogeneous waveforms and beat frequencies on reactivation.

Under favorable conditions, sperm may be stored in the demembranating solution at 0°C for several hours and still show good quality on subsequent reactivation. However, it is safer to limit the duration of the Triton extraction to less than 1 minute if possible. Exposure to high concentrations of Triton X-100 (>1%) does not in itself damage the motile mechanism.

In *Tripneustes,* the flagellar-beat frequency of the live sperm is about 50 Hz at 23°C, whereas in sperm reactivated in 1 m*M* ATP it is about 30 Hz (B. Gibbons, unpublished data; I. R. Gibbons and Gibbons, 1980); in sperm of *Colobocentrotus* the live and reactivated beat frequencies have been reported to be 46 Hz and 32 Hz, respectively, at 25°C (Gibbons and Gibbons, 1972). Reactivated sperm of *Lytechinus pictus* often show poor motility at 1 m*M* ATP, but at 0.1 m*M* ATP give a stable, homogeneous frequency of about 25 Hz, whereas the live sperm beat at 30 Hz at this temperature (16°C) (Brokaw, 1965, 1975). The beat frequency of reactivated sperm flagella decreases with decreasing ATP concentration, and double reciprocal plots of beat frequency against ATP concentration are linear at ATP concentrations above about 0.05 m*M* (Brokaw, 1967, 1975; Holwill, 1969; Gibbons and Gibbons, 1972; Gibbons *et al.,* 1978; Okuno and Brokaw, 1979). A threshold concentration of about 10 μM ATP is required to produce flagellar beating (Gibbons and Gibbons, 1972).

As a consequence of certain differences between the properties of the

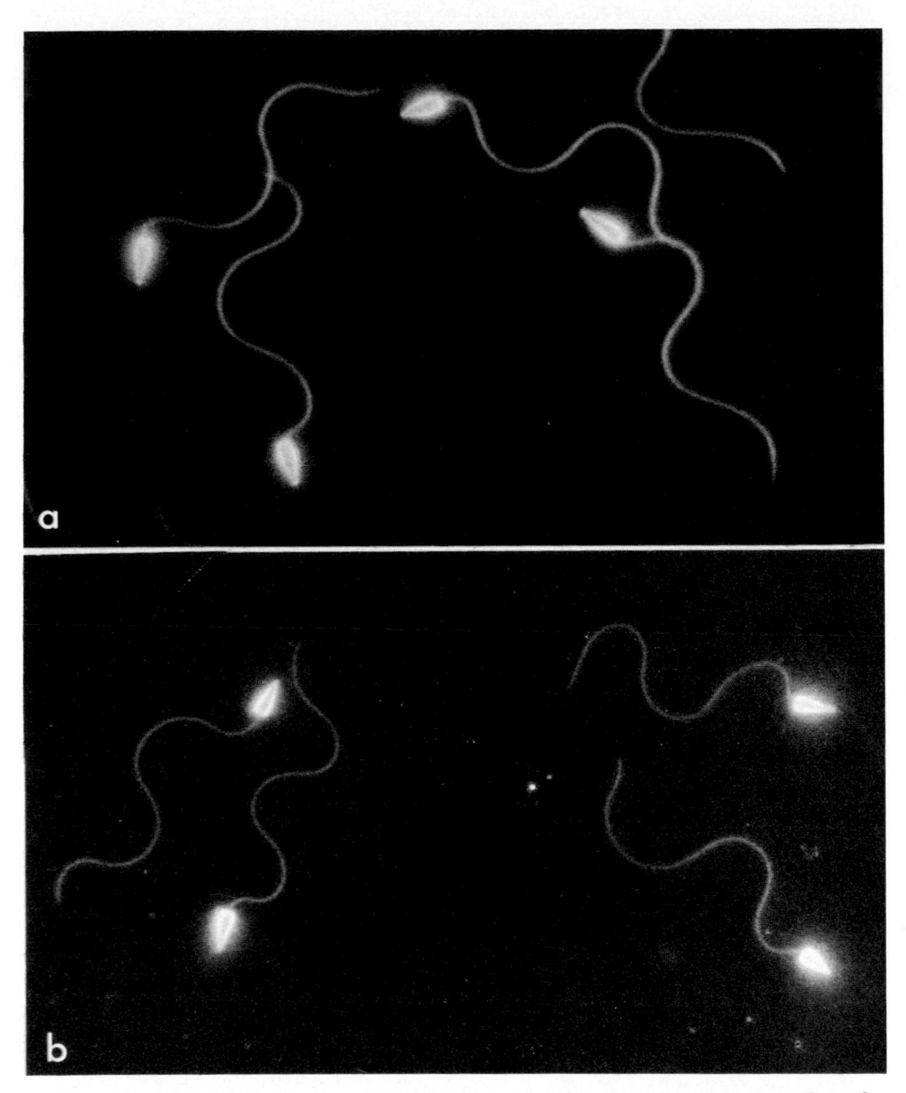

FIG. 1. Dark-field micrographs of *Tripneustes* sperm swimming at the bottom surface of an observation dish. (a) Live sperm in EDTA–seawater. (b) Reactivated sperm prepared according to the standard procedure (Section III, A). These appear lighter than the live sperm because demembranated flagella scatter less light than do intact ones. Taken with ×40 0.75 NA water-immersion objective, 1.2–1.4 NA oil-immersion dark-field condenser, flash exposure. ×1100.

axonemes of sperm from *Colobocentrotus* (on which the original reactivation studies based on demembranation by Triton X-100 were performed) and those of sperm from the majority of other sea urchin species, two different basic demembranating-solution compositions have evolved. These stem from the discovery by Brokaw *et al.* (1974) of a critical requirement for the presence of 4–5 m*M* Ca^{2+} in the demembranating solution in order to obtain symmetrical flagellar bending waves in reactivated preparations of *Strongylocentrotus* sperm, and from the observation (B. Gibbons, unpublished results) that lowering the salt concentration improves the subsequent quality of the beating waveforms and longevity of reactivated sperm of *Tripneustes,* perhaps by decreasing the susceptibility of certain axonemal proteins to solubilization during reactivation. Therefore, whereas the composition of the demembranating solution originally devised for *Colobocentrotus* sperm (Gibbons and Gibbons, 1972, 1974) contains 0.04% Triton X-100–0.15 *M* KCl–2 m*M* $MgSO_4$–1 m*M* DTT–0.5 m*M* EDTA–10 m*M* Tris HCl (pH 8.1), the standard one given above (Section III, A) contains $CaCl_2$, and the KCl and $MgSO_4$ are omitted. It should be pointed out that in this standard procedure, the 25 μl of seawater transferred with the sperm contribute about 0.04 *M* NaCl, 4 m*M* Mg^{2+}, and 0.8 m*M* Ca^{2+} to the demembranating mixture. Most workers continue to use satisfactorily a demembranating solution containing the fuller list of ingredients as originally described for *Colobocentrotus,* usually with addition of $CaCl_2$ (cf. Shingyogi *et al.,* 1977; Okuno *et al.,* 1976; Brokaw, 1979; Goldstein, 1976, 1979). It is to be expected that slight changes in pH, ionic composition, or temperature of reactivation may be required to obtain optimal results in different species. *Lytechinus* sperm, for example, are usually reactivated with 0.1–0.2 m*M* ATP at 16–18°C (Brokaw, 1975; Brokaw and Simonick, 1977).

The replacement of EDTA in the reactivating solution devised for *Colobocentrotus* by EGTA in the standard procedure just given serves to favor the formation of symmetrical flagellar waveforms by maintaining the free-Ca^{2+} concentration at a low level as described by Brokaw *et al.* (1974).

C. Variations That Produce Asymmetric Bending

The finding by Brokaw *et al.* (1974) that the presence or absence of 4–5 m*M* Ca^{2+} in the demembranating solution influences the asymmetry of the flagellar bending waves in reactivated sperm of *Stronglycentrotus* [whereas *Colobocentrotus* sperm are relatively insensitive (Gibbons and Gibbons, 1972)] is also applicable to sperm of *Lytechinus* and *Tripneustes* (Brokaw, 1975; B. H. Gibbons and Gibbons, 1980). Recently, Okuno and Brokaw (1979) have reported that the required concentration of Ca^{2+} is greatly reduced (to about 0.1 m*M*) when Mg^{2+} is omitted from the demembranating solution.

If the demembranating solution contains 2 m*M* EGTA instead of 4 m*M* $CaCl_2$, highly asymmetric waveforms result when the sperm are subsequently reactivated with 1 m*M* ATP (Brokaw *et al.*, 1974). For the sake of brevity, sperm demembranated in the presence of 4 m*M* Ca^{2+} have been termed potentially symmetrical, and sperm demembranated in a solution in which the Ca^{2+} concentration is maintained at a very low level by the presence of 2 m*M* EGTA have been termed potentially asymmetric (B. H. Gibbons and Gibbons, 1980). Study of both types of preparation is proving to be useful (Brokaw, 1979; B. H. Gibbons and Gibbons, 1980) as a means of increasing understanding of the mechanisms by which microtubule sliding is converted into coordinated bending waves.

Potentially symmetrical sperm of *Tripneustes* and *Strongylocentrotus* can be induced to swim with moderately asymmetric waveforms by adding Ca^{2+} to the reactivating solution at concentrations in excess of the EGTA concentration. The degree of asymmetry increases with increasing concentrations of Ca^{2+} up to the millimolar range (Brokaw *et al.*, 1974; Brokaw, 1979), but is never as great as with potentially asymmetric sperm.

D. Discussion

When *Tripneustes* sperm are prepared according to the standard procedure (Section III, A) with Ca^{2+} in the demembranating solution, they swim with bending waves that closely resemble the waveforms of live sperm swimming in seawater, except that they are significantly more symmetrical (Fig. 1). This suggests that the swimming behavior of reactivated sperm prepared by the potentially symmetrical demembranation procedure may be artifactual, and that waveforms with a slight degree of asymmetry are more physiological. However, it has not as yet been possible to reproduce the asymmetry of live sperm in reactivated preparations under conditions that are thought to represent the *in vivo* situation—that is, an intracellular concentration of free Ca^{2+} of about 10^{-8} *M*.

The reactivated motility of cilia and flagella of several other organisms is also sensitive to Ca^{2+}, and in these cases it is possible to mimic physiological behavior more closely with a presumptive physiological concentration of free Ca^{2+}. Specifically, changes in the direction and form of beating have been demonstrated in cilia of *Paramecium* (Naitoh and Kaneko, 1972), in flagella of *Chlamydomonas* (Hyams and Borisy, 1978; Bessen *et al.*, 1980), and in *Crithidia* (Holwill and McGregor, 1976). An effect of Ca^{2+} in inducing arrest of movement has been studied in demembranated mussel gill cilia by Walter and Satir (1978) and by Tsuchiya (1977), and in sperm flagella by B. H. Gibbons and Gibbons (1980). In most of these cases, the free-Ca^{2+} concentration required to effect the switch in behavior is about 1 μM.

IV. Preparation of Dynein-1-Depleted Reactivated Sperm

A. Method

Standard preparations of dynein-l-depleted sperm (Gibbons and Gibbons, 1973) are made by adding 25 μl of the diluted sperm suspension in seawater to 0.3 ml of demembranating solution containing 0.04% Triton X-100–0.5 *M* KCl–2 m*M* $MgSO_4$–1 m*M* DTT–0.5 m*M* EDTA–10 m*M* Tris HCl buffer (pH 8.1), at room temperature (23°C). The extraction is terminated after 50–60 seconds by diluting about 10 μl of the demembranated-sperm suspension into a dish containing 2.5 ml of reactivating solution, and the sperm motility is observed by dark-field microscopy.

The above procedure applied to sperm of the species *Colobocentrotus* results in a uniform reduction of about 50% in flagellar beat frequency, whereas the percentage of motile sperm and their swimming motion and waveform resemble that of standard reactivated sperm (Gibbons and Gibbons, 1973). This decrease in frequency is paralleled by a selective solubilization of the outer dynein arms of the axoneme, while leaving the inner arms apparently intact. So far, this selective solubilization of the outer arms by brief extraction to give homogeneous preparations of reactivated sperm with their flagella beating at half-frequency has been achieved only with sperm of *Colobocentrotus*. For reasons that are not yet clear, the same brief extraction applied to sperm of other species of sea urchin, including *Tripneustes*, results in poor motility with incomplete propagation of bends in the distal region of the flagellum and heterogeneous beat frequencies ranging from 14 to 28 Hz within a preparation.

B. Discussion

It is necessary to use a brief extraction time in these preparations, for when the extraction time exceeds about 1 minute at room temperature, the waveform of the dynein-l-depleted sperm deteriorates (Gibbons and Gibbons, 1973).

Preparations of dynein-l-depleted *Colobocentrotus* sperm are very useful for assaying the functional activity of dynein 1 ATPase preparations as described by Gibbons and Gibbons (1979) and by Bell and Gibbons (1982), and in other species the technique has been used to demonstrate ATP-induced microtubule sliding in axonemes depleted of outer dynein arms (Hata *et al.*, 1980).

V. Preparation of Rigor-Wave Sperm

A. Method

Two procedures are available, both based on a rapid decrease in the concentration of Mg ATP^{2-} to below the level required to sustain motility (Gibbons and Gibbons, 1974, 1978). This decrease can be achieved either by diluting the swimming sperm into reactivating solution lacking ATP, or by diluting them into a reactivating solution that lacks Mg^{2+} and contains sufficient EDTA to chelate essentially all the Mg^{2+} carried over with the sperm from the demembranating solution. In both cases, the motile axonemes become set into stationary waves whose forms resemble those present at different phases of the normal beat cycle.

The protocol for preparing rigor-wave sperm by dilution from ATP is as follows. Twenty-five microliters of the dilute suspension of sperm in seawater is added to 0.3 ml of demembranating solution containing 20 μM ATP–0.15 M KCl–4 mM $CaCl_2$–2 mM $MgSO_4$–1 mM DTT–0.5 mM EDTA–0.04% Triton

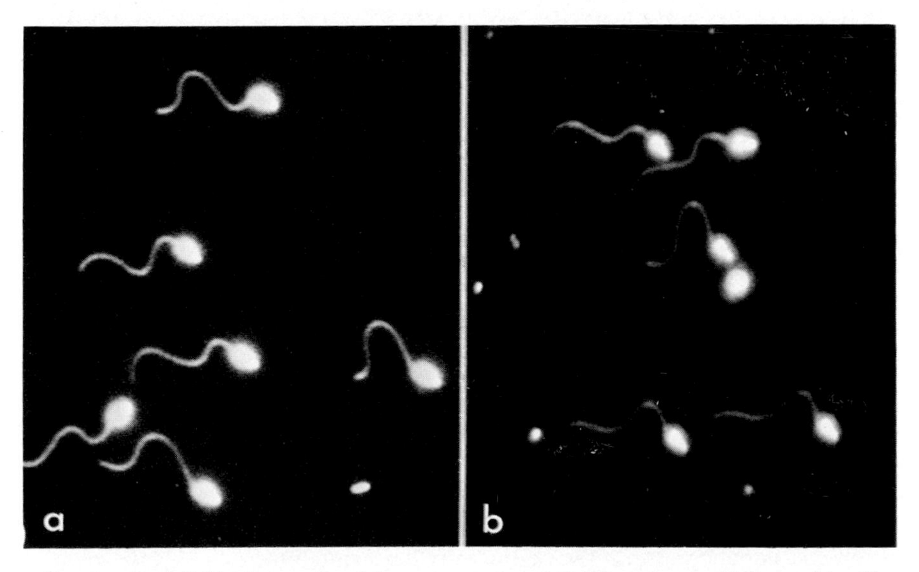

FIG. 2. Dark-field micrographs of rigor-wave sperm of *Colobocentrotus*. (a) Prepared by dilution from $MgATP^{2-}$ into suspension solution lacking ATP. (b) Prepared by dilution from $MgATP^{2-}$ into suspension solution lacking added Mg^{2+} and containing extra EDTA. Taken after sperm had settled to bottom of observation dish, with ×10 0.22 NA objective and 0.8–0.95 NA dark-field condenser. [Image (a) reprinted from Gibbons and Gibbons, 1974.] ×450.

X-100 (w/v)–10 m*M* Tris HCl buffer (pH 8.1) at room temperature. After gentle mixing for 20 to 90 seconds, 100 μl of the demembranated-sperm suspension is withdrawn into a capillary pipet and added immediately to 5 ml of suspension solution (0.15 *M* KCl–2 m*M* $MgSO_4$–1 m*M* DTT–0.5 m*M* EDTA–10 m*M* Tris HCl buffer, pH 8.1) that is being stirred in a 10-ml beaker with a small magnetic bar turning at about 60 rpm. The resulting suspension of rigor-wave sperm may be stored at 0°C for up to 2 hours without the sperm losing their potential for motility on readdition of ATP.

Alternatively, rigor-wave sperm may be prepared by abruptly lowering the concentration of Mg^{2+} (Gibbons and Gibbons, 1978). In this case the protocol is changed so as to avoid carrying over too much Mg^{2+} into the suspension solution either by reducing the volumes transferred at each step, or by shedding the sperm initially into artificial seawater lacking Mg^{2+}. Ten microliters of dilute sperm suspension in seawater is added to 0.3 ml of demembranating solution (composition as just given). Following an extraction period of about 60 seconds, 5–10 μl of the demembranated-sperm suspension is withdrawn with a capillary pipet and immediately added to 5 ml of suspension solution (20 μM ATP–0.15 *M* KCl–1 m*M* DTT–2 m*M* EDTA–10 m*M* Tris HCl buffer, pH 8.1) that is being stirred. The resulting rigor-wave sperm closely resemble those prepared by the first method (Fig. 2).

B. Discussion

The $MgATP^{2-}$ concentration in the demembranating solution must be low (only slightly above the minimum needed to sustain complete reactivation) at the moment that the sperm are diluted into the suspension solution. If the concentration of $MgATP^{2-}$ is too high, the resulting rigor waveforms are of abnormally low amplitude (Gibbons and Gibbons, 1974). Since the sperm flagella are beating and hydrolyzing ATP in the demembranating solution, the ATP concentration is continually dropping during demembranation, and it is usually necessary to adjust the duration of the demembranation period of individual preparations to optimize the quality of rigor waveforms obtained. Depending on the sperm concentration, this period will vary between 30 and 90 seconds. It may be noted that in this procedure the endogenous ATP of the sperm makes a significant contribution to the total ATP in the demembranating solution, and, of course, the sperm concentration affects the rate of ATP consumption.

It is best to omit PEG from the suspension solution because the increase in viscosity that it causes slows the mixing process and is detrimental to the formation of normal-looking rigor waveforms. The presence of DTT and EDTA during storage of rigor-wave sperm is required to prevent slow changes in the waveforms that appear to result from a twisting of the axonemes.

VI. Ca^{2+}-Induced Quiescence in Sperm Flagella

Sperm of the sea urchin genus *Tripneustes* are demembranated by the protocol for the preparation of potentially asymmetric sperm previously described, and transferred to reactivating solution containing 0.15 *M* KCl–2 m*M* $MgSO_4$–1 m*M* DTT–0.1 m*M* EGTA–10 m*M* Tris HCl (pH 8.1)–1 m*M* ATP (2% PEG may be added but is not necessary). After the swimming of the sperm has been checked under the microscope, 0.2–0.3 m*M* Ca^{2+} is added from a stock solution of 0.1 *M* $CaCl_2$ in water. The majority of the sperm immediately stop swimming, with their flagella uniformly bent into the quiescent waveform seen in Fig. 3 (B. H. Gibbons and Gibbons, 1980).

Alternatively, the ATP may be omitted initially from the reactivating solution and replaced with 0.2 m*M* Ca^{2+}. Then, after the sperm are added, quiescence is induced by the addition of ATP, giving waveforms identical to those described above. However, much lower percentages of quiescent sperm flagella are obtained if the free-Ca^{2+} concentration is above about 0.5 m*M*, or if the $MgATP^{2-}$ concentration is below about 0.7 m*M*.

Quiescent sperm are mechanically fragile and their flagellar form is unstable on standing at room temperature. Gradually over a period of 10–30 minutes, a majority of the sperm flagella exhibit drastic shape changes that often include sliding and looping out of some microtubules from the main axonemal bundles

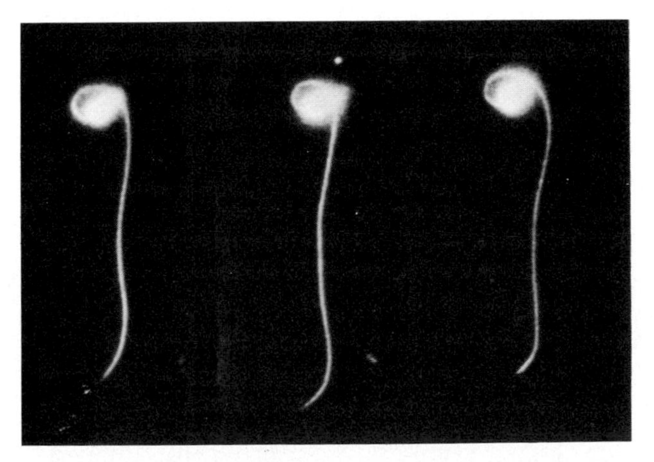

FIG. 3. Dark-field micrographs of reactivated sperm of *Tripneustes* that have become quiescent after addition of 0.1 m*M* free Ca^{2+} as described in Section VI. Taken at high magnification showing flagellar waveforms of individual quiescent reactivated sperm. Optics same as Fig. 1. Flash exposure. (Reprinted from B. H. Gibbons and Gibbons, 1980.) ×930.

and a concomitant diminution in the proximal-bend angle (B. H. Gibbons and Gibbons, 1980). In the brief period of up to 1–2 minutes after their formation, quiescent sperm may be induced to resume motility by treatment with EGTA to lower the free-Ca^{2+} concentration, by dilution to lower the ATP concentration, or by lowering the temperature of the preparation to below 20°C.

VII. Reactivation of Mammalian Sperm

Bishop and Hoffmann-Berling (1959) treated sperm from several mammalian species with an EDTA–digitonin extraction medium and demonstrated that motility could be reactivated by the addition of ATP, although the cells did not show any pronounced forward progression. Reactivated motility of a more lifelike nature was later achieved by Lindemann and Gibbons (1975) by using Triton X-100. In their procedure, sperm are first suspended in a storage medium that consists of 0.2 *M* sucrose, 70 m*M* K_2SO_4, 5 m*M* $MgSO_4$, 3–5 μM $CaCl_2$, and 2 m*M* phosphate buffer (pH 7.4). In the case of frozen bull semen, the samples are thawed rapidly, diluted 10-fold with storage medium, and washed twice by centrifugation and resuspension in fresh storage medium. The sperm are then stored at 0°C. The same procedure is used for human sperm in order to remove the seminal fluid. The condition of the sperm in the stock suspensions is always inspected immediately before use by examining the motility of a small sample diluted into storage medium at 25°C. The percentage of sperm showing normal motility is usually about 90% in fresh samples of epididymal bull sperm, 30–50% in frozen bull sperm, and 50% in fresh human sperm.

Reactivated sperm of optimal quality are obtained by using the following procedure. Fifty microliters of washed sperm suspension is added to 0.5 ml of extracting solution containing 0.2% Triton X-100–0.2 *M* sucrose–70 m*M* potassium glutamate–1 m*M* DTT–30 m*M* Tris HCl buffer (pH 7.9) at room temperature. After about 30 seconds in extraction solution, 50 μl of the extracted sperm suspension is transferred to a dish containing 2 ml of reactivating solution (0.13 *M* sucrose–45 m*M* potassium glutamate–5 m*M* $MgSO_4$–0.6 m*M* DTT–20 m*M* Tris HCl buffer, pH 7.9). After the sperm have been mixed with reactivating solution, ATP is added from a 0.1 *M* stock solution to give a final concentration of 5 m*M*.

The percentage of motility in reactivated preparations of bull sperm is generally 60–70%; the comparable figure for human sperm is 30%. In bull sperm, the relative concentrations of Mg^{2+} and ATP appear to be highly important in obtaining coordinated motility, and the best results were obtained when the ATP concentration was 3–5 m*M* and the Mg^{2+} concentration was 4 m*M* (Lindemann and Gibbons, 1975).

VIII. Reversible Inhibition of Motility

Several agents have been found that inhibit reversibly the motility of reactivated sperm. These include CO_2, vanadate, and lowered pH.

For treatment with CO_2, demembranated sperm are suspended in an open drop of reactivating solution on a microscope slide without a cover glass, and a CO_2 + N_2 gas mixture is blown over the surface of the drop, or alternatively, the sperm may be enclosed in a covered well slide in reactivating solution containing $KHCO_3$ in the range up to about 0.1 *M*. Studies on sperm from *Lytechinus,* reactivated with 0.2 m*M* ATP (Brokaw and Simonick, 1976; Brokaw, 1977), indicate that the CO_2 molecule exerts a specific effect of reducing the amplitude of flagellar bending, with relatively little effect on beat frequency, although at sufficiently high concentrations motility ceases. At concentrations of ATP below 20 μM there is a reduction of beat frequency without a reduction in bend amplitude, and following exposure to a relatively high CO_2 concentration the flagellar bending may be set into stationary rigor waves similar to those described in Section V, A.

The inhibitor vanadate may be added to reactivated sperm from a 1 m*M* or 0.1 m*M* stock solution of sodium orthovanadate ($NaVO_3$) buffered near pH 8 to prevent polymerization of the vanadate ion to the yellow-colored decavanadate ion (Pope and Dale, 1968). For convenience, a more concentrated, stable stock solution of 0.1 *M* sodium orthovanadate is first prepared in water. The critical concentration of vanadate that is needed to inhibit completely the motility of reactivated sea urchin sperm depends on the ATP concentration, and is also sharply dependent on the pH and KCl concentration of the reactivating medium (Gibbons *et al.*, 1978; Kobayashi *et al.*, 1978). A study of the dependence on pH and salt concentration has been carried out by this author on reactivated sperm of *Tripneustes* at 0.1 m*M* ATP, and the results are presented in Fig. 4. The motility of reactivated cilia of sea urchin embryos, and of mammalian sperm flagella (Gibbons *et al.*, 1978; Kobayashi *et al.*, 1978; Gibbons, 1979), as well as the reactivated motility of the lateral cilia of mussel gills (Wais-Steider and Satir, 1979) and bracken fern spermatozoids (Wolniak and Cande, 1980), is also inhibited by vanadate, with only micromolar concentrations or less being required.

Vanadate inhibition can be reversed either by dilution to reduce the vanadate concentration or by addition of 1–2 m*M* catechol or norepinephrine (from a solution prepared freshly in 1 m*M* HCl and kept in the dark). The latter compounds reverse inhibition by complexing and reducing the vanadate from the +5 oxidation state to the inactive +4 state. This reversal of inhibition occurs slowly over a period of about 1 minute. Preliminary experiments indicate that vanadate forms a complex with oxidized DTT that reduces its effectiveness as an inhibitor

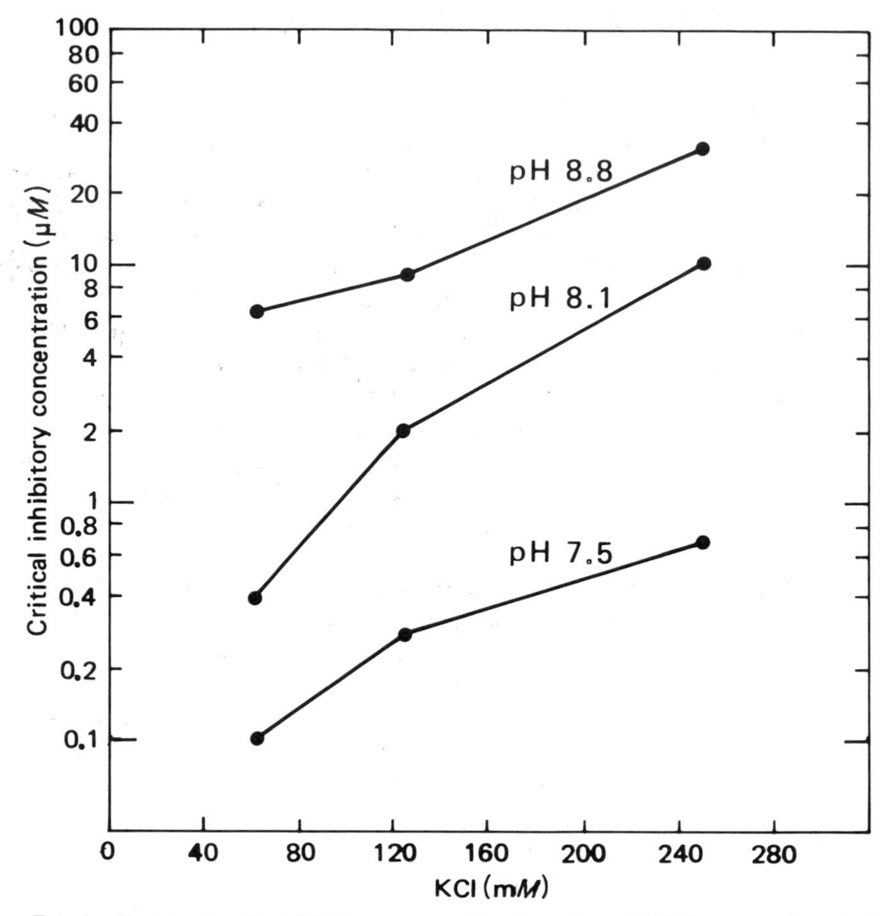

FIG. 4. Variation in critical inhibitory concentration of vanadate with KCl concentration and pH of reactivating solution. Reactivated sperm of *Tripneustes* in 0.1 m*M* ATP, prepared as in Section III, A, except for changes indicated. Concentration of vanadate is that needed to inhibit completely the motility of all sperm in the sample.

of reactivated sperm motility, but that the complex formed with EDTA appears not to interfere with inhibition (B. H. Gibbons and I. R. Gibbons, unpublished results).

Study of the inhibition of reactivated motility at low pH is performed simply by transferring demembranated sperm to reactivating solution buffered at the desired pH (Goldstein, 1979). The inhibition of motility in *Lytechinus* and *Strongylocentrotus* sperm at pH 7.3–7.5 can be reversed by local addition of buffer at pH 8.0 from a micropipet. As the pH is raised, a series of principal bends form near the base and are propagated to the tip. With a further increase in pH, reverse bends also begin to develop. These results suggest that the principal and reverse bends have different sensitivities to pH (Goldstein, 1979).

IX. Trypsin-Treated Axonemes for Study of ATP-Induced Microtubule Sliding

In a slight modification of the procedure originally developed by Summers and Gibbons (1971), sperm are demembranated and broken by homogenization (Dounce glass homogenizer) in a solution containing 1% (w/v) Triton X-100–5 m*M* $MgSO_4$–0.1 *M* KCl–1 m*M* DTT–0.5 m*M* EDTA–10 m*M* imidazole buffer (pH 7.4). The axonemal fragments are isolated as free of sperm heads as possible by differential centrifugation and resuspended at a concentration of about 0.25 mg protein/ml in Tris-Mg solution (2.5 m*M* $MgSO_4$–0.25 m*M* EDTA–0.1 m*M* DTT–30 m*M* Tris HCl buffer, pH 8.0) at room temperature. Digestion is carried out by addition of trypsin to the suspension of axonemes to give a trypsin:protein ratio of about 1:1500. Treatment proceeds to the desired extent in a few minutes and its course may be monitored by measurement of the turbidity of the suspension at 350 nm, but contamination by sperm heads must be strictly avoided or a false indication of the degree of digestion will be obtained. Preparations of axonemal fragments digested to about 80% of their initial turbidity have been found best to demonstrate uniform disintegration by sliding (Summers and Gibbons, 1971; Sale and Gibbons, 1979). The course of the digestion may also be followed by observing the response of the axonemes to ATP. At successive time intervals of a few minutes, a sample is removed, placed on a trough slide, and covered with a coverslip. A small volume of ATP is applied to one end, usually from a 0.5 m*M* stock solution, and evaporation and diffusion are then allowed to draw the ATP slowly through the suspension. The partially digested axonemes are observed by dark-field microscopy, which allows visualization of individual doublet tubules (Gibbons, 1975); when the digestion has proceeded far enough, the axonemes should be seen to slide apart uniformly and in high percentage (Summers and Gibbons, 1971). Digestion of the stock suspension may then be stopped by addition of an excess of soybean-trypsin inhibitor. Trypsin-treated axonemes are mechanically fragile and must be pipetted gently to avoid damage.

For homogenization on a smaller scale, the sperm may be forced through 26- or 30-gauge plastic tubing with a syringe. The trypsin treatment and ATP-induced sliding may be done without separation of the sperm heads, but in this case a much higher ratio of trypsin to protein (about 1:500) must be used.

Microtubule sliding disintegration induced by ATP in axonemes from cilia of *Tetrahymena* has also been studied and has been reported to require (Sale and Satir, 1977) or not to require (Warner and Mitchell, 1978) prior trypsin digestion.

X. Optical and Photographic Techniques

In as much as dark-field microscopy is the method of choice for observation of the swimming motion of sperm, it is appropriate to describe here optical systems that give good results. Lenses are by Zeiss unless otherwise specified.

Two types of illumination are useful: a steady source for motionless sperm, such as rigor preparations; and brief pulses of light whose frequency can be varied to strobe the flagellar waveform and measure the beat frequency of motile sperm. Steady illumination may be provided by a XBO 150 xenon lamp (Osram, Berlin) with a DC power supply. For stroboscopy, Brokaw (1963, 1977) developed a system consisting of a xenon flashtube (Strobex Model 71B) and a power unit (Model 136, Chadwick-Helmuth, Monrovia, California). For routine observation at low magnification, sperm may be observed swimming at the bottom surface of an open dish with a ×10 0.22 NA achromatic objective, and either a 0.7–0.85 or 0.8–0.95 NA dark-field condenser. For higher magnification, a particularly satisfactory and easy arrangement is to observe and photograph the sperm in an open dish with a ×40 0.75 NA water-immersion achromatic objective and a 1.2–1.4 oil-immersion dark-field condenser. With motionless sperm it is preferable to omit PEG from the reactivating solution to reduce swirling of the sperm near the objective when the dish is moved on the microscope stage. In combination with a 1-mm well slide and coverslip, a ×25 0.65 NA planapochromatic objective gives good results; somewhat improved resolution is obtainable with ×60 0.8–1.2 NA and ×100 0.8–1.2 NA oil-immersion planapochromatic objectives at a slight loss of convenience.

Elegant cinematography on 16-mm film has been performed by Hiramoto and Baba (1978) using phase-contrast objective lenses ×40 (Nikon BM 40/0.65) and ×100 (Nikon BN 100/1.25 and Plan BN 100/1.30 oil immersion).

For dark-field photomicrography Kodak Tri-X, 4-X, and 2475 recording film are all satisfactory, and they may be developed in Diafine to improve sensitivity and resolution. For some types of use, Polaroid Type 107C is suitable (cf. Gibbons, 1980). A multiple-exposure technique employing repeated flash illumination and a 35-mm camera modified so that film transport occurs continuously while the shutter is open has been used very effectively by Goldstein (1976) and by Brokaw (1979).

XI. Future Utility of Reactivation

Studies of demembranated and reactivated cilia and flagella have already yielded much information about the mechanism of microtubule-mediated motility (for review, see Blum and Hines, 1979; Gibbons, 1981), and many lines of approach now being pursued promise to add further to our knowledge.

An approach that is receiving much attention at present concerns the role of Ca^{2+} in determining the direction, form, and arrest of ciliary and flagellar beating (see Section III, C,D), and generally its involvement in the coordination of microtubule sliding to produce bending. The recent discovery of the presence of calmodulin in *Tetrahymena* cilia (Jamieson *et al.*, 1979), and of a calmodulin-like protein in *Chlamydomonas* flagella (Van Eldik *et al.*, 1980), add to the interest in this topic.

Steps in the dynein-arm cross-bridge cycle may be further clarified by experiments such as those using vanadate (Sale and Gibbons, 1979; Wais-Steider and Satir, 1979; Okuno, 1980), or measurements of flagellar stiffness under various conditions (Okuno and Hiramoto, 1979; Okuno, 1980). Studies from the laboratory of Miki-Noumura (Hata *et al.*, 1979, 1980; Yano and Miki-Noumura, 1980) of the ATP-concentration dependence of the tubule-extrusion velocity from dynein-depleted and normal axonemes, and those from Warner's laboratory (Zanetti *et al.*, 1979; Warner and Zanetti, 1980) of the effects of divalent cations on ciliary-microtubule sliding, constitute other new approaches to investigating the tubule-sliding mechanism.

Use of various techniques to alter or interrupt normal flagellar beating, and study of their effect on various wave parameters, are beginning to provide information about the mechanisms controlling movement. A few years ago, Gibbons (1974) noted that beat frequency and flagellar waveform are perturbed nearly independently by a number of different agents, such as ATP concentration (Gibbons and Gibbons, 1972). Since then, other studies have extended this observation, including the work of Brokaw (1977) on the effect of CO_2 on the amplitude of bending in reactivated sperm; of Brokaw and Simonick (1977) and of Brokaw (1980) on the effects of protease digestion during movement; of Shingyogi *et al.* (1977) on local reactivation with ATP; of Gibbons *et al.* (1976) and of Ogawa *et al.* (1977) on the effects of an antiserum to dynein 1-ATPase; of Asai and Brokaw (1980) on the effects of antisera to tubulin; of Goldstein (1976, 1979) on bend development during starting transients in sperm flagella; and of Okuno and Brokaw (1979) on the inhibitory effects of Mg^{2+}, and ADP and P_i on several parameters of the movement of demembranated sperm. Lindemann *et al.* (1980) have recently reported a selective inhibition of bend initiation by Ni^{2+} in reactivated bull sperm flagella.

The reports that separated bundles of demembranated microtubules have been observed to beat in the presence of ATP (Nakamura and Kamiya, 1978; B. H. Gibbons and Gibbons, 1980; Hata *et al.*, 1980) suggest a new, simplified system for studying the minimal attributes and components essential for axonemal-bend propagation, although at present it appears that the observations may be limited by the difficulty of identifying the components that are present in these bundles. However, a comparison with the behavior of motile cilia and flagella having atypical structure, such as those with 3 + 0 and 6 + 0 patterns of microtubules (Goldstein *et al.*, 1979; Prensier *et al.*, 1980), may be profitable.

Agents such as vanadate or CO_2 are important in their usefulness in studying the active bending of flagella under conditions in which oscillatory beating is inhibited (B. H. Gibbons and Gibbons, 1980). The exploitation of this technique can be expected to yield important information about the factors regulating the activity of dynein cross-bridges at different positions on the flagellum, as well as about the properties of the structural components that resist active sliding between tubules and convert it into bending.

REFERENCES

Asai, D. J., and Brokaw, C. J. (1980). *J. Cell Biol.* **87,** 114–123.

Bell, C. W., and Gibbons, I. R. (1982). *In* "Muscle and Non-Muscle Motility (A. Stracher, ed.), in press. Academic Press, New York.

Bessen, M., Fay, R. B., and Witman, G. B. (1980). *J. Cell Biol.* **86,** 446–455.

Bishop, D. W., and Hoffmann-Berling, H. (1959). *J. Cell. Comp. Physiol.* **53,** 445–466.

Blum, J. J., and Hines, M. (1979). *Q. Rev. Biophys.* **12,** 104–180.

Brokaw, C. J. (1963). *J. Exp. Biol.* **40,** 149–156.

Brokaw, C. J. (1965). *J. Exp. Biol.* **43,** 155–169.

Brokaw, C. J. (1967). *Science* **156,** 76–78.

Brokaw, C. J. (1975). *J. Exp. Biol.* **62,** 701–719.

Brokaw, C. J. (1977). *J. Exp. Biol.* **71,** 229–240.

Brokaw, C. J. (1979). *J. Cell Biol.* **82,** 401–411.

Brokaw, C. J. (1980). *Science* **207,** 1365–1367.

Brokaw, C. J., and Simonick, T. F. (1976). *Cold Spring Harbor Conf. Cell Proliferation* **3** [Book C], 933–940.

Brokaw, C. J., and Simonick, T. F. (1977). *J. Cell Biol.* **75,** 650–665.

Brokaw, C. J., Josslin, R., and Bobrow, L. (1974). *Biochem. Biophys. Res. Commun.* **58,** 795–800.

Gibbons, B. H. (1979). *In* "The Spermatozoon" (D. W. Fawcett and J. M. Bedford, eds.), pp. 91–97. Urban & Schwarzenberg, Munich.

Gibbons, B. H. (1980). *J. Cell Biol.* **84,** 1–12.

Gibbons, B. H., and Gibbons, I. R. (1972). *J. Cell Biol.* **54,** 75–97.

Gibbons, B. H., and Gibbons, I. R. (1973). *J. Cell Sci.* **13,** 337–357.

Gibbons, B. H., and Gibbons, I. R. (1974). *J. Cell Biol.* **63,** 970–985.

Gibbons, B. H., and Gibbons, I. R. (1978). *J. Cell Biol.* **79,** 285a.

Gibbons, B. H., and Gibbons, I. R. (1979). *J. Biol. Chem.* **254,** 197–201.

Gibbons, B. H., and Gibbons, I. R. (1980). *J. Cell Biol.* **84,** 13–27.

Gibbons, B. H., Ogawa, K., and Gibbons, I. R. (1976). *J. Cell Biol.* **71,** 823–831.

Gibbons, I. R. (1974). *In* "The Functional Anatomy of the Spermatozoon" (B. A. Afzelius, ed.), pp. 127–140. Pergamon, Oxford.
Gibbons, I. R. (1975). *In* "Molecules and Cell Movement" (S. Inoué and R. E. Stephens, eds.), pp. 207–231. Raven, New York.
Gibbons, I. R. (1981). *J. Cell Biol.* **91,** 1075–1245.
Gibbons, I. R., and Gibbons, B. H. (1980). *J. Muscle Res. Cell Motil.* **1,** 31–59.
Gibbons, I. R., Cosson, M. P., Evans, J. A., Gibbons, B. H., Houck, B., Martinson, K. H., Sale, W. S., and Tang, W.-J. Y. (1978). *Proc. Natl. Acad. Sci. U.S.A.* **75,** 2220–2224.
Goldstein, S. F. (1976). *J. Exp. Biol.* **64,** 173–184.
Goldstein, S. F. (1979). *J. Cell Biol.* **80,** 61–68.
Goldstein, S. F., Besse, C., and Schrével, J. (1979). *Acta Protozool.* **18,** 131.
Hata, H., Yano, Y., and Miki-Nomura, T. (1979). *Exp. Cell Res.* **122,** 416–419.
Hata, H., Yano, Y., Mohri, T., Mohri, H., and Miki-Nomura, T. (1980). *J. Cell Sci.* **41,** 331–340.
Hiramoto, Y., and Baba, S. A. (1978). *J. Exp. Biol.* **76,** 85–104.
Hoffmann-Berling, H. (1955). *Biochim. Biophys. Acta* **16,** 146–154.
Holwill, M. E. J. (1969). *J. Exp. Biol.* **50,** 203–222.
Holwill, M. E. J., and McGregor, J. L. (1976). *J. Exp. Biol.* **65,** 229–242.
Hyams, J. S., and Borisy, G. G. (1978). *J. Cell Sci.* **33,** 235–253.
Jamieson, G. A., Jr., Vanaman, T. C., and Blum, J. J. (1979). *Proc. Natl. Acad. Sci. U.S.A.* **76,** 6471–6475.
Kobayashi, T., Martensen, T., Nath, J., and Flavin, M. (1978). *Biochem. Biophys. Res. Commun.* **81,** 1313–1318.
Lindemann, C. B., and Gibbons, I. R. (1975). *J. Cell Biol.* **65,** 147–162.
Lindemann, C. B., Fentie, I., and Rikmenspoel, R. (1980). *J. Cell Biol.* **87,** 420–426.
Lowry, O. H., Rosebrough, N. J., Farr, A. L., and Randall, R. J. (1951). *J. Biol. Chem.* **193,** 265–275.
Naitoh, Y., and Kaneko, H. (1972). *Science* **176,** 523–524.
Nakamura, S., and Kamiya, R. (1978). *Cell Struct. Funct.* **3,** 141–144.
Nakaoka, Y., and Toyotama, H. (1979). *J. Cell Sci.* **40,** 207–214.
Ogawa, K., Asai, D. J., and Brokaw, C. J. (1977). *J. Cell Biol.* **73,** 182–192.
Okuno, M. (1980). *J. Cell Biol.* **85,** 712–725.
Okuno, M., and Brokaw, C. J. (1979). *J. Cell Sci.* **38,** 105–123.
Okuno, M., and Hiramoto, Y. (1979). *J. Exp. Biol.* **79,** 235–243.
Okuno, M., Ogawa, K., and Mohri, H. (1976). *Biochem. Biophys. Res. Commun.* **68,** 901–906.
Pope, M. T., and Dale, B. W. (1968). *Q. Rev., Chem. Soc.* **22,** 527–548.
Prensier, G., Vivier, E., Goldstein, S. F., and Schrével, J. (1980). *Science* **207,** 1493–1494.
Sale, W. S., and Gibbons, I. R. (1979). *J. Cell Biol.* **82,** 291–298.
Sale, W. S., and Satir, P. (1977). *Proc. Natl. Acad. Sci. U.S.A.* **74,** 2045–2049.
Shingyogi, C., Murakami, A., and Takahashi, K. (1977). *Nature (London)* **265,** 269–270.
Summers, K. E., and Gibbons, I. R. (1971). *Proc. Natl. Acad. Sci. U.S.A.* **68,** 3092–3096.
Tsuchiya, T. (1977). *Comp. Biochem. Physiol. A* **56A,** 353–361.
Van Eldik, L. J., Piperno, G., and Watterson, D. M. (1980). *Proc. Natl. Acad. Sci. U.S.A.* **77,** 4779–4783.
Wais-Steider, J., and Satir, P. (1979). *J. Supramol. Struct.* **11,** 339–347.
Walter, M. F., and Satir, P. (1978). *J. Cell Biol.* **79,** 110–120.
Warner, F. D., and Mitchell, D. R. (1978). *J. Cell Biol.* **76,** 261–277.
Warner, F. D., and Zanetti, N. C. (1980). *J. Cell Biol.* **86,** 436–445.
Wolniak, S. M., and Cande, W. Z. (1980). *J. Cell Sci.* **43,** 195–207.
Yano, Y., and Miki-Noumura, T. (1980). *J. Cell Sci.* **44,** 169–186.
Zanetti, N. C., Mitchell, D. R., and Warner, F. D. (1979). *J. Cell Biol.* **80,** 573–588.

Chapter 13

Stereological Analysis of Microtubules in Cells with Special Reference to Their Possible Role in Secretion

EVE REAVEN

Department of Medicine, Stanford University School of Medicine
and
Geriatric Research, Education, and Clinical Center, Veterans Administration Medical Center
Palo Alto, California

I. Introduction

Reliable methodology now exists for the assessment of total soluble tubulin in cells and tissues through the use of either colchicine binding (Bamburg *et al.*, 1973) or radioimmunotechniques (Van De Water *et al.*, 1982). However, for the investigator it is often of more interest to know what proportion of the total tubulin pool is in the form of assembled microtubules, or where the microtubules are located in cells. Various types of questions have been asked. (1) Do different cell types within the same organism have similar microtubule content? Does the cell's content of microtubules change

ISBN 0-12-564125-7

under different physiological or growth conditions? (2) Do agents such as colchicine, vinblastine, D_2O, taxol, chlorpromazine, glycerol (to name just a few) change the content of microtubules in cells, and in so doing can one observe related functional changes in cells? (3) Do certain cells have polarity with regard to the distribution of microtubules? Is this polarity related to a function? Finally, (4) Is the location of microtubules in cells strictly random or do microtubules have specific associations with certain organelles of cells? If a specific association exists, does it change under differing physiological conditions?

One way to approach these questions is through the use of light-microscopic-fluorescent techniques (Osborn and Weber, 1982) or electron-microscopic-cytochemical techniques (Sternberger, 1979). However, even under the best circumstances, immunofluorescence has limited resolution, and subtle changes in the microtubule content of cells would be lost. Reliable immunocytochemical techniques at the electron-microscopic level are still being developed, and with the unsolved problems involving the penetration of antibodies, quantitation of microtubules within cells or tissues by this methodology does not seem imminent.

It is possible, however, to obtain quantitative information regarding the content and location of microtubules within cells and tissues using stereological techniques applied to standard thin sections prepared for electron microscopy. The technique is simple and reproducible; it can be used to examine the details of microtubule distribution within cells, or to observe alterations in microtubule content or distribution in cells undergoing physiological changes. This chapter will describe the use of this technique as it has been applied to the estimation of cytoplasmic microtubules during secretion and, as such, will deal with those cytoplasmic microtubules that are unstable (sensitive to cold or antimicrotubule drugs) and that do not form organized structures (such as nerves). In addition, the limitations of the technique will be described, as well as its relative advantages and disadvantages when compared to biochemical methods aimed at isolating and quantitating the assembled microtubules of cells.

II. Quantitation of Asssembled Cytoplasmic Microtubules in Thin Sections

A. Methods Currently in Use

1. Number of Microtubules/Area Cytoplasm

The simplest technique for quantitating microtubules in thin sections is to *count* the number of microtubules per unit area (or volume) of cytoplasm. Cytoplasm containing microtubules is quantified by stereological techniques in thin sections using the methods of Weibel (1973), or by cutting the cytoplasmic region out of each micrograph and weighing it. These techniques have been used in several laboratories (Porter *et al.*, 1974; Rubin and Weiss, 1975; Murphy and Tilney, 1974; Luftig *et al.*, 1977), and can provide reasonable estimates of differences in microtubule content when cells are subjected to changing conditions. The technique suffers from the fact that the longest and shortest longitudinal segments of microtubules are assessed equally, and, as such, counts of microtubule may not necessarily reflect cytoplasmic content. If, for example, certain fixatives cause microtubules to undulate, then a microtubule of a given length may appear as more segments after one fixative than after another.

2. Volume of Microtubules/Volume Cytoplasm

To avoid this kind of problem we have used the stereological technique of Weibel (1973) to estimate the fraction of the cytoplasm occupied by microtubules (Reaven, 1977; Reaven and Reaven, 1975, 1977, 1978, 1980; Reaven *et al.*, 1978). Stereological techniques permit the making of three-dimensional (or volume) estimates from two-dimensional (or area) measurements on sections of cells and tissues. In using these techniques to estimate microtubules on electron micrographs, transparent grids are placed over the area of interest, and the line intercepts or points (P) that fall on microtubules or on other areas of the cytoplasm are recorded. Thus the percentage of the cytoplasm occupied by microtubules on a given micrograph would be

$$\frac{\text{P microtubules}}{\text{P cytoplasm}} \times 100$$

In our laboratory, the procedure of quantitating microtubules by this method is as follows.

1. Cells or areas of cells to be quantitated are selected at the electron-microscopic level at a magnification too low for microtubules to be identified.

2. Photography of the selected area is carried out at a higher magnification (16,000×) and these micrographs are further enlarged three times for a final print magnification of 48,000×.

3. All microtubule lengths and cross sections are subsequently identified on these prints with the aid of a magnifying lens.

4. For measurement of cytoplasmic volume density, plastic sheets inscribed with 1-cm grid squares are placed over each photograph and the number of grid intersections falling on cytoplasmic regions are scored; for microtubule volume-density measurements, the plastic sheets contain tiny dots 1 mm apart. Because of this small grid dimension, most microtubules in the micrographs are accounted for; whereas a microtubule cross section could conceivably hit only one point on

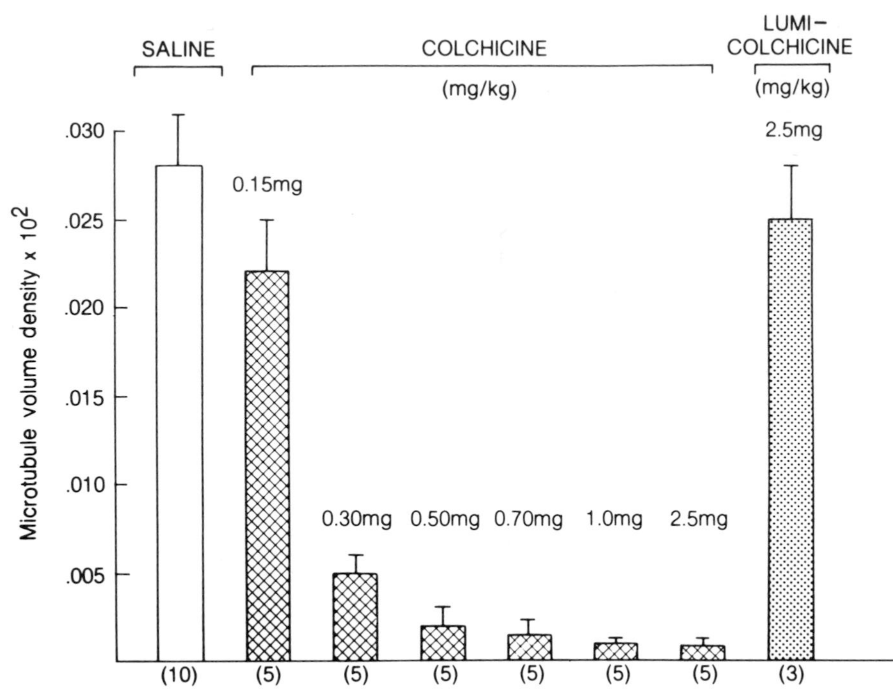

FIG. 1. Hepatic very-low-density-lipoprotein secretion rates observed after treating rats with increasing doses of cholchicine. Data expressed are as mean (± S.E.M.) secretion rates 3–5 hours after the administration of saline (clear bar), colchicine (hatched bars), or lumicolchicine (stippled bar). Figures in parentheses represent the number of livers perfused at each drug dose. (Reaven and Reaven, 1980. Reprinted with permission from *Journal of Cell Biology*.)

the millimeter grid, a longitudinal segment of a microtubule could hit several points depending on its length.

5. The scored values from several micrographs of each cell are averaged and these figures are corrected for differences in the dimensions of the grid used. In stereological techniques, the grid points used for measurement should be infinitely small relative to the structure being measured. This is clearly not possible with microtubules, and care must be taken to construct the grid so as to minimize the problem (Reaven, 1977).

In our hands the volume-density method for estimating the microtubule content of cells is reproducible [the coefficient of variation is less than 5% (Reaven, 1977)] and remarkably sensitive, as illustrated in Fig. 1. In this experiment, differing amounts of colchicine were given to rats 5 hours before liver perfusion with glutaraldehyde. Volume-density measurements of hepatocyte microtubule content showed a clear stepwise decrease in the amount of microtubules in cells that correlated perfectly with the small incremental increases in the amount of colchicine given to the rats.

B. Technical Considerations for Quantitative Stereology

1. Fixation

To ensure good quantitation of microtubules, tissue samples must be fixed optimally. This means that the samples should be very small (e.g., tissue culture cells or small organs such as islets of Langerhans), or the tissues should be perfused with fixative under controlled-pressure conditions that do not alter the size of the cells. The choice of fixatives for quantitation of microtubules is of some moment. Since the objective is to assay the content and distribution of microtubules in cells, it is important that the cellular microtubules "survive" the fixation procedure, that they are neither increased nor decreased in number, and that the size of the cells themselves is not altered by the procedure. As such, the pH, osmolality, and temperature of the fixative should be considered. Whereas gluteraldehyde is the fixative of choice, it is not yet clear whether the buffer of choice should include agents such as glycerol, DMSO, sucrose, GTP, Mg^{2+}, and EGTA, which aid in the stabilization or polymerization of microtubules. Rubin and Weiss (1975) reported that buffers containing 50% glycerol and 10% DMSO did not alter the number or length of microtubules in cultured Chinese Hamster ovary cells. On the other hand, Luftig *et al*. (1977) reported that polymerization buffers (containing glycerol, sulfonic acid buffer, EGTA, GTP, $MgSO_4$) increased the content of microtubules in HeLa cells over that measured with phosphate or cacodylate buffer. It is not certain whether the observed increase was due to superior preservation of existing microtubules or new polymerization of microtubules, an issue of some importance when quantitating microtubules. Since, in our own hands, perfusion of the aforementioned polymerization buffer in rat liver did not alter hepatocyte microtubule content (E. Reaven, unpublished results), we have continued to use cacodylate buffer for fixing samples.

2. Identification of Microtubules

In order to identify microtubules in micrographs, the magnification of the print has to be high enough to visualize even small longitudinal segments and cross sections, and yet low enough so that a reasonable number of microtubules can be measured per print. We use the same final print magnification of 48,000× regardless of the tissue being examined, so that all technical problems are equalized and we can confidently compare the microtubule content of one cell type with another. With a print magnification of 48,000×, it is possible to identify microtubules with some confidence when the length of the microtubule is 2–3 mm or more. Slightly off-center cross sections of microtubules pose more of an identification problem because there appear to be other small vesicular structures in cells that resemble this aspect of microtubules. However, since the identification of microtubules is made using blindfold techniques, our opinion is that arbitrary decisions regarding microtubule identity are not biased, and that results from different tissue samples can be compared with each other.

A more serious problem concerns the fact that longitudinally sectioned microtubules are thinner (250 Å) than the plastic sections (500 Å) that house them. As a result, microtubule volume is overestimated when related to cytoplasmic volume. One cannot effectively correct for this problem because the level of overestimation varies depending on the chance orientation of the microtubules. Again, this problem will not affect comparisons between samples, but will affect values for the absolute volume of microtubules in cells (Reaven, 1977).

3. Sampling of Cells

No strict rule can be made regarding the optimal number of cells to sample for cellular quantitation of microtubules. In general, we find that averaged measurements of several prints of each of six different cells chosen at random from tissues of each of 10 animals (or cell preparations) result in standard errors of the mean within 10%. However, the number of micrographs required of each of the six cells depends on the size, shape, and polarity of the cell, as well as the number and distribution of microtubules within the cells.

4. Evaluation of the Cytoplasmic Compartment

In order to compare the cytoplasmic content of microtubules in cells undergoing physiological or environmental changes, one must know that the overall volume of the cells has not been altered. We routinely check for this when experimental conditions change. When it is clear that the cells have not swollen or shrunk, then relating microtubule volume to cytoplasmic volume is generally sufficient. However, when the microtubule density of different types of cells is being compared, it may be more meaningful to relate microtubule volume to the cytosolic volume of the cells—that is, the portion of the cytoplasm that lies between membrane-bound organelles.

III. Advantages and Disadvantages of Stereological Methods for Estimating Assembled Microtubules in Cells

Biochemical methods utilizing glycerol stabilization and centrifugation of microtubules are now being used to estimate changes in the amount of assembled microtubules in cells (Patzelt *et al.*, 1977; Pipeleers *et al.*, 1977; Ostlund *et al.*, 1979; Sherline *et al.*, 1974; Jennett *et al.*, 1980; Matsuda *et al.*, 1979). Except for one study (Matsuda *et al.*, 1979), stereological and biochemical methods have not been compared on identical cells. This comparison is important so that one can begin to evaluate the effect of technical problems associated with either method (e.g., inadvertent polymerization of microtubules by the biochemical method or lack of preservation of microtubules by the stereological method) on the final results. It may turn out that the biochemical method assesses all aggregated microtubule protein regardless of the extent of aggregation, whereas the stereological method identifies only those microtubules that have attained a visible length, and as such, the methods might give very different kinds of information.

In any event, stereological measurements provide certain advantages over biochemical methods. Most important is the fact that in complex tissues containing several different cell types it is not necessary to separate the cells physically before evaluating microtubule content. In addition, one can see the distribution of microtubules within the cells and evaluate their relationship to other organelles of the cell.

The disadvantages of the stereological approach are obvious. It is a time-consuming technology that permits only limited sampling of cells.

IV. Assessment of Microtubule Associations with Other Cellular Organelles

Two statistical methods have been used in an attempt to relate cytoplasmic microtubules to other cellular structures. Murphy and Tilney (1974) have analyzed "nearest-neighbor distances" to show that pigment granules are closely associated with microtubules. In this approach the authors computed the mean squared distance between the centers of pigment granules with their nearest neighboring microtubules. The values were analyzed to determine if the observed distances were consistent with a random distribution of microtubules. The approach used in our laboratory (Reaven and Reaven, 1978; Reaven *et al.*, 1978) evaluates a defined microenvironment of each microtubule cross section in micrographs, and determines whether the different structures found within this microenvironment have a nonrandom relationship to the microtubules. This is done by comparing the numerical density of different structures found in the microenvironment of observed microtubules with the numerical density of these same structures in the microenvironment of randomly selected points in the cytosol of the identical photographic prints. Nonparametric or paired t-tests can be used to test the possibility that certain organelles appear in the microenvironment of the real microtubules more often than they do in the microenvironment of the random points.

These methods are extremely time-consuming, and are not recommended when the intracellular distribution of microtubules is obvious. However, it is quite enlightening to discover that in many secretory cells there is surprisingly little space available for structures like microtubules in the cytosolic compartment, and what appears to be a close association with organelles may, in fact, be the only place where microtubules can fit.

V. Utilization of Stereological Measurements in Secretory Systems

Given these efforts at quantifying microtubules, what have we learned about the importance of assembled microtubules to the secretory process? Some answers are now available to the questions posed in our Introduction. First, different secretory cells (even from the same animal) can have very different amounts of cytoplasmic or cytosolic assembled microtubules. For example, the rat hepatocyte, which packages and secretes a large number of different substances, has a much lower cytoplasmic content of microtubules [0.03% (Reaven, 1977)] than the chief cells of the parathyroid gland [0.07% (Reaven and Reaven, 1975)] or the β cell of islets of Langerhans [0.10% (E. Reaven, unpublished results)], which package and secrete only one secretory protein. This kind of variation in the microtubule content between cells can be seen even within the same cell if the cell exhibits functional secretory polarity, such as the absorptive cell of the intestine (Reaven and Reaven, 1977). In this cell type, microtubules in the apical, Golgi, and basal regions of the cell average 0.1%, 0.07%, and 0.01%, respectively, of the cytoplasm.

The question as to whether the basal content of microtubules is altered in association with secretory changes in cells, is more complex. In some studies, in which significant secretory changes were documented on the same tissues from which samples were taken for morphology, it appears that parenchymal cell microtubule content is not altered, even though secretory rates do increase. Thus in livers perfused with high levels of fatty acids, lipoprotein secretion rates increase 3-fold, but the content of microtubules in hepatocytes remains the same as in the unstimulated state (Reaven and Reaven, 1980); in *in vitro* studies of islets of Langerhans where glucose-stimulated insulin-secretion rates are increased 15- to 20-fold, the cytoplasmic content of microtubules in β cells is the same in the unstimulated and stimulated states (unpublished data). Parenthetically, when biochemical methods are used to estimate a change in assembled microtubules in these same tissues, an increase in the ratio of assembled tubulin seems to occur (Pipeleers *et al.*, 1976, 1977). It should be mentioned, however, that intact tissues (rather than isolated hepatocytes or β cells) were examined in the biochemical studies—and it is not yet clear whether the differences observed between the biochemical and stereological studies represent methodological problems or biological differences relating to parenchymal cell versus total organ analysis.

Morphometric analysis of microtubules in tissues has been particularly useful in answering the question whether antimicrotubule agents such as colchicine do, in fact, lower the microtubule content of cells (Reaven, 1977; Reaven and Reaven, 1975, 1977, 1978, 1980; Reaven *et al.*, 1978; Sheterline *et al.*, 1977). The apparent relationship between a colchicine-induced reduction of cellular microtubules with a functional change can be illustrated in hepatocyte lipoprotein secretion, as seen in Fig. 2 (Reaven and Reaven, 1980). This figure makes several points. (1) A close relationship exists between hepatocyte microtubule content and lipoprotein secretion. (2) Microtubule content reaches its nadir in advance of the nadir in lipoprotein secretion. However, (3) even when microtubule content is essentially zero in hepatocytes, lipoprotein secretion continues to occur at 20–30% of control values.

Finally, an illustration of a presumed association between microtubules and another organelle of the secretory cell can be shown by the correlation coefficient in Fig. 3. In this figure, measurements of hepatocyte microtubule volume density from various rat colchicine experiments are plotted against measurements of

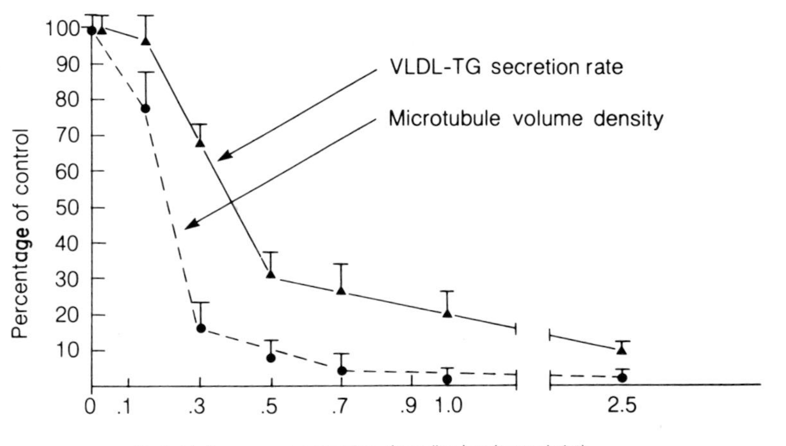

FIG. 2. Comparison of hepatic very-low-density-lipoprotein secretory response and hepatocyte microtubule response to increasing doses of administered colchicine. Data are plotted as percentages of the control (saline) value. Triangles represent mean (± S.E.M.) secretory rates; circles represent values for mean (± S.E.M.) microtubule content. (Reaven and Reaven, 1980. Reprinted with permission of *Journal of Cell Biology*.)

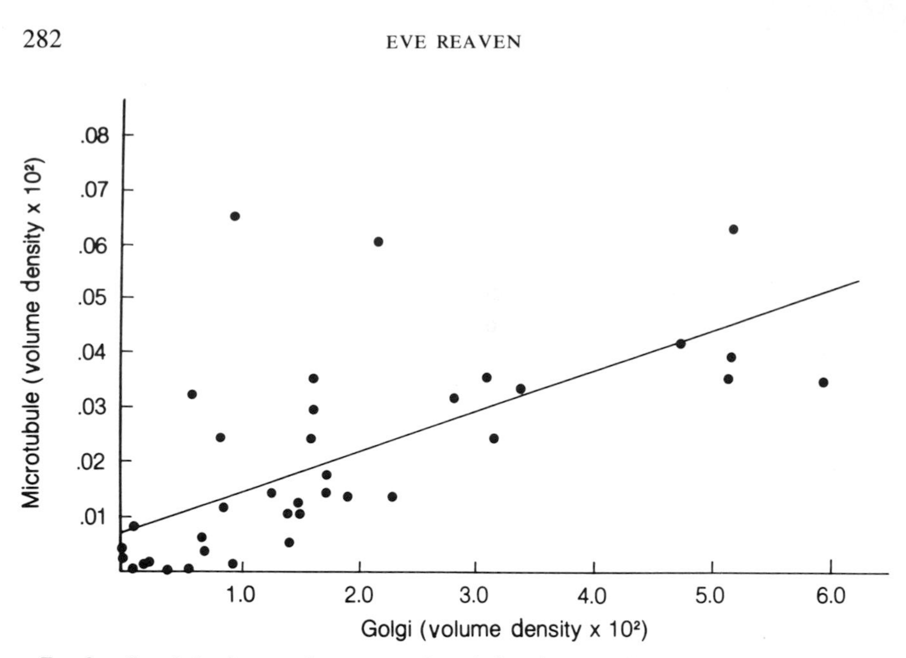

FIG. 3. Correlation between hepatocyte microtubule volume density (x axis) and Golgi volume density (y axis). $n = 36$; $r = 0.63$; $p < 0.01$. Each point represents averaged data from five to six cells obtained from the liver of a rat given saline or colchicine (0.7 mg/kg body weight) 5–24 hours earlier. In general, the data points that show the highest Golgi and highest microtubule content are from control animals or animals given colchicine 24 hours earlier; data points showing the lowest Golgi and microtubule content are from rats given colchicine 5 hours earlier; the middle cluster of points (with Golgi content ≅ 1–1.5% and microtubule content between 0.005 and 0.01% of the cytoplasm) are from animals given colchicine 12 hours earlier.

Golgi volume density in the same cells. In some of the experiments rats were given saline and tissues were taken 5–24 hours later; in other experiments, cells were examined 5 hours, 12 hours, or 24 hours after the rats were given colchicine (0.7 mg/kg body weight). Although there are obvious outlying points on this curve, the relationship between microtubule content and the content of intact Golgi complexes in the same hepatocytes is clear; this point has been made also in various other cell systems in which colchicine was used (De Brabander *et al.*, 1978; Moskalewski *et al.*, 1976; Patzelt *et al.*, 1977; Thyberg *et al.*, 1977; Thyberg *et al.*, 1978). Whether this association with microtubules is unique for the Golgi complex or exists for other organelles as well, is obviously an issue of importance in current investigations of the secretory process.

VI. Comment

As indicated in this chapter, the stereological analysis of cytoplasmic microtubules can be a time-consuming task. Nonetheless, stereological techniques are worthy of attention because they provide sensitive and reproducible methods for assessing the content and distribution of assembled microtubules within cells. The major advantages such methods have over biochemical methods is that multiple analyses can be carried out on specific cell types of complex tissues without the need of physically separating the cells. The major limitation of the stereological method, as applied to microtubule research, is that those aggregated forms of tubulin that are not assembled into microtubule lengths cannot be identified and quantified. It is hoped that this limitation can be overcome in the future with the combined use of antibody probes and ultrastructural stereological analyses.

REFERENCES

Bamburg, J. R., Shooter, E. M., and Wilson, L. (1973). *Neurobiology* **3,** 162–175.

De Brabander, M., Wanson, J.-C., Mosselmans, R., Geuens, G., and Drochmans, P. (1978). *Biol. Cell.* **31,** 127–140.

Jennett, R. B., Tuma, D. J., and Sorrell, M. F. (1980). *Arch. Biochem. Biophys.* **204,** 181–190.

Luftig, R. B., McMillan, P. N., Weatherbee, J. A., and Weihing, R. R. (1977). *J. Histochem. Cytochem.* **25,** 175–187.

Matsuda, Y., Baraona, E., Salaspuro, M., and Lieber, C. S. (1979). *Lab. Invest.* **41,** 455–463.

Moskalewski, S., Thyberg, J., and Friberg, U. (1976). *J. Ultrastruct. Res.* **54,** 304–317.

Murphy, D. B., and Tilney, L. G. (1974). *J. Cell Biol.* **61,** 757–779.

Osborn, M., and Weber, K. (1982). *Meth. Cell Biol.* **24,** 97–132.

Ostlund, R. E., Leung, J. T., and Hajek, S. V. (1979). *Anal. Biochem.* **96,** 155–164.

Patzelt, C., Brown, D., and Jeanrenand, B. (1977). *J. Cell Biol.* **73,** 578–593.

Pipeleers, D. G., Pipeleers-Marichal, M. A., Sherline, P., and Kipnis, D. H. (1977). *J. Cell Biol.* **74,** 341–350.

Pipeleers, D. G., Pipeleers-Marichal, M. A., and Kipnis, D. M. (1976). *Science* **191,** 88–89.

Porter, K. R., Puck, T. T., Hsie, A. W., and Kelley, D. (1975). *Cell* **2,** 145–162.

Reaven, E. P. (1977). *J. Cell Biol.* **75,** 731–742.

Reaven, E. P., and Reaven, G. M. (1975). *J. Clin. Invest.* **56,** 49–55.

Reaven, E. P., and Reaven, G. M. (1977). *J. Cell Biol.* **75,** 559–572.

Reaven, E. P., and Reaven, G. M. (1978). *J. Cell Biol.* **77,** 735–742.

Reaven, E. P., and Reaven, G. M. (1980). *J. Cell Biol.* **84,** 28–39.

Reaven, E. P., Maffly, R., and Taylor, A. (1978). *J. Membr. Biol.* **40,** 251–267.

Rubin, R. W., and Weiss, G. D. (1975). *J. Cell Biol.* **64,** 42–53.

Sherline, P., Bodwin, C. K., and Kipnis, D. M. (1974). *Anal. Biochem.* **62,** 400–407.

Sheterline, P., Schofield, J. G., and Mira-Moser, F. (1977). *Exp. Cell Res.* **104,** 127–134.

Sternberger, L. A. (1979). "Immunocytochemistry." (2nd ed.). Wiley, New York.

Thyberg, J., Axelsson, J., and Hinek, A. (1977). *Brain Res.* **137,** 323–332.

Thyberg, J., Moskalewski, S., and Friberg, U. (1978). *Cell Tissue Res.* **193,** 247–257.

Van De Water, L., Guttman, S., Gorovsky, M. A., and Olmsted, J. B. (1982). *Meth. Cell Biol.* **24,** 79–96.

Weibel, E. R. (1973). *Princ. Tech. Electron Microsc.* **3,** 237–291.

Chapter 14

Chromatophores: Their Use in Understanding Microtubule-Dependent Intracellular Transport

MANFRED SCHLIWA

Department of Molecular, Cellular, and Developmental Biology
University of Colorado
Boulder, Colorado

I. Introduction

Changes of animal coloration have been the subject of scientific interest since earliest record; as in many other areas of science, Aristotle was the first to give an accurate account of adaptive color changes more than 2000 years ago. It was not,

METHODS IN CELL BIOLOGY, VOLUME 25

ISBN 0-12-564125-7

however, until 1852 that a cellular basis for changes in body coloration was established by Bruecke, who recognized the movement of pigment granules as cause for the color change of chameleons. Since then numerous reports have been published devoted primarily to the study of problems of pigment cell development and differentiation, their neural and hormonal control, and malignancy. Until recently the cellular mechanisms of color change have attracted comparatively little attention, but this situation has significantly changed during the past 15 years in conjunction with our increased interest in cytoskeletal organization and cellular motility. As will be demonstrated in this chapter, pigment cells evidently have some unique properties that make them appear suitable as model systems for the study of phenomena related to intracellular transport.

As a rule, two entirely different color-change phenomena are distinguished. *Morphological* color change is a slow process based on an increase or decrease in the number of pigment cells or the content of pigment, which may require weeks. *Physiological* color change, on the other hand, may be quite rapid. It results from an intracellular rearrangement of pigment-containing organelles that, in some species, requires only seconds. In response to certain stimuli, pigments are gathered at the cell center where they aggregate into a small spherical mass, thereby leaving a substantial cell area unpigmented and causing a blanching in the respective body regions. This process is referred to as aggregation. The reverse process, dispersion, is the more or less uniform redistribution of pigment throughout the entire cell, resulting in darkening of the animal. The cells in which these processes take place are collectively referred to as chromatophores, with certain cell types being named after the color of the pigment they contain. For example, melanophores (Greek; *melanos* = black, *phorein* = to carry) bear pigment granules (melanosomes) of brown or black color; erythrophores contain red pigments, and xanthophores those of a yellowish color. Unfortunately, this nomenclature is not useful for all pigment cells. Although, as a rule, a particular chromatophore contains only one pigment, there are several cases in which one cell bears pigment organelles of different size, shape, chemical composition, and color (e.g., Arnott *et al.*, 1970; Bagnara, 1972). In some invertebrates, as many as four different pigments may be found in one cell type (Robison and Charlton, 1973; Schliwa and Euteneuer, 1979).

To achieve aggregation and dispersion of pigment granules, two entirely different mechanisms are established in different groups of organisms. The first is found in many cephalopods such as *Sepia* and *Octopus*. Pigment effectors of these species are actually chromatophore organs, since they are composed of several cells: a pigment cell surrounded by an elastic sacculus to which radially oriented muscle fibers are connected (Cloney and Florey, 1968; Florey, 1969). Contraction of the muscle fibers, which are under neural control, passively expands the pigment cell plus sacculus and distributes the pigment over a large area. Their relaxation allows the elastic sacculus to return to a collapsed state, resulting in passive bunching of pigment granules. Since the muscle fibers associated with the pigment cells can contract within fractions of a second, the extremely rapid and fascinating fluctuations in color of these animals are easily explained.

The second color-change mechanism is based on active intracellular movement of pigment organelles within the chromatophore. The velocities with which redistribution of pigment is achieved vary considerably among different cell types and species. In the polychaete *Platynereis* it may take hours (Fischer, 1965), whereas some fish species accomplish aggregation within a few seconds (Porter, 1973). Pigment transport is an exceptional logistical problem in that thousands of granules have to be translocated at the same time in a coordinated fashion. The characteristics of transport (notably in some fish chromatophores) are not matched by any other cell type, which possibly is one of the reasons for the attractiveness of the pigment cell model. Those who have studied chromatophores can hardly miss the esthetic quality of this phenomenon, particularly in cells with rapid transport. The precise radial orientation of granule movements, their uniformity, and their manipulability are features quite unique to pigment cells. Beyond the esthetic quality, there is an obvious connection of a cellular activity to the macroscopic appearance of an organism. Relatively minute cellular changes can drastically alter the exterior of the individual, which in turn may be of great consequence for its successful interaction with its environment and possibly its survival. Thus color change based on intracellular translocation of granules is of vital importance for a wide variety of organisms as part of camouflage, courtship behavior, territorial display, temperature regulation, protection from harmful radiation, and warning signals to enemies.

This chapter will first give a brief overview of the techniques applied to the study of pigment cell motility and then will summarize recent advances in our understanding of the motile behavior of these cells. A comparison will then be made with other cellular systems in which intracellular transport has been studied.

II. Methods

A. Cell Systems Used

Although chromatophores are present in virtually every multicellular organism from cnidaria to mammals, representatives from only a few classes are suitable for experiments and have been used for studies of granule motility: amphibia, fish, crustacea, and to some extent, sea urchins and insects. Each system has its own virtues and special requirements, hence methods applicable to the study of one cell system or organism are not always transferrable to another. However, all

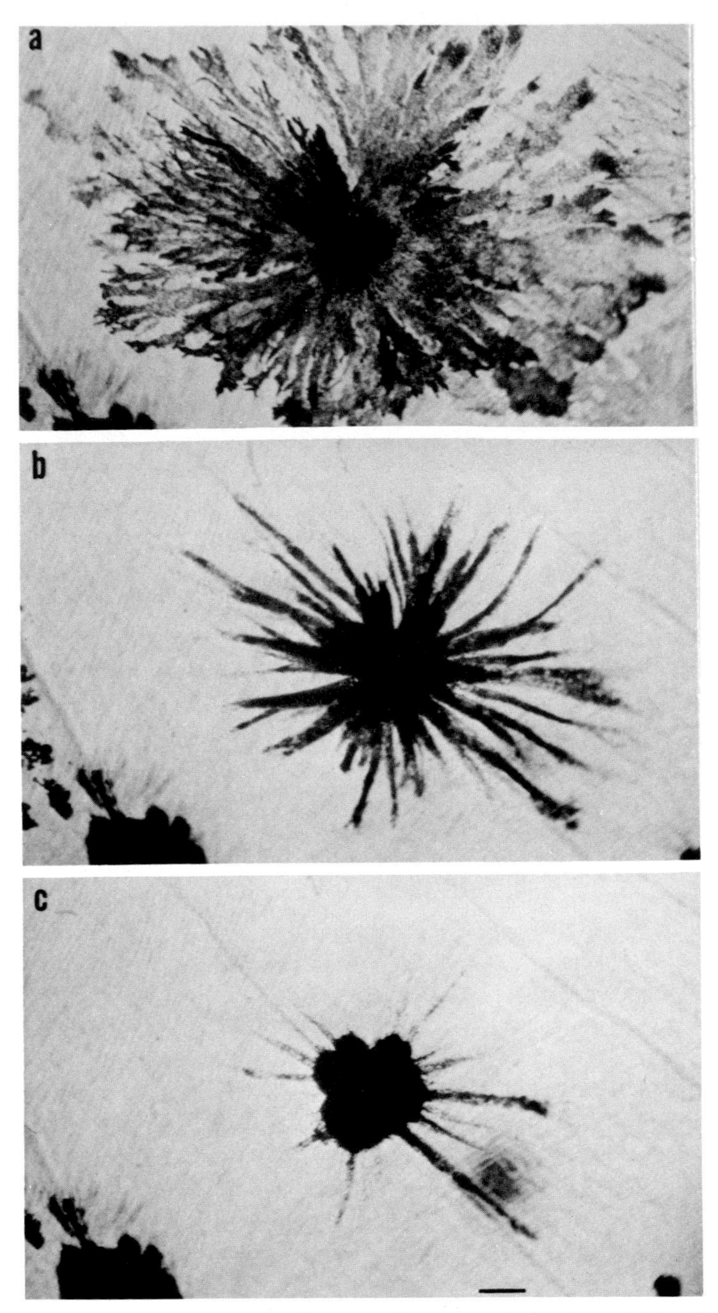

FIG. 1. Melanophore of *Gymnocorymbus ternetzii* on an excised scale maintained in Ringer's solution. (a) Completely dispersed state; (b) 30 seconds after stimulation with 10^{-4} adrenalin; (c) 60 seconds after stimulation with adrenalin. Bar = 10 μm.

systems require a method to determine, preferably in a quantitative way, the responses of pigment cells to external stimuli or drugs, which usually involves light-microscopic observation of excised tissue pieces. Whole animals can only be used if they are small enough to fit at least under a dissecting microscope. Therefore, sample preparation for *in situ* studies of chromatophores under *in vitro* conditions involve one of the following procedures: removal of portions of the carapace with adhering chromatophores or dissection of internal organs (crustacea), excision of scales or fin tissue (fish), or skin biopsies (amphibia). With the use of appropriate saline solutions, the preparations usually remain viable and active for hours (Fig. 1).

B. Measurement of Chromatophore Responses

The activity of chromatophores either under physiological conditions or in response to certain stimuli, drugs, or hormones has to be assessed in a quantitative fashion before conclusions concerning the effects of experimental treatments can be drawn. In 1931, before automated recording systems were available, Hogben and Slome developed a simple scheme for the visual determination of the degree of dispersion of a chromatophore. This scheme is widely known as the chromatophore index. In their system, maximal pigment aggregation represents stage 1 and maximal pigment dispersion stage 5, with stages 2, 3, and 4 describing increasing degrees of dispersion. This system provides, however, only a rough classification scheme for the determination of chromatophore responses; subtle differences are left out of consideration. Furthermore, the classification of intermediate stages of dispersion is a matter of the subjective impression of the experimenter, and the system is almost inapplicable to chromatophores with fast responses (e.g., certain fish cells). Although this was a simple and convenient method, useful mainly as a qualitative description, more accurate methods were desirable.

A step forward toward accurate determination of chromatophore responses was undertaken by Kamada and Kinosita (1944), who used a semiautomatic recording system to measure the velocity of single pigment granules. An ocular screw micrometer was adjusted in such a way that its cross mark was kept on a moving granule, while the advance of the screw was recorded on a kymograph. The real breakthrough came with the studies of Hill *et al.*, (1935) and Smith (1936), who used recording systems that allowed the continuous automatic registration of motile responses of chromatophores. These photoelectric recording systems take advantage of the fact that changes in pigment distribution cause immediate alterations in the amount of light transmitted through, or reflected from, a chromatophore-bearing tissue sample. In principle, the device consists of a photoelectric cell hooked to the ocular of a microscope. Photoelectric changes in resistance are monitored by a chart recorder to which the photocell is connected. According to the special needs of the organism or tissue sample under examination, either transmittance or reflectance measurements are made. The former have widely been applied to the study of fish chromatophores (e.g., Fujii and Novales, 1969; Wikswo and Novales, 1969; Novales and Fujii, 1970; Schliwa and Bereiter-Hahn, 1973b; Junqueira *et al.*, 1974; Iga, 1975), whereas reflectance measurements were used with amphibian (e.g., Novales and Davis, 1967; Malawista, 1971a; Novales and Novales, 1972; Magun, 1973) and sea urchin (Dambach and Weber, 1975; Dambach, 1968) chromatophores. In most instances, these measurements represent the summarized response of several pigment cells in the field of observation. Also, differential responses of different pigment cell types are not distinguished, a difficulty only partially overcome by the use of color filters. To photoelectrically record activities of a single cell, Iga (1975) have placed a diaphragm in the eyepiece that restricted the area under investigation to just one cell. Luby and Porter (1980) first took a 16-mm movie and then made photoelectric measurements directly from the running film.

Single-frame analysis of time-lapse films has only rarely been employed to assess the activities of pigment cells (Egner, 1971; Schliwa and Bereiter-Hahn, 1974), mainly because it is a time-consuming and tedious procedure. This technique, however, undoubtedly gives the most accurate results as to the velocity of moving pigment granules. By taking a time-lapse film of pigment movements in an isolated fish scale and simultaneously making photoelectric recordings of the same area, Schliwa and Bereiter-Hahn (1974) have demonstrated significant differences in the velocities of pigment movements monitored by the two methods. These studies call for some caution in estimations of granule velocities from photoelectric recordings. Nevertheless, the general usefulness of the photoelectric method as a fast and reliable tool for the assay of chromatophore responses is beyond doubt.

C. Attachments to the Light Microscope for *in Vitro* Observations

Light-microscopic observations and photoelectric recordings of chromatophore responses require devices that firmly hold tissue pieces in place and yet allow exchange of solutions for drug treatment or stimulation. Several such attachments have been described for the study of amphibian (McGuire *et al.*, 1972; Iga and Bagnara, 1975; Magun, 1973; Geschwind *et al.*, 1977), fish (e.g., Schliwa and Bereiter-Hahn, 1974; Kinosita, 1953; Morimoto and Iwata, 1976), and sea urchin tissues (Dambach and Weber, 1975). In principle, they are perfusion chambers with clamps to maintain a fixed position of the tissues. These chambers have to meet the special requirements of the tissue under investigation and therefore *the* chamber does not exist. The interested reader is referred to the original literature cited above.

For studies with hydrostatic pressure, Marsland (1944), Marsland and Meisner (1967), and Murphy and Tilney (1974) have used specially designed pressure chambers.

D. *In Vitro* Culture of Pigment Cells

Most studies on chromatophores have been carried out on cells *in situ* using either intact organisms or excised tissue pieces. However, attempts have also been made to culture pigment cells *in vitro*. Such efforts were considered worthwhile for several reasons: The cells become more accessible to various experimental treatments due to the absence of a diffusion barrier of overlying epidermal cells; chromatophores become disconnected from control systems of the intact organisms (hormones, nervous system) that might interfere with experimental treatments; they are better suited for microscopic observations; and they become amenable to limited biochemical study. So far, only amphibian, fish, or sea urchin chromatophores have been studied in culture. Novales and collaborators (Novales, 1960; Novales and Davis, 1967; Lyerla and Novales, 1972) used embryonic tissue derived from different species of amphibia in either Niu-Twitty's complex salt solution (Flickinger, 1949) or Barth's medium supplemented with calf serum (Barth and Barth, 1969). Pieces of tissue from the caudal fin of the angelfish were cultured by Egner (1971) in a medium consisting of 64% calf serum and diluted Ringer's solution. Under these conditions, melanophores migrate from the explant after 3–4 days. Other authors have used various methods of dissociating intact tissue from adult animals (Obika, 1976; Byers and Porter, 1977; Ozato, 1977; Schliwa and Euteneuer, 1978a; Matsumoto *et al.*, 1978). The methods used vary to some extent, but enzyme digestion of skin pieces (scales, fin tissue) with collagenase, trypsin, hyaluronidase, or different combinations of these enzymes is involved. Dissection is followed by collection of cells with a micropipet or by centrifugation. The cell suspensions, which usually contain a variable number of nonchromatophore cells, are then plated on a substrate and allowed to settle and attach in the presence of culture medium (Fig. 2). The media used differ considerably, but apparently each works well in the hands of the different authors. As a rule, the cells can be kept viable for several days or even weeks (e.g., Ozato, 1977). We have recently cultured goldfish melanophores in amphibian culture medium supplemented with 10% calf serum for 8 weeks, during which the cells went through four cycles of detachment and replating (M. Schliwa, unpublished). With time, however, the number of viable cells decreased. In case of goldfish erythrophores, mitotic activity of cultured cells has been reported (Ozato, 1977), but usually fully differentiated chromatophores do not divide. Ide (1974, 1979), however, has succeeded in culturing melanophores isolated from bullfrog tadpoles for nearly 2 years. These cells have not been used for studies of pigment

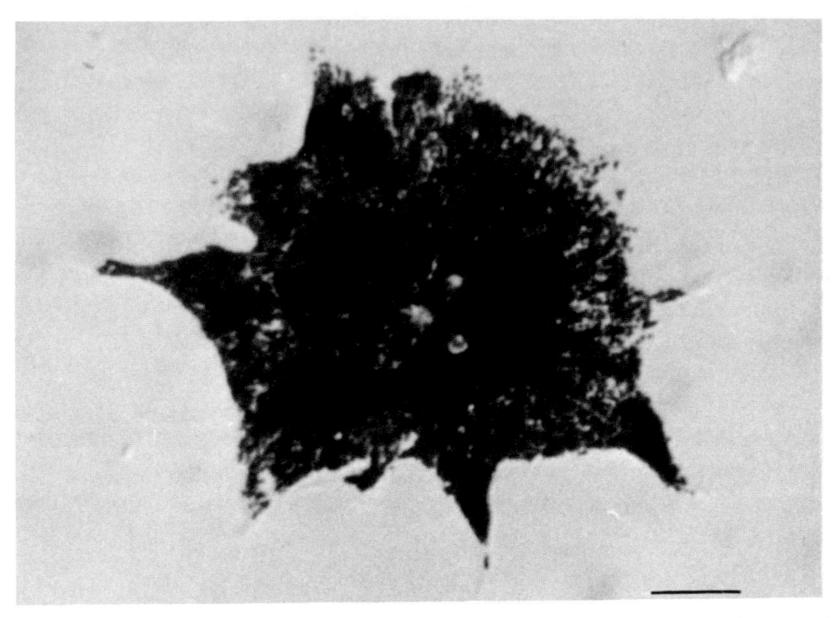

FIG. 2. Isolated melanophore of the angelfish, *Pterophyllum scalare*, settled on a glass coverslip. The cell was kept in a 3:1 mixture of Ringer's solution and amphibian culture medium for 3 hours. Pigment granules are dispersed. Bar = 10 μm.

transport because their motile activities are very low. It would be of great value to culture a cell type that displays all the characteristics of a fully differentiated cell, particularly rapid granule transport, and yet could be propagated *in vitro*. As yet, no such system exists.

Pigment cells *in vitro* have been useful for the study of various aspects of pigment cell physiology. With respect to cytoskeletal organization and intracellular motility, the use of isolated chromatophores has allowed the application of such powerful tools as immunofluorescence microscopy and high-voltage electron microscopy, and has also facilitated quantitative electron-microscopic analysis of thin sections.

E. Electron Microscopy

Ultrastructural studies of pigment cells are numerous and devoted to a variety of problems, including those related to cytoskeletal organization. Few of these studies have, however, been carried out in a sufficiently quantitative way to permit drawing of conclusions as to the involvement and significance of cytoskeletal structures (microtubules, microfilaments) in transport. Following the report by Bikle, Tilney, and Porter (1966) on the presence of microtubules in fish melanophores, numerous pigment cell types were screened for the presence of these organelles. Most of these studies are purely descriptive and few are really thorough. Attempts to characterize the microtubule system and analyze possible changes associated with the redistribution of pigment are confined to some fish chromatophores. These studies report the sectioning of chromatophores perpendicular to the plane in which pigment is moved in order to obtain transverse sections of cell processes that permit the counting of microtubule profiles. Two earlier studies following this approach are in disagreement, perhaps because of variations in the species used, perhaps because of technical difficulties (Schliwa and Bereiter-Hahn, 1973a; Murphy and Tilney, 1974). A more recent study has employed a more accurate and reliable procedure to obtain precise counts at known positions in isolated angelfish melanophores settled and spread on a glass coverslip. Cells were fixed and embedded on the coverslip, which was removed after hardening of the resin by a cold shock in liquid nitrogen. The cells were then re-embedded and oriented for sectioning precisely perpendicular to the substrate surface and micrographs were taken of sections every 5 μm. This study has confirmed earlier findings (Schliwa and Bereiter-Hahn, 1973a) of a decrease by about 60% in the number of microtubules in the cell processes on pigment aggregation. In addition, it allowed a quantitative morphometric analysis of some other parameters, such as relative volume of pigment granules and cytoplasm and their redistribution in the course of pigment aggregation, and a rough estimation of the content of tubulin of these cells.

Using a similar but much more tedious approach—namely, tracking of individual microtubules in transverse serial thin sections—it has been possible to characterize the prominent microtubule apparatus of a melanophore in more detail (Schliwa, 1978). Thus information has been obtained about the course of individual microtubules along the length of a cell process; their spatial relationship to other microtubules, pigment granules, and the cell membrane; and their presumptive point of origin. Such studies, although undoubtedly time-consuming, can provide useful information about a cell's microanatomy and have proved to be indispensable for the study of the organization of the mitotic apparatus (e.g., Tippit *et al.*, 1978; McIntosh *et al.*, 1979; McDonald *et al.*, 1979).

F. Whole-Mount Electron Microscopy

Now that functional chromatophores can be kept *in vitro* in an active state, they are amenable to procedures developed recently for whole-mount electron microscopy. Instead of analyzing and interpreting thin sections, it is possible to look at the structure of whole cells without a resin matrix that might interfere with the proper visualization of delicate cytoplasmic structures. By employing techniques for stereo viewing, the three-dimensional organization of a whole cell or part of a cell is faithfully depicted. This might be of particular importance for the study of highly active cells such as chromatophores, in order to reveal the structural relationship of moving granules to other structural elements of importance (e.g., microtubules). The procedures for whole-mount electron microscopy have been described (e.g., Byers and Porter, 1977; Wolosewick and Porter, 1979) and need not be outlined in detail here. Briefly, the following steps are involved:

1. A thin Formvar film is floated onto a water surface, and a gold grid precleaned in nitric acid is placed on the floating film.
2. Formvar film and grid are picked up with a precleaned coverslip in such a way that the grid is sandwiched between Formvar and glass.
3. A thin carbon film is evaporated onto the Formvar. If necessary, the film may also be coated with an adhesion-promoting material (poly-L-lysine, collagen). The grids are then sterilized under UV light.
4. A drop of a cell suspension containing isolated pigment cells is placed on top of the grid and the cells are allowed to settle in the presence of culture medium.
5. At the proper distribution of pigment, cells are fixed with glutaraldehyde (1–2%, 5–20 minutes), washed with buffer, and postfixed with 0.5–1% osmium tetroxide for 1–5 minutes.
6. Following dehydration in ethanol, specimens are critical-point-dried from CO_2 (Anderson, 1951).
7. Grids are covered with a thin film of carbon either on the cell side alone or on both sides, and stored in a desiccator under a slight vacuum to prevent remoistening and collapse of cellular structures.

Studies using whole-mount preparations of isolated chromatophores have taken advantage of the 1-MeV electron microscope installation at the University of Colorado at Boulder, which allows penetration of even relatively thick cell areas. However, whole-mount electron microscopy is in principle possible with conventional microscopes. If equipped with a goniometer stage, stereo pairs will reveal almost as much detail, at least in relatively thin cell areas (Byers and Porter, 1977; Luby and Porter, 1979; Schliwa, 1979).

G. Immunofluorescence Microscopy

Isolated pigment cells allow the application of yet another powerful structural technique. Immunofluorescence microscopy using antibodies directed against a wide variety of proteins involved in cytoskeletal organization and cell motility has become one of the most useful tools for the study of localization and organization of these proteins in cells. It provides an excellent overview of the distribution of a particular antigen at the light-microscopic level, and it allows the rapid

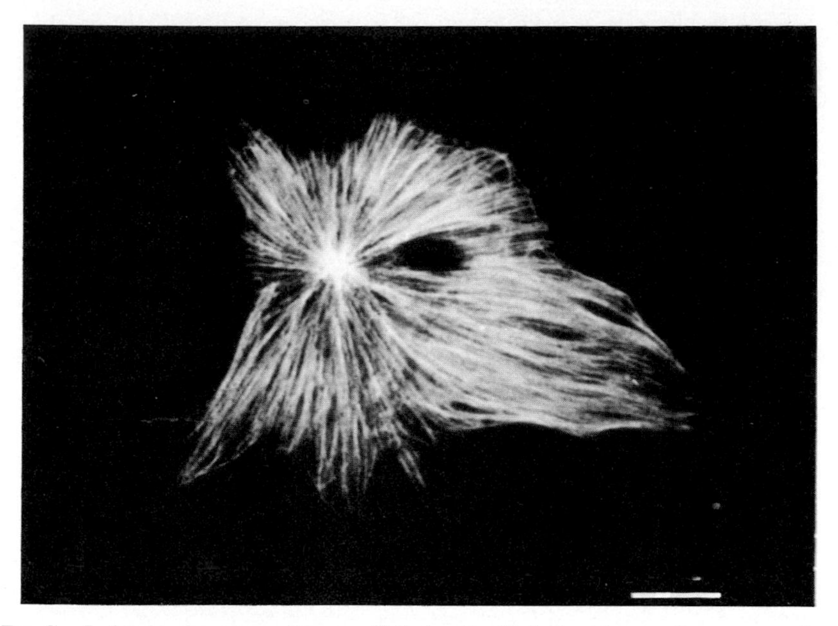

FIG. 3. Isolated melanophore of the angelfish visualized in indirect-immunofluorescence microscopy after decoration with antibodies against tubulin. The cell was fixed with glutaraldehyde and reduced with tetraborohydride according to the method of Weber *et al.* (1978). Bar = 10 μm.

screening of large cell populations. The distribution of both tubulin and actin has been studied in some detail in fish chromatophores (Schliwa *et al.*, 1978, 1980; Beckerle *et al.*, 1979; Fig. 3). The methods used resemble those applied to the study of other cell types and include:

1. Fixation with 3.7% formalin in phosphate-buffered saline (PBS), followed by treatment with methanol at −20°C. Experiments that require good preservation of microtubules will involve glutaraldehyde fixation, followed by reduction of free aldehyde groups with tetraborohydride as described by Weber *et al.* (1978).
2. The incubation at 37°C with the respective antibodies for 30–45 minutes. Ordinarily the antibodies used were rabbit anti-porcine brain tubulin made monospecific by passage over tubulin coupled to Sepharose-4B; rabbit anti-bovine brain cytoplasmic actin was purified by antigen-affinity chromatography (Weber *et al.*, 1978; Aubin and Weber, 1979).
3. Following several washes with PBS, a commercially available fluorescein-isothiocyanate-labeled goat-anti-rabbit antibody is applied for 30–45 minutes. If necessary, this second antibody can be absorbed with tissue homogenates (e.g., rat liver powder) to reduce nonspecific staining.

4. The preparations are viewed in a good microscope equipped with epifluorescence illumination.

Immunofluorescent studies of chromatophores have to deal with one major disadvantage: light absorption by pigment granules. Some pigments like those present in erythrophores of *Holocentrus* (probably carotenoid droplets) will also autofluoresce, a problem that can usually be overcome by the use of special filters. Light absorption of melanin is so effective that photographs of the upper and lower cell side will reveal different microtubule patterns, namely, those associated with the upper and lower cortices, respectively (Schliwa *et al.*, 1978). Pigment granules in between the two will act as a screen.

It is hoped that this technique will be applied to a variety of pigment cells using a number of different antibodies. The power of immunocytochemistry can even be increased if antibody techniques are combined with either conventional or high-voltage electron microscopy. Methods for the localization of antigen–antibody complexes in the electron microscope are already available and are currently being improved in several laboratories. Their application to pigment cells will, it is hoped, give some new insights into the nature of structural connections between pigment granules and skeletal elements such as microtubules.

III. Overview

A. Theories before the Use of Electron Microscopy

Until the 1960s, published interpretations of motile behavior in pigment cells were based on careful light-microscopic observations of living normal or experimentally manipulated cells, as well as fixed material. During the second half of the nineteenth century it was unclear whether pigment cells (notably of lower vertebrates) had a fixed outline or whether they displayed some sort of amoeboid movement by extending and retracting cell processes filled with pigment granules. From his studies of fish chromatophores, Zimmermann (1893) inferred that pulsations of cell processes were responsible for aggregation and dispersion. Kahn and Lieben (1907) demonstrated that in amphibian melanophores, pigment granules return into the same cell processes during successive phases of aggregation and dispersion. This observation was later confirmed by Matthews (1931) for fish melanophores. The alignment of pigment granules in parallel rows was taken as evidence for the existence of intracellular tubes or channels (Ballowitz, 1914; Behre, 1935) that, by their peristaltic contraction and relaxation, propel pigment granules. Both Spaeth (1916) and Franz (1939) proposed a contractile mechanism with rodlike structures forming a "cytoskeleton" (Franz, 1939). On the basis of his pressure and centrifugation experiments, Marsland (1944) considered sol-gel transformations of the cytoplasm to be responsible for the propulsion of the pigment granules. According to his theory, aggregation involves proximally directed gelation of the cytoplasm.

Most of the theories just cited envisaged the existence of some sort of contractile machinery in the chromatophore cytoplasm, although in most instances the nature of this component remained obscure. A completely different mechanism has been proposed by Kinosita (1953, 1963). He assumes the force-producing machinery to reside not in the cytoplasm, but in the chromatophore membrane. According to his electrophoretic theory, negatively charged pigment granules of melanophores migrate along a radial gradient of an electric potential generated at the plasma membrane. This potential is assumed to change with an aggregating or dispersing stimulus, respectively, being more positive in the cell periphery than near the cell center during dispersion, the situation reversing during aggregation. Kinosita's theory received much attention in the subsequent years because, despite its simplicity, it can explain several experimental results. It is, however, in disagreement with others. Changes in the chromatophore membrane potential are not necessarily accompanied by the movement of pigment granules (Fujii and Novales, 1969), and movements can be elicited without altering the membrane potential (Martin and Snell, 1968). Kinosita's own observation (1963) that the onset of pigment dispersion trails several minutes behind the completion of the change in membrane potential is also somewhat at variance with the proposed mechanism. Green (1968) also suspects that the gradients reported by Kinosita are not enough to allow the granules to penetrate the structural resistance of the cytoplasm. Taken together, the evidence accumulated so far has led many authors to believe that the electrophoresis theory can no longer be considered likely.

Of the older theories, only that of Marsland (1944) appears to be relevant in the light of our knowledge about the structural and biochemical basis of cell motility accumulated in the 1960s and 1970s. Whether it is compatible with recent observations on pigment cell physiology and ultrastructure may be judged after considering the following sections, which will summarize findings obtained in the past 15 years.

B. Components of the Pigment Cell Cytoskeleton

1. Microtubules

Bikle, Tilney, and Porter (1966) were the first to describe in detail the ultrastructure of a fish chromatophore. They noted the presence of numerous microtubules positioned parallel to the direction of movement of pigment granules

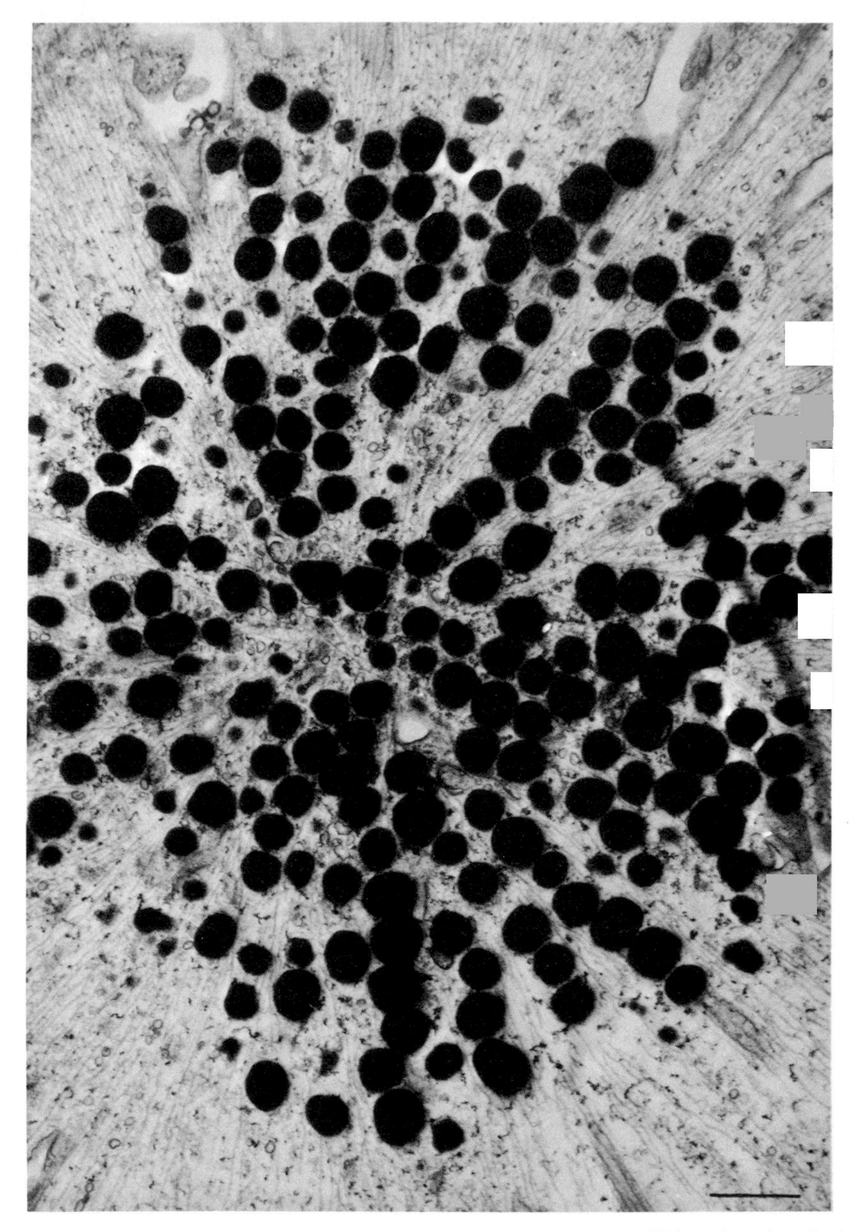

FIG. 4. Thin section parallel to the substrate surface of an isolated angelfish melanophore in the aggregated state. Only the central portion of the cell is shown. Numerous microtubules are arranged in a precise radial fashion relative to the cell center. Bar = 1 μm.

and suspected that microtubules define radial pathways for their movement (Fig. 4). The presence of microtubules has subsequently been confirmed for a number of fish chromatophores (e.g., Green, 1968; Porter, 1973; Egner, 1971; Murphy and Tilney, 1974; Takeuchi, 1975; Lamer and Chavin, 1975) and pigment cells of other species (e.g., Wise, 1969; Gartz, 1970; Moellmann *et al.*, 1973; Elofsson and Kauri, 1971; Robison and Charlton, 1973; Berthold and Seifert, 1977). Their exact distribution has, however, been studied only in a few fish chromatophores. Microtubules extend parallel to the long axis of cell processes and, in favorable sections, may be followed over long distances. When cell processes are viewed in transverse sections, a subcortical arrangement of a considerable portion of the microtubules is a conspicuous feature. In angelfish melanophores, the vast majority of the microtubules is aligned just underneath the cell membrane, and this position is maintained over long distances (Schliwa, 1978). The observation of two distinct locations of microtubules in chromatophores led early to the hypothesis that there are two sets of microtubules, one associated with the cell cortex, the other occupying more central regions in a cell process (Bikle *et al.*, 1966; Porter, 1973; Schliwa and Bereiter-Hahn, 1973a). The two sets were believed to originate in different parts of the cytoplasm. Recently, however, Schliwa (1978) has shown by microtubule tracking that all microtubules of a particular process insert in, or end near, dense aggregates associated with the centrioles in the central cell area (centrosphere). This serial-section analysis has also demonstrated continuity of microtubules over long distances, suggesting that the vast majority extends between the cell center and the periphery. This contention has also been confirmed by stereo high-voltage electron microscopy of whole cells (Byers and Porter, 1977; Schliwa, 1979), which allows one to follow the course of microtubules without tedious reconstruction procedures.

Possible changes in the number and/or arrangement of microtubules in conjunction with different phases of pigment distribution are a matter of controversy. Schliwa and Bereiter-Hahn (1973a) were the first to report a decrease in the number of microtubules in the aggregated state on the basis of microtubule counts by electron microscopy of angelfish melanophores *in situ*. Working with *Holocentrus* erythrophores, Porter (1973) came to a similar conclusion, although quantitative data were not presented. Murphy and Tilney (1974), on the other hand, claimed that in *Fundulus* melanophores, the number of microtubules stayed unchanged in the two extreme states of pigment distribution. These authors also maintained that both shape and volume of cell processes remained constant, whether they were filled with pigment granules or not. This view is difficult to reconcile with numerous observations in different cell types that demonstrate flattening of the cell periphery on withdrawal of pigment granules (e.g., Wise, 1969; Gartz, 1970; Schliwa and Bereiter-Hahn, 1973a; Porter, 1973; Obika, 1975, 1976; Byers and Porter, 1977; Schliwa and Euteneuer,

1978a). Since changes in cell shape might in some way be causally related to the reorganization of microtubules, Murphy and Tilney's observation is difficult to evaluate. To explain the proposed constancy of cell process volume, one would also have to assume either counter-transport of cytoplasm during aggregation and dispersion of pigment, or swelling and shrinkage of cell processes in association with pigment movements. As yet, supporting evidence for one or the other of these possibilities is not available. On the contrary, a morphometric study of isolated angelfish melanophores has shown that approximately 50% of the cytoplasmic matrix moves in conjunction with the pigment granules, causing even more pronounced flattening of the cell periphery than that expected from the withdrawal of pigment granules alone (Schliwa and Euteneuer, 1978a).

In an attempt to characterize further aggregation- and dispersion-related fluctuations in the number of microtubules and to identify underlying control mechanisms, Schliwa *et al.* (1979) studied the morphology and properties of the central apparatus (Porter, 1973), a prominent complex in the cell center comprising a multitude of ill-defined electron-dense aggregates from which all the cell's microtubules appear to arise (Schliwa, 1978). There are reasons to believe that this structure is responsible for the control of microtubule initiation and perhaps orientation, and thus might be a true "microtubule-organizing center" (MTOC), according to the terminology of Pickett-Heaps (1969). A similar organelle has been observed in other chromatophores (Bikle *et al.*, 1966; Gartz, 1970; Porter, 1973) but not yet studied in detail. The structure of the central apparatus undergoes dramatic changes on pigment movements in angelfish melanophores. It is composed of a multitude of electron-dense aggregates with which numerous microtubules are associated in cells with pigment dispersed, whereas in the aggregated state, the amount of dense material plus associated microtubules is greatly reduced. There are marked differences in the capacity of the central apparatus of cells with pigment either aggregated or dispersed to initiate the assembly of microtubules from exogenous pure porcine brain tubulin in lysed cell preparations (Schliwa *et al.*, 1979). It has been concluded that it is the activity of the central apparatus that basically regulates the expression of microtubules in melanophores. The importance of these changes in microtubule number for the mechanism of granule transport is not yet clear. Possibly the pathways of assembly and disassembly of part of the microtubules help to provide directionality for the action of other components that cooperate with the remaining microtubules to propel pigment granules.

Since microtubules are polar structures, it is theoretically possible that their polarity plays an important role in defining the direction of granule transport. According to such a hypothesis, different sets of microtubules having opposite polarities would be responsible for either the centripetal or centrifugal transport of granules. Thus microtubule polarity would be as important for intracellular movement as the polarity of actin filaments for muscle contraction. Until recently, a test of this hypothesis was not possible, since a polarity marker for microtubules (comparable to heavy meromyosin in the case of actin filaments) was not available. Using a rapid and simple method developed by Heideman and McIntosh (1980) for the visualization of microtubule polarity in thin sections, it has now been shown that the overwhelming majority of microtubules in angelfish melanophores are of uniform polarity (Euteneuer and McIntosh, 1981). With respect to the central apparatus, it is identical to that of microtubules growing from a mitotic center or the distal end of a basal body. This observation rules out microtubule polarity as a determinant for the directionality of granule transport. It further strengthens the view that the central apparatus functions as an organizing center for the initiation of virtually all microtubules in this cell type.

None of the electron-microscopic studies has demonstrated the existence of direct connections or "cross bridges" between microtubules and pigment granules, although Murphy and Tilney (1974) statistically proved an "affinity" of the two for each other in *Holocentrus* erythrophores. Byers and Porter (1977) later suggested that this affinity is mediated by structural components of the cytoplasmic ground substance.

2. "Actin-Like" Microfilaments

Until recently, the presence of 5- to 8-nm filaments in chromatophores has not been demonstrated in conventionally embedded specimens. Not only do they lack filament bundles (stress fibers) as observed in various cell types, but even a less ordered network arrangement is not easy to detect. It has not yet been possible to decide whether this is due to a lack or paucity of microfilaments or to an arrangement that makes detection in thin sections very difficult. Several authors have described "filaments" or "filamentous structures," but a clear distinction between thin and intermediate filaments (described below) has not been made. Obika *et al.* (1978a) have used the heavy-meromyosin labeling technique to identify actin filaments in a fish melanophore. They claim the presence of an extensive actin-filament network throughout the cell processes, but their conclusion is not based on sufficient published evidence. The involvement (hence, the presence) of actin filaments in chromatophore motility has in the vast majority of other published reports been inferred from the effects of cytochalasin B on granule movement. Unfortunately, only a few authors have exercised adequate caution in the interpretation of inhibitor studies in the absence of supportive ultrastructural evidence. Accordingly, the value of many of these studies is not clear at the moment.

A close structural association of pigment granules and actin filaments has been demonstrated in only one instance. Melanosomes in frog retinal pigment epithelial cells move within small finger-like cell processes, which have been shown to contain bundles of thin filaments composed of actin (Murray and Dubin, 1975). Since microtubules are completely absent from these processes, granules are thought to travel along the actin filaments.

In isolated angelfish melanophores, Schliwa *et al.* (1981) employed a variety of methods—including immunofluorescence microscopy, conventional transmission, and high-voltage electron microscopy—to identify the presence of a loose cortical filament mesh composed of actin filaments. Detergent extraction under conditions that stabilize both microtubules and microfilaments has further shown that no such filaments are found in association with pigment granules. The role of actin in granule transport of these cells remains unclear, since the actin-specific substances, phalloidin and DNase I, do not significantly interfere with granule transport (see Section III, C).

An unusual filament organization has been described in vertebrate iridophores (Rohrlich and Porter, 1972; Rohrlich, 1974). In addition to intermediate filaments, these cells contain 6-nm filaments in an intriguing array linking pigment crystals in crystalline sheets. It has been speculated that they are composed of actin (Rohrlich, 1974), but direct evidence is lacking.

3. Intermediate-Sized Filaments (10-nm Filaments)

All vertebrate chromatophores studied so far (with the exception of *Holocentrus* erythrophores) display variable amounts of filaments with a diameter of 8–11 nm summarized under the term intermediate-sized filaments. Their number, orientation, and distribution vary greatly among different cell types, and possibly also with the state of pigment distribution. As a rule, these filaments are oriented in a more or less radial fashion, parallel to the direction of pigment movements, and often they are found to surround the nucleus. Apparent changes in the number of intermediate-sized filaments as a function of the distribution of pigment have been described. Junqueira, Reinach, and Salles (1977) observed an increase in filament number after treatment with adrenalin (which causes pigment aggregation) and two other sympathomimetic drugs in fish melanophores. In frog melanocytes, a dramatic increase in the number of filaments in cells with pigment dispersed has been noted (McGuire and Moellmann, 1972; Moellmann *et al.,* 1973). Cells with pigment aggregated display microtubules instead. This correlation of filaments and microtubules with the dispersed and aggregated state, respectively, has led to the conclusion (Moellmann and McGuire, 1975) that microtubules and 10-nm filaments are interconvertible, a view that, according to our present knowledge of intermediate-sized filaments, appears to be erroneous (Lazarides, 1980). The results of Moellmann and co-workers are difficult to reconcile with the generally extreme stability of intermediate-sized filaments; their observation may in part be attributed to the fixation protocol employed (warming of frog skin pieces to 35°C 15 minutes before fixation, glutaraldehyde fixation at 35°C).

An apparent increase in the number of filaments after colchicine treatment, originally thought to result from the breakdown of microtubules (e.g., Wikswo and Novales, 1972), may simply be the result of enhanced visibility due to aggregation and bundling of the filaments.

Intermediate-sized filaments are the least-understood filament type in pigment cells. Their composition is largely unknown; immunofluorescent evidence suggests that in angelfish melanophores, they may be of the vimentin type (M. Schliwa and K. Weber, unpublished observation). Their function in pigment cells remains a mystery. They seem not to be intimately involved in transport, since the cell type with the very fastest transport system (*Holocentrus* erythrophores) does not contain intermediate-sized filaments.

4. Microtrabeculae

The application of stereo high-voltage electron microscopy to the study of whole critical-point-dried cells has revealed a structural component of the cytoplasmic matrix that forms an extensive three-dimensional system of interconnected filamentous strands or microtrabeculae (Wolosewick and Porter, 1976). This system undergoes conformational changes in association with pigment movements in erythrophores of the squirrelfish (Byers and Porter, 1977) and melanophores of the angelfish (Schliwa, 1979). The microtrabecular lattice, which interconnects pigment granules, microtubules, and membranes, has been proposed to be a causal agent for granule transport, "contracting" during aggregation and re-extending during dispersion (Byers and Porter, 1977). The biochemical nature of microtrabeculae is unknown, and so far they are only defined as a structural entity. Their functional and biochemical relationship to the better-known filamentous components of the cytoplasm (actin filaments, intermediate-sized filaments) remains to be determined.

C. Mechanism of Granule Transport

The understanding that microtubules play an essential role in granule translocations in pigment cells is based on both morphological and physiological evidence. Their orientation parallels the direction of movement and defines physical constraints for the arrangement of granules, which, as a consequence, often appear aligned in parallel rows. Microtubule disruption results in a disorganization of this alignment (Murphy and Tilney, 1974; Schliwa and Euteneuer, 1978b), and a decrease in the rate of granule movements in fish chromatophores (Wikswo and Novales, 1969; Schliwa and Bereiter-Hahn, 1973b; Murphy and Tilney, 1974; Junqueira *et al.*, 1974). In amphibian chromatophores, antimicrotubular agents inhibit centripetal transport (Malawista, 1965, 1971a), but also enhance or cause dispersion (Malawista, 1965; Fisher and Lyerla, 1974). However, inhibitor studies on amphibian chromatophores have not been combined with ultrastructural observations, thus our knowledge of microtubule distribution and the effects of drugs on them is limited. The same applies to most invertebrate pigment cells, where colchicine has been shown to inhibit pigment concentration (Lambert and Crowe, 1973, 1976; Fingerman *et al.*, 1975). Contrary to the vast majority of other cell types to which colchicine has been applied, unusually high concentrations (up to 25 m*M*) are necessary to observe measurable effects. Since the same concentration of lumincolchicine can exert a similar effect (Lambert and Fingerman, 1976; Obika *et al.*, 1978b), the specificity of colchicine inhibition is entirely unclear.

Recent studies with fish chromatophores have shown that melanophores depleted of their microtubules by cold (0°C) treatment in the presence of colchicine can, after rewarming to room temperature with the drug still present, undergo aggregation and dispersion-like reactions (Schliwa and Euteneuer, 1978b; Obika *et al.*, 1978b). Since these responses are slow and incomplete, the contention is supported that microtubules are an absolute requirement for normal, rapid granule movements. On the other hand, the experiments indicate the presence of another component able to translocate pigment granules in the total absence of microtubules. Does this component involve contractile proteins?

Numerous studies have employed cytochalasin B to test for the possible involvement of actin filaments in pigment movements. There is considerable variation as to the effect of this drug on pigment cells of different species. Fish chromatophores are not appreciably affected (Ohta, 1974); in amphibian cells, dispersion is inhibited, while aggregation may be enhanced (Malawista, 1971b; McGuire and Moellmann, 1972; Novales and Novales, 1972; Magun, 1973; Fisher and Lyerla, 1974); both aggregation and dispersion are affected in crustacean and echinoderm pigment cells (Lambert and Crowe, 1973; Robison and Charlton, 1973; Lambert and Fingerman, 1978; Dambach and Weber, 1975). However, the value of cytochalasin B experiments is unclear, since the systems studied lack unequivocal morphological evidence for an actin–myosin system. Cytochalasin B experiments might be taken as a preliminary indication for the involvement of a contractile actomyosin system, but supporting evidence is absolutely necessary; the conclusions drawn from the results of cytochalasin experiments might be premature.

Based on immunofluorescent and electron-microscopic evidence, a loose cortical meshwork of actin filaments has been demonstrated in angelfish melanophores (Schliwa *et al.*, 1981). Whether these filaments are involved in transport remains unclear, since their location apparently does not change on pigment aggregation or dispersion, and since microinjection of the actin-specific substances phalloidin and DNase I does not interfere with transport. Extraction of isolated cells with the nonionic detergent Triton X-100 has shown that although the cortical filament system is easily detectable, no such filaments can be demonstrated in association with pigment granules and microtubules. In fact, more central regions of the extracted cell appear virtually "empty." Intact cells, however, display a dense cytoplasmic ground substance that, in high-voltage electron microscopy of whole critical-point-dried cells is resolved into a prominent system of slender microtrabeculae interconnecting granules, microtubules, and membranes. Since this lattice undergoes profound structural changes in association with different phases of pigment movements (Byers and Porter, 1977; Schliwa, 1979), it is believed to be responsible for mediating granule transport in association with microtubules.

The demonstration of a dynamically changing matrix component supports Green's original observation (1968) that the granules behave as if they were embedded in an elastic continuum that expands during dispersion and collapses during aggregation. That actin is a major component in the form of a finely divided network is still a possibility; given its different behavior on Triton extraction, though, its physiological properties can be expected to differ from those of the cortical mesh. This observation suggests a chemical difference between the Triton-stable actin mesh and the microtrabeculae, whose elucidation will undoubtedly contribute to an understanding of granule motility.

Whether all pigment cells employ identical mechanochemical mechanisms to translocate granules is not at all clear at the moment. Although one would anticipate a common mechanism on the basis of their equivalent ontogeny (all vertebrate chromatophores stem from the neural crest), fully differentiated cells may show fundamental differences with respect to their motile activity. For instance, the slower migration of pigment granules in amphibian chromatophores is not just a slow-motion version of the fast fish chromatophore with its fast concentration/slower re-expansion behavior; rather, aggregation and dispersion take place at comparable rates, or dispersion requires even less time (e.g., Malawista, 1971b). It is plausible, therefore, that future studies will establish a structural and/or biochemical basis for differences in the motile behavior of "fast" and "slow" pigment cells.

D. Chromatophore Control

The regulation of chromatophore activities has been the subject of numerous investigations extensively reviewed by Bagnara and Hadley (1973). Basically, three control mechanisms operate in different pigment cell types: light, hormones, and neurotransmitters. Sea urchin chromatophores undergo dispersion on illumination, even when isolated from the animal (Weber and Dambach, 1974). Amphibian and crustacean pigment cells are under humoral control, whereas those of most fish species are controlled by the nervous system. Although hormones do also influence fish cells, the release of transmitters from nerve endings in contact with chromatophores can, as a rule, override hormone levels (e.g., melanophore-stimulating hormone, melatonin) normally present in the blood.

Interestingly, under conditions of *in vitro* culture, fish chromatophores may respond to light directly (Wakamatsu *et al.*, 1980), as do larval *Xenopus* melanophores (Bagnara, 1957). These observations suggest the existence of a hierarchy of control mechanisms for any given chromatophore where nerve endings (if present) exert the most direct and strongest effect; hormonal influence and illumination rank in second and third position, respectively.

It is now well established that cyclic AMP acts as a second (intracellular) messenger for the action of hormones and/or neurotransmitters on chromatophores. Externally applied or iontophoretically injected cyclic AMP, as well as phosphodiesterase inhibitors, cause pigment dispersion in both amphibian and fish melanophores (e.g., Novales and Davis, 1967; Van de Veerdonk and Konijn, 1970; Novales and Fujii, 1970; Geschwind *et al.*, 1977). The intracellular target of cyclic AMP is not known, but direct interaction with cytoskeletal or motility-related components is possible. Evidence implicating cyclic AMP in the regulation of microtubules in a variety of cell types has recently been reviewed by Dedman, Brinkley, and Means (1979).

E. Energy Requirements

Cellular motility is an active process that requires metabolic energy in the form of ATP. Energy requirements of the motile activity of pigment cells has been studied only in a few instances. Teleost chromatophores appear to require ATP for the process of pigment dispersion only (Junqueira *et al.*, 1974; Saidel, 1977; Luby and Porter, 1980), supporting Green's original suggestion (1968) that dispersion is an energy-storage process. Stored energy is then converted into kinetic energy in the process of pigment aggregation. Luby and Porter (1980) further correlated the effects of different metabolic inhibitors (DNP, NaCN, oligomycin) with morphological alterations in the microtrabecular lattice in erythrophores of *Holocentrus* and concluded that ATP is required to establish and maintain the elongated morphology of individual microtrabeculae. Whether an energy requirement for the dispersion process is unique only to the "fast" chromatophores studied thus far remains to be determined.

IV. Hypothetical Evolution of the Motile Apparatus of Pigment Cells

Without indulging in oversimplification, pigment cells may be grouped into three categories according to the characteristics of their granule transport: (1) "Fast" pigment cells, which require only seconds to accomplish pigment aggregation and which usually have a very prominent microtubule system (some fish chromatophores). These cells have a fixed outline regardless of the state of pigment distribution, although their thickness may change considerably on the redistribution of pigment granules. (2) "Slow" chromatophores, which require at least several minutes for pigment aggregation (most amphibian and crustacean, as well as some fish chromatophores). From the limited evidence we have, it seems fair to state that the microtubule system of these cells is only fairly well developed. The outline of these cells, too, remains unchanged on aggregation or dispersion. (3) Thus far, two differentiated pigment cells have been described that change their outline in association with the migration of pigment organelles: sea urchin chromatophores (Weber and Dambach, 1972, 1974) and melanophores of *Corethra* larvae (Kopenec, 1949). Dispersing sea urchin chromatophores, which are initially rounded, extend small pseudopods that carry pigment granules with them. On aggregation (in the dark), these pseudopods are retracted into the cell body. Apart from these two examples described in the literature, only undifferentiated chromatophores of vertebrates are known to change their outline in association with translational movements during embryogenesis. This applies even to those cells that maintain a fixed position and outline in the differentiated state.

Attempts to deduce the phylogenetic development of chromatophores from the morphological and physiological properties of this cell type as expressed in existing species may be misleading, but they may also help us to understand the variability of motility characteristics observed in pigment cells from different animals. It is plausible that primitive chromatophores were amoeboid cells that changed their shape and outline to achieve redistribution of pigmentary organelles contained in their cell bodies. This activity can be assumed to have been based on a contractile actomyosin system; the reactions were probably very slow, as they in fact are in sea urchin chromatophores, a cell type displaying amoeboid motility (dispersion takes about 20 minutes). Possibly these characteristics of primitive pigment cells are also still reflected in the amoeboid motility of undifferentiated vertebrate chromatophores (migrating neural crest cells). With time, cells might have developed a fixed outline stabilized by linear elements such as microtubules and an association with extracellular matrices. The major advantage of this achievement appears to be the establishment of a system of preformed cytoplasmic channels within which the pigment can readily be redistributed. No time (and possibly energy) is wasted for the *de novo* formation of a cell process; thus rapid color change became possible. The transport mechanisms might then have become intimately associated with the cytoskeletal component (the microtubule). This view is based on the hypothesis that granule transport is essentially microtubule dependent, though possibly generated by another component. In fact, that microtubules are indispensable for rapid and directed granule movements is about the only safe statement one can make about the mechanism of pigment transport in "fast" chromatophores. The precise nature of the mechanochemical link is still unclear. The hypothetical scheme outlined here views the rapid fish chromatophore with its prominent microtubule apparatus as the most advanced system in this evolution. It would be of great interest to know whether the fast color change observed in some reptiles (e.g., chameleon) is based on similar cell-anatomical principles.

V. Comparison with Other Examples of Intracellular Motility

Transport of structural components (particles, organelles) is a conspicuous feature of virtually all eukaryotic cells, although it may be expressed to varying degrees. Bulk translocation of cytoplasm (e.g., during establishment of a cell process), protoplasmic streaming (plant cells, slime molds), axonal transport (neurons), mitosis, and saltatory movement (cultured cells, heliozoa) are other examples from the broad spectrum of intracellular-motility phenomena that have been studied in some detail. Of these, the first two can clearly be distinguished from other forms of intracellular transport according to their phenomenology and probably the mechanisms involved. The other examples of particle transport have two main features in common with chromatophores: Transport is bidirectional and it is spatially associated with microtubules. Whether these similarities also imply similar mechanisms is not clear at the moment, although the possibility is recognized by many authors.

Considerable variation exists as to the size of transported particles, ranging from ribosomes in ovarioles (Macgregor and Stebbings, 1970) to nuclei in virus-induced syncytia (Holmes and Choppin, 1968); the same is true for the distances travelled (up to 1 m in some axons) and the rate of transport, which ranges from fractions of a micrometer per second to up to 20 μm/second. Bidirectionality is a conspicuous feature of particle transport. This is perhaps best illustrated by the mass movements of pigment granules in chromatophores where inward and outward motion are separated temporally. However, bidirectional transport of granules is also observed at any given time in cells with pigment dispersed, where granules are in constant motion back and forth. Here, as in neurons and heliozoan axopodia, particles may travel in opposite directions even in very close proximity to one another. The idea that transport in opposite directions is associated with microtubules having opposite polarities has already been dismissed for the angelfish melanophore (see Section III,B,1). It probably also does not apply to particle transport along heliozoan axopodia (Euteneuer and McIntosh, 1981), since the microtubules of heliozoan axonemes are of uniform polarity.

It therefore seems probable that the direction in which a particle travels is not determined by the intrinsic polarity of a microtubule.

As in chromatophores, inhibitor studies in other cell systems also have yielded divergent results. Let us take a brief look at the phenomenon of axonal transport, which has some features in common with granule transport in pigment cells. Although colchicine undoubtedly inhibits axonal transport in many systems, electron microscopy has shown an almost normal complement of microtubules in some cells where transport was blocked with this drug (Fernandez *et al.*, 1970; Karlsson *et al.*, 1971; Flament-Durand and Dustin, 1972). Conversely, particle transport in cultured regenerating chick sensory neurons is only slightly affected with 10 μM colchicine, whereas the number of microtubules is reduced by as much as 80% (Chang, 1972).

Both results again call for caution in the interpretation of inhibitor studies on transport phenomena. It cannot be concluded unambiguously that microtubules are the organelles providing the motive force for transport. It should also be borne in mind that the proportion of microtubules in relation to neurofilaments may vary considerably, and yet, in axons with a few microtubules and a well-developed neurofilament array, particle transport appears identical to that in microtubule-rich axons. Perhaps, then, microtubules and/or neurofilaments provide linear guidelines on which another, as yet unidentified, component acts to propel the particles.

Probably the most direct evidence thus far in support of the contention that microtubules (or other linear elements) act only as guidelines comes from a simple and brilliant experiment by Edds (1975) in a system long thought to represent a paradigm for microtubule-dependent granule movements. By pushing a microneedle through the cell body of the heliozoan *Actinospherium,* he created an artificial axopod with the microneedle acting as an axoneme. Transport of particles along the microneedle-axopod proceeded normally even in concentrations of colchicine known to cause retraction of natural axopodia. What this experiment and the indirect evidence from diverse studies of neurons and chromatophores tell us is that transport requires a backbone of linear cytoskeletal structures that may be an array of microtubules, a bundle of filaments, or a glass rod; the motive force, however, is delivered by another component that acts on, or in conjunction with, this framework. Future studies will show whether this hypothesis is correct and, if so, what the nature of the force-generating component is.

Acknowledgments

I thank Drs. K. R. Porter and J. R. McIntosh for helpful discussions and critical reading of the manuscript. This article was written while the author was supported by a Heisenberg fellowship from the Deutsche Forschungsgemeinschaft.

References

Anderson, T. (1951). *Trans. N.Y. Acad. Sci.* [2] **13,** 130.

Arnott, H. J., Best, H. C. G., and Nicol, J. A. C. (1970). *J. Cell Biol.* **46,** 426.

Aubin, J. E., Weber, K., and Osborn, M. (1979). *Exp. Cell Res.* **124,** 93.

Bagnara, J. T. (1957). *Proc. Soc. Exp. Biol. Med.* **94,** 527.

Bagnara, J. T. (1972). *In* "Pigmentation: Its Genesis and Biologic Control" (V. Riley, ed.), pp. 68–75. Appleton, New York.

Bagnara, J. T., and Hadley, M. E. (1973). "Chromatophores and Color Change." Prentice-Hall, Englewood Cliffs, New Jersey.

Ballowitz, E. (1914). *Pfluegers Arch. Gesamte Physiol.* **157,** 165.

Barth, L. G., and Barth, L. J. (1969). *J. Embryol. Exp. Morphol.* **7,** 210.

Beckerle, M., Byers, H. R., Fujiwara, K., and Porter, K. R. (1979). *J. Cell Biol.* **83,** 352a.

Behre, K. (1935). *Science* **31,** 292.

Berthold, G., and Seifert, G. (1977). *Entomol. Ger.* **3,** 303.

Bikle, D., Tilney, L. G., and Porter, K. R. (1966). *Protoplasma* **61,** 322.

Bruecke, E. (1852). *Denkschr. Akad. Wiss. Math. Nat. Kl.* **4,** 179.

Byers, H. R., and Porter, K. R. (1977). *J. Cell Biol.* **75,** 541.

Chang, C. M. (1972). *J. Cell Biol.* **55,** 37a.

Cloney, R. A., and Florey, E. (1968). *Z. Zellforsch. Mikrosk. Anat.* **89,** 250.

Dambach, M. (1968). *Z. Vergl. Physiol.* **64,** 400.

Dambach, M., and Weber, W. (1975). *Comp. Biochem. Physiol. C* **50C,** 49.

Dedman, J. R., Brinkley, B. R., and Means, A. R. (1979). *Adv. Cyclic Nucleotide Res.* **11,** 131.

Edds, K. T. (1975). *J. Cell Biol.* **66,** 145.

Egner, O. (1971). *Cytobiologie* **4,** 262.

Elofsson, R., and Kauri, T. (1971). *J. Ultrastruct. Res.* **36,** 263.

Euteneuer, U., and McIntosh, J. R. (1981). *Proc. Nat. Acad. Sci. U.S.A.* **78,** 372.

Fernandez, H. L., Huneeus, F. C., and Davison, P. F. (1970). *J. Neurobiol.* **1,** 395.

Fingerman, M., Fingerman, S. W., and Lambert, D. T. (1975). *Biol. Bull.* (*Woods Hole, Mass.*) **149,** 165.

Fischer, A. (1965). *Z. Zellforsch. Mikrosk. Anat.* **65,** 290.

Fisher, M., and Lyerla, T. A. (1977). *J. Cell. Physiol.* **83,** 117.

Flament-Durand, J., and Dustin, P. (1972). *Z. Zellforsch. Mikrosk. Anat.* **130,** 440.

Flickinger, R. A. (1949). *J. Exp. Zool.* **12,** 465.

Florey, E. (1969). *Am. Zool.* **9,** 429.

Franz, V. (1939). *Z. Zellforsch. Mikrosk. Anat.* **30,** 194.

Fujii, R., and Novales, R. R. (1969). *Am. Zool.* **9,** 453.

Gartz, R. (1970). *Cytobiologie* **2,** 220.

Geschwind, I. I., Horowitz, J. M., Mikuckis, G. M., and Dewey, R. D. (1977). *J. Cell Biol.* **74,** 928.

Green, L. (1968). *Proc. Natl. Acad. Sci. U.S.A.* **59,** 1179.

Heideman, S. R., and McIntosh, J. R. (1980). *Nature (London)* **286,** 570.

Hill, A. V., Parkinson, J. L., and Solandt, D. Y. (1935). *J. Exp. Biol.* **12,** 397.

Hogben, L., and Slome, D. (1931). *Proc. R. Soc. London, Ser. B* **108,** 10.

Holmes, K. V., and Choppin, P. W. (1968). *J. Cell Biol.* **39,** 526.

Ide, H. (1974). *Dev. Biol.* **41,** 380.

Ide, H. (1979). *Pigment Cell* **4,** 28.

Iga, T. (1975). *Mem. Fac. Lit. Sci., Shimane Univ.* **8,** 75.

Iga, T., and Bagnara, J. T. (1975). *J. Exp. Zool.* **192,** 331.

Junqueira, L. C. U., Raker, E., and Porter, K. R. (1974). *Arch. Histol. Jpn.* **36,** 339.

Junqueira, L. C. U., Reinach, F., and Salles, L. M. M. (1977). *Arch. Histol. Jpn.* **40,** 435.
Kahn, R. H., and Lieben, S. (1907). *Arch. Anat. Physiol., Physiol. Abt.* **17,** 104.
Karlsson, J. O., Hansson, H. A., and Sjöstrand, J. (1971). *Z. Zellforsch. Mikrosk. Anat.* **115,** 265.
Kinosita, H. (1953). *Annot. Zool. Jpn.* **26,** 115.
Kinosita, H. (1963). *Ann. N.Y. Acad. Sci.* **100,** 992.
Kopenec, A. (1949). *Z. Vergl. Physiol.* **31,** 490.
Lambert, D. T., and Crowe, L. H. (1973). *Comp. Biochem. Physiol. A* **46A,** 11.
Lambert, D. T., and Crowe, L. H. (1976). *Comp. Biochem. Physiol. C* **54C,** 115.
Lambert, D. T., and Fingerman, M. (1976). *Comp. Biochem. Physiol. C* **53C,** 25.
Lambert, D. T., and Fingerman, M. (1978). *Biol. Bull. (Woods Hole, Mass.)* **155,** 563.
Lamer, H. I., and Chavin, W. (1975). *Cell Tissue Res.* **163,** 383.
Lazarides, E. (1980). *Nature (London)* **283,** 249.
Luby, K. J., and Porter, K. R. (1980). *Cell* **21,** 13.
Lyerla, T. A., and Novales, R. R. (1972). *J. Cell. Physiol.* **80,** 243.
McDonald, K., Edwards, M. K., and McIntosh, J. R. (1979). *J. Cell Biol.* **83,** 443.
McGuire, J., and Moellmann, G. (1972). *Science* **175,** 642.
McGuire, J., Moellmann, G., and McKeon, F. (1972). *J. Cell Biol.* **52,** 754.
McIntosh, J. R., McDonald, K., Edwards, M. K., and Ross, B. M. (1979). *J. Cell Biol.* **83,** 428.
Macgregor, H. C., and Stebbings, H. (1970). *J. Sci.* **6,** 431.
Magun, B. (1973). *J. Cell Biol.* **57,** 845.
Malawista, S. E. (1965). *J. Exp. Med.* **122,** 361.
Malawista, S. E. (1971a). *J. Cell Biol.* **49,** 848.
Malawista, S. E. (1971b). *Nature (London)* **234,** 354.
Marsland, D. A. (1944). *Biol. Bull. (Woods Hole, Mass.)* **87,** 252.
Marsland, D. A., and Meisner, D. (1967). *J. Cell. Physiol.* **70,** 209.
Martin, A. R., and Snell, R. S. (1968). *J. Physiol. (London)* **195,** 755.
Matsumoto, J., Watanabe, Y., Obika, M., and Hadley, M. (1978). *Comp. Biochem. Physiol.* **61,** 509.
Matthews, S. A. (1931). *J. Exp. Zool.* **58,** 471.
Moellmann, G., and McGuire, J. (1975). *Ann. N.Y. Acad. Sci.* **253,** 711.
Moellmann, G., McGuire, J., and Lerner, A. B. (1973). *Yale J. Biol. Med.* **46,** 337.
Morimoto, K., and Iwata, K. S. (1976). *Biol. J. Okayama Univ.* **17,** 23.
Murphy, D. B., and Tilney, L. G. (1974). *J. Cell Biol.* **61,** 757.
Murray, R. L., and Dubin, M. W. (1975). *J. Cell Biol.* **64,** 705.
Novales, R. R. (1960). *Trans. Am. Microsc. Soc.* **79,** 25.
Novales, R. R., and Davis, W. J. (1967). *Endocrinology* **81,** 283.
Novales, R. R., and Fujii, R. (1970). *J. Cell. Physiol.* **75,** 133.
Novales, R. R., and Novales, B. J. (1972). *Gen. Comp. Endocrinol.* **19,** 363.
Obika, M. (1975). *J. Exp. Zool.* **191,** 427.
Obika, M. (1976). *Annot. Zool. Jpn.* **49,** 157.
Obika, M., Menter, D. G., Tchen, T. T., and Taylor, J. D. (1978a). *Cell Tissue Res.* **193,** 387.
Obika, M., Turner, W. A., Negishi, S., Menter, D. G., Tchen, T. T., and Taylor, J. D. (1978b). *J. Exp. Zool.* **205,** 95.
Ohta, T. (1974). *Biol. Bull. (Woods Hole, Mass.)* **146,** 258.
Ozato, K. (1977). *J. Cell Sci.* **26,** 93.
Pickett-Heaps, J. D. (1969). *Cytobios* **3,** 257.
Porter, K. R. (1973). *Ciba Found. Symp.* [N.S.] **14,** 149.
Robison, W. G., and Charlton, J. S. (1973). *J. Exp. Zool.* **186,** 279.
Rohrlich, S. T. (1974). *J. Cell Biol.* **62,** 295.
Rohrlich, S. T., and Porter, K. R. (1972). *J. Cell Biol.* **53,** 38.

Saidel, W. M. (1977). *Experientia* **33,** 1573.
Schliwa, M. (1978). *J. Cell Biol.* **76,** 605.
Schliwa, M. (1979). *Exp. Cell Res.* **118,** 323.
Schliwa, M., and Bereiter-Hahn, J. (1973a). *Z. Zellforsch. Mikrosk. Anat.* **147,** 107.
Schliwa, M., and Bereiter-Hahn, J. (1973b). *Z. Zellforsch. Mikrosk. Anat.* **147,** 127.
Schliwa, M., and Bereiter-Hahn, J. (1974). *Microsc. Acta* **75,** 235.
Schliwa, M., and Euteneuer, U. (1978a). *J. Supramol. Struct.* **8,** 177.
Schliwa, M., and Euteneuer, U. (1978b). *Nature (London)* **273,** 556.
Schliwa, M., and Euteneuer, U. (1979). *Cell Tissue Res.* **196,** 541.
Schliwa, M., Osborn, M., and Weber, K. (1978). *J. Cell Biol.* **76,** 229.
Schliwa, M., Euteneuer, U., Herzog, W., and Weber, K. (1979). *J. Cell Biol.* **83,** 623.
Schliwa, M., Wehland, J., Weber, K., and Porter, K. R. (1981). Submitted for publication.
Smith, D. C. (1936). *J. Cell. Comp. Physiol.* **8,** 83.
Spaeth, R. A. (1916). *J. Exp. Zool.* **20,** 193.
Takeuchi, I. K. (1975). *Annot. Zool. Jpn.* **48,** 242.
Tippit, D. H., Schultz, D., and Pickett-Heaps, J. D. (1978). *J. Cell Biol.* **79,** 737.
Van de Veerdonk, F. C. G., and Konijn, T. M. (1970). *Acta Endocrinol. (Copenhagen)* **64,** 364.
Wakamatsu, Y., Kawamura, S., and Yoshizawa, T. (1980). *J. Cell Sci.* **41,** 65.
Weber, K., Rathke, P., and Osborn, M. (1978). *Proc. Natl. Acad. Sci. U.S.A.* **75,** 1820.
Weber, W., and Dambach, M. (1972). *Z. Zellforsch. Mikrosk. Anat.* **133,** 87.
Weber, W., and Dambach, M. (1974). *Cell Tissue Res.* **148,** 437.
Wikswo, M. A., and Novales, R. R. (1969). *Biol. Bull. (Woods Hole, Mass.)* **137,** 228.
Wikswo, M. A., and Novales, R. R. (1972). *J. Ultrastruct. Res.* **41,** 189.
Wise, G. E. (1969). *J. Ultrastruct. Res.* **27,** 472.
Wolosewick, J. J., and Porter, K. R. (1976). *Am. J. Anat.* **147,** 303.
Wolosewick, J. J., and Porter, K. R. (1979). *In* "Practical Tissue Culture Applications" (K. Maramorosh and H. Hirami, eds.), pp. 59–85. Academic Press, New York.
Zimmermann, K. W. (1893). *Arch. Mikrosk. Anat.* **41,** 367.

Chapter 15

Acanthamoeba castellanii: *Methods and Perspectives for Study of Cytoskeleton Proteins*

EDWARD D. KORN

Laboratory of Cell Biology, National Heart, Lung, and Blood Institute
National Institutes of Health
Bethesda, Maryland

I. Introduction

The study of the cytoskeletal proteins of *Acanthamoeba* developed from earlier investigations of endocytosis in this amoeba. For those studies, a eukaryote was sought that might be totally dependent on endocytosis (to maximize the fraction of cell protein and membrane engaged in endocytic processes), that could be grown axenically on soluble medium (to eliminate complications de-

METHODS IN CELL BIOLOGY, VOLUME 25

ISBN 0-12-564125-7

rived from mixed cultures), and that could be grown in essentially unlimited quantities (to obtain sufficient material to isolate membranes and to purify proteins to homogeneity for biochemical studies). In the early 1960s, *Acanthamoeba castellanii* (Neff) seemed an appropriate choice. It can be grown axenically on soluble medium in aerated carboys or fermenters, is large enough for easy visualization of phagocytosis in the light microscope but small enough that good fixation for electron microscopy can be obtained, has a plasma membrane unencumbered by a cell wall or glycocalyx, and, as it happened, seems not to have mechanisms for the active transport of nutrients and is thus totally dependent on endocytosis. Subsequent experience has shown that *Acanthamoeba* was not an unreasonable choice (Weisman and Korn, 1967; Korn and Weisman, 1967; Wetzel and Korn, 1969; Ulsamer *et al.*, 1969, 1971; Korn *et al.*, 1974; Bowers and Olszewski, 1972; Bowers, 1977). *Acanthamoeba* may, in fact, provide a more useful system than has generally been recognized for the study of many important problems in cell biology (Byers, 1979).

From our earliest morphological studies of *Acanthamoeba* (Bowers and Korn, 1968, 1969), we recognized the abundance of cytoplasmic microfilaments particularly in the ectoplasmic region underlying the plasma membrane, in pseudopodal areas, and totally filling the microspikes that give *Acanthamoeba* its name. Actin microfilaments are specifically enriched at the site of formation of phagocytic vesicles (Korn *et al.*, 1974), and are enriched in preparations of highly purified plasma membranes (Korn and Wright, 1973), with which they seem to be specifically associated (Pollard and Korn, 1973a). These circumstantial observations were sufficient to convince us of what we already suspected: Actin and myosin are the proteins primarily responsible for force generation in amoeboid movement and phagocytosis. Encouraged by the prior isolation of actin (Hatano and Oosawa, 1966) and myosin (Adelman and Taylor, 1969) from *Physarum*, we undertook the purification of these proteins from *Acanthamoeba* in the belief that the cell biology of actomyosin-dependent motile processes could be fully comprehended only when the underlying biochemistry was understood.

I am now certain of the presence in *A. castellanii* (Neff) of the following 12 proteins, each of which probably functions in one or another motile process: actin; profilin; gelactins I, II, III, and IV; myosin IA, IB, and IC; myosin II; myosin I heavy-chain kinase; and two myosin II heavy-chain kinases. Some of these proteins have been purified to homogeneity, others have been only partially purified. In addition, at least two other proteins must be assumed to be present: myosin I and myosin II heavy-chain phosphatases. If the system is similar to others that have been studied, there will be additional cytoskeletal proteins, but even without postulating the existence of proteins yet to be discovered, the cytoskeletal and motility proteins probably account for about 25 to 30% of the total proteins of *A. castellanii* (Neff).

II. Biochemistry of the Cytoskeletal Proteins

A. Actin and Related Proteins

From amoebas to humans, actins are remarkably similar, although not identical, in amino acid sequence and properties (Korn, 1978). All actins have a molecular weight of about 42,000 and contain one mole of N^{τ}-methylhistidine. Of the 228 residues of *Acanthamoeba* actin that have been sequenced (Elzinga and Lu, 1976; M. Elzinga, personal communication), only 14 (6%) are different from the corresponding positions in rabbit skeletal-muscle actin, which contains a total of 374 residues (Elzinga *et al.*, 1973). In marked contrast to the actins of vertebrate nonmuscle and smooth-muscle cells, which are composed of two isoelectric species owing to differences in sequence of the N-terminal tetrapeptide (Vandekerckhove and Weber, 1978a,b,c), *Acanthamoeba* [and also *Dictyostelium* (Uyemura *et al.*, 1978; Rubenstein and Deuchler, 1979) and *Physarum* (Vandekerckhove and Weber, 1978a)] contains only one type of actin molecule (Gordon *et al.*, 1977). Actin from *Acanthamoeba* has a more alkaline isoelectric point than any other actin (Gordon *et al.*, 1977), perhaps because it may be the only isolated actin that contains one mole of N^{ε}-methyllysines (Weihing and Korn, 1970, 1971; Gordon *et al.*, 1976). The N-terminal sequence of *Acanthamoeba* actin has yet to be determined but it is likely to contain, as do all actins thus far sequenced, an N-acetylated aspartate or glutamate as the first residue. The remarkable conservation of amino acid sequence is such that the actins of amoebas are more similar to actins of vertebrate nonmuscle cells than the latter are to muscle actins of the same species (Vandekerckhove and Weber, 1978a,b).

Actin functions in nonmuscle and muscle cells not as the monomeric globular G-actin but as two-stranded double-helical polymeric F-actin. In nonmuscle cells, at least, the temporal and spatial regulation of actin polymerization will exert profound control of its cytoskeletal and motile activities. It is useful to recall that the polymerization of actin *in vitro* is induced by raising the ionic strength (e.g., to 50 or 100 mM KCl) or increasing the concentration of Mg^{2+} (e.g., to 0.5 or 1 mM); that is, purified actin is polymeric under physiological conditions. The first step in polymerization is the slow, rate-limiting formation of nuclei (three or four subunits) from monomeric G-actin, followed by rapid elongation to filaments of F-actin and, finally, the redistribution of filament lengths (Oosawa and Asakura, 1975). Approximately one molecule of actin-bound ATP is hydrolyzed to actin-bound ADP for every molecule of actin converted from the monomeric to the polymeric state. More recent elaboration of this model (Wegner, 1976) recognizes that the actin filament can elongate at both ends (Woodrum *et al.*, 1975), although growth is apparently more rapid at one end than at the other. Also, one end of the filament (probably the end that elongates more rapidly) has a lower critical concentration (the concentration of monomer in equilibrium with polymer when polymerization ceases) than does the other end of the filament (Brenner and Korn, 1979, 1980a,b). At steady state, therefore, when the concentrations of polymer and monomer are constant, the actual monomer concentration will lie between the theoretical critical concentrations at the two ends of the filament, and there will be net addition of actin monomers at the low-critical-concentration end of the filament and equivalent net loss of actin monomers from the high-critical-concentration end. Although actin can polymerize in the absence of ATP, this directionality of polymerization (an actin treadmill) depends on the hydrolysis of ATP.

In *Acanthamoeba* (Gordon *et al.,* 1977), as in other nonmuscle cells, about 50% of the actin is nonpolymerized at a monomer concentration greatly in excess of the critical concentration of purified actin, under similar ionic conditions. The explanation seems to be the presence of a 14,000-dalton protein, profilin, that forms a 1:1 molar complex with monomeric actin and inhibits the nucleation step (Reichstein and Korn, 1979). More may be involved, however, because the reconstituted complex is not as resistant to polymerization as is the nonpolymerizable actin in the crude extract. It is possible to imagine that actin polymerization might also be affected by one or more proteins that interact with either end of the elongating filaments, but such modulators have not yet been found or sought in *Acanthamoeba*.

Once formed, actin filaments associate into larger structures in all cells. In *Acanthamoeba,* filament–filament interaction seems to occur by the cross-linking of F-actin by any one of four relatively low-molecular-weight proteins: gelactins I, II, III, and IV, with subunit molecular weights of 23,000, 28,000, 32,000, and 38,000, respectively (Maruta and Korn, 1977a). All but gelactin I may be dimers.

Other than its putative cytoskeletal role, the only known function of filamentous actin (monomeric actin has no known function) is the activation of the Mg^{2+}-ATPase activity of myosin in such a way that the energy of hydrolysis of ATP is made to perform work (Korn, 1978). In this regard, all actins are qualitatively similar, but there are significant quantitative differences between the affinities of *Acanthamoeba* (Gordon *et al.,* 1976, 1977) and other nonmuscle actins (Gordon *et al.,* 1977; Korn, 1978) and rabbit skeletal-muscle actin. In addition, the effects of muscle tropomyosin on actin-activated myosin ATPase activity are qualitatively different (Yang *et al.,* 1977, 1979a,b) with *Acanthamoeba* actin (activation) from with muscle actin (inactivation). The significance of this for *Acanthamoeba,* which is not known to contain tropomyosin, is unclear, but it may provide useful clues to the tropomyosin sites on actin, since *Acanthamoeba* and muscle actin are so similar in amino acid sequence. The ease with which *Acanthamoeba* and muscle actin form (apparently random) copolymers (Gordon *et al.*, 1976; Yang *et al.*, 1977, 1979a) in which the specific properties of each are still expressed suggests that similar experiments with the isoactins of vertebrate nonmuscle cells may be informative. A more general discussion of the properties of actin and its polymerization will have been presented in other contributions to this volume.

B. Myosins

Except for those from *Acanthamoeba,* all known myosins consist of a pair of heavy chains of MW ~200,000 and two pairs of light chains of MW ~20,000 and 16,000 (Korn, 1978). The heavy chains contain an α-helical rod segment and a globular head region: The rods associate to form a rigid intramolecular double-stranded helix, and intermolecular association of the rod portions of molecules results in the formation of bipolar filaments that are certainly the functional states *in situ,* at least for muscle. The two pairs of light chains are associated with the globular heads of the heavy chain, which project from the bipolar filament. These myosin heads contain the actin-binding site and catalytic site. The Mg^{2+}-ATPase activity of actomyosin is generally regulated by Ca^{2+}. In skeletal muscle. For example, this occurs through interaction of the tropomyosin-troponin complex with the actin filament in such a way that the actomyosin Mg^{2+}-ATPase activity can be expressed only when Ca^{2+} is present. In vertebrate smooth-muscle and nonmuscle cells, regulation occurs through Ca^{2+}-dependent phosphorylation of the 20,000-dalton light chains of the myosin; only the phosphorylated form of the myosin can express its actomyosin Mg^{2+}-ATPase activity. To all of these generalizations, *Acanthamoeba* is an exception—and, thus far, the only known exception—for it contains three myosin isoenzymes, each of atypical molecular weight and subunit composition and each with unusual regulatory mechanisms.

1. *Acanthamoeba* MYOSINS IA AND IB

Two of the three *Acanthamoeba* myosin isoenzymes are single-headed molecules of MW ~150,000 (Pollard and Korn, 1973b; Maruta *et al.*, 1979). Myosin IA has a single heavy chain of 130,000 daltons, and a single light chain of 17,000 daltons, as well as a variable (and apparently less than equimolar) associated peptide of 14,000 daltons (Maruta *et al.,* 1979). Myosin IB has a single heavy chain of 125,000 daltons, a single light chain of 27,000 daltons, and, again, a variable amount of an associated peptide of 14,000 daltons (Maruta *et al.,* 1979). As isolated, both enzymes possess Ca^{2+}-ATPase and (K^+, EDTA)-ATPase activities similar to those of rabbit skeletal-muscle myosin, but their low Mg^{2+}-ATPase activities are not actin-activated. Extensive activation by actin occurs, however, when the heavy chains of these myosins are phosphorylated (1 mol/mol) by the action of a specific myosin I heavy-chain kinase that has purified about 800-fold (Maruta and Korn, 1977c). This is the only known example of regulation of a myosin through phosphorylation of its heavy chain. The kinase may have a molecular weight of about 100,000; it is not affected by either Ca^{2+} or cAMP.

Myosin IA can also be isolated with a tightly associated peptide of 20,000 daltons (Maruta *et al.*, 1979). This peptide seems to modify the ATPase activities of myosin IA specifically (Maruta *et al.*, 1979), and may be a second regulatory mechanism for this enzyme. This form of the enzyme has been called myosin IC; however, it is now known not to be a third myosin I isoenzyme, but rather, another form of myosin IA.

Myosin IB heavy chain can be isolated free of any light chains and shown to possess all of the enzymatic activities of the holoenzyme (Maruta *et al.*, 1978). Therefore, the heavy chain must contain the ATP-binding site and the actin-binding site in addition to the phosphorylation site. Controlled proteolysis allows the isolation of a molecule that is actin-activatable without phosphorylation, indicating that the activation of the holoenzyme by phosphorylation is a derepression. Recently, it has been possible to photoaffinity-label the heavy chains of myosin IA and IB (as well as myosins from other sources) with ATP in a substrate-specific manner (Maruta and Korn, 1981a,b).

Myosin IA and IB are true isoenzymes (i.e., products of different genes), as demonstrated by peptide maps (Gadasi *et al.*, 1979) and immunochemistry with antibodies raised against the whole enzymes (Gadasi and Korn, 1979) and the electrophoretically homogenous heavy chains (Gadasi and Korn, 1980). The two isoenzymes do show some immunochemical cross-reactivity. Myosin IC heavy chain is the same as myosin IA heavy chain and it is, therefore, not another isoenzyme. Despite the very low molecular weight of their heavy chains, the isolated forms of myosin IA and IB probably represent their true size, since immunochemically cross-reactive polypeptides of higher molecular weight have not been detected in whole-cell homogenates (Gadasi and Korn, 1979, 1980).

Immunocytochemistry has also demonstrated that every cell contains myosin I and that the myosin I isoenzymes are preferentially concentrated near the plasma membrane of the cell. Myosin IA and IB can be detected immunochemically on electrophoretic gels of highly purified plasma membranes (Gadasi and Korn, 1980), as well as by immunofluorescence. The apparent inability of these enzymes to form bipolar filaments (Pollard and Korn, 1973b), probably because they have very short rod segments (H. Gadasi and E. D. Korn, unpublished data), presents an intriguing problem in their mechanism of action.

2. *Acanthamoeba* Myosin II

This third *Acanthamoeba* myosin isoenzyme physically more closely resembles other known myosins. It is a double-headed molecule consisting of a pair of heavy chains of 170,000 daltons and two pairs of light chains of about 17,500 and 17,000 daltons, respectively (Maruta and Korn, 1977b; Pollard *et al.*, 1978). The heavy chain of myosin II is variably phosphorylated as isolated. Myosin II prepared by the procedure of Pollard *et al.* (1978) is not actin-activatable. Actin-activatable myosin II can be prepared by the procedure described in this article (Collins and Korn, 1980). Both preparations have the same native molecular weight of about 400,000; that is, they elute from gel-filtration columns at the identical position, and have the same subunit composition by polyacrylamide-gel electrophoresis in sodium dodecyl sulfate (SDS) and urea and by isoelectric-gel focusing (Collins and Korn, 1980).

The only difference between the two molecules is that myosin II isolated by the procedure of Pollard *et al.* (1978) contains 4 moles of phosphate—that is, 2 moles of P per mole of heavy chain (Collins and Korn, 1980); whereas myosin II isolated by the procedure of Collins and Korn (1980) contains only about 2 moles of phosphate—that is, about 1 mole of P per mole of heavy chain. When incubated with any one of several phosphatases (Collins and Korn, 1980), the actin-activated ATPase activities of both myosin preparations increase in direct proportion to the loss of phosphate maximal activity being reached with the almost completely nonphosphorylated enzyme (Collins and Korn, 1980). *Acanthamoeba* myosin II, therefore, is the first example of a myosin that can contain as many as 2 moles of P per mole of heavy chain and that is inactive in the phosphorylated and active in the nonphosphorylated form.

Both by peptide maps and immunochemistry, myosin II is totally unrelated to either myosin IA or myosin IB (Gadasi *et al.*, 1979; Gadasi and Korn, 1980). As with myosin IA and IB, immunochemical analysis of electrophoretic gels of whole amoebas reveals no polypeptide of MW $>$170,000 that reacts with antibodies raised against the isolated heavy chain of myosin II. All cells contain myosin II but, in contrast to myosin I isoenzymes, it is generally distributed throughout the cytoplasm but seems to be excluded from regions near the plasma membrane (Gadasi and Korn, 1980), where myosin I isoenzymes are enriched. Myosin II is almost undetectable in preparations of isolated plasma membranes (Gadasi and Korn, 1980).

3. Implications of Multiple Myosin Isoenzymes

A. castellanii (Neff) and related amoebas are the only known examples of a nonmuscle cell with multiple myosin isoenzymes. Indirect-immunofluorescence studies show that all cells of *A. castellanii* (castellanii) (ATCC #30234), *A. astronyxis* (ATCC #30137), and *Nigleria gruberi* (ATCC #30133) react with antibodies to *Acanthamoeba* myosins IA, IB, and II. Both of the other *Acanthamoeba* strains contain polypeptides of approximately the same molecular weights as the heavy chains of *Acanthamoeba* I and II that cross-react with the appropriate antibodies, and the homologue of *Acanthamoeba* myosin I in *A.*

astronyxis seems to have the same native molecular weight as the enzyme from A. castellanii (H. Gadasi and E. D. Korn, unpublished data). *Dictyostelium discoideum, Entomoeba histolytica* human platelets, and SV40-transformed 3T3 cells (as well as smooth and skeletal muscle) do not contain immunologically cross-reactive proteins, but this does not rule out the possibility that structurally and functionally isoenzymes might exist in these or other cells.

The occurrence of multiple isoenzymes, the fact that they are differently regulated, and their apparently different localization within the cell all imply that there must be significantly different motile activities in which they function. Additionally, the apparent inability of myosin IA and IB to form bipolar filaments raises the possibility that structural organizations other than interacting, sliding filaments may be responsible for some of these motile activities.

III. Methods

A. Cell Cultures

Amoebas are grown in one of three ways: aerated 25-ml cultures in 2 × 20 cm screw-cap test tubes, 1-liter cultures in rotating 2.5-liter low-form culture flasks, and aerated 15-liter cultures in 5-gallon carboys. Medium is prepared by mixing 15 gm of glucose and 15 gm of Difco proteose peptone with 25 ml each of D,L-methionine (12 gm/liter) and KH_2PO_4 (160 gm/liter), and 5 ml each of $CaCl_2$ (3 gm/liter), $FeCl_3$ (320 mg/liter), $MgCl_2$ (24 gm/liter), thiamine (2 gm/liter), biotin (0.4 gm/liter), and vitamin B_{12} (20 mg/liter) in a total volume of 1 liter. Medium is autoclaved for 15 minutes in the test tubes, 20 minutes in the low-form culture flasks, and 1 hour in the carboys.

For purposes of aeration, a hole is drilled in the test tube caps and a glass tube fixed in place so that it will reach to the bottom of the tube. The other end is fitted with a short piece of rubber tubing and connected to a drying tube filled with cotton. The tubing is closed with a screw clamp during autoclaving. In use, air enters through the filter and escapes from around the loosened cap. The carboys are similarly fitted with a rubber stopper (clamped in place), through which a glass tube with an attached fritted-glass filter extends to the bottom of the carboy (this tube is clamped shut during autoclaving) and a second glass tube that extends only into the air space at the top of the carboy (this tube is left open during autoclaving to allow for volume change). Air enters the carboys through a cotton filter attached to the first tube and leaves through a cotton filter attached to the second tube. Simple aquarium pumps provide sufficient pressure, which can be regulated through a bleeder valve. Normally, six test tubes or two carboys are attached to one pump. It is necessary to add a small amount of Dow antifoam A to the carboys so as to prevent excessive foaming, and to pass the air through a water trap before it enters the 25-ml cultures so as to avoid excessive evaporation of the medium.

Normally, 0.2 to 0.4 ml of inoculum is transferred twice weekly from one 25-ml culture to another. Inocula of any size can be used to inoculate the 1-liter flasks and, normally, one flask is used to inoculate a carboy. Generation time is about 8 hours and about 15 gm (2×10^9) of cells are obtained from 1 liter of medium. The optimal temperature for growth is about 28°–30°C; below this temperature they will grow much more slowly and at higher temperatures the cells will encyst or die.

For all of the purposes to be described in this paper, the cells are harvested by low-speed centrifugation (about 1000–3000 rpm), washed twice in the appropriate buffer, suspended in 2 ml of buffer per gram of cells, and ruptured by release of the suspension from a Parr bomb equilibrated with nitrogen at 400 psi.

B. Actin

1. Assay

The simplest method for assaying for the presence of actin is to convert it to polymeric F-actin by adding 0.1 *M* KCl and 2 m*M* $MgCl_2$, and examining for birefringence by gently stirring the solution and viewing through crossed polaroids. Quantitative assays require scanning of stained SDS–polyacrylamide gels.

2. Purification (Gordon et al., 1976)

Approximately 100 gm of cells are washed in 10 m*M* imidazole chloride (pH 7.5), and suspended and ruptured in 3 m*M* imidazole containing 0.1 m*M* $CaCl_2$–0.5 m*M* ATP–0.75 m*M* β-mercaptoethanol and adjusted to pH 7.5. The homogenate is centrifuged at 100,000 *g* for 90 minutes and the pellet is discarded. About 200–300 ml of extract is obtained, which has a protein concentration of about 12 to 16 mg/ml, of which about 15% is actin. This extract is then fractionated on a column of DEAE–cellulose to remove proteins that interfere with actin polymerization.

Approximately 300 ml (wet volume) of Whatman DE-52 is washed with 1 *N* NaOH, 1 *N* HCl, and water, and then suspended in about 200 ml of buffer containing 10 m*M* imidazole–0.1 m*M* $CaCl_2$–0.5 m*M* ATP–0.75 m*M* β-mercaptoethanol (pH 7.5). ATP is added until its concentration (calculated from the A_{260}) is 0.5 m*M* at equilibrium. The pH is adjusted to 7.5 before and after addition of ATP. This requires about 1.5 gm of ATP and it takes about 15 minutes to reach equilibrium. The DEAE–cellulose is then packed into a 2.5 × 45 cm column and equilibrated by flowing through 500 ml of the same buffer in which the DEAE–cellulose was originally suspended. The amoeba extract (200–300 ml) is applied to this column and washed in with about 50 ml of the extraction buffer. The column is then washed with 200–300 ml of the equilibrating buffer and eluted with 2 liters of a linear gradient from 0.1 to 0.5 *M* KCl in the same buffer. Flow rate is maintained at about 50 ml/hour under gravity. Protein elution is monitored at A_{290} (to avoid interference by the absorbance of ATP), and the KCl concentration is monitored by measuring conductance. Actin is eluted between approximately 0.19 and 0.24 *M* KCl. All fractions are adjusted to 2 m*M* $MgCl_2$, warmed to room temperature, and those fractions that contain F-actin detectable by flow birefringence under crossed polaroids are pooled. The pooled fraction contains about 15% of the applied protein and is about 40% actin.

The polymerized F-actin is centrifuged at 100,000 *g* for 13.5 hours at 20°C, and the clear gelatinous pellets are homogenized in 20 to 30 ml of the original extracting buffer to which 0.02% NaN_3 is added to prevent growth of bacteria. Actin is depolymerized by dialysis for 60 to 80 hours against two to three changes of the same buffer, and material that has not depolymerized is removed by centrifugation at 100,000 *g* for 90 minutes. The supernatant solution contains about 5% of the original protein and is about 70% actin.

About 20 ml of the solution of depolymerized actin is applied to a column of Sephadex G-150 (2.5 × 70 cm) equilibrated and eluted with the same buffer against which the actin was dialyzed. About 100 mg (i.e., about 3% of the original protein) is obtained as about 99% pure actin. This material can be concentrated by ultrafiltration on an Amicon PM-10 membrane (or by other methods) and polymerized by adding 0.1 *M* KCl or 2 m*M* $MgCl_2$, or both.

This purification procedure is equally applicable to the isolation of actin from other nonmuscle cells, for example human platelets and embryonic chick brain (Gordon *et al.*, 1977). Its advantages over some other procedure is that chromatography on DEAE–cellulose immediately separates the actin from several other proteins that, on the one hand, interact with G-actin and prevent its polymerization, and, on the other hand, interact with F-actin and retard its depolymerization. The concentrating feature of anion-exchange chromatography also facilitates the processing of large volumes of crude extracts.

C. Profilin

1. Assay

Profilin has the property of inhibiting the rate of polymerization of actin apparently by inhibiting the nucleation step. The most direct assay would be to test the ability of protein fractions to inhibit the increase in viscosity when the ionic strength of a solution of G-actin is raised by addition of 0.1 *M* KCl and 2 m*M* $MgCl_2$. Unfortunately, this is not possible with crude extracts or partially purified fractions, because of the presence of contaminating actin, but mostly because of the presence of gelactins, which interact with any F-actin that is formed and increase the viscosity. Therefore, only qualitative evidence for the presence of profilin can be obtained until the material is highly purified.

2. Purification (Reichstein and Korn, 1979)

Cells are extracted exactly as described for the isolation of actin and the extract is applied to an identical column of DEAE–cellulose, which is eluted in the same way. Profilin is contained in the unbound fraction of protein that elutes before the gradient of KCl is begun. Proteolysis can be inhibited by the addition of 0.5 m*M* phenylmethylsulfonyl fluoride to all buffers. The DEAE-unbound fraction is fractionally precipitated with ammonium sulfate. The material recovered between 1.2 and 2.4 *M* is collected and dissolved in and dialyzed against 5 m*M* potassium phosphate–0.5 m*M* dithiothreitol (pH 7.5). The dialyzed protein is then applied to a 20-ml column of hydroxyapatite equilibrated in and eluted by the same buffer (flow rate of 7 to 15 ml/hour). The unbound protein (detected by A_{280}) is pooled, concentrated to 6 ml by dialysis against solid sucrose, and applied to a column of Sephadex G-75 (80 ml), which is also equilibrated with and eluted by 5 m*M* potassium phosphate–0.5 m*M* dithiothreitol (pH 7.5). The central portion of the peak is pooled and concentrated by dialysis against solid sucrose to a protein concentration of at least 2 mg/ml. The purified profilin can be assayed for its ability to inhibit the polymerization of actin, but the assay is interfered with at earlier stages of purification. Approximately 25 mg of pure profilin is recovered from about 10 gm (wet weight) of cells.

D. Gelation Factors or Gelactins I, II, III, and IV

1. Assay (Maruta and Korn, 1977a)

The gelactins can be assayed in any of several ways, each of which is a measure of their interaction with F-actin to form higher-ordered structures. None of the assays is rigorously quantitative, and thus all are less than satisfactory for comparison of the four gelactins or of the gelactins with proteins from other sources that similarly seem to cross-link filaments of F-actin. In the gelation assay, fractions to be assayed are incubated at 30°C, in glass tubes with inside diameters of 7.5 mm, with F-actin (2 mg/ml) in a total volume of 0.4 ml of 15 m*M* imidazole chloride (pH 8.0)–1 m*M* EGTA–1 m*M* dithiothreitol–1 m*M* ATP–5 m*M* $MgCl_2$. The tubes are gently inverted at intervals of 30 seconds to determine the time required to form a solid gel. The time required is a nonlinear function of the concentration of the gelactin. In the viscosity assay, fractions are incubated for 15 minutes at 30°C with actin at 1.5 mg/ml in the same buffer as in the gelation assay, and the viscosities are then measured in conventional Cannon–Manning semimicro viscometers. The concentration of gelactin necessary to double the viscosity is determined. Finally, the ability of the gelactin fraction to inhibit the actin-activated Mg^{2+}-ATPase of myosin can be measured. The relative order of activities of the four gelatins are different in the three assays.

2. Purification (Marata and Korn, 1977a)

Highly purified, but not homogeneous, gelactins can be purified as follows. About 10 gm of cells are washed in 0.1 *M* NaCl and ruptured in two volumes of 15 m*M* imidazole chloride–1 m*M* dithiothreitol–1 m*M* EGTA containing 0.34 *M* sucrose. The supernatant solution after centrifugation at 100,000 *g* for 90 minutes is fractionated on a DEAE–cellulose column (5 × 25 cm), with a 0 to 0.15 *M* KCl gradient in the same buffer but without sucrose. Most of the gelation activity (i.e., the ability to increase the viscosity, or gel, a solution of F-actin) is recovered in two fractions: Gelactins I and III are eluted together before the gradient is started; gelactins II and IV are eluted together between 0.03 and 0.06 *M* KCl. The two active fractions are concentrated by dialysis against solid sucrose, then dialyzed against the original extracting buffer without sucrose. They are then separately fractionated on hydroxyapatite columns (1.6 × 2.5 cm) with a 0 to 0.12 *M* KH_2PO_4 gradient in the extracting buffer containing 0.1 *M* KCl. From the one column, gelactins I and III elute at about 0.06 and 0.08 *M* KH_2PO_4, respectively, whereas gelactins II and IV elute from the other column at about 0.04 and 0.08 *M* KH_2PO_4, respectively. The four partially purified gelactins are concentrated by dialysis against solid sucrose, dialyzed against 5 m*M* imidazole (pH 7.0)–1 m*M* dithiothreitol–0.5 *M* KCl, and separately chromatographed on a column of Bio-Gel A-1.5m (2 × 100 cm) equilibrated and eluted with the same solution. Approximately 5 mg of gelactin II and 1 mg each of gelactins I, III, and IV will be obtained. The approximate concentrations required to form a gel in 1 minute when added to F-actin (2 mg/ml) are as follows: I, 7 μg/ml; II, 25 μg/ml; III, 7 μg/ml; IV, 3 μg/ml. This procedure has not been carried out many times and its reproducibility is not known.

E. Myosins IA, IB, and IC

1. Assays (Pollard and Korn, 1973b)

Myosins are characterized by their ATPase activities under three different ionic conditions: (1) K^+ and EDTA; (2) Ca^{2+}; and (3) Mg^{2+} with and without the addition of F-actin. To monitor the purification of the myosin I isoenzymes, the (K^+, EDTA)-ATPase assay is used because it is less interfered with by other proteins and ions, and because under these assay conditions the myosin I isoenzymes have their highest specific activities. (K^+, EDTA)-ATPase activity is measured in buffer containing 15 m*M* imidazole chloride (pH 7.5)–0.5 *M* KCl–2 m*M* EDTA; Ca^{2+}-ATPase activity is measured in 15 m*M* imidazole chloride (pH 7.5)–1 m*M* EGTA–10 m*M* $CaCl_2$; Mg^{2+}-ATPase activity is measured in 15 m*M* imidazole chloride (pH 7.5)–1 m*M* EGTA–2 m*M* $MgCl_2$, with and without excess F-actin (usually 0.05 to 1 mg/ml). All solutions contain 2 m*M* [γ-^{32}P]ATP in a total volume of 1 ml and are incubated at either 25° or 30°C. The reaction is stopped by adding 2 ml of 1:1 2-butanol:benzene and 0.5 ml of 4% silicotungstic acid in 3 *N* H_2SO_4 and vortexing for several seconds. Then 0.2 ml of 10% ammonium molybdate is added and the vortexing is continued for 15 seconds. After phase separation has occurred, 0.2 ml of the upper, organic phase, which contains the phosphomolybdate complex, is added to an appropriate scintillation cocktail and its radioactivity measured. The rate of formation of inorganic phosphate can also be measured colorimetrically, without the need for radioactive ATP, by transferring 1 ml of the organic phase to a tube containing 2 ml of 0.73 *N* H_2SO_4 in ethanol followed by 0.1 ml of a solution prepared by mixing 1 ml of 10% stannous chloride in 12 *N* HCl with 25 ml of 1 *N* H_2SO_4. The blue color is quantified by its A_{720}. The radioactive assay is much simpler for routine use.

2. Purification (Maruta et al., 1979)

Approximately 1 kg of cells is extracted with 2 volumes of buffer containing 75 m*M* KCl–12 m*M* sodium pyrophosphate–5 m*M* dithiothreitol–30 m*M* imidazole chloride (pH 7.0). The supernatant obtained after centrifugation at 100,000 *g* for 3 hours is adjusted to pH 8.0 with Tris and dialyzed overnight against 10 volumes of 7.5 m*M* sodium pyrophosphate–0.5 m*M* dithiothreitol–10 m*M* Tris chloride (pH 8.0). The solution is then applied to a DEAE–cellulose column (5 × 80 cm), previously equilibrated with 10 m*M* KCl–1 m*M* dithiothreitol–10 m*M* Tris chloride (pH 8.0), and eluted with a linear gradient of KCl between 10 and 250 m*M* in 4.5 liters of the same buffer. Fractions are analyzed for (K^+, EDTA)-ATPase activity, and the peak of activity that is eluted at about 10 m*M* KCl is pooled. This contains a mixture of myosins IA, IB, and IC but is entirely free of myosin II, which elutes later from the column. All of the ATPase activity is concentrated by adding ammonium sulfate to 2 *M* and collecting the precipitate and dissolving it in 40 ml of a solution containing 15 m*M* imidazole chloride (pH 7.5)–1 m*M* EGTA–1 m*M* dithiothreitol–0.34 *M* sucrose. The solution is dialyzed overnight against the same buffer and fractionated by addition of ammonium sulfate. The precipitate obtained between 0.9 and 1.5 *M* ammonium sulfate is dissolved in 20 ml of the buffer just described and dialyzed overnight against the same buffer.

Myosin IC can be separated from a mixture of myosins IA and IB by affinity chromatography on ADP–agarose. The dialyzed solution containing myosins IA, IB, and IC is applied to an ADP–agarose affinity column (1.6 × 25 cm) equilibrated with 15 m*M* Tris chloride (pH 7.5)–50 m*M* KCl–1 m*M* dithiothreitol. The column is washed with 100 ml of the equilibrating buffer and myosin IC is eluted with 150 ml of equilibrating buffer adjusted to 200 m*M* KCl and 2 m*M* EDTA. This fraction will also contain myosin I heavy-chain kinase activity.

A mixture of myosins IA and IB is then eluted with the same buffer adjusted to 1.0 *M* KCl and 2 m*M* EDTA. This separation may be better after the ADP–agarose column has been used several times. The fractions containing myosin IC and the mixture of myosins IA and IB are separately dialyzed overnight against a buffer of 15 m*M* imidazole chloride (pH 7.5)–1 m*M* dithiothreitol–42 m*M* KCl–0.34 *M* sucrose.

The final purification of myosins IA and IB and their separation from each other is obtained by adsorption chromatography on Bio-Gel A-1.5m followed by chromatography on hydroxyapatite. The dialyzed mixture of myosins IA and IB obtained from the ADP–agarose column is adsorbed onto a column of Bio-Gel A-1.5m (200–400 mesh; 1.5 × 85 cm), equilibrated with 10 m*M* imidazole chloride (pH 7.5)–42 m*M* KCl–1 m*M* dithiothreitol. The column is then eluted with a linear gradient of KCl between 42 and 300 m*M* in 200 ml of the same buffer. Myosin IB is eluted at about 60 m*M* KCl and myosin IA is eluted at about 120 m*M* KCl. Each of these fractions is then applied separately to a column of hydroxyapatite (2.5 × 25 cm) equilibrated with 10 m*M* Tris chloride (pH 7.5)–1 m*M* dithiothreitol–0.5 *M* KCl. The column is washed with the same buffer and then eluted with 200 ml of a gradient of potassium phosphate, pH 7.5 (0 to 0.2 *M*) at a flow rate of 10 ml/hour. Myosin IA is eluted at about 80 m*M* potassium phosphate and myosin IB is eluted at about 160 m*M* potassium phosphate. The column can be regenerated for further use by washing it with 0.4 *M* potassium phosphate in the same buffer. Both myosin IA and IB are about 90% pure as judged electrophoretically.

Myosin IC is further purified, following its separation from myosins IA and IB on ADP–agarose, by chromatography on hydroxyapatite exactly as just described for the separation of myosin IA and IB. Myosin IC is eluted at about 125 m*M* potassium phosphate. Myosin IC will be about 70% pure. All additional efforts to increase the purity of myosin IC have failed because they have led to the loss of the 20,000-dalton light chain from myosin IC, which thereby converts it to myosin IA.

Purified in this way, the yields of the myosin I isoenzymes from 1 kg (wet weight) of amoebas are about 5 mg of myosin IA, 10 mg of myosin IB, and 15 mg of partially purified myosin IC. However, since myosin IC is identical to myosin IA, except for the presence of an additional 20,000-dalton peptide (Maruta *et al.*, 1979; Gadasi *et al.*, 1979), for many purposes it is desirable to increase the yield of myosin IA at the expense of myosin IC. This can be done by a slight change in the purification steps just described. Instead of eluting myosin IC separately from myosins IA and IB on the ADP–agarose column, a mixture of all three enzymes is eluted by omitting the elution step with 200 m*M* KCl and proceeding immediately to 1.0 *M* KCl. The mixture of myosins IA, IB, and IC is then applied to the hydroxyapatite column rather than the Bio-Gel column. When all three enzymes are present, elution of the hydroxyapatite column in the way described previously will result in one broad peak that contains myosins IA, IB, and IC. The three enzymes are usually too incompletely resolved to justify separating them into different fractions. The mixture of the three enzymes is then applied to the Bio-Gel A-1.5m column, which is eluted in the usual way. On this column, the 20,000-dalton peptide is removed from myosin IC and only two peaks are obtained: about 10 mg of myosin IB and about 15 mg of myosin IA, each purified to about 95%.

The detailed characterization of myosins IA, IB, and IC is presented in the several papers that have already been referenced. It may be useful here, however, to record the specific enzymatic activities that are usually obtained for the highly purified enzymes. Myosin IA has a (K^+, EDTA)-ATPase activity of 3 $\mu mol \cdot min^{-1} \cdot mg^{-1}$; a Ca^{2+}-ATPase activity of 1.1 $\mu mol \cdot min^{-1} \cdot mg^{-1}$; a Mg^{2+}-ATPase activity, in the absence of F-actin, of 0.11 $\mu mol \cdot min^{-1} \cdot mg^{-1}$; and an actin-activated Mg^{2+}-ATPase activity, when phosphorylated, of 1.7 $\mu mol \cdot min^{-1} \cdot mg^{-1}$. The corresponding values for myosin IB are 4, 1.2, 0.1, and 2.4 $\mu mol \cdot min^{-1} \cdot mg^{-1}$; and for myosin IC, which is less pure, 1.1, 0.02, 0.01, and 0.5 $\mu mol \cdot min^{-1} \cdot mg^{-1}$, respectively.

F. Myosin I Heavy-Chain Kinase

1. Assay

This protein was originally referred to as "cofactor protein" (Pollard and Korn, 1973c) because its presence in the assay mixture was required in order to demonstrate actin activation of the Mg^{2+}-ATPase activity of the original preparations of myosin IA. Later work (Maruta and Korn, 1977c) demonstrated that the protein is actually a kinase that specifically phosphorylates the heavy chains of myosin IA, IB, and IC, and that the phosphorylated myosins have actin-activated Mg^{2+}-ATPase activities in the absence of the kinase. The enzyme, therefore, can be assayed by its ability to phosphorylate myosin IA or IB using [γ-^{32}P]ATP, but it is simpler merely to assay the actin-activated Mg^{2+}-ATPase activites of myosin IA or IB in the presence and absence of the kinase.

2. Purification (Marata and Korn, 1977c)

The enzyme has not yet been purified to homogeneity, but it can be obtained free of interfering substances as a by-product of the purification of myosin IC.

The fraction eluted from ADP–agarose that contains myosin IC (200 m*M* KCl) also contains the kinase. The two are separated on hydroxyapatite by the procedure just described for the preparation of myosin IC. the kinase is eluted at about 40 m*M* potassium phosphate, whereas myosin IC is not eluted until a concentration of about 125 m*M* potassium phosphate is reached. Of course, if it is desired to isolate myosin I heavy-chain kinase the modified procedure suggested for eluting myosins IA, IB, and IC together in one peak should not be used. It should be possible, however, to elute the kinase with a solution containing 40 m*M* potassium phosphate and then to elute the myosin isoenzymes together by proceeding directly to 1 *M* potassium phosphate. Kinase activity cannot be assayed in the crude amoeba extract but it is about 60-fold purified beginning from the DEAE–cellulose fraction. Assuming it was recovered with no loss when the extract was separated by chromatography on DEAE–cellulose, the purification of the kinase would be about 1000-fold from the crude extract. Approximately 1 μg/ml of the partially purified kinase is sufficient to phosphorylate completely (1 mol/mol of heavy chain) about 200 μg/ml of myosin IA, IB, or IC. It has no activity toward myosin II.

G. Myosin II

1. Assay

In contrast to the myosins I, myosin II is more active as a Ca^{2+}-ATPase than as a (K^+, EDTA)-ATPase, and therefore, its activity should be monitored in that way during its purification (Section III, E, 1). The Ca^{2+}-ATPase activity is inhibited by pyrophosphate and so this anion must be removed by dialysis to obtain quantitatively meaningful results.

2. Purification

Two published procedures exist for the purification of myosin II. The first method that was described (Maruta and Korn, 1977b) separates the high-molecular-weight myosin II (about 400,000 daltons) from most other proteins in the amoeba extract by molecular-seive chromatography, removes contaminating RNA by incubation with RNase and DEAE chromatography, and obtains final purification by affinity chromatography on ADP–agarose. About 10 mg of highly purified enzyme is obtained from about 200 gm of cells. The second procedure (Pollard *et al.*, 1978) utilizes DEAE–cellulose chromatography in a procedure identical to that described in Section III, B, 2, for the purification of actin from *Acanthamoeba*. Myosin II, which elutes with the front half of the actin peak, is precipitated as an actomyosin complex, by dialysis against 0.1 *M* KCl, dissolved in 0.6 *M* KI, and isolated by chromatography on agarose at that stage. About 20 mg (twice the yield of the procedure of Maruta and Korn) of at least equally pure myosin II is obtained from 200 gm of cells. The enzymes prepared by both procedures have subunit compositions and enzymatic activities identical to a Ca^{2+}-ATPase, Mg^{2+}-ATPase, and (K^+, EDTA)-ATPase. Disappointingly, neither preparation shows more than 1.5- to 2-fold activation of its Mg^{2+}-ATPase activity by F-actin, nor has it yet been found possible to convert either enzyme to an actin-activatable form.

Recently, we (Collins and Korn, 1980) have developed a third purification procedure that produces good yields of a very highly purified myosin II that is physically indistinguishable from the myosin II obtained by either of the published procedures. This enzyme has the same Ca^{2+}-ATPase, (K^+, EDTA)-ATPase, and Mg^{2+}-ATPase activities as the other two preparations, but—for reasons that are still unclear—its Mg^{2+}-ATPase activity can be at least 30-fold activated by F-actin. The procedure that results in the actin-activatable myosin II is described here.

The cells (150 gm) are washed twice with 10 m*M* triethanolamine hydrochloride (pH 7.5) in 0.1 *M* NaCl, and homogenized in two volumes of buffer (pH 7.5) containing 10 m*M* triethanolamine hydrochloride, 40 m*M* sodium pyrophosphate, 30% sucrose, 0.4 m*M* dithiothreitol, and the following inhibitors of proteolysis: 0.5 m*M* phenylmethylsulfonyl fluoride, pepstatin A (10 μg/ml), and leupeptin (0.8 μg/ml). The extract obtained after centrifuging for 3 hours at 100,000 *g* is dialyzed for 24 hours against two changes of 15 liters each of 10 m*M* triethanolamine hydrochloride–0.1 *M* KCl–1 m*M* EDTA–1 m*M* dithiothreitol (pH 7.5), to allow actomyosin to form. The actomyosin pellet is collected by centrifugation for 1 hour at 100,000 *g* and is homogenized in 75 ml of buffer containing 3 m*M* triethanolamine hydrochloride–0.1 m*M* ATP–0.1 m*M* $CaCl_2$–1 m*M* dithiothreitol (pH 7.5), and dialyzed for 20 hours against two changes of 1 liter each of the same buffer. This procedure allows the excess F-actin to depolymerize to G-actin. The myosin-enriched actomyosin pellet is recovered by centrifugation for 1 hour at 100,000 *g* and homogenized in 5 ml of buffer containing 10 m*M* TES–20 m*M* sodium pyrophosphate–5% sucrose–1 m*M* dithiothreitol–0.5 m*M* ATP (pH 7.5); the suspension is left on ice for 2 hours, and then mixed with an equal volume of a solution containing 1.2 *M* KI–10 m*M* $MgCl_2$–10 m*M* ATP–2 m*M* $CaCl_2$–1 m*M* EDTA–1 m*M* dithiothreitol (pH 7.5), to dissolve the actomyosin. The solution is immediately centrifuged for 30 minutes at 200,000 *g* and the supernatant is immediately applied to a column of Bio-Gel A-15m (200–400 mesh, 2.5 × 90 cm), previously equilibrated with 10 m*M* TES–20 m*M* sodium pyrophosphate–5% sucorse–1 m*M* dithiothreitol–0.5 m*M* ATP (pH 7.5). The column is eluted with the same buffer and the Ca^{2+}-ATPase peak that elutes just after the void volume of the column is pooled, adjusted to 40 m*M* in sodium pyrophosphate, and applied to a column of DEAE-cellulose (5 ml in volume) equilibrated with 10 mM TES–40 mM sodium pyrophosphate–5% sucrose–1 mM dithiothreitol (pH 7.5). After washing the column with the same buffer, myosin II is eluted with the same buffer containing 0.2 M KCl. This step removed RNA, which remains on the DEAE-cellulose. The myosin II is then dialyzed for 48 hours against two changes of 1 liter of 10 mM imidazole chloride (pH 7.5)–0.1 M KCl–1 mM dithiothreitol, to remove the pyrophosphate, and concentrated by dialysis for 12 hours against solid sucrose. About 5 to 10 mg of highly purified myosin II is obtained. It has the following specific enzymatic activities: (K^+, EDTA)-ATPase, 0.15–0.2 $\mu mol \cdot min^{-1} \cdot mg^{-1}$; Ca^{2+}-ATPase, 0.5–0.7 $\mu mol \cdot min^{-1} \cdot mg^{-}$; Mg^{2+}-ATPase, without F-actin, <0.01 $\mu mol \cdot min^{-1} \cdot mg^{-1}$; actin-activated Mg^{2+}-ATPase, 0.15–0.35 $\mu mol \cdot min^{-1} \cdot mg^{-1}$ (as high as 0.88 $\mu mol \cdot min^{-1} \cdot mg^{-1}$ when dephosphorylated).

IV. Conclusion

To the biochemist it is axiomatic that our understanding of cell biology depends on, and will be limited by, our understanding of molecular events, and that such biochemical information depends, in the last analysis, on the detailed study of highly purified molecules. This is no less true for cell motility than, for example, for protein synthesis or energy metabolism. We will understand actomyosin-dependent cell motility only after we have defined all of the proteins involved and the nature of their interactions, for no model of a motile process can be correct that is incompatible with the biochemical properties of the proteins involved. Purification procedures are necessary tools in this process. As they continue to improve, as better sources for some of the proteins become available, and as the full complement of components in each system is defined, progress in our understanding of the biological problems will accelerate exponentially. For the moment, the discovery of multiple actins in some cells, of multiple myosins in other cells, and of a multiplicity of interacting proteins in many cells has expanded the range of questions more than it has provided answers. But possibly some of the broader outlines can be discerned.

Clearly, the state of actin polymerization in the cell will be regulated not only by the inherent properties of the actin molecule, but also by the interaction of monomeric actin with one or more proteins that greatly modify the polymerization process and the interaction of the actin filaments with other proteins that will similarly affect the steady-state situation. Actin filaments themselves will be interconnected by yet other proteins into still more organized structures. Actin filaments will interact with one or more myosin molecules in ways that will be regulated at least in part by the state of phosphorylation of the myosins, which will, in turn, be a function of the relative rates of phosphorylation and dephosphorylation by kinases and phosphatases. The enzymatic activities of these latter enzymes will also be subjected to regulation. The actomyosin system must also be integrated into the total cell organization through specific associations with at least plasma membranes and probably other organelles. Many of the details of these interactions have already been described for the erythrocyte cytoskeleton. Just a few years ago none of these statements would have had any experimental basis. The considerable progress in the last 10 years bodes well for the future.

REFERENCES

Adelman, M. R., and Taylor, E. W. (1969). *Biochemistry* **8,** 4976–4988.
Bowers, B. (1977). *Exp. Cell Res.* **110,** 409–417.
Bowers, B., and Korn, E. D. (1968). *J. Cell Biol.* **39,** 95–111.
Bowers, B., and Korn, E. D. (1969). *J. Cell Biol.* **41,** 786–805.
Bowers, B., and Olszewski, T. E. (1972). *J. Cell Biol.* **53,** 681–694.
Brenner, S. L., and Korn, E. D. (1979). *J. Biol. Chem.* **254,** 9982–9985.
Brenner, S. L., and Korn, E. D. (1980a). *J. Biol. Chem.* **255,** 841–844.
Brenner, S. L., and Korn, E. D. (1980b). *J. Biol. Chem.* **255,** 1670–1676.
Byers, T. J. (1979). *Int. Rev. Cytol.* **61,** 283–338.
Collins, J. H., and Korn, E. D. (1980). *J. Biol. Chem.* **255,** 8011–8014.
Elzinga, M., and Lu, R. C. (1976). *In* "Contractile Systems in Non-Muscle Tissues" (J. V. Perry, A. Margreth, and R. S. Adelstein, eds.), pp. 29–37. North-Holland Publ., Amsterdam.
Elzinga, M., Collins, J. H., Kuehl, W. M., and Adelstein, R. S. (1973). *Proc. Natl. Acad. Sci. U.S.A.* **70,** 2687–2691.
Gadasi, H., and Korn, E. D. (1979). *J. Biol. Chem.* **254,** 8095–8098.
Gadasi, H., and Korn, E. D. (1980). *Nature (London)* **286,** 452–456.
Gadasi, H., Maruta, H., Collins, J. H., and Korn, E. D. (1979). *J. Biol. Chem.* **254,** 3631–3636.
Gordon, D., Eisenberg, E., and Korn, E. D. (1976). *J. Biol. Chem.* **251,** 4778–4786.
Gordon, D., Boyer, J. L., and Korn, E. D. (1977). *J. Biol. Chem.* **252,** 8300–8309.
Hatano, S., and Oosawa, F. (1966). *Biochim. Biophys. Acta* **127,** 488–498.
Korn, E. D. (1978). *Proc. Natl. Acad. Sci. U.S.A.* **75,** 588–599.
Korn, E. D., and Weisman, R. A. (1967). *J. Cell Biol.* **34,** 219–227.
Korn, E. D., and Wright, P. L. (1973). *J. Biol. Chem.* **248,** 439–447.
Korn, E. D., Bowers, B., Batzri, S., Simmons, S. R., and Victoria, E. J. (1974). *J. Supramol. Struct.* **2,** 517–528.
Laemmli, U. K. (1970). *Nature (London)* **227,** 680–685.
Maruta, H., and Korn, E. D. (1977a). *J. Biol. Chem.* **252,** 399–402.
Maruta, H., and Korn, E. D. (1977b). *J. Biol. Chem.* **252,** 6501–6509.
Maruta, H., and Korn, E. D. (1977c). *J. Biol. Chem.* **252,** 8329–8332.
Maruta, H., and Korn, E. D. (1981a). *J. Biol. Chem.* **256,** 499–502.
Maruta, H., and Korn, E. D. (1981b). *J. Biol. Chem.* **256,** 503–506.
Maruta, H., Gadasi, H., Collins, J. H., and Korn, E. D. (1978). *J. Biol. Chem.* **253,** 6297–6300.
Maruta, H., Gadasi, H., Collins, J. H., and Korn, E. D. (1979). *J. Biol. Chem.* **254,** 3624–3630.
Oosawa, F., and Asakura, S. (1975). "Thermodynamics of Protein Polymerization." Academic Press, New York.
Pollard, T. D., and Korn, E. D. (1973a). *J. Biol. Chem.* **248,** 448–450.

Pollard, T. D., and Korn, E. D. (1973b). *J. Biol. Chem.* **248,** 4682–4690.
Pollard, T. D., and Korn, E. D. (1973c). *J. Biol. Chem.* **248,** 4691–4697.
Pollard, T. D., Stafford, W. F., and Porter, M. E. (1978). *J. Biol. Chem.* **253,** 4798–4808.
Reichstein, E., and Korn, E. D. (1979). *J. Biol. Chem.* **254,** 6174–6179.
Rubenstein, P., and Deuchler, J. (1979). *J. Biol. Chem.* **254,** 11142–11147.
Ulsamer, A. G., Smith, F. R., and Korn, E. D. (1969). *J. Cell Biol.* **43,** 105–114.
Ulsamer, A. G., Wright, P. L., Wetzel, M. G., and Korn, E. D. (1971). *J. Cell Biol.* **51,** 193–215.
Uyemura, D. G., Brown, S. S., and Spudich, J. A. (1978). *J. Biol. Chem.* **253,** 9088–9096.
Vandekerckhove, J., and Weber, K. (1978a). *Nature (London)* **276,** 720–721.
Vandekerckhove, J., and Weber, K. (1978b). *Eur. J. Biochem.* **90,** 451–462.
Vandekerckhove, J., and Weber, K. (1978c). *Proc. Natl. Acad. Sci. U.S.A.* **75,** 1106–1110.
Wegner, A. (1976). *J. Mol. Biol.* **108,** 139–150.
Weihing, R. R., and Korn, E. D. (1970). *Nature (London)* **227,** 1263–1264.
Weihing, R. R., and Korn, E. D. (1971). *Biochemistry* **10,** 590–600.
Weisman, R. A., and Korn, E. D. (1967). *Biochemistry* **6,** 485–497.
Wetzel, M. G., and Korn, E. D. (1969). *J. Cell Biol.* **43,** 90–104.
Woodrun, D. T., Rich, S. A., and Pollard, T. D. (1975). *J. Cell Biol.* **97,** 231–237.
Yang, Y.-Z., Gordon, D. J., Korn, E. D., and Eisenberg, E. (1977). *J. Biol. Chem.* **252,** 3374–3378.
Yang, Y.-Z., Korn, E. D., and Eisenberg, E. (1979a). *J. Biol. Chem.* **254,** 2084–2088.
Yang, Y.-Z., Korn, E. D., and Eisenberg, E. (1979b). *J. Biol. Chem.* **254,** 7137–7140.

Chapter 16

Biochemical and Immunocytological Characterization of Intermediate Filaments in Muscle Cells

ELIAS LAZARIDES

Division of Biology
California Institute of Technology
Pasadena, California

ISBN 0-12-564125-7

I. Introduction

For some years electron microscopy has identified a morphological class of filaments with an average diameter of 100 Å in the cytoplasm of many higher eukaryotic cells. These filaments are referred to as intermediate filaments, since their diameter is intermediate to those of the 60 Å actin filaments and the 250 Å microtubules in nonmuscle cells, and intermediate to those of the 60 Å actin filaments and the 150 Å myosin filaments in muscle cells. Intermediate filaments were at first widely regarded as disaggregation products of either microtubules or myosin filaments, and as a result were neglected. New biochemical and immunofluorescence data have recently established intermediate filaments as a distinct fibrous system composed of unique yet chemically heterogeneous subunits. On the basis of biochemical and immunological criteria, we can define five major classes of intermediate filaments (for a review, see Lazarides, 1980):

1. Keratin (tono) filaments, found in epithelial cells and cells of epithelial origin, and composed of a heterogeneous class of polypeptides with molecular weights between 45,000 and 68,000.
2. Desmin filaments, found predominantly in smooth, skeletal, and cardiac muscle cells, have a subunit molecular weight of 50,000.
3. Vimentin filaments, found in all cells of mesenchymal origin, have a subunit molecular weight of 52,000.
4. Neurofilaments, found in neurons. Three major polypeptides with molecular weights of 210,000, 160,000, and 68,000 consistently copurify with neurofilaments (for a review, see Lazarides, 1980). Whether indeed any one of these three polypeptides or all of them function as bona fide subunits of neurofilaments is presently unknown, since it has not been determined whether they can polymerize to intermediate filaments *in vitro*.
5. Glial filaments, found predominantly in glial cells. The major subunit of these filaments, termed the glial fibrillary acidic (GFA) protein has a subunit molecular weight of 51,000. As we will discuss later on, this classification is rather arbitrary, since two, three, or even more intermediate-filament subunits can be expressed in a given cell type.

This chapter will review our knowledge of the function, structure, expression, and cytoplasmic localization of intermediate filaments in muscle cells and during myogenesis. Furthermore, it will summarize our present-day methodology for purifying the major subunit of these filaments, desmin, from smooth muscle; the production and characterization of antibodies against this protein; and the related intermediate-filament subunit, vimentin. The methodology for the immunofluorescent localization of these proteins during myogenesis and in mature myofibrils will also be described.

II. Desmin and Vimentin in Both Muscle and Nonmuscle Cells

An indispensable tool for the unambiguous characterization of the subunits of intermediate filaments in both muscle and nonmuscle cells has been the technique of two-dimensional isoelectric focusing (IEF)/sodium dodecyl sulfate (SDS)-polyacrylamide slab gel electrophoresis (PAGE). One-dimensional SDS–PAGE is not sufficient to resolve adequately the different filament subunits, since (a) they vary only slightly in molecular weight, and (b) some of the higher-molecular-weight subunits are rapidly, but only partially, proteolyzed during extraction, generating peptides that coincide in molecular weight with other lower-molecular-weight intermediate-filament subunits. For example, vimentin is rapidly proteolyzed to a number of lower-molecular-weight peptides, one of which has the same molecular weight as desmin (50,000), a second with a molecular weight close to that of one of the keratins (45,000), and another with the molecular weight of actin (42,000). Similarly, desmin is partially degraded during extraction to a small number of lower-molecular-weight peptides, one of which has the same molecular weight as a proteolytic fragment of vimentin, and another the same molecular weight as actin. However, all these peptides and the parent molecules can be resolved clearly by two-dimensional IEF/SDS–PAGE. Immunofluorescence alone is not adequate in establishing the presence (or absence) of different intermediate-filament subunits in the cytoplasm of various cell types. Thus the absence of fluorescence using antibodies specific for a given intermediate-filament subunit in a given cell type cannot be taken as evidence for the absence of the antigen. This is, for example, the case with the presence of desmin in nonmuscle cells. Antibodies to muscle desmin do not display much cytoplasmic fluorescence, which can be clearly attributed to the presence of desmin-containing filaments (Lazarides, 1978; Gard *et al.*, 1979); yet the presence of desmin can be demonstrated using two-dimensional IEF/SDS–PAGE in some nonmuscle cells (Gard *et al.*, 1979; Tuszynski *et al.*, 1979).

A. Expression of Desmin

As determined by two-dimensional IEF/SDS–PAGE, adult chicken smooth, skeletal, and cardiac muscle desmin exists as two isoelectric variants with approximate isoelectric points of 5.70 (termed β-desmin) and 5.65 (termed

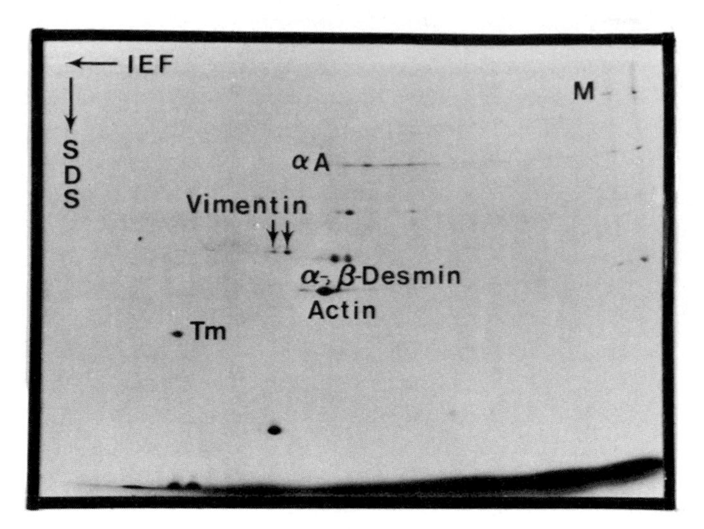

FIG. 1. Two-dimensional gel electropherogram of briefly extracted myofibrils from chicken pectoralis muscle. Isoelectric focusing (IEF) was performed from right to left (acid on the left) and SDS–polyacrylamide gel electrophoresis (SDS–PAGE) was performed from top to bottom. Major proteins include vimentin, α- and β-desmin, actin, and tropomyosin (Tm; the $'\alpha$ variant predominates in this muscle). Myosin (M) and α-actinin (αA) do not focus well in this system.

α-desmin) (Fig. 1). Mammalian-muscle desmin exists only as one major variant with a molecular weight slightly higher than that of avian desmin ($d_m = 51{,}000$; $d_a = 50{,}000$) and with a slightly more acidic pI than α-desmin (d_m pI = 5.60). In all three avian-muscle types, the two variants maintain the same charge and molecular weight (Izant and Lazarides, 1977; Lazarides and Balzer, 1978). However, comparative one-dimensional peptide mapping of total (α plus β) smooth- and skeletal-muscle desmin has revealed the presence of peptide differences in desmin within a given species and among species, indicating that although desmin is a conserved molecule, it may be coded by different genes in the different muscle types and species (Gard *et al.*, 1979; O'Shea *et al.*, 1979). The two desmin variants copurify with desmin purified from chicken gizzard through a number of cycles of polymerization–depolymerization, indicating that they are both integral components of desmin filaments and of the desmin molecule. The ratio of α- to β-desmin is variable; since α-desmin is phosphorylated (see later on), this variability is probably due to the presence of phosphatase activity in cell extracts. The presence of phosphatase inhibitors during the preparation of cell extracts greatly enhances the presence of α-desmin, suggesting that dephosphorylation of desmin occurs during cell fractionation.

Extraction of chicken embryo fibroblasts (CEF), baby hamster kidney (BHK) cells, and mouse 3T3 cells with 1% Triton X-100 and 0.6 *M* KCl leaves an insoluble cytoskeletal residue composed predominantly of intermediate filaments and nuclear remnants (Brown *et al.*, 1976). Analysis of these cytoskeletons by two-dimensional IEF/SDS–PAGE shows that they are composed primarily of the 52,000 MW subunit of intermediate filaments (vimentin). In addition, CEF cytoskeletons exhibit a minor component with MW 50,000 that coelectrophoreses and has indistinguishable peptide maps with α-desmin; β-desmin appears to be absent in these cells. The mammalian cells contain the mammalian form of desmin (Gard *et al.*, 1979). However, desmin is not present in easily detectable amounts in actively proliferating cells, such as a number of virally transformed cells, Chinese hamster ovary (CHO) cells, and Novikoff hepatoma cells (D. L. Gard, unpublished observations; Tuszynski *et al.*, 1979). It appears that the induction of desmin synthesis may be a function of the cessation of DNA replication (D. L. Gard, unpublished observations). Detailed analysis of a number of nonmuscle cells, including epithelial cells and differentiated epidermal cells as well as of the intermediate filaments that copurify with tubulin during its cyclic polymerization–depolymerization from mammalian brain, has revealed the presence of desmin. The quantity of desmin varies from as little as 20% (CEF) to as much as 90% (BHK) of the quantity of vimentin.

B. Expression of Vimentin

All cell types examined thus far contain vimentin filaments, so called because immunofluorescence studies with antibodies to the major subunit of these filaments reveal a wavy (Latin, *vimentus*) cytoplasmic filamentous system (Hynes and Destree, 1978; Franke *et al.*, 1978, 1979; Bennett *et al.*, 1978; Blose, 1979; for a review, see Lazarides, 1980). The wide cross-reactivity of vimentin antibodies with cells of mammalian, avian, and amphibian origin indicates that this protein is conserved in evolution (Franke *et al.*, 1979). Vimentin is similarly conserved by charge and electrophoretic mobility, since two-dimensional IEF/SDS–PAGE analysis of this molecule from a variety of avian and mammalian cell types, including muscle cells, shows a consistent molecular weight (52,000) and isoelectric point (pI = 5.35) (Gard *et al.*, 1979). In general, vimentin from fibroblast, myoblast, and myotube cell extracts is rapidly degraded to a family of lower-molecular-weight peptides that form a diagonal line toward the acidic end of the parent molecule (Gard *et al.*, 1979). The lowest-molecular-weight product has an apparent molecular weight of 43,000 and a pI of 5.1. The presence of these degradation products is unaffected by inclusion of *p*-tosyl-L-arginine methyl ester (TAME), phenylmethylsulfonyl fluoride (PMSF), or *O*-phenanthroline in the 9.2 *M* urea IEF sample buffer, but is reduced or eliminated by heating the samples to 100°C in 1% SDS and 0.5% β-mercaptoethanol for 3 minutes prior to the addition of urea to 9.2 *M* and subsequent IEF. Similarly, proteolysis of vimentin can be avoided if the cells are first cross-linked with the reversible cross-linker dimethyl-3,3′-dithiobispropionimidate, then lysed with 0.5% NP-40, and finally denatured and reduced with 9.2 *M* urea–0.5% β-mercaptoethanol prior to IEF. Prolonged cell manipulation and denaturation of the cytoskeleton with urea concentrations below 8 *M* can result in the complete conversion of vimentin to the 43,000-dalton polypeptide.

In chicken skeletal myotubes differentiating in tissue culture, vimentin represents approximately 2.0% of the total protein-synthetic activity of the cell and, unlike desmin, it is expressed continuously throughout myogenesis (Gard and Lazarides, 1980). Thus the synthesis of desmin and vimentin appear to be noncoordinately regulated. In adult muscle, two-dimensional IEF/SDS–PAGE analysis of purified chicken skeletal myofibrils indicates that both desmin and vimentin are components of these structures. Desmin and vimentin remain associated with the myofibril cytoskeletons after extraction of the actomyosin with 0.6 *M* KCl or 0.6 *M* KI (Granger and Lazarides, 1978, 1979; see Fig. 1).

One of the most interesting and characteristic, but as yet unexplained, properties of vimentin- or desmin-containing intermediate filaments is their aggregation into filamentous bundles in the form of a perinuclear cap or cytoplasmic ribbons in cells exposed to colcemid. Similar structures exist transiently during normal cell adhesion and spreading onto a substrate during mitosis or in nonmotile tumor cells (for a review, see Lazarides, 1980). Isolated filament caps from BHK cells contain two predominant polypeptides, vimentin and desmin, in approximately equimolar amounts (see above). Vimentin and desmin can be purified from the colcemid-induced caps by repeated cycles of disassembly–reassembly, induced by lowering and raising the ionic strength of the extracting solution (Starger and Goldman, 1977; Starger *et al.*, 1978; Zackroff and Goldman, 1979). Reassembled filaments have an average diameter of 100 Å. Similarly, vimentin copurifies with desmin through cycles of polymerization–depolymerization of the latter molecule from chicken gizzard (see later on). It is not presently known whether vimentin and desmin copolymerize *in vitro* into the same filaments or whether they polymerize into distinct filaments with common solubility properties.

Thus, in summary, both vimentin and desmin are coexpressed in a variety of cell types. In muscle cells, in particular, desmin and vimentin are expressed throughout myogenesis, and in adult muscle they are components of myofibrils.

III. Phosphorylation of Desmin and Vimentin

Desmin and vimentin are phosphorylated in both chicken muscle and nonmuscle cells (O'Connor *et al.*, 1979). $^{32}PO_4^{2-}$ is incorporated *in vivo* into α-desmin, as well as into two or three more acidic desmin variants, but is not incorporated into β-desmin. Similarly, $^{32}PO_4^{2-}$ is incorporated into a number of acidic vimentin variants, but it is not incorporated into the main vimentin variant.

Desmin and vimentin exhibit the same pattern of phosphorylation in chicken fibroblasts, chicken smooth, skeletal, and cardiac muscle cells, and embryonic chicken myotubes differentiating in tissue culture. Desmin and vimentin are also phosphorylated in mammalian muscle and nonmuscle cells. Since chicken fibroblasts apparently lack β-desmin, all the desmin variants are phosphorylated in these cells. In adult muscle cells, the main phosphorylated desmin variant is α-desmin with the more acidic desmin variants greatly diminished or absent (Gard *et al.*, 1979). From *in vivo* labeling experiments, it is evident that only a fraction of the total amount of desmin or vimentin is phosphorylated. The exact ratio has been difficult to estimate, since all extracts contain phosphatase activities that rapidly remove $^{32}PO_4^{2-}$ from desmin and vimentin. As has been noted above, it is presently unknown whether the phosphorylated forms of desmin and vimentin arise from phosphorylation of the major nonphosphorylated form of the two proteins or whether they are distinct gene products. Cytoskeletons that are enriched in intermediate filaments, prepared from myogenic cultures by extraction with 0.5% Triton X-100 and 0.6 *M* KCl, contain all of the phosphorylated and nonphosphorylated desmin and vimentin variants. Similarly, the desmin and vimentin that are associated with myofibrils (Z disks, see later on) are also phosphorylated. These observations indicate that the desmin and vimentin assembled in intermediate filaments and Z disks are phosphorylated.

The chicken muscle-kinase activities that phosphorylate desmin and vimentin *in vivo* copurify with the two major cAMP-dependent protein kinases from this tissue. Purified preparations of the chicken skeletal-muscle or beef heart catalytic subunits of the cAMP-dependent protein kinases phosphorylate purified chicken smooth-muscle desmin, or desmin and vimentin in filamentous cytoskeletons. The pattern of phosphorylation of the two proteins by the catalytic subunit is indistinguishable from that obtained for desmin and vimentin *in vivo*. One-dimensional peptide mapping of *in vivo* and *in vitro* phosphorylated desmin (and vimentin) has indicated that the two types of phosphorylated peptides are indistinguishable, attesting to the specificity of the reaction *in vitro* (O'Connor *et al.*, 1981). These experiments indicate that desmin and vimentin phosphorylation *in vivo* may be mediated by cAMP-dependent protein kinases. At present, the function of desmin or vimentin phosphorylation remains unclear.

IV. Methods

A. Purification of Desmin and Its Polymerization to Intermediate Filaments

The most convenient starting material for the purification of desmin is gizzard. The protein is purified as follows (Hubbard and Lazarides, 1979). All operations are performed at 4°C.

1. 500 gm of fresh or frozen-thawed gizzards are dissected free of mucosa and connective tissue.

2. 200 gm of muscle is homogenized in a Waring blender at low speed in 200 ml of 100 m*M* Tris HCl (pH 7.5) and 10 m*M* EDTA. The mixture is increased to 1 liter in volume with the same buffer and blended further for 45 seconds at an intermediate speed. The material is then collected at 10,000 *g* for 20 minutes.

3. Pellets are suspended in 500 ml of 10 m*M* Tris HCl (pH 7.5) and 10 m*M* EGTA, and stirred for 2–3 hours. The extract is filtered through cheesecloth and insoluble material is collected at 10,000 *g* for 20 minutes. The pellets are extracted in this buffer for 2 hours, filtered, and centrifuged. These three steps are repeated two more times. These latter two EGTA extracts contain a considerable amount of desmin along with a small number of other proteins including actin, tropomyosin, α-actinin, filamin, and the 130,000-dalton protein (vinculin) (Geiger, 1979). Since this desmin appears to be tightly complexed with actin, these extracts have not been used for the routine purification of desmin.

4. EGTA pellets are resuspended in 500 ml of 1.0 m*M* KCl–10 m*M* Tris HCl (pH 7.5)–1 m*M* EGTA. The material is stirred for 3 hours and the extract is filtered through cheesecloth to remove large gelatinous aggregates. After filtration, insoluble material is collected at 10,000 *g* for 20 minutes and the high-salt extraction process is repeated twice. The final insoluble material contains numerous intermediate filaments, occasionally attached to dense bodies, but no detectable actin or myosin filaments. One-dimensional SDS–gel electrophoresis reveals an enrichment in desmin (MW 50,000), with actin and myosin still the major contaminants.

5. The final high-salt insoluble material is washed with distilled H_2O until free of salt and is then treated in one of two ways: (a) suspended directly in 1 *M* acetic acid (Small and Sobieszek, 1977); or (b) extracted twice with equal volumes of acetone. Each acetone extraction is carried out for 2 hours at 4°C and the final residue is air dried in a fume hood. The air-dried residue is then suspended in distilled H_2O for 1 hour at room temperature, and the residue is rinsed well with distilled H_2O on a Buchner funnel. The residue is then suspended in five volumes of 1 *M* acetic acid on ice. The material from (a) or (b) is stirred at 0°C for 1 hour and centrifuged at 30,000 *g* for 15 minutes. The pellet is extracted once more, then the supernatants are pooled together and filtered through glass wool.

6. Final purification of desmin is accomplished by polymerization–depolymerization. The 1 *M* acetic acid extract is treated in one of two ways: (a) It is neutralized to pH 7.5 by the slow dropwise addition of 0.2 *M* NaOH; or (b) it is dialyzed with repeated buffer changes against 10 m*M* Tris HCl, pH 8.0. The precipitates are collected on a table-top centrifuge at 1000 *g* for 5 minutes. They are then suspended in five volumes of 1 *M* acetic acid and the process is repeated once or twice more. Desmin cycled two to three times is 90–95% pure. Actin is an occasional contaminant, varying from 0 to 4%. Desmin purified in the manner

described above is slightly proteolyzed, especially on storage, to two polypeptides with molecular weights of 47,000 and 45,000. Desmin is usually stored frozen in 1 m*M* HCl or lyophilized. Alternatively, it can be dialyzed extensively against 1 m*M* Tris HCl, pH 8.5. In this case, a small amount (< 30%) of the molecule stays in solution and it can be stored at 4°C in the presence of 10 m*M* NaN_3, or lyophilized. Desmin dialyzed against 10 m*M* Tris HCl (pH 8.0) forms numerous intermediate filaments with an average diameter of 120 Å (Fig. 2; whole mounts). More than 90% of the molecule is converted into filaments under these conditions. The rate of polymerization can be accelerated appreciably by the addition of 0.1 *M* NaCl or KCl, 5 m*M* $MgCl_2$, or 1 m*M* $CaCl_2$ in the dialysis medium. Nucleoside triphosphate also enhances the rate of polymerization. However, the molecule does not exhibit any specific triphosphate and monovalent or divalent cation requirements for polymerization.

Two-dimensional IEF/SDS–PAGE of highly purified and polymerized desmin reveals the presence of both α- and β-desmin as integral parts of the molecule. As mentioned above, the ratio of α- to β-desmin is difficult to estimate with accuracy, due to the presence of phosphatase activities in the extracts. The molecular weight of desmin using known molecular-weight standards and the Tris–glycine system of Laemmli (1970) is 50,000. Slightly higher molecular-weight values are obtained using different electrophoresis systems in SDS.

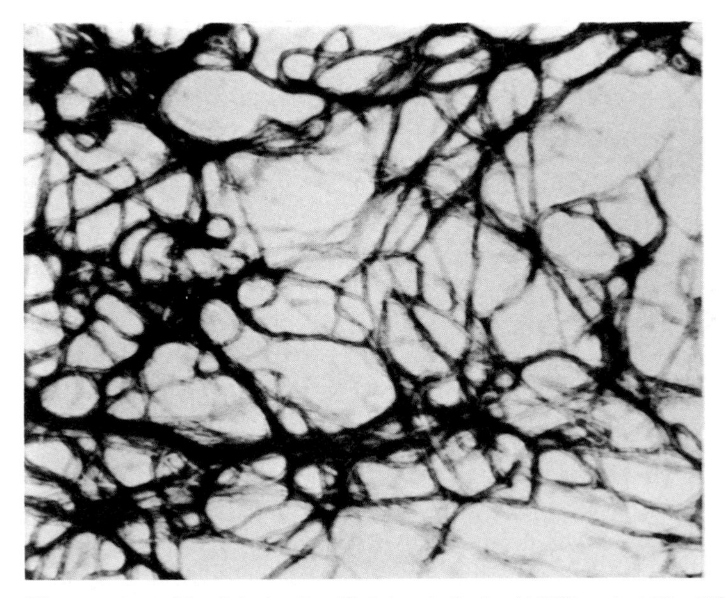

FIG. 2. Filaments formed by dialysis of purified desmin (in 1 m*M* HCl) against 10 m*M* Tris HCl (pH 8.0). The filaments measure approximately 120 Å.

Two-dimensional IEF/SDS–PAGE of high concentrations (10–20 μg) of purified desmin reveals the following:

1. A small amount of vimentin (MW 52,000) (approximately 10% of the quantity of desmin) copurifies with desmin through all the purification steps and cycles of polymerization–depolymerization. The cellular origin of this vimentin is difficult to assess. It may be a component of smooth-muscle cells, as is the case with skeletal and cardiac muscle (Granger and Lazarides, 1979; Gard and Lazarides, 1980); or it may originate in the nonmuscle cells that contaminate the gizzard tissue. However, since desmin and vimentin coexist and copurify in skeletal and cardiac muscle cells and in BHK21 cells (see above), it is likely that the vimentin present in smooth-muscle extracts is an intermediate-filament component also in smooth-muscle cells.
2. A number of more basic desmin variants are present. These variants appear to be α- and β-desmin aggregates that are not dissociated with 9 *M* urea. On two-dimensional IEF/SDS–PAGE reelectrophoresis, these variants electrophorese at the positions of α- and β-desmin.
3. A minor component with molecular weight of approximately 230,000 is present. For the copurification of vimentin and desmin from baby hamster kidney (BHK) cells and their *in vitro* polymerization properties, the reader is referred to the work of Goldman and his associates (Starger and Goldman, 1977; Starger *et al.*, 1978; Zackroff and Goldman, 1979).

B. Preparation of Antibodies to Desmin and Vimentin

As mentioned above, desmin purified by acetic acid cycling is contaminated by small amounts of vimentin, actin, and a high-molecular-weight protein. Similarly, vimentin preparations are contaminated by desmin in a number of cell types. Thus to ensure the highest purity of the antigen to be used for immunization, the two proteins are purified by preparative one-dimensional SDS–PAGE before immunization (Lazarides and Hubbard, 1976; Granger and Lazarides, 1979). Desmin, purified to near homogeneity by cycles of polymerization–depolymerization, is further purified by preparative SDS–gel electrophoresis before immunization. Vimentin is purified from the cytoskeletons of embryonic muscle. Desmin is not purified from the same material, since vimentin tends to be partially proteolyzed in these extracts to a lower-molecular-weight polypeptide, which comigrates with desmin. Vimentin to be used as an antigen is prepared from Triton–KCl cytoskeletons of embryonic chicken muscle as follows (Granger and Lazarides, 1979). Leg and breast muscles from 14- to 16-day-old embryos are dissected out and homogenized in a Dounce homogenizer in 0.1 *M* KCl–50 m*M* Tris HCl (pH 8.0)–5 m*M* EGTA–5 m*M* 2-mercaptoethanol (2-

ME)–1 m*M* *O*-phenanthroline–0.5 m*M* phenylmethylsulfonyl fluoride (PMSF). The muscle homogenate is washed (by pelleting and resuspension) twice in this buffer, filtered through cheesecloth, and extracted three times in 0.6 *M* KCl–2 m*M* $Na_4P_2O_7$–0.5% Triton X-100–20 m*M* Tris HCl (pH 7.5)–1 m*M* EGTA–10 m*M* 2-ME–0.5 m*M* *O*-phenanthroline–0.1 m*M* PMSF. The final insoluble residue is washed with water, dissolved in hot solubilization buffer (2% SDS–50 m*M* Tris HCl, pH 6.8–0.5% 2-ME–10% glycerol, bromophenol blue) for 3 minutes, and loaded onto preparative slab gels.

Bands are visualized by staining 15 minutes in 0.25% Coomassie Brilliant Blue R-250–47.5% ethanol–10% acetic acid, and destaining for 1–2 hours in 12.5% ethanol–5% acetic acid. The same procedure is followed for the final purification of desmin, which has been purified by cycles of polymerization–depolymerization. The desmin and vimentin bands are carefully excised from the two types of preparative gels (desmin from the gizzard-cycled preparation and vimentin from the embryonic-muscle cytoskeletons), and neutralized for 5–10 minutes in 0.15 *M* sodium phosphate, pH 7.4. The gel is homogenized in a motor-driven Teflon glass homogenizer, and is then coagulated by the addition of a few drops of 10% $AlCl_3$. To this material 0.5–1 volume of adjuvant is added and the mixture is emulsified by repeated passage between two syringes. Subcutaneous dorsal injections into female New Zealand white rabbits are made with complete adjuvant on days 0 and 14, and with incomplete adjuvant on days 28, 47, 65, and 106. Protein (200–400 μg) in a volume of 5–10 ml is injected at two sites each time. Blood is collected from the marginal ear vein on days 34, 40, 53, 72, 82, 94, and 112. After clot formation and contraction, the serum is clarified by centrifugation. Gamma globulins are partially purified by precipitation at 0°C with ammonium sulfate at 50% saturation, dialyzed against 150 m*M* NaCl–10 m*M* Tris HCl (pH 7.5)–10 m*M* NaN_3–1 m*M* ϵ-amino-*n*-caproic acid (to inhibit plasminogen activation), and stored at −70°C. Antiserum specificities and titers are usually tested by double-immunodiffusion analysis and indirect immunofluorescence; antidesmin is routinely tested for by the labeling of Z lines in isolated chicken myofibrils, whereas antivimentin is tested for by the staining of intermediate filaments in cultures of chicken fibroblasts. Since double immunodiffusion is not a sensitive test of specificity, the antisera are tested by immunoautoradiography for cross-reaction with any other protein other than the homologous antigen.

C. Assessing the Specificity of Desmin and Vimentin Antibodies by Immunoautoradiography

Even if desmin- and vimentin-specific antibodies are positive by immunofluorescence, it is generally desirable to use a technique that can unequivocally demonstrate the reaction of antidesmin or antivimentin antibodies with only desmin or vimentin in a whole-cell extract. This is particularly desirable, since the desmin- or vimentin-specific antibodies may carry a weak cross-reactivity toward antigens unrelated to them, or toward each other for the reasons discussed earlier on. For this purpose, the technique of immunoautoradiography has been used successfully on one-dimensional SDS–polyacrylamide gels to show that actin antibodies react only with actin in cell extracts (Burridge, 1976). We have used immunoautoradiography on two-dimensional IEF/SDS–polyacrylamide gels as a means of demonstrating not only the specificity of a given antibody, but also (in the case of desmin) to show whether all three isoelectric variants of desmin react with equal efficiency. This technique is lengthy (2 weeks), but offers extreme sensitivity (detection of less than 50 ng of protein on the gel) and reliability, providing convincing evidence of the specificity of the antisera (Granger and Lazarides, 1979). Furthermore, it allows the detection of cross-reacting proteins even if the antibodies are nonprecipitating for these proteins. Since it is applicable to a variety of antigens, it is described here in detail, and it is recommended as an ultimate test for the specificity of any of the structural antigens. This technique is also described in another section of this volume (Lazarides, 1981).

For testing the specificity of antibodies to desmin and vimentin, a variety of cell extracts from embryonic tissues, cells grown in tissue culture, or isolated myofibrils can be used. To obtain the maximum number of cytoskeletal proteins on the gel, we routinely use embryonic-muscle extracts. Leg and breast muscle of a 14-day chick embryo is disrupted in a Dounce homogenizer in 4 ml of 20 m*M* Tris HCl (pH 8.0)–5 m*M* EGTA–1 m*M* *O*-phenanthroline–0.5 m*M* PMSF, and spun for 1 hour at 150,000 *g*. The supernatant is made 10 *M* in urea, 2% in NP-40, and 0.05% in 2-ME, and 75–100 μg of protein is loaded onto two-dimensional gels. Two-dimensional IEF/SDS–polyacrylamide gel electrophoresis is carried out as described by O'Farrell (1975), with modifications to maximize the resolution of the various isoelectric variants of actin and desmin (Hubbard and Lazarides, 1979). These extracts contain ample quantities of tubulin, α-actinin, myosin, the various actin variants, all the variants of desmin and vimentin, the light chains of myosin, the various troponin subunits, the tropomyosin subunits, and so on. (Granger and Lazarides, 1979).

After electrophoresis, the gels are fixed for 6 hours in 50% ethanol–10% acetic acid (all incubations are carried out at room temperature in rocking Pyrex baking dishes), and then neutralized for 1 day in several changes of 50 m*M* Tris HCl (pH 7.5)–0.1 *M* NaCl–10 m*M* NaN_3. After another day of washing in buffer A [0.01 *M* Tris HCl, pH 7.5–0.140 *M* NaCl–0.01 *M* NaN_3–0.1% gelatin (Difco; dissolved by heating the buffer to 60°C)], the gels are incubated for 1 day in 100 ml of buffer to which 100 μl of antiserum has been added (1:1000 dilution). The actual final dilution of the antiserum depends on the titer of the antibodies. Good

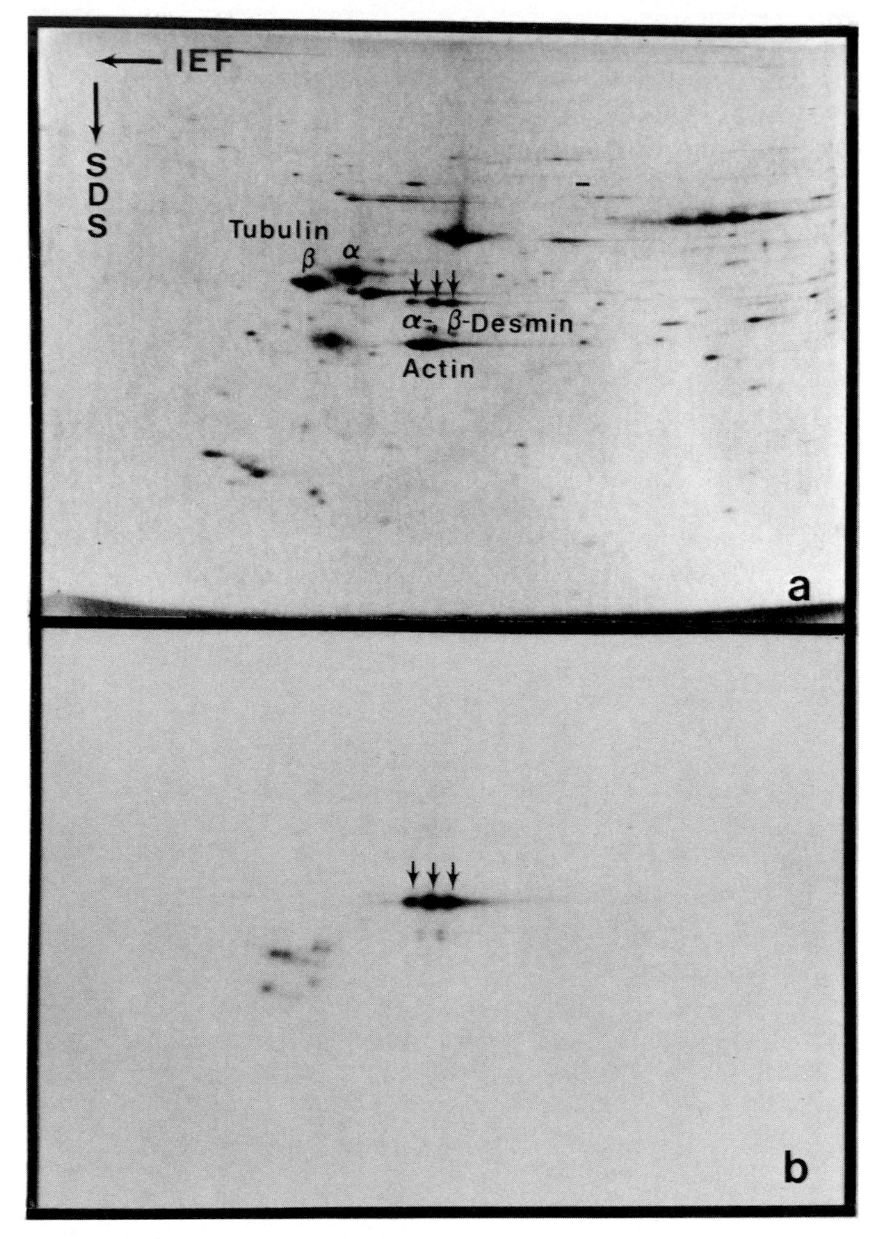

FIG. 3. Two-dimensional immunoautoradiography using antidesmin antiserum. (a) Dried Coomassie Blue-stained two-dimensional gel of a high-speed supernatant of a low-ionic-strength extract of embryonic chicken muscle. (b) Corresponding autoradiogram showing specific labeling of desmin and its degradation products with this antiserum. (From Granger and Lazarides, 1979.)

titer antisera can be diluted 1:500–1:1000, but lower or higher dilutions need to be tested. If too low a dilution is used, nonspecific staining of a number of proteins results. If too high an antibody dilution is used, no staining is obtained.

After incubation with the antibody solution, the gels are washed for 3 days in several changes of buffer A. They are then incubated for 1 day in 100 ml of buffer A containing ^{125}I-labeled protein A (see later on). Two days of washing in buffer A and one day in buffer A *without* the gelatin are followed by staining in 0.1% Coomassie Brilliant Blue R–47.5% ethanol–10% acetic acid, and destaining in 12.5% ethanol–5% acetic acid. The destained gels are dried, photographed (using a Polaroid setup and P/N 55 film), and autoradiographed for 12–25 hours at −70°C on Kodak X-Omat R XR5 film with a DuPont Cronex Lightning-Plus intensifying screen. Autoradiograms are developed in Kodak X-ray developer after 1 day (short) or 3–4 days (long) of exposure (longer if necessary).

Protein A (from Pharmacia) is iodinated by the chloramine T method, as described by Greenwood, Hunter, and Glover (1963), with the exception that the reaction is terminated by adding an excess of tyrosine. One hundred microliters of 0.5 *M* potassium phosphate (pH 7.5) is added to 1 mCi of Na-^{125}I (ICN; in 2 μl of NaOH, pH 8.8); 20 μl of protein A (5 mg/ml) and 20 μl of chloramine T (2.5 mg/ml) are added next. After 2 minutes, 150 μl of tyrosine (0.4 mg/ml) is added. The mixture is passed centrifugally through a 3-ml bed of Sephadex G-25 and the void fraction is used for labeling. Specific activities on the order of 0.1 μCi/μg protein are obtained, and each gel is incubated with approximately 1 μCi. The iodinated protein A is used within a week.

The entire procedure requires approximately 14 days to complete (including exposure of the gels on film). Using this technique, a number of proteins that react with the antibodies and that do not stain with Coomassie Blue can be detected. An example of the use of this technique for the demonstration of the specificity of antibodies to desmin is shown in Fig. 3. The peptides below desmin are proteolytic products of desmin and are present in this gel in quantities less than 50 ng.

D. Affinity Purification of Desmin and Vimentin Antibodies

For the affinity purification of desmin, chicken smooth-muscle desmin, purified by cycles of polymerization–depolymerization, is coupled to CNBr-activated Biogel A15 in the presence of 0.1% SDS. The presence of SDS is necessary in order to keep desmin in solution during the coupling. Since, however, even this preparation of desmin has a small vimentin contamination, desmin and vimentin are purified by preparative SDS–gel electrophoresis and subsequent electrophoretic elution before coupling. This method, as well as the affinity purification of the antibodies, is described in another article in this volume (Lazarides, 1981).

E. Preparation of Skeletal Myofibrils and Z-Disk Sheets

Lower leg or breast muscles are cut in strips from adult male leghorn chickens, tied to wooden applicator sticks to prevent contraction, and stored at −15°C in a solution of 50% glycerol–50% PBSA (163 mM Na^+–140 mM Cl^-–4 mM K^+–10 mM PO_4^{2-}–10 mM NaN_3^-, pH 7.4–1 mM EGTA–0.5 mM O-phenanthroline) for 1–6 months. Myofibrils are prepared by vigorous blending of glycerinated muscle in a Lourdes MM-1B homogenizer for 3–5 minutes at near top speed. Blending and storage is carried out in the 50% glycerol solution. The myofibrils are stored for periods up to 12 months.

For immunofluorescence, coverslips are covered with a drop of the 50% glycerol–myofibril suspension. Numerous myofibrils rapidly adhere to the glass and remain adherent throughout subsequent buffer washings and antibody incubations.

For the preparation of Z-disk sheets (Granger and Lazarides, 1978), very small strips of well-glycerinated muscle are tied to glass micropipets and swirled in several changes of 0.6 M KI–10 mM $Na_2S_2O_3$ (to prevent free-iodine formation)–20 mM Tris HCl (pH 7.4)–0.1 mM EGTA–0.1 mM O-phenanthroline for about 1–2 weeks at 4°C or for 1 day at room temperature. Subsequent blending in the KI solution in a Virtis "45" microhomogenizer produces Z-disk sheets that will adhere tenaciously to glass coverslips. A drop of homogenate is normally placed on a coverslip and then immediately rinsed off.

F. Immunofluorescence of Myofibrils and Z-Disk Sheets (Granger and Lazarides, 1978, 1979)

All immunofluorescence is performed on myofibrils and Z-disk sheets adhering to glass coverslips. Incubations and washes are performed in either PBSA or in PBSA containing 0.5% Triton X-100 and 0.1% SDS (in the case of Z-disk sheets), which gives the same results but generally results in cleaner backgrounds. Incubations with primary antisera are typically for 1 hour at 37°C; incubations with fluorescein-conjugated goat anti-rabbit IgG are for 0.5 hour at 37°C. Washes are for 0.5 to 2 hours at room temperature. The rest of the details are described in another section of this volume (Lazarides, 1981).

Indirect immunofluorescence for the detection of a single antigen and double direct-indirect immunofluorescence for the detection of two antigens in the same cell is carried out as described in detail elsewhere in this volume (Lazarides, 1981; see also Gard and Lazarides, 1980). For the double immunofluorescence, antibodies to vimentin (or α-actinin) are coupled directly to tetramethyl rhodamine isothiocyanate as follows. The IgG fraction is prepared from immune sera by ammonium sulfate precipitation (to 37% saturation at 4°C) and DEAE–cellulose chromatography in 10 mM phosphate buffer, pH 7.5. The resulting IgG (flowthrough) is concentrated by ultrafiltration and reacted with tetramethyl rhodamine isothiocyanate (25–50 μg/ml protein) in a buffer containing 0.15 *M* NaCl and 0.1 *M* sodium carbonate (pH 9.5) for 1 hour. At this time, an additional one-fifth volume of 0.5 *M* sodium carbonate (pH 9.5) is added, and the reaction is continued overnight at 4°C. Unconjugated rhodamine is removed by chromatography on Sephadex G-25 in 10 m*M* phosphate (pH 7.5), followed by DEAE chromatography, as described by Cebra and Goldstein (1965) and Brandtzaeg (1973). The 280/560 ratios of the conjugates used are usually between 1.3 and 4.3.

G. Preparation of Myogenic Cultures

Primary cultures of embryonic chicken thigh muscle are prepared from 10-day-old embryos according to the methods of Konigsberg *et al.* (1978). A modification of their procedure is used in the preparation of secondary cultures (Gard and Lazarides, 1980). Primary cells (18 hours, at 10^7 cells/100-mm plate) are washed twice with Earle's balanced salt solution (EBSS) and trypsinized with 5 ml trypsin in EBSS (0.025 mg/ml) at 37°C for 3–5 minutes. Trypsin digestion is terminated by the addition of 5 ml growth medium, consisting of Eagle's minimal essential medium (MEM) supplemented with 15% horse serum, nonessential amino acids, 5% chick embryo extract, and antibiotics (100 μg of streptomycin and 100 U of penicillin per ml). The cells are then gently pipetted from the plates, pelleted at 500 *g*, and washed once with growth medium. These cells are then pre-plated twice, according to the procedure of O'Neill and Stockdale (1972). For immunofluorescence, cells are plated at a density of $2.5–4 \times 10^5$ cells/60-mm petri plate, with four or five collagen-coated coverslips per plate. For [^{35}S]methionine labeling, cells are plated at a density of $1–4 \times 10^6$ cells/100-mm collagen-coated plate. These procedures yield cultures that are more than 95% free of contaminating fibroblasts. Within the first 3 days after plating of secondary cultures, more than 80% of the cells fuse into multinucleated myotubes. After 4 days in culture, more than 95% of the cells fuse. To prevent overgrowth of fibroblasts during long-term cultures, cells are fed on day 3 with complete medium, Eagle's MEM containing 10% horse serum, and 2% embryo extract, containing 10 μM cytosine arabinofuranoside (ARA-C). Cultures are then fed at 3- or 4-day intervals with complete medium without ARA-C.

V. Immunofluorescent Localization of Desmin and Vimentin in Adult Skeletal Muscle

The small size of smooth-muscle cells and their complicated morphology makes *in situ* studies difficult, and for this reason, we have used skeletal muscle in most of our studies on the immunocytological localization of desmin and vimentin. Although common in embryonic striated muscle, intermediate filaments become progressively less visible as the sarcomeres condense and become laterally registered. Intermediate filaments are rarely seen in adult skeletal muscle, but substantial amounts of both desmin and vimentin can be detected in isolated adult skeletal myofibrils both by two-dimensional IEF/SDS–PAGE and by indirect immunofluorescence (Granger and Lazarides, 1979; Lazarides and Hubbard, 1976; Lazarides, 1978). Using immunofluorescence on isolated myofibrils, we observed that desmin and vimentin are components of both skeletal and cardiac myofibril Z lines as well as of intercalated disks (Lazarides and Hubbard, 1976; Granger and Lazarides, 1979). Thus the Z disk is conclusively known to contain the following proteins: actin, α-actinin, filamin, desmin, and vimentin (Granger and Lazarides, 1978, 1979; Bechtel, 1979). To obtain information on the spatial relationship of desmin and vimentin to actin and α-actinin within the plane of a single Z disk, it was necessary to develop a method that allowed a face-on view of Z disks by light, as well as by electron microscopy. Normally, when a muscle fiber is sheared (as previously described), it is cleaved lengthwise, between laterally associated Z disks, to produce myofibrils. However, if most of the actin and myosin filaments are first extracted from the fibers with a high-ionic-strength buffer, as described earlier in this chapter,

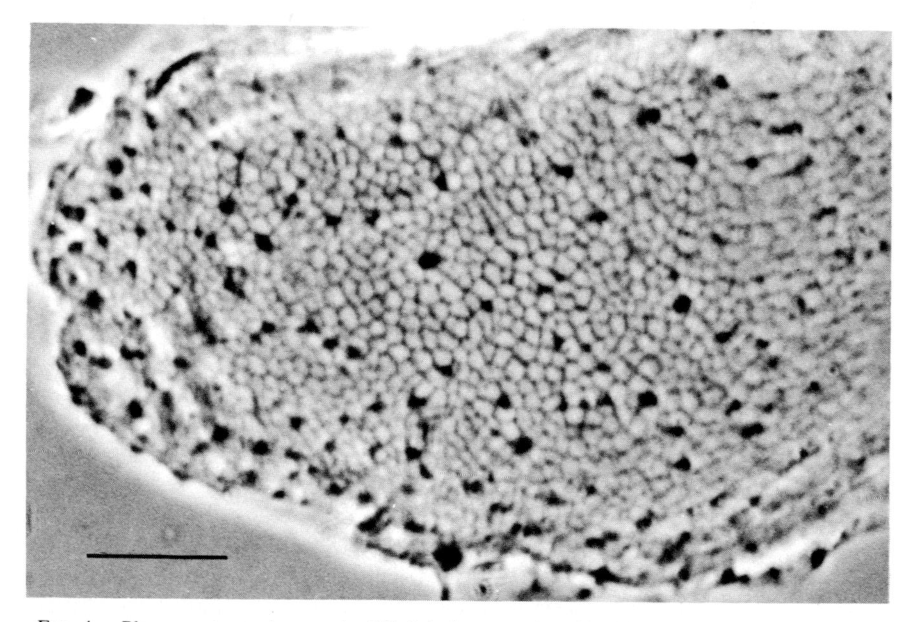

FIG. 4. Phase-contrast micrograph of Z-disk sheets produced by blending of KI-extracted muscle fibers. (From Granger and Lazarides, 1978.) Bar = $10 \mu m$.

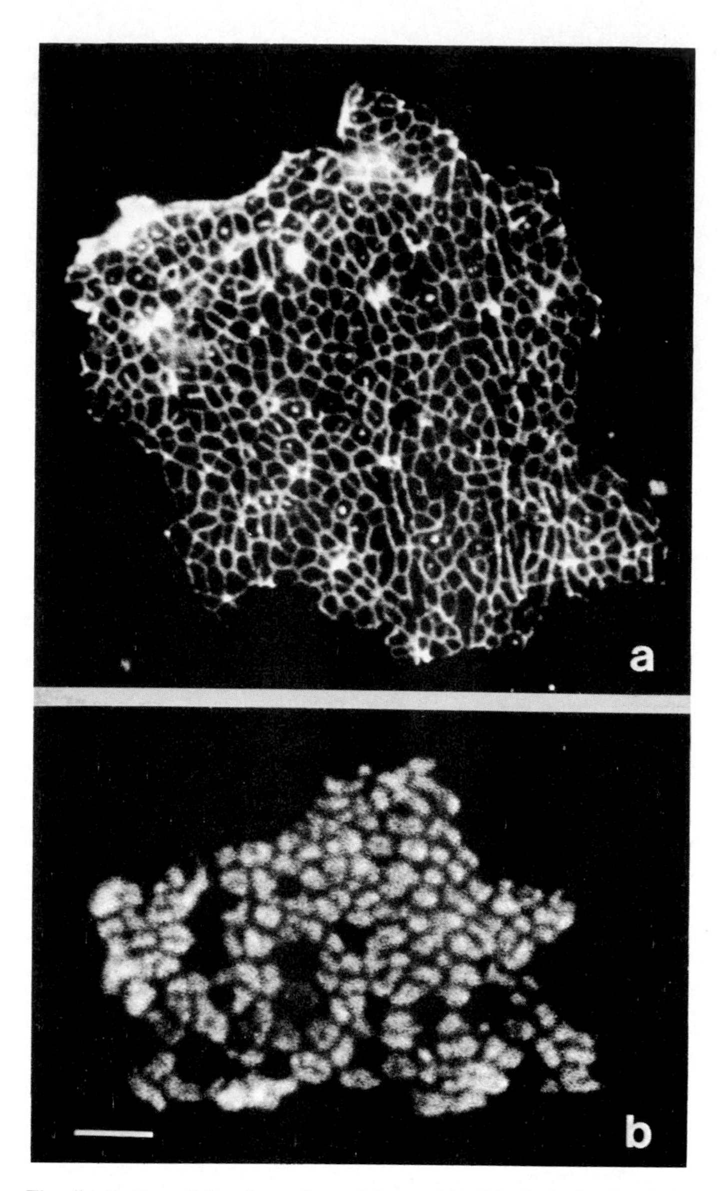

FIG. 5. The distribution of desmin at the periphery of Z disks in isolated Z-disk sheets from chicken skeletal muscle. (a) Indirect immunofluorescence using antibodies to chicken smooth-muscle desmin. Vimentin shows the same distribution as desmin in these sheets. The fluorescent-desmin foci in the interior of some Z disks represent the areas of Z-disk subdivision. (b) The distribution of α-actinin in the interior of Z disks in isolated Z-disk sheets from chicken skeletal muscle. Indirect immunofluorescence using antibodies to chicken skeletal-muscle α-actinin. Bar = 2.5 μm.

the fibers' longitudinal strength is reduced and subsequent blending shears the fibers transversely between Z planes, producing honeycomb-like sheets of Z disks (Fig. 4) (Granger and Lazarides, 1978, 1979). The generation of sheets of interconnected Z disks provides unambiguous evidence that the Z disks within a Z plane are actually connected to one another. Two-dimensional IEF/SDS–PAGE of these Z-disk sheets reveals actin as the major component, desmin and vimentin as two other prominent components, and lesser quantities of tropomyosin and α-actinin. α-Actin appears to be the predominant actin variant in these Z-disk sheets.

Indirect immunofluorescence performed on a single Z-disk sheet reveals that desmin and vimentin are present at the periphery of each Z disk, forming an interconnecting network across the fiber (Fig. 5a). α-Actinin is localized within each disk, giving a face-on fluorescence pattern that is complementary to that of desmin (Fig. 5b). Actin is present throughout the Z plane, apparently coexisting with both desmin and vimentin at the periphery of the Z disk, and α-actinin in the interior of the Z disk (Granger and Lazarides, 1978, 1979).

VI. Association of Membranous Organelles with Isolated Z-Disk Sheets and Their Relationship to Desmin and Vimentin

Electron microscopy has revealed an abundance of vesicularized membrane near the level of the Z line in extracted muscle fibers. These vesicles are most likely remnants of transverse (T) tubules and sarcoplasmic reticulum (SR) that have survived the disruptive effects of glycerol and KI. Cross sections of the Z disks in extracted fibers show that these vesicles actually surround each Z disk, as does the T system. In the fibers we have used so far for the extraction, the T-system membrane is located at the periphery of the Z disk. Sarcoplasmic reticulum membranes that normally occupy the region adjacent to the A band probably coalesce at the peripheries of the Z disks during extraction. Mitochondria also survive complete extraction and are generally aligned in rows along the length of the muscle fiber. On shearing, mitochondria fragments remain tightly associated with isolated Z-disk sheets (Fig. 6). In phase-contrast micrographs of Z-disk sheets, they appear as dark patches (Fig. 4). Longitudinal thin sections of Z-disk sheets clearly indicate a firm association and a continuity between the peripheries of Z disks and mitochondria. Immunofluorescence reveals that the areas of association between mitochondria, the T-system vesicular material, and the peripheries of the Z disks contain both desmin and vimentin (Granger and Lazarides, 1978) (Fig. 5a).

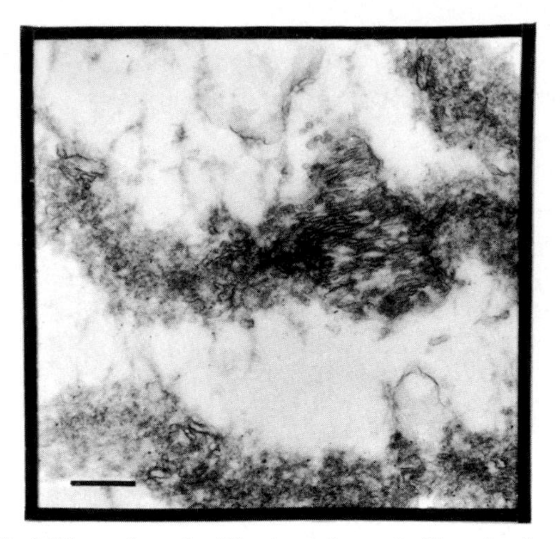

FIG. 6. Longitudinal thin section of a KI-extracted muscle fiber showing the association of a residual mitochondrion and membrane vesicles with the Z plane. (From Granger and Lazarides, 1978.) Bar = 0.5μm.

VII. Relationship of Desmin and Vimentin to Myofibril Subdivision

Unlike the way most cells proliferate, in muscle cells proliferation is a direct result of radial growth of myofibrils, which ultimately results in a longitudinal subdivision of the myofibril after a maximum size is attained (Goldspink, 1970, 1971; Shear, 1978). This division allows the SR and T-system to expand into the myofibril and complete the formation of new, smaller myofibrillar assemblies. Electron micrographs show that this process probably starts with the invagination of these membranous organelles toward the center of the disk, or with the formation of a small fissure at the center of the Z disk, which proceeds outward radially (Goldspink, 1970, 1971; Shear, 1978). Immunofluorescence reveals occasional small patches of desmin and vimentin at the center of the disk, as well as short radial segments of desmin and vimentin reaching from the center to the edge. Conversely, α-actinin is absent from these locations (Granger and Lazarides, 1978, 1979) (see Fig. 5a,b). Thus desmin and vimentin, SR, and T tubules are most probably invading the larger myofibrils as they split into smaller units. These observations and those described in the previous section indicate a close structural relationship between desmin and vimentin and membranous organelles, such as the T–SR membrane system and mitochondria.

VIII. Steps in the Assembly of Z Disks

The biochemical and immunofluorescent electron-microscopic evidence summarized above demonstrates that the Z disk is composed of at least two major molecular domains: a peripheral one that contains at least desmin, vimentin, actin, and membranous organelles (T system–SR, mitochondria); and a central one that contains at least α-actinin and actin (and presumably filamin). The central domain functions to link together the sarcomeric actin filaments, which emerge from these sites with opposite polarities. In the peripheral domain, we have proposed that desmin- and vimentin-containing filaments function to link Z disks laterally, as well as to link membranous organelles such as the T system–SR and mitochondria to the periphery of the Z disk (Granger and Lazarides, 1978, 1979; Lazarides, 1980; Gard and Lazarides, 1980).

Do these two domains assemble at different times during myogenesis? In the early stages of myogenesis, electron microscopy reveals that numerous intermediate filaments exist scattered in the cytoplasm, not associated with the newly forming Z disks and sarcomeres (Ishikawa *et al.*, 1968; Kelly, 1969). Similarly, double-immunofluorescence techniques reveal that the distribution of desmin and vimentin is indistinguishable; both proteins are found in an intricate network of free cytoplasmic filaments (Fig. 7a). During this time, desmin- and vimentin-containing intermediate filaments can be induced to aggregate into bundles by exposing the cells to colcemid (Gard and Lazarides, 1980). Later in myogenesis, several days after the appearance of α-actinin containing Z-line striations, both filament proteins become associated with the Z lines of newly assembled myofibrils, with a corresponding decrease in the number of cytoplasmic filaments (Fig. 7b). This transition corresponds to the time when the α-actinin containing Z lines become aligned laterally. After the transition, Z disk-associated desmin and vimentin can no longer be induced to aggregate in cells exposed to colcemid (Gard and Lazarides, 1980). These results indicate that the two molecular domains of the Z disk are indeed assembled sequentially at distinct time intervals. Thus the Z disk or the desmin- and vimentin-containing intermediate filaments must undergo some sort of molecular maturation during myogenesis that allows their association. The exact nature of this molecular maturation is presently unknown. It is similarly unclear how desmin- and vimentin-containing filaments eventually integrate the T system–SR membranes and mitochondria at the peripheries of Z disks.

One puzzling observation is that in adult skeletal-muscle fibers, intermediate filaments are only rarely seen by conventional electron-microscopic techniques (Kelly, 1969; Page, 1969). The presence of high concentrations of membranous material in the intermyofibrillar space may obscure the detection of intermediate-filament connections. However, in cardiac muscle, where the

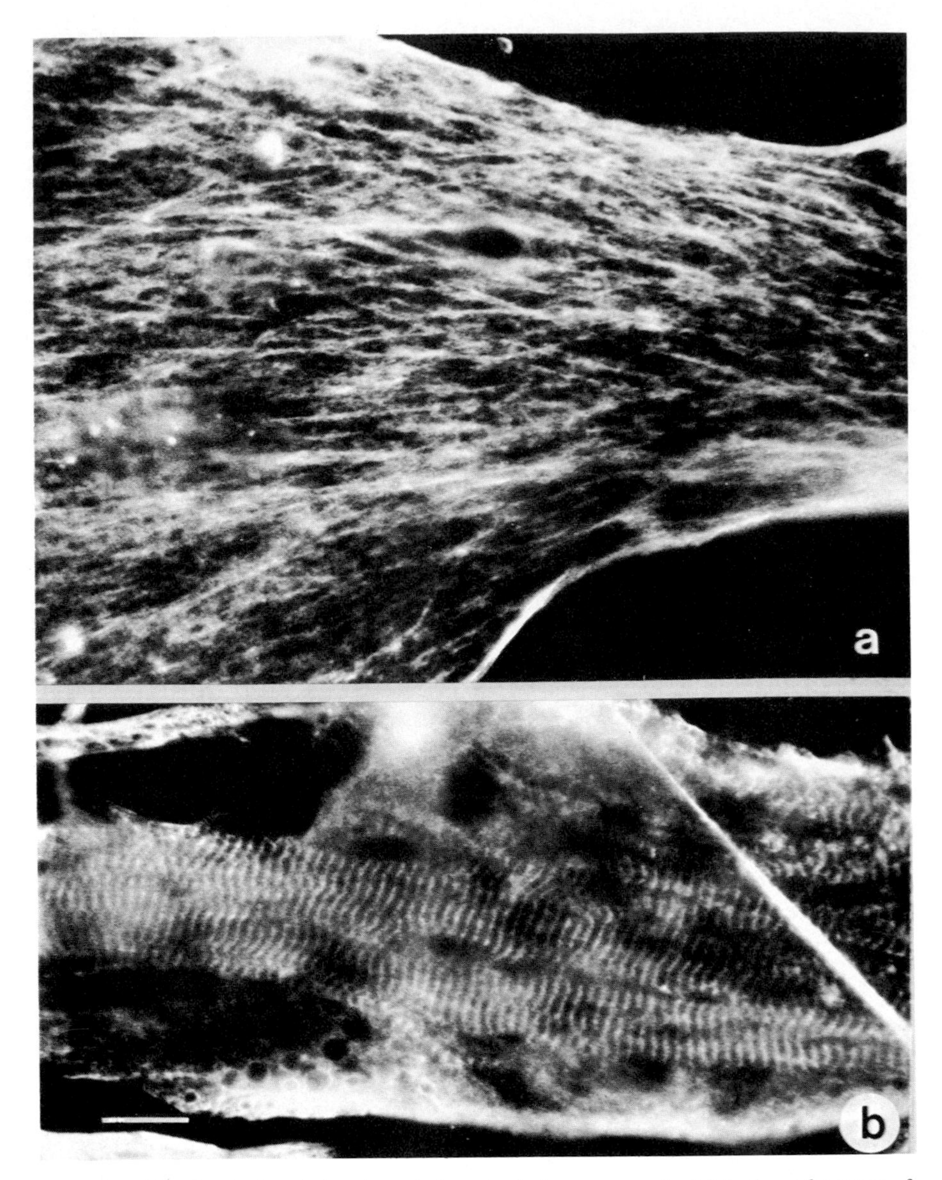

FIG. 7. The distribution of desmin and vimentin during myogenesis. (a) In the early stages of differentiation (3–4 days postfusion) of chicken embryonic skeletal myotubes grown in tissue culture, desmin and vimentin are deposited in filamentous structures that remain unassociated with the newly forming Z disks. (b) Later in differentiation and about the time when myofibrils become laterally aligned (7 days), desmin and vimentin become associated with Z disks. Indirect immunofluorescence using antibodies to chicken smooth-muscle desmin. Bar = 10 μm.

myofibrils are not as tightly packed, bundles of intermediate filaments appear to surround Z disks and to connect them to each other. Occasionally, these filaments extend from the Z disk to the nuclear and plasma membranes, as well as to mitochondrial membranes (Ferrans and Roberts, 1973; Oliphant and Loewen, 1976; Behrendt, 1977; Junker and Sommer, 1977).

IX. Function of Intermediate Filaments in Muscle

The observations on the structure, expression, and localization of desmin- and vimentin-containing filaments in muscle cells that are summarized here lead us to propose the following functions of these filaments in muscle cells (for a review, see Lazarides, 1980). The two intermediate-filament systems, desmin and vimentin, play an important role in the lateral organization and registration of myofibrils. They may mechanically integrate all contractile actions of a muscle fiber by linking individual myofibrils laterally and to other cytoplasmic membranous organelles. This arrangement ensures the maintenance of the proper alignment of cellular structures during the contraction–relaxation cycle of muscle. Thus, during the biogenesis and assembly of myofibrils, these two intermediate filament classes may have an important role in the generation of the striated appearance of muscle by bringing the Z disks into lateral register. The two intermediate filament classes may also function in the biogenesis of membranous systems. Some mechanism is required to direct the deposition and attachment of the T–SR membrane system and of mitochondria to the periphery of the Z disk and the myofibril in general. A protein system with both membrane and Z-disk affinities could be vital to this process. We can further hypothesize that this transverse mechanical integrating system of desmin filaments may be an essential feature of all muscle types regardless of their position in the evolutionary scale.

X. Prospects for the Future

The methodology and results summarized here about intermediate filaments have provided us with a new way of looking at muscle structure and muscle differentiation. However, many new and unexpected questions have arisen and await to be answered in the future. What are the biochemical events that culminate in the lateral registration of myofibrils and how do intermediate filaments change their distribution to associate with myofibrils? How do intermediate filaments interact with membranous organelles? How do the actin-filament bundles usually found in fibroblasts and early myotubes become transformed into contractile sarcomeres during myogenesis? A great deal more experimental evidence and yet new ways of looking at muscle structure are needed before we can begin to unravel the molecular complexity of this system.

Acknowledgments

I thank Janet Sauer for her help in the preparation of this manuscript. This work was supported by a grant from the National Institutes of Health (PHS-GM-06965), a grant from the Muscular Dystrophy Association of America, and a grant from the National Science Foundation. The author is a recipient of a National Institutes of Health Research Career Development Award.

References

Bechtel, P. J. (1979). *J. Biol. Chem.* **254,** 1755–1758.
Behrendt, H. (1977). *Cell Tissue Res.* **180,** 303–315.
Bennett, G. S., Fellini, S. A., Croop, J. M., Otto, J. J., Bryan, J., and Holtzer, H. (1978). *Proc. Natl. Acad. Sci. U.S.A.* **75,** 4364–4368.
Blose, S. H. (1979). *Proc. Natl. Acad. Sci. U.S.A.* **76,** 3372–3376.
Brandtzaeg, P. (1973). *Scand. J. Immunol.* **2,** 273–290.
Brown, S., Levinson, W., and Spudich, J. A. (1976). *J. Supramol. Struct.* **5,** 119–130.
Burridge, K. (1976). *Proc. Natl. Acad. Sci. U.S.A.* **73,** 4457–4461.
Cebra, J. J., and Goldstein, G. (1965). *J. Immunol.* **95,** 230–245.
Ferrans, V. J., and Roberts, W. C. (1973). *J. Mol. Cell. Cardiol.* **5,** 247–257.
Franke, W. W., Schmid, E., Osborn, M., and Weber, K. (1978). *Proc. Natl. Acad. Sci. U.S.A.* **75,** 5034–5038.
Franke, W. W., Schmid, E., Osborn, M., and Weber, K. (1979). *J. Cell Biol.* **81,** 570–580.
Gard, D. L., and Lazarides, E. (1980). *Cell* **19,** 263–275.
Gard, D. L., Bell, P. B., and Lazarides, E. (1979). *Proc. Natl. Acad. Sci. U.S.A.* **76,** 3894–3898.
Geiger, B. (1979). *Cell* **18,** 193–205.
Goldspink, G. (1970). *J. Cell Sci.* **6,** 593–603.
Goldspink, G. (1971). *J. Cell Sci.* **9,** 123–137.
Granger, B. L., and Lazarides, E. (1978). *Cell* **15,** 1253–1268.
Granger, B. L., and Lazarides, E. (1979). *Cell* **18,** 1053–1063.
Greenwood, F. C., Hunter, W. M., and Glover, J. S. (1963). *Biochem. J.* **89,** 114–123.
Hubbard, B. D., and Lazarides, E. (1979). *J. Cell Biol.* **80,** 166–182.
Hynes, R. O., and Destree, A. T. (1978). *Cell* **13,** 151–163.
Ishikawa, H., Bischoff, R., and Holtzer, H. (1968). *J. Cell Biol.* **38,** 538–555.
Izant, J. G., and Lazarides, E. (1977). *Proc. Natl. Acad. Sci. U.S.A.* **74,** 1450–1454.
Junker, J., and Sommer, J. R. (1977). *Proc.—Annu. Meet., Electron Microsc. Soc. Am.* **35,** 582–583.
Kelly, D. E. (1969). *Anat. Rec.* **163,** 403–426.
Konigsberg, I. R., Sollmann, P. A., and Mixter, L. O. (1978). *Dev. Biol.* **63,** 11–26.

Laemmli, U. K. (1970). *Nature (London)* **227,** 680–685.
Lazarides, E. (1978). *Exp. Cell Res.* **112,** 265–273.
Lazarides, E. (1980). *Nature (London)* **283,** 249–256.
Lazarides, E. (1981). *Methods Cell Biol.* **24,** 313–331.
Lazarides, E., and Balzer, D. R., Jr. (1978). *Cell* **14,** 429–438.
Lazarides, E., and Hubbard, B. D. (1976). *Proc. Natl. Acad. Sci. U.S.A.* **73,** 4344–4348.
O'Connor, C. M., Balzer, D. R., and Lazarides, E. (1979). *Proc. Natl. Acad. Sci. U.S.A.* **76,** 819–823.
O'Connor, C. M., Gard, D. L., and Lazarides, E. (1981). *Cell* **23,** 135–143.
O'Farrell, P. H. (1975). *J. Biol. Chem.* **250,** 4007–4021.
Oliphant, L. W., and Loewen, R. D. (1976). *J. Mol. Cell. Cardiol.* **8,** 679–688.
O'Neill, M. C., and Stockdale, F. E. (1972). *J. Cell Biol.* **52,** 52–63.
O'Shea, J. M., Robson, R. M., Huiatt, T. W., Hartzer, M. K., and Stromer, M. H. (1979). *Biochem. Biophys. Res. Commun.* **89,** 972–890.
Page, S. G. (1969). *J. Physiol. (London)* **205,** 131–145.
Shear, C. R. (1978). *J. Cell Sci.* **29,** 297–312.
Small, J. V., and Sobieszek, A. (1977). *J. Cell Sci.* **23,** 243–268.
Starger, J. M., and Goldman, R. D. (1977). *Proc. Natl. Acad. Sci. U.S.A.* **74,** 2422–2426.
Starger, J. M., Brown, W. E., Goldman, A. E., and Goldman, R. D. (1978). *J. Cell Biol.* **78,** 93–109.
Tuszynski, G. P., Frank, E. D., Damsky, C. H., Buck, C. A., and Warren, L. (1979). *J. Biol. Chem.* **254,** 6138–6143.
Zackroff, R. V., and Goldman, R. D. (1979). *Proc. Natl. Acad. Sci. U.S.A.* **76,** 6226–6230.

Chapter 17

Dictyostelium discoideum: *Methods and Perspectives for Study of Cell Motility*

JAMES A. SPUDICH

Department of Structural Biology
Stanford University School of Medicine
Stanford, California

I. Introduction

In the last decade the field of cell motility has advanced considerably. During this period many key questions have been defined by pooling a large amount of data on numerous cell types. It is now necessary to be able to isolate in large quantities not only actin and myosin but also accessory components that occur in cells in much smaller quantities and interact with actin or myosin to affect their function and organization. The next decade will require an extensive effort to identify what could be a large number of such components, to purify them, and to study their properties. To accomplish this it will be important to have a focus of effort on a single cell type. We have seen the effects of such focus when *E. coli* was established as an organism of choice to sort out much of classical biochemistry and molecular biology.

ISBN 0-12-564125-7

In choosing an organism on which to focus, several important criteria must be met. It must be possible to clone the cells in order to assure a homogeneous population. In addition, it must be feasible to grow the cells in very large quantities. A kilogram or more of wet cell pellet will be required to purify and characterize many of the accessory components of the contractile system. Preferably the cells should grow in a totally defined medium to allow radioactive components to be taken up by the cell with defined specific radioactivities. The cells should be easily manipulated and extracts readily prepared. Proteolysis, which occurs in extracts of various cell types, should be easily inhibited. Genetic manipulation of the organism should be possible. Ultimately, the definition of the physiological relevance of accessory components that have been characterized biochemically will depend on the identification of appropriate conditional lethal mutants. Also, the identification of new components will in many instances be made by genetic techniques.

An organism that fulfills the criteria outlined above is *Dictyostelium discoideum* (Loomis, 1975). Consequently, there has been a natural growth in the number of investigators who study motility in this organism. As described here, *Dictyostelium* is very easy to work with, can be grown in large quantities, exists as a stable haploid organism, and can be studied genetically. It exists as individual amoebas as long as bacteria or other food supply is available. In this state the amoebas form pseudopods and phagocytize bacteria as a source of food. Axenic strains (not needing bacteria as a food supply) are available (Loomis, 1971) that grow in a totally defined medium (Franke and Kessin, 1977). Under starvation conditions the cells undergo a chemotactic response for 3′,5′-cyclic AMP, and aggregate into multicellular bodies known as pseudoplasmodia. Each of these moves as a unit and is phototactic. The driving force for pseudoplasmodium movement appears to be the amoeboid movement of its individual cells. Each pseudoplasmodium gives rise to a fruiting body consisting of a long, thin stalk supporting a mass of about 10^5 spores collected together in a balloon-like structure known as a sorus. Germination of the spores gives rise to amoebas and the cycle is completed. Thus, in addition to pseudopod formation that leads to general movements, this organism displays fascinating chemotactic and phototactic responses. How the contractile machinery that causes movement is controlled by cyclic AMP and light will be very interesting to elucidate.

II. Methods

A. Strains and Where to Get Them

There are a number of axenic strains of *Dictyostelium discoideum*. The strain we have used is called Ax-3, which we obtained from Dr. William Loomis

(Biology Department, University of California, San Diego). I am happy to supply this strain to anyone who requests it.

B. Spores—The Starting Material

The cells are easily stored in the form of spores. For short-term storage (one to several weeks), fruiting bodies are formed on agar in petri dishes. SM agar (Table I) is used. For sporulation of the *Dictyostelium* on SM agar plates, use the following procedure. A loopful of *Klebsiella aerogenes,* maintained on an SM agar plate, is used to inoculate a small flask of medium (we generally use HL5, to be described). The culture is allowed to grow overnight. About 0.1 ml of the bacterial culture is deposited onto a fresh SM agar plate. One sorus is picked with a sterile wire loop from a preexisting sporulated plate and mixed with the *Klebsiella*. The mixture is then spread evenly and allowed to develop at about 23°C (do not exceed 26°C). Within one day the bacterial lawn matures. Within the next day the amoebas eat the bacteria, clearing the lawn. This results in starvation of the amoebas and they enter their developmental cycle. Within the next 24 hours the amoebas form fruiting bodies, which consist of a long, thin stalk supporting a sorus, which contains about 10^5 spores. These sori can then be used as a source of spores for more than a week. We routinely prepare a fresh plate at least once each week.

For long-term storage (one to several years), spores can be adsorbed to silica gel and stored at 4°C. This is accomplished as follows. A freshly sporulated SM

TABLE I

COMPOSITION OF SM AGAR[a]

Component[b]	Amount (gm/liter)
Bactopeptone	10
Yeast extract[c]	1
KH_2PO_4	1.9
K_2HPO_4	0.6
Glucose	10
$MgSO_4$	1
Agar	20

[a] From Sussman (1966).

[b] The first four components can be combined to make a 10-fold concentrated stock solution. The pH of the final solution is adjusted to 6.0–6.4 with KOH.

[c] From Oxoid Limited; distributed by K. C. Biological, Inc., P. O. Box 5441, Lenexa, Kansas 66215.

agar plate is turned over abruptly onto a counter top. Most of the sori are dislodged from their stalks and fall into the upper chamber of the SM agar plate. These sori are then suspended in about 1 ml of a powdered-milk suspension (5% nonfat dry milk in cold, sterile H_2O). A small vial (5 ml) is filled with silica gel (6–16 mesh, SX144-01, Matheson Coleman and Bell) and autoclaved. About 0.5 ml of the spore suspension is added to the silica gel. The solution is absorbed by the gel and the spores remain in a dried state on the surface of the silica. The spores are stable and viable under these conditions at 4°C for several years. To germinate these spores, an SM agar plate is coated with *Klebsiella* as described above and several pieces of the silica gel are sprinkled onto the surface of the plate. The spores then germinate and feed on the bacterial lawn, giving rise to fresh fruiting bodies within a few days.

C. Growth of Vegetative Cells

We routinely grow cells in HL5 medium (Table II) in order to harvest large amounts of cells for biochemical work. In this medium they have a doubling time of about 9 hours. We use three sizes of flasks: small flasks (250 ml) containing 25 ml of medium; medium-sized flasks (2 liters) containing 700 ml; and large flasks (6 liters) containing 2 liters. The procedure we use is as follows. A small flask is inoculated with spores of *Dictyostelium* by picking a sorus off its stalk, taking care to avoid picking up bacteria from the surface of the agar. The small flask is allowed to shake at 23°C. (*Dictyostelium* cells should not be allowed to reach a temperature above 26°C or they do not grow properly and begin to die. This is true at all steps, including the maintenance of plates.) In about 7 days the

TABLE II

COMPOSITION OF HL5 MEDIUM[a]

Component[b]	Amount (gm/liter)
Proteose peptone[c]	10
Yeast extract[c]	5
$Na_2HPO_4 \cdot 7H_2O$	0.35
KH_2PO_4	0.35
Glucose	10

[a] From Loomis (1971).

[b] The final solution (pH 6.5) is sterilized in an autoclave for 15 minutes and removed as soon as possible to prevent caramelization.

[c] From Oxoid Limited; distributed by K. C. Biological, Inc., P. O. Box 5441, Lenexa, Kansas 66215.

optical density at 660 nm reaches 0.5 to 1.0. Do not allow the amoebas to enter their stationary phase. We routinely transfer cells to fresh HL5 medium after they reach between 0.5 and 1.0 OD_{660}. The small flask is used to inoculate a medium-sized flask and cells are grown to late logarithmic growth once more. In this laboratory we maintain cells in medium-sized flasks continuously in order to have a supply of cells at any moment. When large amounts of amoebas are needed, each medium-sized flask is used to inoculate two large flasks. These, like the small and medium-sized flasks, are shaken at 250 rpm on a New Brunswick platform shaker. At about 1 OD_{660} the cells are harvested by centrifugation at about 500 *g* for 2 minutes. The yield of cells is 5–10 gm/liter of medium.

In 1977 a chemically defined medium for growing *Dictyostelium* axenic strains was described (Franke and Kessin, 1977). Cells grow in this medium with a doubling time of about 10 hours at 23°C. This medium is extremely useful for studies involving radioactivity where one wishes to know the specific radioactivity of a particular nutrient after it has entered the cell constituents. Thus ^{35}S-labeled actin is prepared in this laboratory at about 4000 cpm/μg (Simpson and Spudich, 1980). Also, ^{32}P-labeled myosin, phosphorylated *in vivo,* is easily obtained using this medium (Kuczmarski and Spudich, 1980; Peltz *et al.*, 1981).

Cells can also be obtained in reasonable quantities by growing them on *Klebsiella* in large petri dishes (14 cm diameter). The SM agar dishes are treated as previously described and the cells are harvested at the point that the bacterial lawn is cleared by the amoebas. This is accomplished by adding 10 ml of medium (such as HL5) to each dish and suspending the cells with a glass spreader. Rinse each dish with an additional 10 ml of medium. The amoebas can be separated from the bacteria by sedimentation at 500 *g* for 2 minutes. Several washes with the medium or buffer of choice essentially eliminates free bacteria. We obtain approximately 1 gm of wet cell pellet from one large petri dish.

D. Preparation of Extracts and Isolation of Actin and Myosin

Methods for preparation of extracts of *Dictyostelium* amoebas that minimize proteolysis are presented by Uyemura *et al.* (1978). The same report details one method of preparation of *Dictyostelium* actin with about 30% yield. *Dictyostelium* myosin purification was first carried out by Clarke and Spudich (1974). The procedure has been modified to eliminate RNA that remains associated with the myosin unless purified by DEAE chromatography (Mockrin and Spudich, 1976), and further improved to obtain better yields by other slight modifications (Kuczmarski and Spudich, manuscript in preparation). References dealing with these and other aspects of isolation of the cytoskeletal proteins can be found in recent reviews (Taylor and Condeelis, 1979; Uyemura and Spudich, 1980).

III. Concluding Remarks

This brief review is intended to emphasize the need for some degree of focus on a single cell type to further elucidate mechanisms of cell motility. Many of the advantages of *Dictyostelium discoideum* are briefly outlined. Some basic methods and materials are discussed and a few relevant references are given. Let me emphasize my willingness to send the *Dictyostelium* strains we use to anyone requesting them. It is my hope that this article will encourage additional investigators to work on *Dictyostelium discoideum* as a model system to study cell motility.

References

Clarke, M., and Spudich, J. A. (1974). *J. Mol. Biol.* **86,** 209–222.

Franke, J., and Kessin, R. (1977). *Proc. Natl. Acad. Sci. U.S.A.* **74,** 2157–2161.

Kuczmarski, E. R., and Spudich, J. A. (1980). *Proc. Natl. Acad. Sci. U.S.A.* **77,** 7292–7296.

Kuczmarski, E. R., and Spudich, J. A. (1981). In preparation.

Loomis, W. F. (1971). *Exp. Cell Res.* **64,** 484–486.

Loomis, W. F. (1975). "Dictyostelium Discoideum: A Developmental System." Academic Press, New York.

Mockrin, S. C., qnd Spudich, J. A. (1976). *Proc. Natl. Acad. Sci. U.S.A.* **73,** 2321–2325.

Peltz, G., Kuczmarski, E. R., and Spudich, J. A. (1981). *J. Cell Biol.* **89,** 104–108.

Simpson, P. A., and Spudich, J. A. (1980). *Proc. Natl. Acad. Sci. U.S.A.* **77,** 4610–4613.

Sussman, M. (1966). *Methods Cell Physiol.* **2,** 397–410.

Taylor, D. L., and Condeelis, J. S. (1979). *Int. Rev. Cytol.* **56,** 57–144.

Uyemura, D. G., and Spudich, J. A. (1980). *Biol. Regul. Dev.* **2,** 317–338.

Uyemura, D. G., Brown, S. S., and Spudich, J. A. (1978). *J. Biol. Chem.* **253,** 9088–9096.

Chapter 18

Axonal Transport: A Cell-Biological Method for Studying Proteins That Associate with the Cytoskeleton

SCOTT T. BRADY AND RAYMOND J. LASEK

Department of Anatomy
Case Western Reserve University
School of Medicine
Cleveland, Ohio

ISBN 0-12-564125-7

I. Introduction and Overview

Many early studies of the nervous system used one of a number of silver-based histological stains to reveal the shape and extent of neuronal cells (Ramón y Cajál, 1928). These silver stains showed more than just the shape of neurons, however; they also revealed the presence of a fibrous network that arose in the cell body near the nucleus and coursed into the axon and dendrites (Fig. 1). Not only did pioneers like Ramón y Cajál provide the first real insights into the organization of the nervous system, but they also gave us the first pictures of cytoskeletons. More important, even at this early stage of understanding, they recognized that such fibrous structures were the basis of cellular form and organization (Ramón y Cajál, 1928). Such a realization came quickly because, from the very first, two aspects of neuronal form were apparent: Neurons from different regions of the nervous system displayed a remarkable diversity of shape and size, whereas homologous neurons in the same regions of the nervous system in different individuals, or even different species phylogenetically quite distant, maintain an even more remarkable constancy of overall form (Bullock *et al.*, 1977; see esp. p. 458). The form of neuronal cells is as essential to the functioning of a nervous system as the electrical properties of those cells, and the substance of that form is found in the cytoskeletal elements.

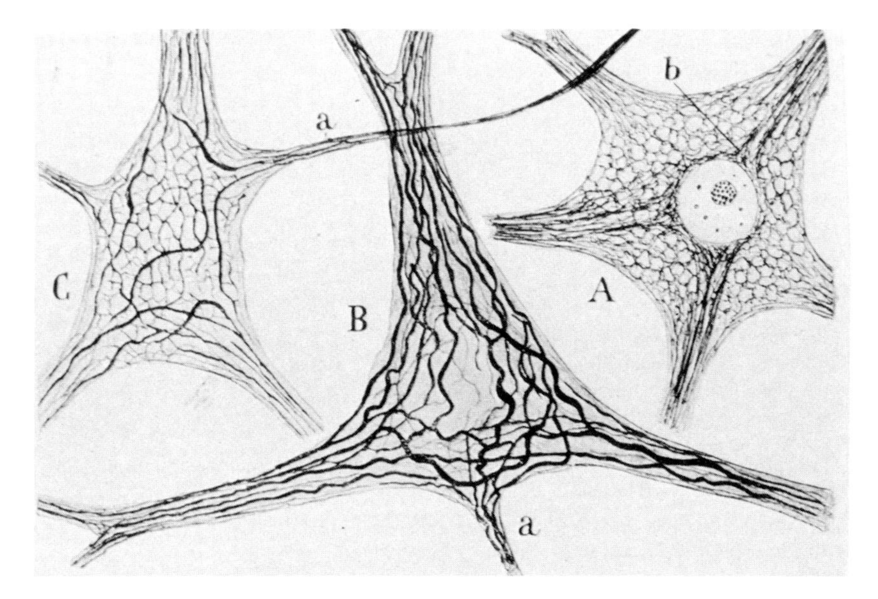

FIG. 1. The fibrous networks of neurons as visualized by classic silver-based histological stains. Shown are neurons from the central nervous system of rabbit prepared and drawn by Ramón y Cajál (1909). Such fibrous structures are now known to correspond to the microtubule-neurofilament network of the neuron, which interacts with certain silver stains in a highly specific manner.

Through the years, neuronal tissue has remained an important source for study of cytoskeletal elements including tubulin (Wuerker and Kirkpatrick, 1972; Karlsson and Sjöstrand, 1971; Hoffman and Lasek, 1975), nonmuscle actomyosin (Berl *et al.*, 1973; Black and Lasek, 1979; Willard *et al.*, 1979), and intermediate filaments (Hoffman and Lasek, 1975; Schlaepfer and Freeman, 1978; Liem *et al.*, 1978). As biochemical and ultrastructural techniques have increased in resolution, it has become apparent that each of these cytoskeletal elements represents a family of polypeptides in which each member has a distinct set of properties. This combination of conservation and variation has made it important to look at the cytoskeletal elements of individual cell types and even at subcellular regions of a single cell type in order to understand cytoskeletons and associated proteins in a physiological context. The neuron, or—more specifically—the cytoplasmic extension of the neuron known as the axon, represents an valuable model system in which to study the composition, organization, and dynamic properties of cytological structures *in situ,* particularly the cytoskeletal elements. The reasons for this can be seen if we examine some of the special properties of the axon.

A. Axon and Cytoskeletons

Axons extend from the postmitotic cell body during embryonic development, representing a major step in the full expression of neuronal differentiation. Once established in connection with a target cell, the axon is effectively maintained for the duration of the organism's life. Cytoskeletal elements have important roles both in the initial elongation process and in the maintenance of the mature axon. The utility of axons as a model for the study of cytoskeletal roles in cell elongation has long been appreciated (Wessels *et al.*, 1971; Yamada *et al.*, 1971; Lasek, 1981).

The cytoskeletal elements of the axon are unusually prominent and distinguished by striking regularities in the organization of the different cytoskeletal structures. These regularities have permitted analyses of the detailed relationships among many of the structural elements in the axon (Wuerker and Palay, 1969; Porter *et al.*, 1979; Hodge and Adelman, 1980). With a few exceptions, axons contain all of the major types of structural elements of the cell, including microtubules, intermediate filaments, and microfilaments (Peters *et al.*, 1976). Microtubules and neurofilaments (the neuronal form of intermediate filaments) parallel the long axis of the axon and appear interconnected by "bridges" to form a lattice (Wuerker and Palay, 1969; Metuzals and Izzard, 1969; Burton and Fernandez, 1973; Hodge and Adelman, 1980). Very short microfilaments of the type associated with plasma membranes have been visualized in the axon (LeBeux and Willemot, 1975), and the microtrabecular networks that have been described in many cells have also been demonstrated in the axon (Porter *et al.*, 1979; Ellisman and Porter, 1980).

In point of fact, it is useful to emphasize that, with one important exception, virtually all of the structures and metabolic processes associated with cytoplasmic regions of cells in general can be demonstrated in the axon. The exception is that no protein synthesis occurs in the axon (Lasek *et al.*, 1977), which is a natural consequence of the absence of cytoplasmic ribosomes in mature axons (Peters *et al.*, 1976). Virtually all proteins making up the axon are synthesized in the cell body and then conveyed to their sites of utilization by a set of processes known as axonal transport (Lasek, 1970; Grafstein and Forman, 1980). Thus cytoskeletal elements and the other macromolecular complexes of the axon are continually being renewed by axonal transport (Lasek and Hoffman, 1976).

Finally, axons are remarkable both for the size that they may reach and for their regularity of organization. Even in a relatively small animal, such as the rat, axons may exceed 10 cm in length; and they increase in size correspondingly in larger animals. At the same time, neuronal cell bodies associated with a nerve or tract are typically grouped together so that their axons travel in parallel over these macroscopic distances. Such regularity means that the molecular processes of a large number of cells have a functional equivalence in a nerve and can in many respects be treated as if they were a single very large neuron. The macroscopic

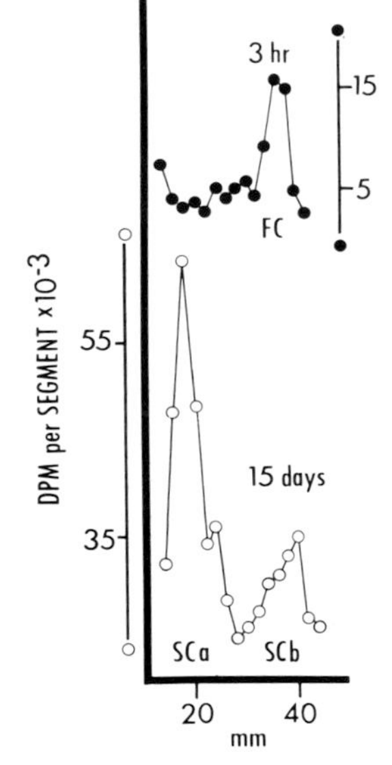

FIG. 2. The major orthograde rate components of axonal transport. ^3H amino acids were injected into the region of the medulla containing the cell bodies for the hypoglossal nerve in adult guinea pigs. At either 3 hours or 15 days after injection, the animals were sacrificed and the amount of radioactive protein in consecutive 2-mm segments of the hypoglossal nerve was plotted. Peaks of radioactivity corresponding to the fast component (FC), slow component a (SCa), and slow component b (SCb) are clearly visible. (Data from studies by Dr. Mark Black.)

dimensions of axons organized into nerves make it possible to study cellular and molecular events in real time, because they occur over distances measurable with an ordinary metric ruler. The most common way of doing this has been to use radioactive precursors to pulse-label proteins in neuronal perikarya in order to produce a series of easily analyzed waves of radioactive proteins moving along the nerve (Fig. 2).

As will be seen later on, the axonal-transport paradigm has a particular value in studies of cytoskeletons, because it allows the investigator to focus on a specialized region of a cell whose physiological roles are relatively well documented. The major goal of this chapter is to describe the methodologies for the study of axonal transport, with an emphasis on those that are of use for the study of cytoskeletal elements in transport. It is hoped that we will be able to serve both those cell biologists who might wish to employ these techniques in their own studies and those biologists who are more interested in a perspective from which to interpret the literature on axonal transport of cell structures. In order for the discussions of technique to be most useful, however, it is necessary to present a brief overview of the axonal transport of cytoskeletons and other macromolecular complexes.

B. Reviews of Axonal Transport

The phenomenon of axonal transport was first demonstrated by Weiss and Hiscoe (1948), though the necessity of such processes had been suggested earlier (notably by Ramón y Cajál, 1928). Not surprisingly, a considerable literature has accumulated since that time dealing with the properties, mechanisms, composition, pharmacology, functions, pathologies, and uses of axonal transport. An exhaustive review of the literature is obviously beyond the scope of this chapter, and fortunately is made unnecessary by the existence of a number of recent reviews dealing with either the general phenomenon or specific issues in the field. The superb and encyclopedic review published last year by Grafstein and Forman (1980) is a model both for completeness and clarity. A number of other reviews could also be noted, including Lasek (1970) and Heslop (1975), for detailed reviews of the earlier literature.

Many reviews dealing with specific issues in axonal transport also illuminate their special areas. Lubinska (1964, 1975) treats the phenomena of streaming and accumulation at a block. Wilson and Stone (1979) focus on the fast component of transport and protein composition, whereas Rambourg and Droz (1980) concentrate on the role of the smooth endoplasmic reticulum (SER) in the fast component. Schwartz (1979) also focuses on the fast component, particularly in invertebrates. McClure (1972), Samson (1976), and Hanson and Edström (1978) focus on the pharmacology of transport. Griffin and Price (1976) examine the role of transport in pathologies associated with the nervous system. Kristensson

(1978) focuses on retrograde transport. Cowan and Cuénod (1975) have edited a book devoted to the use of axonal transport in tracing pathways in the nervous system. Axonal transport also receives considerable attention in many reviews on related topics such as axonal regeneration (Grafstein and McQuarrie, 1978). The reader is directed to these authors for detailed reviews; citations here will be intended only as examples to illustrate relevant points.

C. Axonal Transport and Axonal Compartments

As just indicated, it is important to have some familiarity with contemporary views on the axonal transport of cytoskeletons and other macromolecular assemblies and of the organization of the axon. Four classes of cytological structures are readily apparent in studies of the fine structure of the axon: (1) a tubulovesicular membranous system; (2) mitochondria; (3) the fibrous cytoskeletal elements; and (4) the ground substance of axoplasm. At the same time, pulse-labeling studies of axonal transport have resolved axonally transported material into a number of distinct rate components, each with characteristic properties and protein compositions (Table I and Fig. 3). The correlation of

TABLE I

RATE COMPONENTS OF AXONAL TRANSPORT AND THEIR RELATIONSHIP TO CYTOLOGICAL STRUCTURES

Rate Components			Rate for mammals (mm/day)	Types of materials	Hypothesized structures
Orthograde					
Fast		I	200–400	Glycoproteins, glycolipids, acetylcholinesterase, peptides, serotonin	Vesicles, tubular profiles, neurosecretory granules
Intermediate		II	50	Mitochondrial proteins	Mitochondria
Intermediate		III	15	Myosin-like proteins	
Slow	SC_B	IV	2–4	Actin, clathrin, enolase, creatine phosphokinase, calmodulin, actin-binding protein	Microfilaments, microtrabecular components, matrix
Slow	SC_A	V	0.2–1	Neurofilament triplet proteins, tubulin, certain tau proteins	Microtubule–neurofilament network
Retrograde					
Fast			100–200	Nerve growth factor, polio virus, lysosomal enzymes	Lysosomal and prelysosomal lamellar and multivesicular bodies

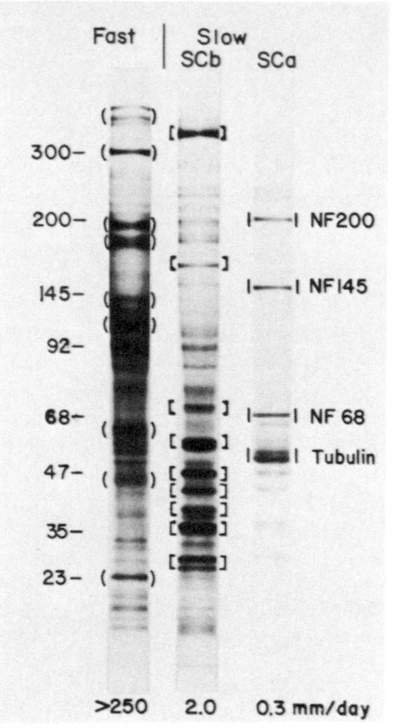

FIG. 3. One-dimensional SDS–gel electrophoresis of the major rate components of axonal transport. Proteins for each of the major rate components were labeled in the optic nerve as described in the text with [^{35}S]methionine, and analyzed on adjacent walls of a slab gel. Fluorography shows the very different patterns of polypeptides associated with each rate component. (From Tytell *et al.*, 1981.)

these distinct rate components with distinct cytological structures (Tytell *et al.*, 1981; Lasek, 1980) has been an important step in placing the processes of axonal transport into a broader cellular context. This correlation has led to the structural hypothesis of axonal transport: each rate component of axonal transport corresponds to a distinct cytological structure.

D. Membranous Structures and Axonal Transport

The membranous structures of the axon include several different types of membranous organelles. Electron micrographs of axons show a broad range of membrane-delimited organelles, including an anastomotic smooth endoplasmic reticulum, small tubules and vesicles, dense core granules, prelysosomal structures (such as multivesicular and lamellar bodies), and mitochondria (Peters *et al.*, 1976). Early in the study of the fast orthograde component (FC) of axonal transport, it was demonstrated that the bulk of the material in FC appeared to be associated with membranes (Grafstein, 1967; McEwen and Grafstein, 1968; DiGiamberardino *et al.*, 1973). At first, it was not certain which membranous organelles were actually moving at a fast (200–400 mm/day) rate. Now, a broad range of evidence from studies of cell fractionation, electron-microscopic autoradiography, protein composition, and direct observations of transported particles has accumulated (Grafstein and Forman, 1980). As a result it is now possible to assign each of these types of organelle to a specific rate and direction of transport.

The elegant studies of Tsukita and Ishikawa (1980) and Smith (1980) graphically illustrate the types of structures that accumulate when fast orthograde transport processes are blocked. Small tubules, vesicles, dense core granules, and related membranous structures (collectively termed the tubulovesicular system) characterize the material that accumulates from a block of FC. These structures are thought to be products of the RER and the Golgi apparatus. A recent report indicates that all FC material passes through the Golgi (Hammerschlag *et al.*, 1981). The SER has also been shown to play a role in the movement of FC materials (Droz *et al.*, 1975; Rambourg and Droz, 1980), either as an active element in the movement of some FC material or as a relatively stationary structure with which the actively moving tubules and vesicles fuse and detach.

The material moving retrogradely (from the axon terminal toward the cell body) differs from membranous elements which move in an orthograde direction. Multivesicular and lamellar bodies are the most prominent structures moving in the retrograde direction (Tsukita and Ishikawa, 1980). These structures are considered to be prelysosomal in nature and they mature into lysosomes on reaching the cell body (Holtzman *et al.*, 1977). A lysosomal function fits with the view that a major function of retrograde transport is to convey segments of membrane to the cell body for degradation and possible recycling. However, retrograde transport has also been shown to convey molecules like nerve growth factor (Stoeckel and Thoenen, 1975) back to the cell body from sites of internalization in the terminals; on reaching the cell body, they function as physiological effectors to modulate cell body processes. Thus retrograde transport has an additional function of conveying information, permitting the cell body to respond to situations in the terminal regions.

Mitochondria fall into a separate category from the other membranous organelles. Light-microscopic observations on living axons show that the mitochondria move by saltations and remain stationary for relatively long periods of time (Cooper and Smith, 1974; Forman *et al.*, 1977a). Studies on the movement of mitochondrial protein (Lorenz and Willard, 1978) indicate that the fastest-moving mitochondria travel at about 40 mm/day (Table I), but the average rate of mitochondrial movement could be slower as a result of the relatively long periods in which mitochondria are not moving.

E. Structures of the Axonal Cytoplast

The bulk of the protein in the axon, however, is not part of any membrane-delimited organelle. Instead, these polypeptides make up the axonal elements that may be collectively termed the cytoplast, including both the cytoskeletal proteins and the soluble proteins of the axon (Porter *et al.*, 1979). Together these proteins constitute the structures corresponding to the two slow components of axonal transport (Black and Lasek, 1980, Table I). The major fibrous cytoskeletal elements of the axon (microtubules and neurofilaments) correspond to the slowest rate component, slow component a (SCa; 0.2–1.0 mm/day) (Hoffman and Lasek, 1975). Actin, clathrin, and more than 100 other polypeptides make up slow component b (SCb; 2–4 mm/day) (Lasek, 1980; Black and Lasek, 1980). Together with FC, these three rate components account for the great majority of labeled proteins that enter the axons after pulse-labeling with radioactive amino acids (Fig. 2).

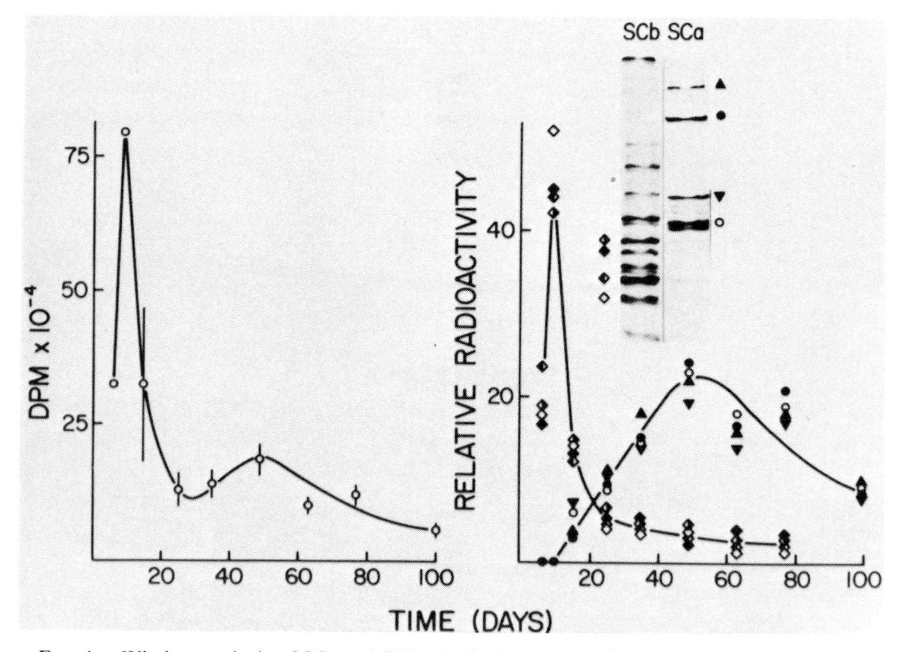

FIG. 4. Window analysis of SCa and SCb. A window segment 3 mm wide and 12–15 mm from the retina was chosen in the optic nerve of a guinea pig. The eye was injected with labeled amino acids as described in the text and animals sacrificed at postinjection times ranging from 6 to 100 days. The amount of radioactivity in the segment at each time was measured both as total radioactivity in the segment (left) and as the amount of radioactivity associated with some of the major polypeptides of SCa and SCb (right). For the graph of radioactivity associated with specific polypeptides, the counts were normalized for comparison. Note the coherent movement of proteins of a given rate component through the segment. (From Black and Lasek, 1980.)

F. Slow Axonal Transport and the Movement of Structures of the Cytoplast

The subunit proteins of microtubules and neurofilaments constitute almost the entire polypeptide composition of SCa. (Hoffman and Lasek, 1975; Tytell *et al.*, 1980). The simplicity of the protein composition for SCa (Fig. 3) is in striking

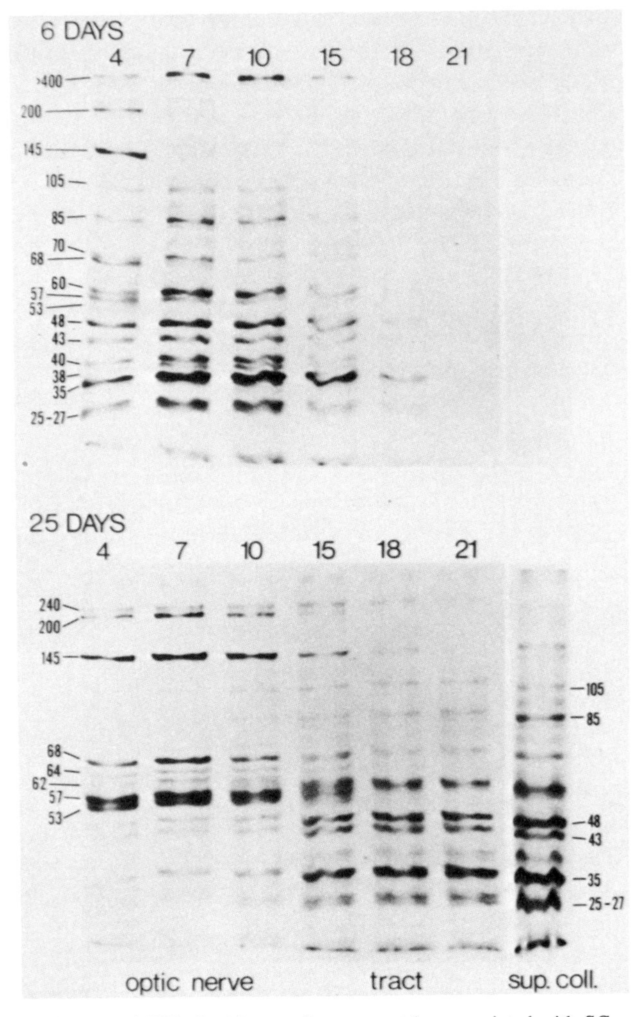

FIG. 5. Distribution analysis of SCa and SCb. In this experiment, proteins associated with SCa and SCb in guinea pig optic nerve were labeled as in Fig. 4, but consecutive 3-mm segments along the nerve and tract were analyzed instead of only a single window segment. Each segment was homogenized in a buffer containing urea and SDS, and run in consecutive wells of a slab-gel apparatus at each time point. Note the movement of a complex group of labeled proteins correspond-

contrast to the more complex polypeptide patterns associated with FC and SCb. In retinal ganglion cells, careful analyses of each of these rate components have demonstrated that the protein composition of each rate component is unique (Tytell *et al.*, 1981). For example, no tubulin, neurofilament proteins, or actin can be detected moving coherently with the FC (Stone *et al.*, 1978; Tytell *et al.*,

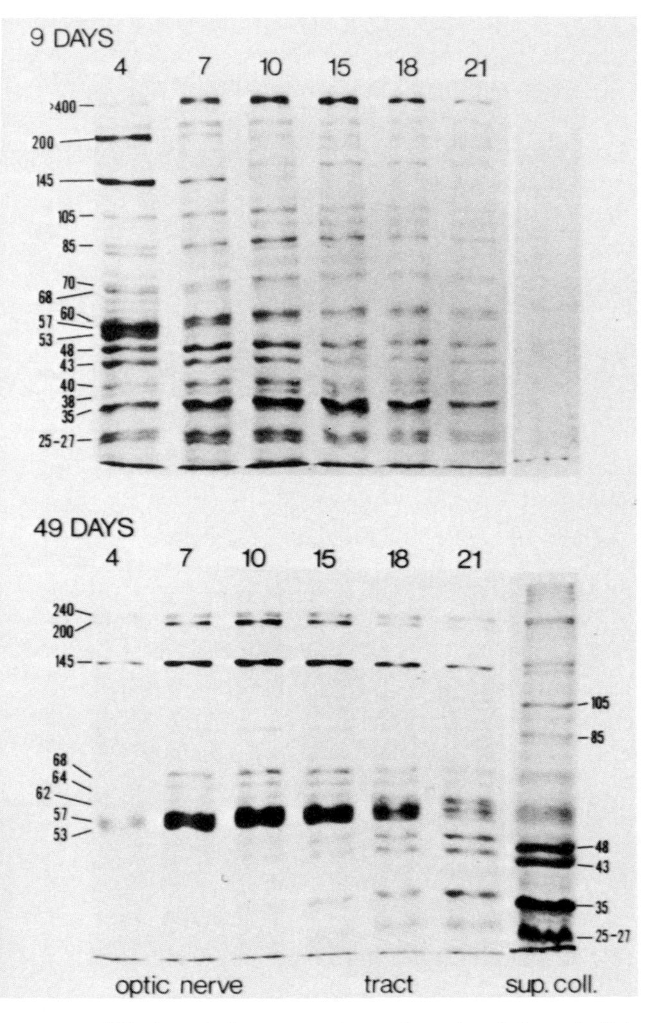

ing to SCb through the nerve and tract from 6 to 9 days. At 9 days, a second, similar group of polypeptides has just begun to enter the first nerve segment. These correspond to SCa and move along the nerve much more slowly, being still present in the nerve at 49 days when most or all of SCb proteins have left the nerve. (Modified slightly from Black and Lasek, 1980.)

1981). The coherent movement of the subunit proteins of microtubules and neurofilaments along the axon (Hoffman and Lasek, 1975; Figs. 4 and 5) is strong evidence in support of the structural hypothesis of axonal transport (Tytell *et al.,* 1981; Lasek, 1980). These proteins behave like elements of a structural complex as they move down the axon, not like individual-subunit proteins. The recent demonstration that certain of the microtubule-associated proteins of the brain also move coherently with the tubulin and neurofilament proteins (Tytell *et al.*, 1980) reinforces this conclusion, since these proteins associate with microtubules, not with tubulin. Just as FC can be seen as the movement of intact organelles, the transport properties of SCa are most consistent with the view that SCa is the movement of microtubules and neurofilaments interacting to form a network (Black and Lasek, 1980).

The coherent transport of the proteins that constitute SCb also suggest that these proteins are organized in a structural complex. However, conventional electron-microscopic techniques have not provided an obvious structural correlate that can take into account the complex polypeptide composition of SCb (Fig. 3 and Tytell *et al.,* 1981). Acknowledged structural proteins such as actin and clathrin make up only a small percentage of the total protein associated with the SCb complex (Black and Lasek, 1979; Willard *et al.,* 1979; Garner and Lasek, 1981). Even more problematic, many of the remaining SCb proteins partition with the soluble fraction in cell-fractionation studies (Lorenz and Willard, 1978; Garner, 1978). Other proteins that have been identified as part of the SCb complex include enzymes of glycolysis and intermediary metabolism (Brady and Lasek, 1981) and other soluble proteins such as calmodulin (Erickson *et al.,* 1980; Brady *et al.*, 1981a). However, all of the proteins associated with SCb move coherently along the axon (Figs. 4 and 5).

Although the problem of a morphological correlate for SCb remains, recent advances in electron microscopy have begun to reveal possible candidates. The description of the microtrabecular network of cells by Porter and his associates (1981a; Porter *et al.*, 1979; Ellisman and Porter, 1980) demonstrates that the cytoplasmic regions of the cell possess a great deal more structure and organization than had been thought. We have suggested that the cytological correlate of SCb be called the axoplasmic matrix (Brady and Lasek, 1981) because of its probable correlation with the ground substance of the axon and the microtrabecular network. The axoplasmic matrix includes actin and clathrin. In addition, many other polypeptides are part of the SCb complex, and these may also be an integral part of the structure of the matrix (Brady and Lasek, 1981). The detailed organization of the complex constellation of proteins that are part of SCb remains to be revealed. However, the axonal-transport studies already demonstrate that those proteins that are generally considered soluble are, in fact, part of a structural complex that may be highly ordered.

G. Axonal Transport and the Study of Cytoskeletons

The size and regularity of organization that are characteristic of axons make them especially attractive as an experimental system for the study of structure, metabolism, composition, and movement of cytoskeletal elements. The axonal-transport paradigm is a particularly powerful approach when dealing with questions about the dynamic properties and organization of cell structures. In the next section we will describe the basic paradigms that are employed in the study of axonal transport. Subsequently, we will explore one application of the paradigm as an example that permits analysis of the long-term associations of proteins with cytological structures in the axon. It is hoped that such an introduction will stimulate interest in the use of axonal transport as a tool to answer similar questions that can be posed by cell biologists.

II. Paradigms: General Principles

Many different paradigms have been used to study the phenomenon of axonal transport. Each of these paradigms has strengths and limitations that must be understood in order to choose the preparation best suited to answer a given question. Before examining specific preparations, however, there are several general considerations to be discussed. First, what types of information can be obtained from studies of axonal transport? In other words, we need to know what the different approaches for analysis of axonal transport can tell us about axonal cytoskeletons. Second, how are the materials in axonal transport to be detected moving in a background of nonaxonal material? In the context of the study of cytoskeletal elements and associated proteins, this general question can be reduced to choosing a way to label specifically the elements of interest.

A. Approaches to Analysis

Three broad categories can be defined in approaches to the analysis of axonal transport, albeit somewhat arbitrarily. In theory, these categories overlap to some extent, but in practice there is generally enough difference in intent and information to be derived to ameliorate any ambiguities. The first category, direct visualization of axonal transport, involves the use of appropriate optics and nerves to follow the movement of certain large structures in the axon. The second approach, damming of axonal transport, involves making a block of the movement of material in the axon and measuring the accumulation at the block. The final category, ''staining'' of axonal transport, analyzes the material in transport during transit along the nerve on the basis of some selective labeling.

B. Direct Visualization

Direct visualization of structures moving in the axon is a technically difficult but interesting approach to the study of axonal transport. Limits on the resolution of light microscopy restrict this approach to the analysis of the movements of large membranous organelles, such as mitochondria and multivesicular bodies (Smith, 1971, 1972, 1980; Forman *et al.*, 1977a,b; Pomerat *et al.*, 1967). An exception is the extension of the growth cone by a neuron during growth or regeneration. The extension of growth cones may be followed under the light microscope and represents the movement of cytoskeletal elements by the neuron (Yamada *et al.*, 1971). This aspect of growth cones is not widely appreciated, but it represents an important model for studying the growth and development of cytoskeletal elements. At present, these approaches are the only ones that permit analysis of the movement of specific structures in the axon without disassembly or destruction of the structures. Observations based on direct visualization may have special relevance to discussions of the mechanisms of transport.

C. Accumulation at a Block

The movement of material in transport may be blocked in a wide variety of ways, both *in vivo* and *in vitro*. Axonal transport was first demonstrated by Weiss and Hiscoe (1948) when they noted a swelling proximal to the compression of a nerve. On relieving the compression, the swelling moved in an anterograde direction corresponding to the slow components of axonal transport (equal to 1 mm/day). Compression is a complex manipulation of the nerve that may elicit complex responses from the nerve tissue (Friede and Miyagishi, 1972). For example, during compression peripheral axons are affected most severely, while axons in the center of the nerve may even be unaffected. Nonetheless, the work of Weiss and Hiscoe demonstrated the value of interrupting axonal transport in order to understand the underlying processes. Since it was relatively difficult to assure a reliable and complete block using simple compression, a ligation or crush that pinches off axons while leaving connective tissue more or less intact largely replaced the use of compression to dam up the moving material in the axon. Even transsection of the nerve has been used, since the truncated axons seal off at the ends. Although the swellings of the nerve that may be produced by the cutting of the axons have been used to demonstrate transport (Zelena *et al.*, 1968), the analysis of axonal transport in these preparations is usually by measurement of radioactivity or enzyme activity accumulating at the block. The use of a crush or some variant has the disadvantage of disrupting not only axons, but also the extraaxonal environment, so it is not specific for axonal transport.

The value in having a specific means of blocking axonal transport led to the use of pharmacological agents to produce a damming of axonal transport. The inhibition of transport by antimitotic agents is the best-documented example (for reviews, see Grafstein and Forman, 1980; Hanson and Edström, 1978; Samson, 1976; McClure, 1972). Colchicine is the most familiar of these agents and it has often been used to demonstrate that a given movement of material is axonal transport rather than diffusion and local incorporation (for examples, see Barondes, 1964; Crothers and McLuer, 1975). This has been a valuable test, although the biological activities of colchicine are far from specific for either axonal transport or microtubules (for examples, see reviews in Wilson, 1975; Brady, 1978; Brady *et al.*, 1980). Since colchicine is slow to act and requires relatively high doses for maximum effectiveness (Paulson and McClure, 1975a), it should not be the antimitotic agent of choice in studies of axonal transport. Podophyllotoxin (Paulson and McClure, 1975b) or nocodazole (Samson *et al.*, 1979) are more appropriate agents to demonstrate that material is being moved by axonal transport. [However, podophyllotoxin also has a wide range of biological activities, such as inhibition of cytochrome oxidase (Kelly and Hartwell, 1954). Nocodazole has not been studied in sufficient detail as yet to determine its specificity.]

Drugs are principally useful in studies of the fast component, where *in vitro* studies are possible. Unfortunately, the cytoskeletal elements (actin, tubulin, neurofilaments, etc.) move only with the slow rate components of transport (Hoffman and Lasek, 1975; Black and Lasek, 1979; Willard *et al.*, 1979). The long time course required in the study of the slow components of transport prevents the use of *in vitro* preparations, and the difficulties in defining the location and concentration of a drug *in vivo* make most pharmacological studies on slow components of transport equivocal. A possible exception is the remarkable finding that the drug β, β iminodiproprionitrile may specifically block SCa, especially the neurofilaments, when administered systemically (Griffin *et al.*, 1978).

Another approach to the damming of axonal transport involves the use of low temperatures applied locally to the nerve, resulting in a reversible cold block of transport. Brimijoin (1975; Brimijoin and Helland, 1976; Brimijoin and Wiermaa, 1978; Brimijoin *et al.*, 1979) has done the most to exploit the use of reversible cold blocks, which may be employed both *in vitro* and *in vivo* (Tsukita and Ishikawa, 1980). Two advantages result from the use of a localized cold block. First, it is a reversible process, hence material may be accumulated at the block and then released. As a result, accurate determinations of rate are made simpler and the signal-to-noise ratio for detecting features such as enzyme activities may be substantially increased. Second, the cold block is the least disruptive method for blocking transport, preserving most morphological features. Microtubules are lost in this treatment (Brimijoin *et al.*, 1979), but compared to the morphological disturbances created by physical manipulations (Zelena *et al.*, 1968; Smith, 1980) or drug treatment (Friede and Ho, 1977), the ultrastructure exhibits minimal disruption.

D. Staining

By calling the final broad category of analysis the "staining" of axonally transported materials, classical histochemistry with its use of organelle-specific staining reagents and procedures is recalled. However, the staining methods employed in axonal-transport studies extend well beyond histochemical methods. In brief, this group of methods involves labeling of neuronal constituents in the cell body during a pulse-label and taking advantage of the normal processes of the neuron to separate the labeled materials into the various rate components of axonal transport. The effect of this pulse-chase procedure is to "stain" elements selectively in each of the rate components in the axonal and terminal regions of the nerve. Staining approaches have been particularly useful methods in recent years to study the slow rate components, which include most or all of the cytoskeletal elements of the axon. By appropriate choices of nerve system and time points for analysis, each rate component and associated proteins may be obtained in a labeled form that is highly selective for that rate component and largely uncontaminated by labeled constituent elements of the other rate components of axonal transport.

The major rate components of axonal transport, in terms of the amount of material carried, are fast component, slow component a, and slow component b (see Table I). In an analysis of total radioactivity along the nerve at various times after the pulse-labeling with amino acids, these three rate components can be visualized as peaks of radioactivity moving at characteristic rates (see Fig. 2). The coherent movement of these peaks along the axon is an unmistakable sign of order in the underlying cellular elements. However, the definition of a rate component in axonal transport can no longer be restricted to the appearance of a peak of radioactivity moving along the nerve. Even when distinct peaks of radioactivity cannot be resolved in analyses of total radioactivity, modern methods of biochemical analysis make it possible to unambiguously identify rate components on the basis of polypeptide composition. Willard and his colleagues (1974) were the first to take full advantage of the sensitivity and resolution of gel electrophoresis and fluorography, but a number of laboratories (for example, Hoffman and Lasek, 1975; Stone *et al.*, 1978; Komiya and Kurokawa, 1978) have confirmed the utility and extended the range of these approaches.

The staining paradigm has become a powerful analytical tool in cell biology, because it can identify the constituent elements of an axonal-transport rate component and place them in a cellular context. The unique polypeptide composition of the major rate components (Tytell *et al.*, 1981) has led to the hypothesis that each rate component corresponds to a cytological structure. As more is understood about the organization of each rate component, it becomes possible to define the molecular interactions of an axonal element by placing it in the context of a rate component. To some extent this is already possible: Recent studies in our laboratory (Tytell *et al.*, 1980) have shown that the proteins associated with SCa include a set of proteins corresponding to the tau class of microtubule-associated proteins, but does not include proteins corresponding to the high-molecular-weight microtubule-associated proteins. The microtubules of the axon thus appear to have long-term associations only with certain of the tau proteins, since only these microtubule-associated proteins maintain an association during transport of tubulin. This ability to define long-term associations of axonal constituents with cytoskeletal structures provides an insight into the microscopic events of the cell not available from other approaches.

There are two primary approaches to analysis that have been employed for the staining paradigm. The first method involves the use of one or more "windows" along the nerve. A "window" may be defined as a segment of nerve (either at the terminal regions or at a convenient interval along the nerve). Axonal transport is monitored by analyzing the transported materials present in the specified nerve segment at various time intervals after labeling. The rate components of transport are identified by the time of arrival at the window segment (Fig. 4). Many classic studies on the kinetics of axonal transport have employed the approach of analysis at a window, including Grafstein (1967), Grafstein and Laureno (1973), Levin (1977, 1978), Karlsson and Sjöstrand (1968), and many others. Recent work by Willard and his associates (1974, 1979; Willard and Hulebak, 1977; Lorenz and Willard, 1978; Willard, 1977) has shown the power of this approach when combined with biochemical analyses. On occasion, two windows on the same nerve may be used to allow more accurate estimations of transport rates (Levin, 1977, 1978). Window analysis often has focused on the leading edges of the rate components in determining the rates of movement, although there is evidence that the velocity of the leading edge of a rate component may differ from the velocity of the peak (Gross and Beidler, 1973; Cancalon, 1979). The window method may not be the most suitable approach for kinetic studies, because interpretation of kinetic data obtained using window analysis may be complicated by degradation of transported proteins in the terminal regions or by trailing of material behind the main peak of a rate component. Window analysis serves best as a means of sampling for studies of the composition of axonally transported proteins.

The second method of analyzing transport using the staining paradigm also has been used in a number of studies of axonal transport. A few notable examples among many are Taylor and Weiss (1965), Lasek (1967, 1968a), Ochs (1972), Gross and Beidler (1973), and Cancalon (1979). Our laboratory (Hoffman and Lasek, 1975; Black and Lasek, 1980; Garner and Lasek, 1981; and others) has emphasized this approach, supplemented with biochemical analyses. In this approach, the entire distribution of a "stained" element along the nerve is analyzed at selected time points, rather than examining only those elements in a single segment. In a sense, this "distribution" analysis can be described as an expansion of the window approach using a large number of small windows to maximize information about the spatial distribution of labeled proteins in the nerve. With careful choices of time points in distribution analysis, pictures not unlike frames from a motion picture can be obtained (Fig. 5). These permit a detailed analysis of the movements of individual elements in the axon (Garner, 1978; Black and Lasek, 1979, 1980). One may study segments down to the limits of detection for the label in order to define the movement and distribution of elements in the axon, which makes distribution analysis especially suitable for kinetic analysis of axonal transport. Perhaps the fullest expression of this approach can be found in the studies of axonal transport using a multiwire proportional chamber (Snyder *et al.,* 1976), which follows the movement of radioactivity in living nerves. Although distribution analysis precisely defines the kinetics of axonal transport, it is more time-consuming than window analysis.

The choice between window analysis and distribution analysis depends on the type of information desired about the elements of axonal transport, and in many cases a combination of the two approaches is most efficient. Distribution analysis can first be used to define the kinetics of axonal transport in a nerve system and to choose both a window along the nerve and an appropriate time point. After these steps have been taken, window analysis can be used most efficiently to characterize the specific elements of a rate component. The combination of the two methods ensures both a maximum amount of labeled material and a minimum of contamination by labeled material from other rate components. Given these analytical approaches, use of the staining paradigm with appropriate choice of label can provide many insights into the organization of the axon.

E. Approaches to Labeling

The process of axonal transport is followed by measuring the movement of labeled material in the neuron, so the labeling of the elements in transport is a critical step. These labels may be extrinsic in the sense that they make use of substances that are not a part of the normal complement of materials available to the neuron. An obvious example is the use of horseradish peroxidase to label retrograde transport (LaVail and LaVail, 1974). More often, the labels used are intrinsic: They are normal constituents of the neuron that may be identified either because they have a characteristic property (fluorescence or an enzyme activity) (Lubinska, 1971; Hökfelt and Dahlström, 1971), or because they have been radioisotopically labeled. The power of intrinsic labels is that, properly utilized, they provide noninvasive and direct access to the interior of the axon. As a result, the most widely employed and generally applicable labeling procedures are based on the introduction of radioactive metabolic precursors into the neuronal cell body. The other labeling procedures, however, do have specific uses that cannot be ignored, and so we shall briefly survey some of the major means of labeling the materials of axonal transport.

In a real sense, the modern era of axonal-transport studies began with the introduction of radioactive tracers and the demonstration of their intraaxonal location (Droz and Leblond, 1963). The radioactive precursor can be introduced into the cell body in several different ways. *In vivo,* the label is suspended in a small volume of carrier (distilled water or physiological saline are commonly used) and injected into the region of the nervous system containing the neuronal cell bodies. Using small volumes and careful injection procedures, the number of cell bodies exposed to labeled precursor may be kept small or restricted to specific class(es) of neurons (e.g., retinal ganglion cells or a specific nucleus of the central nervous system). With certain large neurons, it has even been possible to inject precursors directly into the cell bodies or to iontophorese label onto the cells (Lux *et al.,* 1970; Koike *et al.,* 1972; Schwartz *et al.,* 1975; Isenberg *et al.,* 1980).

Axonal transport may also be labeled *in vitro* either by using tissue culture techniques (Estridge and Bunge, 1978) or by using organ culture of nerves under conditions where only the cell bodies and initial portion of the axons are bathed in a nutrient medium containing the radioactive precursors (Kerkut *et al.,* 1967; Edström and Mattsson, 1972; Theiler and McClure, 1977). In all cases, the labeling efficiency will depend on a number of variables. The precursor employed and the manner in which it is introduced can generally be controlled, but variabilities from animal to animal, uncertainties in the amount of precursor actually available to the cell-synthetic machinery, and the presence of non-neuronal cell bodies prevent the formulation of a simple relationship between the incorporation of radioactivity in the vicinity of the neuronal cell bodies and the amount of material in axonal transport. Once the material has been committed to transport, however, the distribution of material is highly reproducible and may be normalized (Paulson and McClure, 1975b).

For labeling the cytoskeletal elements, the most frequently used radioactive metabolic precursors are the amino acids, because the principal constituents of the cytoskeleton are proteins. The best choices for an amino acid precursor are those that are not synthesized by the experimental animal and, more importantly, are not subject to major metabolic processing. Thus glutamate is a relatively poor choice as a marker, whereas methionine is generally a good one. Ideally, the labeled amino acid should be uniformly distributed throughout all proteins, but in practice this is not possible. As a result, the apparent levels of a given polypeptide in a rate component may vary considerably with respect to the total material in that rate component when different precursors are compared. The high specific activity and relative biological stability of [^{35}S]methionine has made it the precursor of choice for many studies in recent years. Leucine, lysine, and/or proline labeled with ^{3}H have also been widely employed (Elam and Agranoff, 1971; Droz and Leblond, 1962; Kunzle and Cuénod, 1973; Heacock and Agranoff, 1977). For many experiments, there is little advantage to one of these labels over the others, and any of the common precursors may be used. For very long experiments, the long half-life of ^{3}H (12 years) may be an advantage over ^{35}S (half-life of 87.1 days). The higher-energy electrons emitted by ^{35}S do result in a higher efficiency of detection and it is preferable in experiments involving fluorography or autoradiography (Bonner and Laskey, 1974; Laskey and Mills, 1975).

Other types of metabolic precursors are also useful to label classes of proteins or other constituents of the axon. Glycoproteins are labeled effectively with fucose (Forman *et al.*, 1972) or with other amino sugars or precursors (Elam *et al.*, 1970; Forman *et al.*, 1971), though the efficacy may vary. Lipids have been labeled with a variety of precursors including glycerol, cholesterol and precursors of cholesterol, and (^{32}P) inorganic phosphate (Miani, 1963; Abe *et al.*, 1973; Grafstein *et al.*, 1975; Forman and Ledeen, 1972). The transport of lipid is not as well characterized as that of proteins, but appears to be primarily related to the transport of membranous organelles. In the absence of genetic material and protein synthesis in the axon, the nucleic acids and related compounds are not frequently used to follow axonal transport. The small amount of DNA associated with mitochondria may be labeled with thymidine (G. Sheckett and R. J. Lasek, unpublished results) and the tRNA that has been found in low levels in the axon can be detected using uridine (Autilio-Gambetti *et al.*, 1973; Ingoglia *et al.*, 1973).

A new approach to labeling axonal transport with a radioactive tag has recently been described by Fink and Gainer (1979, 1980a,b). We classify this as an extrinsic label because it involves the use of the protein-iodinating agent, Bolton–Hunter reagent, or its tritiated analog, *N*-succinimidyl propionate, to label proteins without utilizing protein synthesis of the cell. These reagents have been used extensively in immunochemistry to label the free amino groups of a wide variety of proteins without loss of activity or change in antigenic properties (Fang *et al.*, 1975; Bolton *et al.*, 1976). Fink and Gainer showed that under proper conditions, these reagents may also be used to label intact cells independent of protein synthesis. The labeled proteins then appear to be processed by the cell normally, hence cell bodies, terminals, or even axonal regions may be labeled in this manner and the subsequent redistribution of labeled species in axonal transport followed. The potential of this technique has not yet been tapped, but it appears to represent an important new tool in cell biology. At present, two primary limitations are apparent. First, both reagents are extremely labile and react with H_2O, thus they have a short effective life span. Second, these reagents tend to label proteins preferentially near the periphery of the cell, especially membrane proteins and those proteins located in the subplasmalemmal cortex (M. J. Katz and R. J. Lasek, personal communication), so not all components of the cell are uniformly labeled. Since the label cannot be reincorporated into the other proteins, this approach is particularly important as a means of studying long-lived proteins independent of protein synthesis.

The enzyme horseradish peroxidase (HRP) is the most widely used extrinsic label of transport, although its primary use is for the study of anatomic pathways rather than the process of axonal transport. When small amounts of HRP are injected near the terminal regions of a nerve, it is taken up by the nerve in pinocytotic vesicles and transported in a retrograde direction to the cell bodies (LaVail and LaVail, 1974). The relative specificity of the uptake and the ease of detection by histochemical means has made this an invaluable tool for the tracing of pathways in the nervous system (LaVail, 1978). Not all materials injected near the terminals of a nerve will be taken up and transported, but several other substances have been employed in this manner, including tetanus toxin (Stoeckel *et al.,* 1975), nerve growth factor (Stoeckel and Thoenen, 1975), and certain viruses (reviewed by Johnson and Griffin, 1978). HRP may also be taken up by neuronal cell bodies and moved in an anterograde direction (Nauta *et al.*, 1975), as may some forms of wheat germ agglutinin (Margolis *et al.,* 1981), but these possibilities have not been exploited to any great extent. Since the uptake of extrinsic material by the neuron can label only retrograde and, to a lesser degree, fast component of anterograde transport, this method is not applicable for the study of the translocation of cytoskeletal elements.

III. Paradigms: Experimental Methods

Having discussed some general principles for study of axonal transport in the previous section, it becomes possible to examine the merits of specific preparations that have proved useful. Many different animal models and nerve systems can be found in the literature. Some of the more widely used animals include rat (Lasek, 1968b; Theiler and McClure, 1977), guinea pig (Black and Lasek, 1978), frog (Edström and Mattsson, 1972), rabbit (Willard *et al.,* 1974), goldfish (Grafstein, 1967), chick (Droz *et al.,* 1973), and Aplysia (Schwartz *et al.,* 1975). Others that have been utilized include mousc (Grafstein and Laureno, 1973), leech (Isenberg *et al.,* 1980), cat (Lasek, 1968a; Ochs *et al.,* 1967), garfish (Gross, 1973; Cancalon and Beidler, 1975), and even human (Brimijoin *et al.,* 1973). In each animal, one, perhaps several nerve systems have been employed. Often the use of a specific animal model is based on the characteristics of a given nerve system, which may be particularly suitable in that animal (i.e., larger, more accessible, better documented, etc.). For example, Aplysia neurons were chosen to take advantage of the extremely large neuronal cell bodies, which permit analysis of axonal transport in a single identified cell, and the garfish is used because of its remarkably long olfactory neurons.

So many different nerve and animal preparations have been used (for a comprehensive listing, see Grafstein and Forman, 1980) that it is neither profitable nor possible to consider them all in this paper. A few preparations have proved particularly useful over the years and will be discussed. We will attempt, therefore, to describe the use of these systems in sufficient detail to permit design and execution of experiments. Other nerve systems will be considered primarily for their general utility and specific advantages, with the reader being referred to the literature for details of methodology. Finally in this section, we will illustrate the use of axonal transport to answer cell-biological questions with an example of an application.

A. Optic Nerve and Optic Tract (ONOT)

Measurement of axonal transport in the ONOT is a rapid, simple means of specifically labeling the proteins associated with major structural elements of the cell, substantially free of proteins associated with other structural components. No surgery is required and the injection is simple, resulting in minimal trauma to the experimental animal. Proteins that are labeled in this manner are produced by a comparatively homogeneous cell population, since only a fraction of the cells in the retina send projections along the optic nerve. This system is one of the best-characterized preparations for study of axonal transport. A substantial literature documents the kinetics and, to some extent, the composition of all five of the anterograde rate components (for example, Willard *et al.,* 1974; Lorenz and Willard, 1978; Hoffman and Lasek, 1975; Black and Lasek, 1980). For studies in which turnover is of interest, the terminal regions of the retinal ganglion cells are well characterized and, for the most part, well localized.

The ONOT does have certain limitations. It can only be used as an *in vivo* preparation, so although the dissection is simple, the nerve is not readily manipulated either physically or pharmacologically during the experiment. The optic system is a relatively poor preparation for doing studies on the pharmacology of axonal transport, because neither concentration nor site of action (cell body, axon, etc.) can easily be determined. Interpretation of pharmacological studies done in systems as complex as the eye is very difficult, and other models should be utilized whenever possible. The relatively short length of the ONOT also makes it a less satisfactory choice for the study of the intermediate rate components because they are not well resolved from the much more prominent slow and fast rate components. These limitations are minor when compared to the overall simplicity and utility of the ONOT preparation.

Axonal transport has been studied on occasion in the ONOT of most common experimental animals. The primary considerations in choosing an animal are first, that there should be a posterior chamber in the eye large enough to permit injections without damaging the retina or anterior chamber; and second, that the optic system is 98–100% crossed at the chiasm. The classic control for nonspecific labeling of neuronal proteins utilizes this effectively complete crossing over of the axons. The amount of label present in the corresponding nerve segment of the uninjected eye represents local incorporation of blood-borne precursor. The guinea pig (Hoffman and Lasek, 1975; Black and Lasek, 1980; Tytell *et al.*, 1981), rabbit (Willard *et al.*, 1974; Lorenz and Willard, 1978), and mouse (Grafstein and Laureno, 1973) are the best-characterized mammalian systems. All have relatively large eyes and are standard experimental animals. The small size of the mouse makes it a more difficult model, but a careful study of the kinetics of transport in mouse ONOT exists (Grafstein and Laureno, 1973; Specht and Grafstein, 1973). The rabbit and guinea pig are considerably larger, and detailed studies of both kinetics and protein composition of the rate components of transport are available (Lorenz and Willard, 1978; Willard *et al.*, 1974; Black and Lasek, 1978; Black and Lasek, 1980; Tytell *et al.*, 1981). Although the rabbit is a considerably larger animal, there is little advantage over the guinea pig, since the head and ONOT do not scale upward in proportion to body weight. Two other ONOT systems have also been extremely valuable: the goldfish (Grafstein, 1967) and the chick (Marchisio *et al.*, 1973). Detailed analyses of the proteins are not yet available with these systems. Rates are much slower in the goldfish, a poikilotherm; however, the ONOT of goldfish regenerate effectively, whereas mammalian systems do not. The chick visual system (Marchisio and Sjöstrand, 1971, 1972; Marchisio *et al.*, 1973) is principally of interest as a developing system.

B. Injecting the Eye

Taking guinea pig as an example, the first step in making injections is to take a length of PE-20 intramedic polyethylene tubing sufficient to hold 110–120 μl and stretch one end to reduce the inner diameter until it forms a tight fit around the base of a 30-gauge needle (broken off from a standard sterile hypodermic needle with a hemostat). The needle of a 100-μl Hamilton microsyringe that has been filled with distilled water is inserted into the other end of the tubing. It is important to control the rate and volume of injections precisely, so we use a repeating dispenser for the microsyringe that will deliver the label in 2-μl aliquots with a 100-μl syringe. For most of our experiments, 200–500 μCi of [^{35}S]methionine or [^{3}H]lysine/proline is injected per eye. The label is concentrated shortly before use by lyophilization in a vacuum centrifuge and resuspension in sufficient distilled water to give a specific activity of 20–50 Ci/μl. Using the syringe, the PE-20 tubing is filled with water to within 5 mm of the 30-gauge needle to minimize dead space. Ten microliters of label are taken up into the tubing and the 30-gauge needle held with a hemostat. Similar amounts of label may be used for injection in other animals, although the volume injected should be reduced for smaller eyes (rat, mouse, etc.).

In the case of the guinea pig, the animal is anesthetized with ether (anesthesia should be sufficiently deep to abolish the corneal reflex). The animal can then be placed on the operating area and the eyelid pulled back to expose the sclera. Holding the needle with a hemostat, the needle is inserted 2–4 mm into the eye on the scleral side of the sclera/cornea border while angled toward the posterior of the eye. For a successful injection, the posterior angle is important to avoid both lens and anterior chamber of the eye. The point on the circumference of the cornea for injection is not critical, though the region most rostral is convenient in allowing the experimentor to brace the eye securely against the medial part of the orbit. The needle can often be seen through the lens when it is in the posterior chamber. Label should be injected without delay in small deliberate aliquots to avoid buildup of pressure or loss of label. The rate of injection is followed by watching the meniscus in the PE-20 tubing. The needle is withdrawn and the animal allowed to recover from anesthesia. In a good injection, there will be no leakage of material from the eye or bleeding at the injection site, and the lens will be intact.

C. Injection–Sacrifice Interval

Sacrifice interval will depend on the rate component of interest and the animal employed. For the guinea pig, a 2- to 4-hour interval is sufficient to isolate FC in the optic nerve and initial segments of the tract. SCb is obtained by sacrifice at 6 days and SCa is present in the optic nerve in essentially pure form from 30 to 60 days. The nerve is harvested by decapitation and dissection. For kinetic studies, the entire nerve and opposing tract should be removed intact (the appropriate lateral geniculate and superior colliculae are included if terminal regions are of interest). The ONOT can be cut into consecutive equivalent segments (1–3 mm) and processed in sequence for kinetic analysis. Types of analysis might include total radioactivity, compositional analysis by electrophoresis, or presence of a specific labeled element by immunological/affinity analysis. Once the kinetics for the rate component of interest have been defined, a "window" segment of the nerve may be chosen (as discussed in the previous section), and only that region of the nerve need be sampled. The times indicated above for guinea pig would correspond to a "window" segment in the distal half of the optic nerve and the proximal region of the contralateral optic tract.

D. Dorsal-Root Ganglion (DRG)

The second most widely used system for study of axonal transport is the DRG and associated sciatic nerve. Sets of kinetics are available in the literature for rat

(Droz and Leblond, 1962) and cat (Ochs, 1972; Lasek, 1968a), both with large sciatic nerves that are accessible after simple surgery. The central branches of the DRG, leading into the spinal cord, have been utilized (Komiya and Kurokawa, 1978; Lavoie *et al.*, 1979), providing a unique opportunity to examine transport in two branches of the same cells—central and peripheral. The length of the axons in the sciatic nerve and dorsal roots permit very long time-course experiments even in rat (more than 4 months is easily studied). A valuable property of the DRG is the option to study transport both *in vivo* and *in vitro*. In both types of studies, the nerve is accessible and may be manipulated either physically or pharmacologically. For FC, this is a preferred system for pharmacological studies, particularly when used *in vitro,* since the length of the nerve permits precise control of parameters and the use of internal controls (Edström and Mattson, 1972; Theiler and McClure, 1977). Even questions about the site of action can be answered with respect to cell bodies and axons (Hammerschlag *et al.*, 1977; Lavoie *et al.*, 1979). Finally, the potential for regeneration in this system permits examination of questions about the growth and development of axons (Bisby, 1978, 1979).

The sciatic-nerve portion of the DRG system does have a heavy investiture of connective tissue sheaths that interfere with pharmacological studies in some cases and can complicate analysis (Crescitelli, 1951; Ochs, 1977). The kinetics of transport in the DRG appear to be more complex than those of the ONOT, giving less well-defined peaks for reasons that are not well understood. To some extent, this difficulty is overcome by the much greater length of this nerve system. The DRG does not have well-defined and localized terminal regions, so detailed analysis of destinations and turnover is not possible. The injection procedures are much more difficult than in the ONOT, but are described in the literature in detail. For *in vitro* studies of transport, FC may be labeled either by injection *in vivo* (Ochs and Smith, 1971) or by incubation of ganglion in labeled media (Edström and Mattson, 1972; Theiler and McClure, 1977). No *in vitro* studies of slow components have been possible to date.

E. Central Motor Neurons

A second set of neurons in the sciatic nerve may also be used in the study of axonal transport: the anterior horn cells that give rise to the motor axons of the sciatic. Since these cell bodies are located in the spinal cord proper, rather than in a discrete ganglion like the DRG, the surgical and injection procedures are more difficult. Most advantages and disadvantages of this system (particularly those of great length) are similar to those of the DRG with a few important exceptions. The motor-neuron cell bodies may not label well by incubation *in vitro* and have been most successfully labeled by injection. It is difficult to resolve clearly the different rate components into simple peaks of radioactivity, particularly the slow components (Hoffman and Lasek, 1980). The reasons for this complexity in kinetics are not known, but may relate to differences in the neurons involved. The primary reasons to study axonal transport in the ventral motor neurons have more to do with neurobiology than with cell biology. These motor axons are cholinergic and regenerate well even in mammals. As a result, they are excellent systems to study the mechanisms of growth and development of axons in the peripheral nervous system. They may also be useful in studies of some neuromuscular pathologies (Griffin and Price, 1976; Mendell *et al.*, 1977; Sahenk and Mendell, 1980).

F. Other Systems

Certain questions about axonal transport may be best answered by use of specific nerve systems. The variety of neuronal forms and functions provide many choices. The large size of certain neuronal cell bodies makes it possible to inject directly into an identified neuron in Aplysia (Koike *et al.*, 1972; Schwartz *et al.*, 1975) and leech (Isenberg *et al.*, 1980), making it possible to ask and answer questions that are impossible with cell populations. The garfish olfactory nerve (Gross, 1973; Gross and Beidler, 1973; Cancalon, 1979) is an exceptionally long and uniform nerve, which has led to remarkably detailed analyses of the kinetics of axonal transport in this system. This length and uniformity of the nerve, when combined with other properties of the garfish, such as being a poikilotherm, permits study of biochemical and biophysical properties of transport (Gross, 1973; Cancalon, 1979). The vagus (McLean *et al.*, 1975), the hypoglossal (Miani, 1963), the saphenous (Tsukita and Ishikawa, 1980), and other nerves have all been used successfully to demonstrate and analyze phenomena associated with axonal transport. Even regions and tracts within the brain have been used on occasion to study axonal transport (Levin, 1977, 1978; Padilla *et al.*, 1979); however, in this case, the difficulty in dissecting out the axonal regions complicates analysis.

IV. Applications: Criteria and Examples

The potential value of the axonal-transport model to the study of cell-biological questions is considerable, particularly with regard to the dynamic properties of structural elements. For example, knowledge about the way a cell partitions its many elements and the dynamic properties of that partitioning are available through the study of axonal transport. In effect, the axonal-transport paradigm can be an assay for protein associations with cell structures (microtubule–intermediate filament network, axoplasmic ground substance, smooth endoplasmic reticulum, etc.). It is beyond the scope of this article to examine all possible applications of axonal transport to cell biology. Thus we will restrict our discussion of applications to the use of axonal transport as an association assay for proteins.

The rate components of axonal transport are now sufficiently well characterized to predict the rate component with which a protein will be associated. The observation that each rate component has a unique polypeptide composition (Tytell *et al.*, 1981) is the basis of a simple rule: A protein generally moves with the rate component whose solubility properties correspond most closely to those of the protein. Membrane-associated proteins are most likely to be associated with the FC (DiGiamberardino *et al.*, 1973; Lorenz and Willard, 1978), except for mitochondrial proteins, which appear to move at an intermediate rate (Lorenz and Willard, 1978). Actin and the bulk of the soluble proteins, including many enzymes of intermediary metabolism, move as elements of SCb (Black and Lasek, 1979; Willard *et al.*, 1979; Garner, 1978; Brady and Lasek, 1981; Brady *et al.*, 1981a), whereas tubulin, neurofilament protein, and their associated proteins make up SCa (Hoffman and Lasek, 1975; Tytell *et al.*, 1980; Brady *et al.*, 1981b).

The identification of specific associations with cytoskeletal and other structural elements is clearly an essential step in the understanding of cellular organization. The axonal-transport paradigm can serve as an association assay for proteins in two ways. It may be used to characterize the elements in association with a specific cytological structure. An example is the recent characterization of the minor components of SCa (Tytell *et al.*, 1980). Two of the minor proteins in SCa were found to correspond to the tau class of microtubule-associated proteins (MAPs), but no high-molecular-weight MAPs could be detected. The alternate approach is to use axonal transport to evaluate the presence or absence of a specific protein in association with the cytological structures of the cell. An example here is the recent identification of calmodulin in SCb (Brady *et al.*, 1981a; Erickson *et al.*, 1980). The analysis of calmodulin in axonal transport will serve to illustrate some general principles for the use of axonal transport as a cell-biological tool.

Although calmodulin may interact with many different cellular components, it is primarily a soluble protein (Cheung, 1980). Because the bulk of the soluble proteins in axonal transport are associated with SCb (Lorenz and Willard, 1978; Garner, 1978), SCb was first examined for calmodulin. For a protein such as calmodulin, which may be obtained in relatively pure form, comigration of the purified protein and the labeled, axonally transported protein in two-dimensional gel electrophoresis (2D PAGE) (O'Farrell, 1975) is the simplest way to identify a transported polypeptide (Stone *et al.*, 1978). Calmodulin does not resolve well on standard 2D PAGE, but runs very well in a modified 2D PAGE system (Brady *et al.*, 1981a; Fig. 6a). A polypeptide in SCb did comigrate in 2D PAGE with purified brain calmodulin (box in Fig. 6b). However, two different proteins may

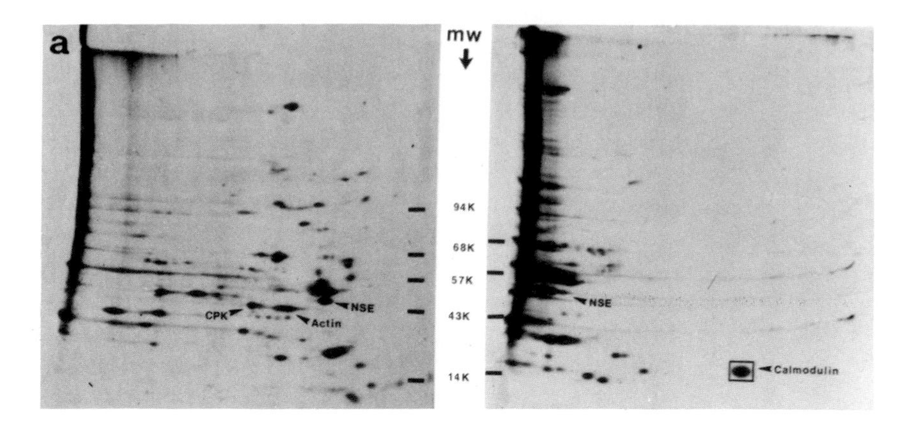
a
mw
94K
68K
57K
43K
14K
NSE
CPK
Actin
Calmodulin

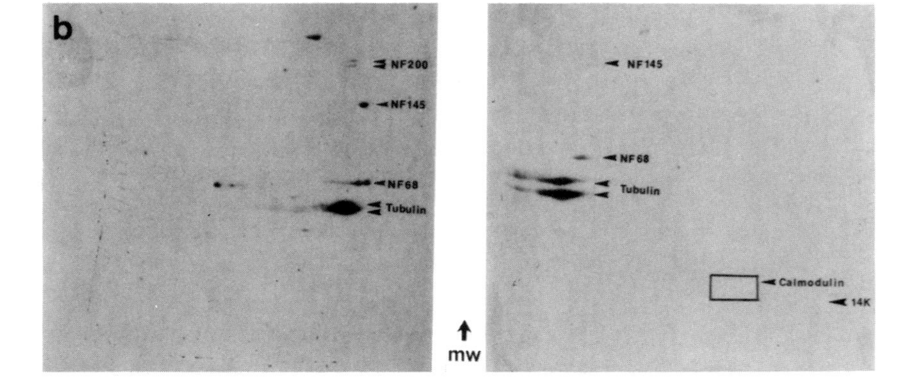
b
NF200
NF145
NF68
Tubulin
Calmodulin
14K
mw

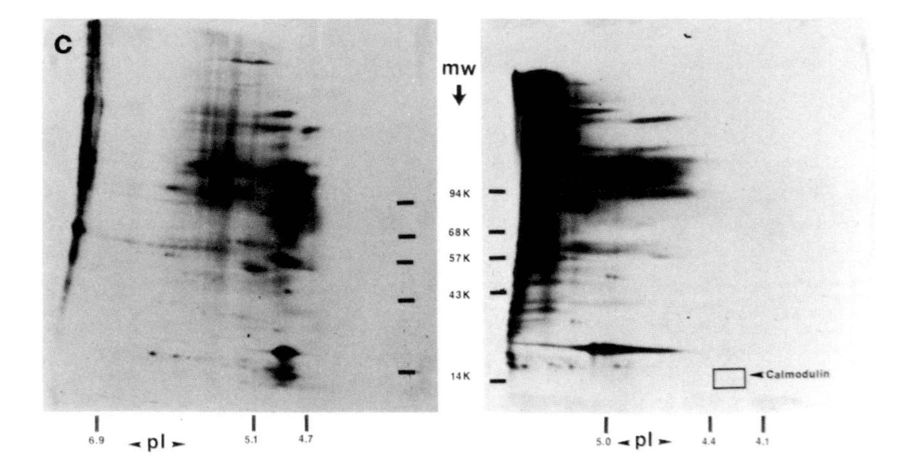
c
mw
94K
68K
57K
43K
14K
Calmodulin
6.9
pI
5.1
4.7
5.0
4.4
4.1

comigrate even in 2D PAGE, so a further independent test of identity is desirable (Henslee and Srere, 1979). Calmodulin is heat stable (Cheung, 1980) and shows a Ca^{2+}-dependent anomalous migration in PAGE (Kobayashi and Field, 1978; Grab *et al.*, 1979), so two independent tests of identity were possible. The labeled protein of SCb that comigrated with purified calmodulin was also heat stable (Brady *et al.*, 1981a), representing a copurification of labeled and unlabeled protein. The SCb protein also migrated anomalously on PAGE in the presence of Ca^{2+}-chelating agents (Brady *et al.*, 1981a), showing a specific ligand interaction for the labeled protein. Examples of other suitable independent tests include immunoprecipitation of a labeled species using monospecific antibody (Marangos *et al.*, 1975), and an identity of peptide maps for pure, identified material and labeled material (Henslee and Srere, 1979; Brady and Lasek, 1981).

Demonstrating the absence of a protein in a specific rate component is generally simpler than showing its presence. The absence of a labeled species comigrating with a pure protein in 2D PAGE can, in most cases, be considered unequivocal demonstration that the protein is not present in that component (see Fig. 6b,c; no calmodulin could be detected in either FC or SCa). One reservation must be noted: It is possible to get a false-negative result with this approach for certain high-affinity interactions, if the solubilization conditions for the isoelectric-focusing step do not disaggregate the constituent proteins into individual subunits. For example, the interaction between calmodulin and phosphorylase kinase is not disrupted by the O'Farrell lysis buffer (Cohen *et al.*, 1981), but is disrupted by SDS-containing buffers. Prior exposure of samples to SDS should, in principle, reduce or eliminate the possibility of a false-negative result.

The axonal-transport paradigm demonstrates long-term interactions between proteins and cellular structures, as well as the intracellular movements of proteins. The example of calmodulin shows that a protein may have hierarchies of

---

FIG. 6. Two-dimensional analysis of the major orthograde rate components of axonal transport. Axonally transported proteins were labeled as before in the optic nerve and the animals sacrificed at 6 hours for FC, 6 days for SCb, and 40–60 days for SCa. The optic nerves were dissected out and the radioactive polypeptides analyzed by two-dimensional electrophoresis and fluorography, using either the O'Farrell (1975) procedure (pH 5–7) or a modified O'Farrell (Brady *et al.*, 1981a) procedure to visualize acidic proteins (pH 2.5–5). The differences in protein composition between FC, SCa, and SCb are even more striking than in Fig. 3. The samples (a–c) were handled identically except for the length of time after injection, so differences in the behavior of proteins in 2D PAGE for different rate components reflect the different properties of those proteins. (a) The complex polypeptide composition of SCb is seen and a few of the identified proteins are indicated, including calmodulin, which focuses reliably only in the modified 2D-PAGE system (box). (b) The polypeptide composition of SCa is much simpler and no calmodulin can be detected in association with SCa. (c) FC has a complex pattern of proteins very different from that of SCb, and like SCa, contains no calmodulin. The difference in focusing reflects the high glycoprotein content of FC (see text). (Taken from Brady *et al.*, 1981a.)

interaction. From biochemistry and traditional cell biological approaches, we know that calmodulin can interact with membrane proteins and microtubules, but axonal-transport studies indicate that these must represent transient and short-term interactions as opposed to the primary interaction with the axoplasmic ground substance. The complementarity of axonal-transport and traditional paradigms is apparent.

V. Conclusion

The phenomenon of axonal transport has not been widely exploited by cell biologists to date, perhaps because it was perceived as an activity peculiar to neuronal cells. It must be remembered, however, that the majority of cytoskeletal and metabolic elements that a neuron uses to establish its form and functions are identical to those of liver cells and fibroblasts. Only a fraction of the neuronal complement of proteins are unique to the nervous system. More importantly, the problems of intracellular transport and subcellular compartmentalization faced by neuronal cells are merely extreme examples of the problems dealt with by all cells. The neuron represents a model for the understanding of cytoskeletal organization and protein association in a dynamic way. Much as striated muscle gave us first insights into the mechanisms of contractility and the erythrocyte provided a model for the study of membrane organization, the neuron and axonal transport represent a powerful paradigm for studies on the organization of the cytoplasm and its constituent elements.

References

Abe, T., Haga, T., and Kurokawa, M. (1973). *Biochem. J.* **136,** 731–740.
Autilio-Gambetti, L., Gambetti, P., and Shafer, B. (1973). *Brain Res.* **53,** 387–398.
Barondes, S. H. (1964). *Science* **146,** 779–781.
Berl, S., Puszkin, S., and Nicklas, W. J. (1973). *Science* **179,** 441–446.
Bisby, M. A. (1978). *Exp. Neurol.* **61,** 281–300.
Bisby, M. A. (1979). *Exp. Neurol.* **65,** 680–684.
Black, M. M., and Lasek, R. J. (1978). *J. Neurobiol.* **9,** 433–443.
Black, M. M., and Lasek, R. J. (1979). *Brain Res.* **171,** 410–413.
Black, M. M., and Lasek, R. J. (1980). *J. Cell Biol.* **86,** 616–623.
Bolton, A. E., Ludlam, C. A., Pepper, D. S., Moore, S., and Cash, J. D. (1976). *Thromb. Res.* **8,** 51–58.
Bonner, W. M., and Laskey, R. A. (1974). *Eur. J. Biochem.* **46,** 83–88.
Brady, S. T. (1978). Ph.D. Dissertation, University of Southern California, Los Angeles.
Brady, S. T., and Lasek, R. J. (1981). *Cell* **23,** 523–531.
Brady, S. T., Crothers, S. D., Nosal, C., and McClure, W. O. (1980). *Proc. Natl. Acad. Sci. U.S.A.* **77,** 5909–5913.

Brady, S. T., Tytell, M., Heriot, K., and Lasek, R. J. (1981a). *J. Cell Biol.* **89,** 607–614.
Brady, S. T., McQuarrie, I. G., Tytell, M., and Lasek, R. J. (1981b). *Trans. Am. Soc. Neurochem.* **12,** 198.
Brimijoin, S. (1975). *J. Neurobiol.* **6,** 379–394.
Brimijoin, S., and Helland, L. (1976). *Brain Res.* **102,** 217–228.
Brimijoin, S., and Wiermaa, M. J. (1978). *J. Physiol. (London)* **285,** 129–142.
Brimijoin, S., Capek, P., and Dyck, P. J. (1973). *Science* **180,** 1295–1297.
Brimijoin, S., Olsen, J., and Rosenson, R. (1979). *J. Physiol. (London)* **287,** 303–314.
Bullock, T. H., Orkand, R., and Grinnell, A. (1977). "Introduction to Nervous Systems," Chapters 2 and 10. Freeman, San Francisco, California.
Burton, P. R., and Fernandez, H. L. (1973). *J. Cell Sci.* **12,** 567–583.
Cancalon, P. (1979). *J. Neurochem.* **32,** 997–1007.
Cancalon, P., and Beidler, L. M. (1975). *Brain Res.* **89,** 225–244.
Cheung, W. Y. (1980). *Science* **207,** 19–27.
Cohen, P., Klee, C. B., Picton, C., and Shenolikar, S. (1981). *Ann. N.Y. Acad. Sci.* **356,** 151–161.
Cooper, P. D., and Smith, R. S. (1974). *J. Physiol. (London)* **242,** 77–97.
Cowan, W. M., and Cuénod, M. (1975). "The Use of Axonal Transport for Studies of Neural Connectivity." Elsevier, New York.
Crescitelli, F. (1951). *Am. J. Physiol.* **166,** 229–240.
Crothers, S., and McLuer, R. H. (1975). *J. Neurochem.* **24,** 209–214.
DiGiamberardino, L., Bennett, G., Koenig, H. L., and Droz, B. (1973). *Brain Res.* **60,** 147–159.
Droz, B., and Leblond, C. P. (1976). *Science* **137,** 1047–1048.
Droz, B., and Leblond, C. P. (1963). *J. Comp. Neurol.* **121,** 325–346.
Droz, B., Koenig, H. L., and DiGiamberardino, L. (1973). *Brain Res.* **60,** 93–127.
Droz, B., Rambourg, A., and Koenig, H. L. (1975). *Brain Res.* **93,** 1–13.
Edström, A., and Mattsson, H. (1972). *J. Neurochem.* **19,** 205–211.
Elam, J. S., and Agranoff, B. W. (1971). *J. Neurochem.* **18,** 375–387.
Elam, J. S., Goldberg, J. M., Radin, N. S., and Agranoff, B. W. (1970). *Science* **170,** 458–460.
Ellisman, M. H., and Porter, K. R. (1980). *J. Cell Biol.* **87,** 464–479.
Erickson, P. F., Seamon, K. B., Moore, B. W., Lasher, R. S., and Minier, L. N. (1980). *J. Neurochem.* **35,** 242–248.
Estridge, M., and Bunge, R. (1978). *J. Cell Biol.* **79,** 138–155.
Fang, V. S., Cho, H. W., and Meltzer, H. Y. (1975). *Biochem. Biophys. Res. Commun.* **65,** 413–419.
Fink, D. J., and Gainer, H. (1979). *Brain Res.* **177,** 208–213.
Fink, D. J., and Gainer, H. (1980a). *J. Cell Biol.* **85,** 175–186.
Fink, D. J., and Gainer, H. (1980b). *Science* **208,** 303–305.
Forman, D. S., and Ledeen, R. W. (1972). *Science* **177,** 630–633.
Forman, D. S., McEwen, B. S., and Grafstein, B. (1971). *Brain Res.* **28,** 119–130.
Forman, D. S., Grafstein, B., and McEwen, B. S. (1972). *Brain Res.* **48,** 327–342.
Forman, D. S., Padjen, A. L., and Siggins, G. R. (1977a). *Brain Res.* **136,** 197–213.
Forman, D. S., Padjen, A. L., and Siggins, G. R. (1977b). *Brain Res.* **136,** 215–226.
Friede, R. L., and Ho, K.-C. (1977). *J. Physiol. (London)* **265,** 507–519.
Friede, R. L., and Miyagishi, T. (1972). *Anat. Rec.* **172,** 1–14.
Garner, J. A. (1978). Ph.D. Dissertation, Dept. of Anatomy, Case Western Reserve University, Cleveland, Ohio.
Garner, J. A., and Lasek, R. J. (1981). *J. Cell Biol.* **88,** 172–178.
Grab, D. J., Berzins, K., Cohen, R. S., and Siekevitz, P. (1979). *J. Biol. Chem.* **254,** 8690–8696.
Grafstein, B. (1967). *Science* **157,** 196–198.
Grafstein, B., and Forman, D. S. (1980). *Physiol. Rev.* **60,** 1167–1283.

Grafstein, B., and Laureno, R. (1973). *Exp. Neurol.* **39,** 44–57.
Grafstein, B., and McQuarrie, I. G. (1978). *In* "Neuronal Plasticity" (C. Cotman, ed.), pp. 155–195. Raven, New York.
Grafstein, B., Miller, J. A., Ledeen, R. W., Haley, J., and Specht, S. C. (1975). *Exp. Neurol.* **46,** 261–281.
Griffin, J. W., and Price, D. L. (1976). *UCLA Forum Med. Sci.* **19,** 33–67.
Griffin, J. W., Hoffman, P. N., Clark, A. W., Carroll, P. T., and Price, D. L. (1978). *Science* **202,** 633–635.
Gross, G. W. (1973). *Brain Res.* **56,** 359–363.
Gross, G. W., and Beidler, L. M. (1973). *J. Neurobiol.* **4,** 413–428.
Hammerschlag, R., Bakhit, C., and Chiu, A. Y. (1977). *J. Neurobiol.* **8,** 439–451.
Hammerschlag, R., Stone, G. C., and Bolen, F. A. (1981). *Trans. Am. Soc. Neurochem.* **12,** 144.
Hanson, M., and Edström, A. (1978). *Int. Rev. Cytol., Suppl.* **7,** 373–402.
Heacock, A. M., and Agranoff, B. W. (1977). *Brain Res.* **122,** 243–254.
Henslee, J. G., and Srere, P. A. (1979). *J. Biol. Chem.* **254,** 5488–5497.
Heslop, J. P. (1975). *Adv. Comp. Physiol. Biochem.* **6,** 75–163.
Hodge, A. J., and Adelman, W. J. (1980). *J. Ultrastruct. Res.* **70,** 220–241.
Hoffman, P. N., and Lasek, R. J. (1975). *J. Cell Biol.* **66,** 351–366.
Hoffman, P. N., and Lasek, R. J. (1980). *Brain Res.* **202,** 317–333.
Hökfelt, T., and Dahlström, A. (1971). *Z. Zellforsch. Mikrosk. Anat.* **119,** 460–482.
Holtzman, E., Schacher, S., Evans, J., and Teichberg, S. (1977). *Cell Surf. Rev.* **4,** 165–246.
Ingoglia, N. A., Grafstein, B., McEwen, B. S., and McQuarrie, I. G. (1973). *J. Neurochem.* **20,** 1605–1615.
Isenberg, G., Schubert, P., and Kreutzberg, G. W. (1980). *Brain Res.* **194,** 588–593.
Johnson, R. T., and Griffin, D. E. (1978). *In* "Handbook of Clinical Neurology" (P. J. Vinken and G. W. Bruyn, eds.), Vol. 34, Part II, 15–37. North-Holland Publ., Amsterdam.
Karlsson, J.-O., and Sjöstrand, J. (1968). *Brain Res.* **11,** 431–439.
Karlsson, J.-O., and Sjöstrand, J. (1971). *J. Neurochem.* **18,** 975–982.
Kelly, M. G., and Hartwell, J. L. (1954). *JNCI, J. Natl. Cancer Inst.* **14,** 967–1010.
Kerkut, G. A., Shapira, A., and Walker, R. J. (1967). *Comp. Biochem. Physiol.* **23,** 729–748.
Kobayashi, R., and Field, J. B. (1978). *Biochim. Biophys. Acta* **539,** 411–419.
Koike, H., Eisenstadt, M., and Schwartz, J. H. (1972). *Brain Res.* **37,** 152–159.
Komiya, Y., and Kurokawa, M. (1978). *Brain Res.* **139,** 354–358.
Kristensson, K. (1978). *Annu. Rev. Pharmacol. Toxicol.* **18,** 97–110.
Kunzle, H., and Cuénod, M. (1973). *Brain Res.* **62,** 213–217.
Lasek, R. J. (1967). *Nature (London)* **216,** 1212–1214.
Lasek, R. J. (1968a). *Brain Res.* **7,** 360–377.
Lasek, R. J. (1968b). *Exp. Neurol.* **21,** 41–51.
Lasek, R. J. (1970). *Int. Rev. Neurobiol.* **13,** 289–321.
Lasek, R. J. (1980). *Trends NeuroSci.* **3,** 87–91.
Lasek, R. J. (1981). *Neurosci. Res. Program Bull.* **19,** 7–32.
Lasek, R. J., and Hoffman, P. N. (1976). *Cold Spring Harbor Conf. Cell Proliferation* **3** [Book C], p. 1021–1049.
Lasek, R. J., Gainer, H., and Barker, J. L. (1977). *J. Cell Biol.* **74,** 501–523.
Laskey, R. A., and Mills, A. D. (1975). *Eur. J. Biochem.* **56,** 335–341.
LaVail, J. H. (1978). *In* "Neuroanatomical Research Techniques" (R. T. Robertson, ed.), pp. 355–384. Academic Press, New York.
LaVail, J. H., and LaVail, M. M. (1974). *J. Comp. Neurol.* **157,** 303–358.
Lavoie, P.-A., Bolen, F., and Hammerschlag, R. (1979). *J. Neurochem.* **32,** 1745–1751.
LeBeux, Y. J., and Willemot, J. (1975). *Cell Tissue Res.* **160,** 1–36.
Levin, B. E. (1977). *Brain Res.* **130,** 421–432.

Levin, B. E. (1978). *Brain Res.* **150,** 55–68.
Liem, R. K. H., Yen, S. H., Salomon, G. D., and Shelanski, M. L. (1978). *J. Cell Biol.* **79,** 637–645.
Lorenz, T., and Willard, M. (1978). *Proc. Natl. Acad. Sci. U.S.A.* **75,** 505–509.
Lubinska, L. (1964). *Prog. Brain Res.* **13,** 1–66.
Lubinska, L. (1971). *Acta Neuropathol., Suppl.* **5,** 136–143.
Lubinska, L. (1975). *Int. Rev. Neurobiol.* **17,** 241–296.
Lux, H. D., Schubert, P., Kreutzberg, G. W., and Globus, A. (1970). *Exp. Brain Res.* **10,** 197–204.
McClure, W. O. (1972). *Adv. Pharmacol. Chemother.* **10,** 185–220.
McEwen, B. S., and Grafstein, B. (1968). *J. Cell Biol.* **38,** 494–508.
McLean, W. G., Frizell, M., and Sjöstrand, J. (1975). *J. Neurochem.* **25,** 695–698.
Marangos, P., Zomzely-Neurath, C., York, C., and Bondy, S. C. (1975). *Biochim. Biophys. Acta* **392,** 75–81.
Marchisio, P. C., and Sjöstrand, J. (1971). *Brain Res.* **26,** 204–211.
Marchisio, P. C., and Sjöstrand, J. (1972). *J. Neurocytol.* **1,** 101–108.
Marchisio, P. C., Sjöstrand, J., Aglietta, M., and Karlsson, J.-O. (1973). *Brain Res.* **63,** 273–284.
Margolis, T. P., Marchand, C., Kistler, H., and LaVail, J. H. (1981). *J. Cell Biol.* **89,** 152–156.
Mendell, J. R., Sahenk, Z., Saida, K., Weiss, H. S., Savage, R., and Couri, D. (1977). *Brain Res.* **133,** 107–118.
Metuzals, J., and Izzard, C. S. (1969). *J. Cell Biol.* **43,** 480–505.
Miani, N. (1963). *J. Neurochem.* **10,** 859–874.
Nauta, H. J. W., Kaiserman-Abramof, I. R., and Lasek, R. J. (1975). *Brain Res.* **85,** 373–384.
Ochs, S. (1972). *J. Physiol.* (*London*) **227,** 627–645.
Ochs, S. (1977). *Nature* (*London*) **270,** 748–750.
Ochs, S., and Smith, C. B. (1971). *J. Neurochem.* **18,** 833–841.
Ochs, S., Johnson, J., and Ng, M.-H. (1967). *J. Neurochem.* **14,** 317–331.
O'Farrell, P. (1975). *J. Biol. Chem.* **250,** 4007–4021.
Padilla, S. S., Roger, L. J., Toews, A. D., Goodrum, J. F., and Morell, P. (1979). *Brain Res.* **176,** 407–411.
Paulson, J. C., and McClure, W. O. (1975a). *J. Cell Biol.* **67,** 461–467.
Paulson, J. C., and McClure, W. O. (1975b). *Ann. N.Y. Acad. Sci.* **253,** 517–527.
Peters, A., Palay, S. L., and Webster, H. (1976). "The Fine Structure of the Nervous System." Saunders, Philadelphia, Pennsylvania.
Pomerat, C. M., Hendelman, W. J., Raiborn, C. W., Jr., and Massey, J. F. (1967). *In* "The Neuron" (H. Hyden, ed.), pp. 119–178. Elsevier, Amsterdam.
Porter, K. R., Byers, H. R., and Ellisman, M. H. (1979). *In* "The Neurosciences: 4th Study Program" (F. O. Schmitt and F. G. Worden, eds.), pp. 703–722. MIT Press, Cambridge, Massachusetts.
Rambourg, A., and Droz, B. (1980). *J. Neurochem.* **35,** 16–25.
Ramón y Cajál, S. (1909). "Histologie du Systeme Nerveux." A. Maloine, Paris, France.
Ramón y Cajál, S. (1928). "Degeneration and Regeneration of the Nervous System" (transl. by R. M. May), Vol. I. Oxford Univ. Press, London and New York.
Sahenk, Z., and Mendell, J. R. (1980). *Brain Res.* **186,** 343–353.
Samson, F. E. (1976). *Annu. Rev. Pharmacol. Toxicol.* **16,** 143–159.
Samson, F. E., Donoso, J. A., Heller-Bettinger, I., Watson, D., and Himes, R. H. (1979). *J. Pharmacol. Exp. Ther.* **208,** 411–417.
Schlaepfer, W. W., and Freeman, L. (1978). *J. Cell Biol.* **78,** 653–662.
Schwartz, J. H. (1979). *Annu. Rev. Neurosci.* **2,** 467–504.
Schwartz, J. H., Goldman, J. E., Ambron, R. T., and Goldberg, D. J. (1975). *Cold Spring Harbor Symp. Quant. Biol.* **40,** 83–92.
Smith, R. S. (1971). *Cytobios* **3,** 259–262.

Smith, R. S. (1972). *Can. J. Physiol. Pharmacol.* **50,** 467–469.
Smith, R. S. (1980). *J. Neurocytol.* **9,** 39–65.
Snyder, R. E., Reynolds, R. A., Smith, R. S., and Kendal, W. S. (1976). *Can. J. Physiol. Pharmacol.* **54,** 238–244.
Specht, S., and Grafstein, B. (1973). *Exp. Neurol.* **41,** 705–722.
Stoeckel, K., and Thoenen, H. (1975). *Brain Res.* **85,** 337–341.
Stoeckel, K., Schwab, M., and Thoenen, H. (1975). *Brain Res.* **99,** 1–16.
Stone, G. C., Wilson, D. L., and Hall, M. E. (1978). *Brain Res.* **144,** 287–302.
Taylor, A. C., and Weiss, P. (1965). *Proc. Natl. Acad. Sci. U.S.A.* **54,** 1521–1527.
Theiler, R. F., and McClure, W. O. (1977). *J. Neurochem.* **28,** 321–330.
Tsukita, S., and Ishikawa, H. (1980). *J. Cell Biol.* **84,** 513–530.
Tytell, M., Brady, S. T., and Lasek, R. J. (1980). *Soc. Neurosci. Abstr.* **6,** 501.
Tytell, M., Black, M. M., Garner, J., and Lasek, R. J. (1981). *Science* **214,** 179–181.
Weiss, P., and Hiscoe, H. B. (1948). *J. Exp. Zool.* **107,** 315–395.
Wessells, N. K., Spooner, B. S., Ash, J. F., Bradley, M. O., Luduena, M. A., Taylor, E. L., Wrenn, J. T., and Yamada, K. M. (1971). *Science* **171,** 135–143.
Willard, M. (1977). *J. Cell Biol.* **75,** 1–11.
Willard, M., Cowan, W. M., and Vagelos, P. R. (1974). *Proc. Natl. Acad. Sci. U.S.A.* **71,** 2183–2187.
Willard, M., Wiseman, M., Levine, J., and Skene, P. (1979). *J. Cell Biol.* **81,** 581–591.
Willard, M. B., and Hulebak, K. L. (1977). *Brain Res.* **136,** 289–306.
Wilson, D. L., and Stone, G. C. (1979). *Annu. Rev. Biophys. Bioeng.* **8,** 27–45.
Wilson, L. (1975). *Ann. N. Y. Acad. Sci.* **253,** 213–231.
Wuerker, R. B., and Kirkpatrick, J. B. (1972). *Int. Rev. Cytol.* **33,** 45–75.
Wuerker, R. B., and Palay, S. L. (1969). *Tissue Cell* **1,** 387–402.
Yamada, K. M., Spooner, B. S., and Wessells, N. K. (1971). *J. Cell Biol.* **49,** 614–635.
Zelena, J., Lubinska, L., and Gutmann, E. (1968). *Z. Zellforsch. Mikrosk. Anat.* **91,** 200–219.

Index

D

E

F

G

H

I

K

L

M

T

U

V

X

Z

CONTENTS OF RECENT VOLUMES

(Volumes I–XX edited by David M. Prescott)

Volume X

Volume XI

Yeast Cells

Volume XII

Yeast Cells

Volume XIII

Volume XIV

Volume XV

Volume XVI

Chromatin and Chromosomal Protein Research. I

Volume XVII

Chromatin and Chromosomal Protein Research. II

Volume XVIII

Chromatin and Chromosomal Protein Research. III

Volume XIX

Chromatin and Chromosomal Protein Research. IV

Volume XX

Volume 21A

Normal Human Tissue and Cell Culture A. Respiratory, Cardiovascular, and Integumentary Systems

Volume 21B

Normal Human Tissue and Cell Culture B. Endocrine, Urogenital, and Gastrointestinal Systems

Volume 22

Three-Dimensional Ultrastructure in Biology

Volume 23

Basic Mechanisms of Cellular Secretion

Volume 24

The Cytoskeleton

DATE DUE